AF506171

Plant Cell and Tissue Culture – A Tool in Biotechnology

Karl-Hermann Neumann · Ashwani Kumar ·
Jafargholi Imani

Plant Cell and Tissue Culture – A Tool in Biotechnology

Basics and Application

Second Edition

 Springer

Karl-Hermann Neumann (deceased)
Hungen, Germany

Ashwani Kumar
Department of Botany
University of Rajasthan
Jaipur, India

Jafargholi Imani
Institute of Phytopathology
Justus-Liebig University Giessen
Giessen, Germany

ISBN 978-3-030-49096-6   ISBN 978-3-030-49098-0   (eBook)
https://doi.org/10.1007/978-3-030-49098-0

This Springer imprint is published by the registered company Springer Nature Switzerland AG.
The registered company address is: Gewerbestrasse 11, 6330 Cham, Switzerland

Preface to First Edition

This book is intended to provide a general introduction to this exciting field of plant cell and tissue culture as tool in biotechnology, without overly dwelling on detailed descriptions of all aspects. It is aimed at the newcomer, but will hopefully also stimulate some new ideas for the "old hands" in tissue culture. Nowadays, with the vast amount of information readily available on the internet, our aim was rather to distill and highlight overall trends, deeming that a complete report of each and every tissue culture investigation and publication was neither possible, nor desirable. For some techniques, however, detailed protocols are given. We have tried to be as thorough as possible, and regret if we have inadvertently overlooked any pertinent literature or specific development that belong in this work.

The three authors have been associated for many years, and have worked together on various aspects in this field. Without this close interaction, this book would not have been possible. At this opportunity, we wish to reiterate our mutual appreciation of this fruitful cooperation. An Alexander von Humboldt Stiftung fellowship to Ashwani Kumar (University of Rajasthan, Jaipur, India) to work in our group at the Institut für Pflanzenernaehrung der Justus Liebig Universität, Giessen, supported this close cooperation and the completion of this book, is gratefully acknowledged.

Such a book takes time to grow. Indeed, its roots lie in a 3–4 week lecture and laboratory course by one of us (K.-H.N.) about 30 years ago as visiting professor at Ain Shams University, Cairo, Egypt, which later led to the development of a graduate training unit at the University of Giessen, Germany, and other universities. So, also older key literature, nowadays risking being forgotten, has been considered, which could be of help for newcomers in this domain.

Thanks are due to our publisher for all the help received, and for patiently waiting for an end product that, we feel, has only gained in quality.

Giessen, Germany   K. -H. Neumann
March 2009   A. Kumar
                           J. Imani

Preface to Second Edition

The first edition of the book was intended to provide a general introduction to this exciting field of plant cell and tissue culture as tool in biotechnology, without overly dwelling on detailed descriptions of all aspects. After a decade of the first edition, and despite the enormous amount of information readily available on the Internet, our aim was to distill and highlight overall trends of tissue culture that currently represent a basic tool in plant biotechnology in the second edition. Thus, a complete report of each and every tissue culture investigation and publication was neither possible nor desirable. The second edition was written to emphasize the latest advances in the developing field of genetic engineering, especially the thoroughly revised Chap. 13. For some techniques, however, detailed protocols are provided in this book. The second edition of this book was written by two authors (Dr. A. Kumar and Dr. J. Imani), because Prof. Neumann, who had played a crucial role in first edition, is unfortunately no longer alive. These two authors have been associated with each other for many years and have worked together on various aspects in this field. Without this close interaction, the second edition of this book would not have been possible. An Alexander von Humboldt Stiftung fellowship to Ashwani Kumar (University of Rajasthan, Jaipur, India) to work in our group at the Institut für Pflanzenernaehrung and Dr. Imani at the Institute for Phytopathology, Justus Liebig Universität, Giessen, which supported this close cooperation and the completion of this book, is gratefully acknowledged. We have tried to be as thorough as possible, and regret if we have inadvertently overlooked any pertinent literature or specific development that belongs in this work.

We also acknowledge numerous publications which have been the source of some pictures in this book.

Acknowledgments are due to our publisher for all the received help and patience while waiting for the manuscript.

Giessen, Germany A. Kumar
Spring 2020 J. Imani

Acknowledgments

Figures 3.2–3.5, 3.8, 3.10, 3.12, 3.13, 3.16, 4.1, 4.4, 5.2, 5.4, 5.5, 5.7, 6.3, 7.3, 7.5–7.9, 7.11, 7.15, 7.16, 7.33, 8.1, 8.3, 8.15, 9.2, 12.1, 13.3 and Tables 2.1, 3.3–3.8, 5.1, 6.1–6.3, 7.1, 7.3, 7.5, 7.8, 12.1 are published with kind permission of Verlag Eugen Ulmer. Further, we are indebted to the following authors and their publishers for getting permissions to use their figures in our book in Chap. 13. Hayta, S., Smedley, M.A., Demir, S.U. et al. (2019) An efficient and reproducible Agrobacterium-mediated transformation method for hexaploid wheat (*Triticum aestivum* L.) Plant Methods 15, 121 (Figs. 13.11 and 13.12); He, L., Hannon, G. (2004) MicroRNAs: small RNAs with a big role in gene regulation. Nat Rev Genet 5, 522–531. https://doi.org/10.1038/nrg1379 (Fig. 13.27); Moises Zotti, Ericmar Avila dos Santos, Deise Cagliari, Olivier Christiaens, Clauvis Nji Tizi Taning, Guy Smagghe (2018) RNA interference technology in crop protection against arthropod pests, pathogens and nematodes. Pest Management Science 74, 1239–1250 (Fig. 13.29); Liu, S., Jaouannet, M., Dempsey, D., Imani, J., Coustau, C. et al. (2020) RNA-based technologies for insect control in plant production. Biotechnology Advances 39 (107463), 1–13 (Fig. 13.30); Karimi, Mansour et al. (2002) GATEWAY™ vectors for Agrobacterium-mediated plant transformation. Trends in Plant Science 7(5), 193–195 (Fig. 13.31); Yuriko Osakabe, Keishi Osakabe. Genome editing with engineered nucleases in plants. Plant and Cell Physiology 56, 389–400 (Fig. 13.34); Yuriko Osakabe, Keishi Osakabe. Genome editing with engineered nucleases in plants. Plant and Cell Physiology 56, 389–400 (Fig. 13.35); Erkes, A., Reschke, M., Boch, J., Grau, J. (2017) Evolution of transcription activator-like effectors in Xanthomonas oryzae. Genome Biol Evol 9(6), 1599–1615 (Fig. 13.36); Joung, J.K., Sander, J.D. (2013) TALENs: a widely applicable technology for targeted genome editing. Nat Rev Mol Cell Biol 14(1), 49–55 (Fig. 13.37); Khatodia, S., Bhatotia, K., Passricha, N., Khurana, S.M., Tuteja, N. (2016) The CRISPR/Cas genome-editing tool: application in improvement of crops. Front Plant Sci 7, 506 (Fig. 13.38); Belhaj, K., Chaparro-Garcia, A., Kamoun, S., Nekrasov, V. (2013) Plant genome editing made easy: targeted mutagenesis in model and crop plants using the CRISPR/Cas system. Plant Methods 9(1), 39 (Fig. 13.39); Belhaj, K., Chaparro-Garcia, A., Kamoun, S., Nekrasov, V. (2013) Plant genome editing made easy: targeted mutagenesis in model and crop plants using the CRISPR/Cas system. Plant Methods 9(1), 39 (Fig. 13.41);

Sharma, M., Schmid, M., Rothballer, M., Hause, G., Zuccaro, A., Imani, J. et al. (2008) Detection and identification of bacteria intimately associated with fungi of the order Sebacinales. Cell Microbiol 10, 2235–2246. https://doi.org/10.1111/j. 1462-5822.2008.01202.x (Fig. 13.42); Reitz, M., Bissue, J.K., Zocher, K., Attard, A., Hückelhoven, R., Becker, K., Imani, J., Eichmann, J., Schäfer, P. (2012) The subcellular localization of Tubby-like proteins and participation in stress signaling and root colonization by the mutualist Piriformospora indica. Plant Physiology 160 (1), 349–364. https://doi.org/10.1104/pp.112.201319 (Fig. 13.43); Sharma, M., Schmid, M., Rothballer, M., Hause, G., Zuccaro, A., Imani, J. et al. (2008) Detection and identification of bacteria intimately associated with fungi of the order Sebacinales. Cell Microbiol 10, 2235–2246. https://doi.org/10.1111/j.1462-5822. 2008.01202.x (Fig. 13.44).

Ashwani Kumar sincerely acknowledges encouragement and continued support of Prof. Dr. Sven Schubert, Director of the Institute of Plant Nutrition, Justus Liebig Universität, Giessen, after the demise of Prof. Dr. K.H. Neumann in 2009, and the award of Alexander von Humboldt Fellowship which enabled his frequent research visits and also helped in shaping the book. Dr. Jafargholi Imani acknowledges the support of Prof. Dr. Karl-Heinz Kogel, Director of the Institute of Plant Pathology, Justus Liebig Universität, Giessen. We thank our research students, fellow workers, and colleagues whose works we have quoted and who have provided support to this work directly or indirectly. We also thank Dr. Andrea Schlitzberger, Project Coordinator, book production Germany and Asia, and Mr. Bibhuti Sharma from Springer. It was pleasure to work with Springer, and we thank them for bringing out the book so nicely.

In Memoriam Prof. Dr. Karl-Hermann Neumann (1936–2009)

Prof. Dr. Karl-Hermann Neumann passed away on October 13, 2009. He was born on 22 May, 1936, in Morgensdorf, near Leitmeritz (Sudetenland), which is now a part of the Czech Republic. After the end of Second World War, his family migrated to Bernburg in Saxonia-Anhaltinia (later German Democratic Republic, GDR). In this place, his father acquired another farm, where he grew up. In 1956, following some problems with Communist administration of the GDR, he had to move to the Federal Republic of Germany, where he finished his schooling. In 1957, he entered Justus Liebig Universität to study agriculture. He spent one semester as a foreign student with a stipend at "Den Kgl. Veterinaer og Landbohojskole" in Copenhagen, Denmark, majoring in agricultural chemistry; he completed his studies again at Giessen, as "Diplomlandwirt" in 1960. In the same year, he received scholarship from Cornell University in Ithaca, NY, USA. He enrolled here in a graduate school, with botany as major and biochemistry and physical chemistry as minor. He also started working for Ph.D. under the supervision of Prof. F.C. Steward, FRS. The topic of his study was "Function of some heavy metals (iron, magnesium and molybdenum) on the growth and metabolism of carrot tissue cultures" (mainly protein metabolism and photosynthesis), which also became major for his Ph.D. thesis later on. Since the scholarship was given to him for only 1 year, he returned to Germany and completed his Ph.D. under the supervision of Prof. H. Linser in 1962 from the University of Giessen. After spending few years with Prof. Linser while doing postdoc, he finished his Habilitation (equivalent to D.Sc.) studies in 1969. He got promoted as "Privatdozent" at the agricultural faculty of Justus Liebig University, Giessen, Germany. In 1972, he became Professor for Biochemistry and Cell Biology of Plants at the Faculty of Nutrition of the Justus Liebig University.

"While working with Prof. Steward and Prof. Linser he was also intensively trained in systemic thinking and multidisciplinary research approaches which influenced his own scientific career to a considerable extent," wrote Arnholdt et al. (2010).

His first major achievement was replacement of coconut milk in White's nutrient medium used in Steward's laboratory at Cornell with an artificial nutrient medium also named as Neumann's Lösung or Neumann's liquid medium of defined chemical composition. "All later investigations can be traced back to the three experiments performed in the 1960s and the medium was named as Neumann's Lösung or Neumann's medium. One of the experiments focussed on the photosynthesis of cultured carrot explants, which was based on work done at Cornell; second on nucleic acid metabolism of carrot cultures; and third on somatic embryogenesis in carrot cells in a defined nutrient medium. The work on photosynthesis was accelerated after Ludwig Bender and Ashwani Kumar joined his group in the 1970s and both held Professorships subsequently." A. Kumar joined as Alexander von Humboldt Fellow while serving at the University of Rajasthan, Jaipur. "The results, covering cytology as well as biochemistry aspects, were published in number of papers. This line of work came to a close with studies on somatic embryogenesis of autotrophic cultures under normal atmosphere (Dr. Eva Plescka)."

"The work on nucleic acid started with studies on metabolic turnover of both DNA and RNA, followed by comparative studies of DNA organization of several plant species by Cot hybridization indicating about 15% identity of unique and repeated DNA. This represents the basic genetic information, which can be used to distinguish higher plants from other biological systems. Here also first results turned up on the occurrence of metabolic DNA localised in repeated fractions and broadly associated with differentiation. This work was done in cooperation with Dr A. Schafer, Dr. E. Duerssen and Professor Savedra of Chilie."

"Later, Dr. B. Arnholdt-Schmidt, now a Professor (Universidade de Évora: Évora, PT) joined the group and was mainly concerned with DNA methylation and amplification as related to differentiation. Based on these early studies in the 1990s, gene technology was taken up resulting in the insertion of the information of a coat protein of Hepatitis B virus into the carrot genome which was also expressed in mature carrot roots at harvest. Dr. Jafargholi Imani took the lead with the cooperation of medical virologist (Professor W. Gerlich, Giessen, Germany) from the University. Clinical studies with respect to immunization after oral application could not be performed till now. Here also the results and experience of many studies on the cell cycle and its synchronization of haploid and diploid Datura cultures and others (Dr. J. Blaschke, Dr. R. Kiebler) was utilized by using synchronised cultures for insertion of foreign DNA into carrot cells, preferably during S-phase," he wrote in his autobiography published in Kumar and Soporty (2008).

He further wrote: "Many studies were performed on somatic embryogenesis, mainly with petiole explant of carrots, including histology, protein, and nucleic acid organization and metabolism. These studies resulted from cooperation of Dr. B. Grieb and Professor Li of the University of Huehot, P.R. China. Carbon metabolism and hormonal system were also studied by Dr. E. Pleschka and F. Schaefer, and published in a number of papers. A broad research programme on the ploidy level and its significance on development and secondary metabolism was initiated by Dr. Forche and Dr B. Zeppernick; both associated with our group for several years. The results of these exciting studies were published in several papers."

"Quite interesting were the cooperative investigations on diurnal variations of the concentrations of several phytohormones in intact plants, in pot experiments, as well as in cultured cells. In cultured cells, in constant environment (including continuous illumination) clear maxima of IAA as well as several cytokinins was observed 24 h a day for several days." "All this work would not have been possible without the dedicated help of my associates especially Frau Christa Lein, who had the same position and function in my laboratory as Mrs M. Mapes had in F.C. Steward's laboratory at Cornell," said Prof. Dr. Neumann in his autobiography.

In 1995, a small book on cell and tissue culture (Pflanzliche Zell—und Gewebekulturen) was published in German by Ulmer Verlag, Stuttgart, Germany. In this book results obtained till then were discussed in context to ideas of the time. This book was subsequently developed into first edition of the present book (Neumann et al. 2009).

He further wrote: "In 1972, I became Professor in the faculty of nutrition at the Justus Liebig University Giessen (Plant nutrition, biochemistry and cell biology of plants) and I worked there till my retirement in 2001. During this period I spent some time abroad mainly as a visiting Professor or in a similar position for several countries of Asia and Africa, from where students came to my laboratory to work for a Ph.D. or Post doctorate. My longest cooperation was with Professor Ashwani Kumar, University of Rajasthan, India (since 1977 till now) as he wrote this in 2009 just before his death."

"One great challenge before me was to establish a research farm in the south of Frankfurt in 1979 to pursue investigations mainly concerned with irrigation and the quality of irrigation water. What a change! After more than 20 years doing basic research, I had to turn to practical problems to continue studies on biochemical and cell biological problems as before. Here Dr. Buno Pauler, a research associate, was a great help, especially in the statistical evaluation of the data obtained from the experimental work of about 15 years. The work on irrigation and salinity was extended to studies on sugar beet cultivation in saline conditions in Egypt. This work was done together with Prof. A. Raafat and Dr. seyed Eisa, Ain Shamps University, including work on biological remediation of saline fields in cooperation with Professor Kumar and Dr. Shekhawat, University of Rajasthan, Jaipur, India. In Egypt, some work was concerned with the control of Orobanche infection of faba beans based on hormonal studies (Prof. N. Al. Gamrawy and Dr Salem, Cairo University); mango malformation (Prof. A. Raafat and Dr El Deep), concentrating on the hormonal system. At the University, I occupied the chair of dean several times, and was also the director of the department as well as the chief of examinations." He has published around 150 research papers and produced over 20 Ph.D. students during his entire career.

Prof. Neumann's special interest in systemic and applied research was probably best mirrored by the small edition dedicated to him by his former scientist colleagues at his retirement with the title "From Soil to Cell—a Broad Approach to Plant Life" (Bender and Kumar 2001).

"Prof. Neumann was truly an international scientist with global vision who cared so well, beyond national boundaries, for his many students, research associates and fellows from all over the world. We are sure that all of them, having had the privilege to work and learn in his team will keep an honourable memory of his generous personality, scientific acumen and lifetime achievement" (Arnholdt-Schmidt et al. 2010).

References

Arnholdt-Schmitt B, Kumar A, Imani R et al (2010) In memoriam Prof. Dr. Karl-Hermann Neumann (1936–2009). Plant Cell Tiss Organ Cult 100:121–122. https://doi.org/10.1007/s11240-009-9644-5

Bender L, Kumar A (2001) From soil to cell: A broad approach to plant life. Giessen Electron. Library GEB, pp 1–5. http://geb.uni-giessen.de/geb/volltexte/2006/3039/pdf/FestschriftNeumann-2001.pdf

Neumann K-H (2008) Professor Dr Karl-Hermann Neumann. In: Kumar A, Sopory SK (eds) Recent advances in plant biotechnology and its applications. IK International, New Delhi

Contents

About the Authors

Ashwani Kumar, FBS, FPSI, FISMPP, FABP, FIFS born (1946) to Mr. Swami Dayal Tewari and Mrs. Shanti Devi Tewari, at Bandikui, Rajasthan, India, received B.Sc. at Agra University and M.Sc. (Botany) at the University of Rajasthan. He was awarded gold medal for standing first in order of merit. His Ph.D. (1971) was under the supervision of Prof. H.C. Arya and postdoc with Prof. Dr. K-H. Neumann and later on with Prof. Dr. Sven Schubert at Justus Liebig Universität, Giessen, Germany, with Alexander von Humboldt Fellowship. His botanist father Prof. Swami Dayal Tewari (M.Sc. in Botany) was his first teacher. Prof. Ashwani Kumar was also selected in the Indian Administrative Services (IAS: IPS) (1972), but he opted for a career in botany being his family subject. His wife Mrs Vijay R. Kumar has also been Professor of Botany at the University of Rajasthan on her own qualifications. He was appointed as Assistant Professor in 1969, Associate Professor in 1985, and Full Professor from 1986 to 2007. Then he was an Adjunct Professor until 2016. He along with members of COC introduced Integrated Biotechnology 5-year course in Rajasthan.

In recognition of his research contributions, Dr. Kumar was awarded Alexander von Humboldt Fellowship in 1977–1979, with resumption of fellowships until 2017, British Council Visitorship, UK (1986); Visiting Professorship at Toyama Medical and Pharmaceutical University in Japan (1999–2000); Toyama Prefectural University Japan (2011); and INSA-DFG visiting Professorship at Germany, 1997. He holds a diploma in German language and has a certificate in French language. His area of research includes photosynthesis in vitro and in vivo, biotic and abiotic resistance, ethnobotany, bioenergy, and presently understanding salinity resistance in maize. He has also carried out research projects granted by UGC, USDA-ICAR, MNES, CSIR, DST, DBT, and FACT. He attended a large number of national and international conferences as invited speaker and served as chair or co-chair in the International Botanical Conference, Berlin, and EU Biomass conferences. He has published 220 research papers in national and international journals and 23 books, of which 10 books are authored and 13 edited from reputed publishers such as Springer and IK. He is a member of the editorial board of *Current Trends in Biotechnology and Pharmacy*. He has guided 39 research students to Ph.D. at the University of Rajasthan, Jaipur, India. He published 220 original research papers in National and Internatioal journals. He is an elected Fellow of Botanical Society, Fellow of Phytopathological Society, Fellow of Indian Society of Mycology and Plant Pathology, Fellow of Mendelian Association, Fellow of Association of Biotechnology and Pharmacology, and Fellow of Indian Fern Society. He received V. Puri Medal as Botanist and Teacher's Excellence award of CEE in 2015. He has been a consultant in the World Bank Project sanctioned to SPRI-HPPI, President Commonwealth Human Ecology Council (India Chapter), and presently President Indian Botanical Society.

Jafargholi Imani (born 1955 in Aliabad-Gorgan, Iran) works as a scientific group leader at the Institute for Phytopathology, Justus Liebig University (JLU) Giessen, Germany. He has graduated in agricultural science and completed his Ph.D. under the supervision of Prof. Dr. Karl-Hermann Neumann at JLU. He worked as postdoctoral fellow in the group of "Plant Tissue Culture and Transformation" at the Institute of Plant Nutrition, JLU, Giessen. Currently, he is administrative head of field station and group leader in Plant Tissue Culture and Transformation at the Institute of Phytopathology, Justus Liebig University, Giessen, Germany. He is continuing scientific works on the Plant Tissue Culture and Genetic Transformation of several plant species. His area of research includes optimization of genome editing procedures in plants especially in cereals and RNA-based technologies for pest control in plant production. He is also specialized in mutualistic interaction between many crop plants and soil-borne and plant root-colonizing fungal endophytes that confers resistance to biotic and abiotic stresses and promotes growth in crop plants. He attended a large number of national and international conferences. He has published 108 research papers in national and international journals and is author of eight books. He is member of Working Group "German in Vitro Cultures e.V." (ADIVK), International Association for Plant Tissue Culture and Biotechnology (IAPTC&B), Society for Plant Biotechnology e.V., and Consultant at the EU Marie Curie Chair, Universidade de Évora, Portugal. In addition to teaching and supervising several bachelor's and master's students in agrobiotechnology, he has guided 17 Ph.D. students at the Justus Liebig University, Giessen, Germany.

Introduction

1

The United Nations predicted that 9.2 billion are likely to occupy our planet by 2050 and 11.2 billion in 2100 (https://www.un.org/development/desa/en/news/popula tion/world-population-prospects-2017.html) (Figs. 1.1 and 1.2). Global warming and resultant climate change poses another threat to the population (Kumar 2013). The food production must increase by 50% under the prevailing circumstances. According to Deroles and Davies (2014), plant cell culture can offer continuous production systems for high-value food and health ingredients, independent of geographical or environmental constrains. Metabolic engineering using plant tissue culture systems has led to increase in yield potential of several metabolites and also helped in understanding the primary and secondary metabolisms. Induction of chloroplast development and carbon fixation studies in autotrophic callus cultures enabled basic understanding of "C4" cycle in C3 callus cultures of *Daucus carota* and *Arachis hypogaea* for the first time (Neumann et al. 1982; Kumar et al. 1977, 1978, 1980; Bender et al. 1985; Neumann et al. 2009). Major efforts are currently underway to introduce carbon concentrating systems into C3 crops plants, such as wheat and rice, with special emphasis on converting C_3 plants to C_4 plants (von Caemmerer et al. 2012).

According to Maurino and Weber (2012), increasing photosynthetic efficiency is of prime importance to increase plant productivity to meet the demands of a growing human population with respect to food, feed, fiber, and energy efficiency. Weber and Brautigam (2013) suggested two major goals for plant metabolic engineering: (1) increasing the yield or quality of plant products or specialized metabolites which have health promoting properties and (2) increasing photosynthetic efficiency to optimize the amount of plant products that can be achieved with a given amount of fertilizer, water, and land. Plant cell culture technology can be used to obtain fundamental metabolic information, supply high-value products, propagate elite species, and promote somatic embryogenesis (Wilson and Roberts 2012). This can only be achieved if conventional breeding is supported by biotechnological applications of plant tissue culture techniques (Dar et al. 2012; Kumar and Sopory 2010; Kumar and Roy 2011; Waltz 2016) (Figs. 1.3 and 1.4). Besides angiosperms

© Springer Nature Switzerland AG 2020

K.-H. Neumann et al., *Plant Cell and Tissue Culture – A Tool in Biotechnology*, https://doi.org/10.1007/978-3-030-49098-0_1

The rise in human population will outrun the growth in food supplies

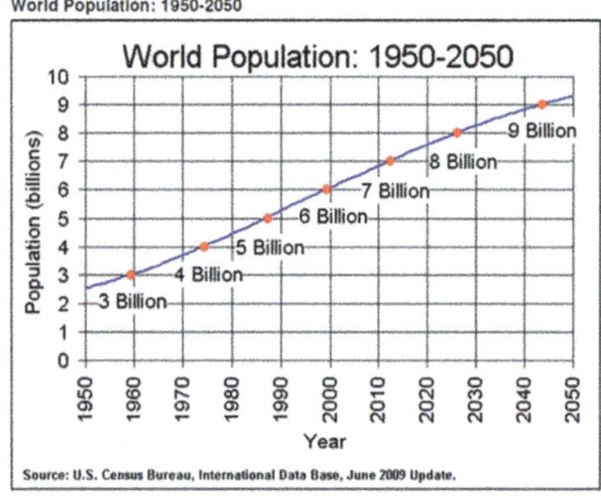

Fig. 1.1   The rise of human population will outrun the food supply

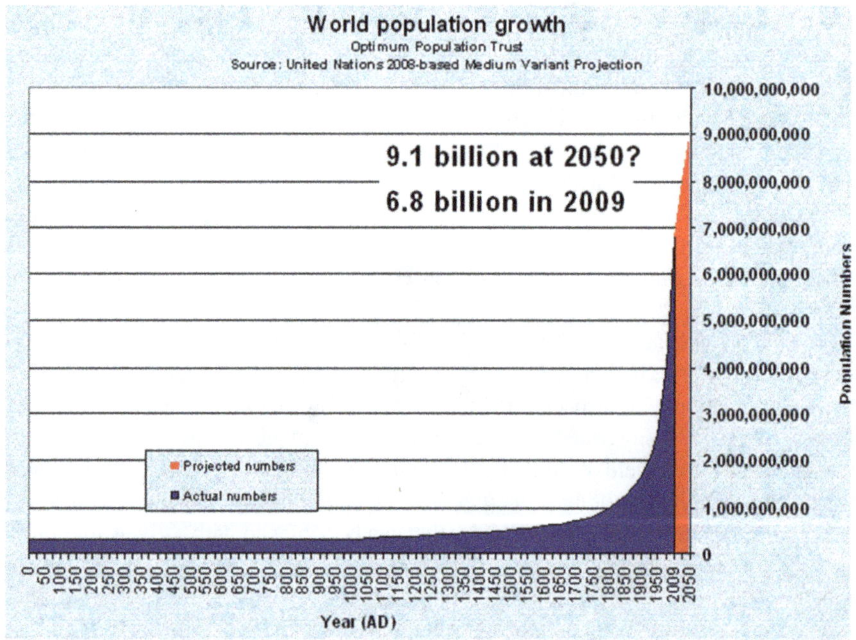

Fig. 1.2   World population growth

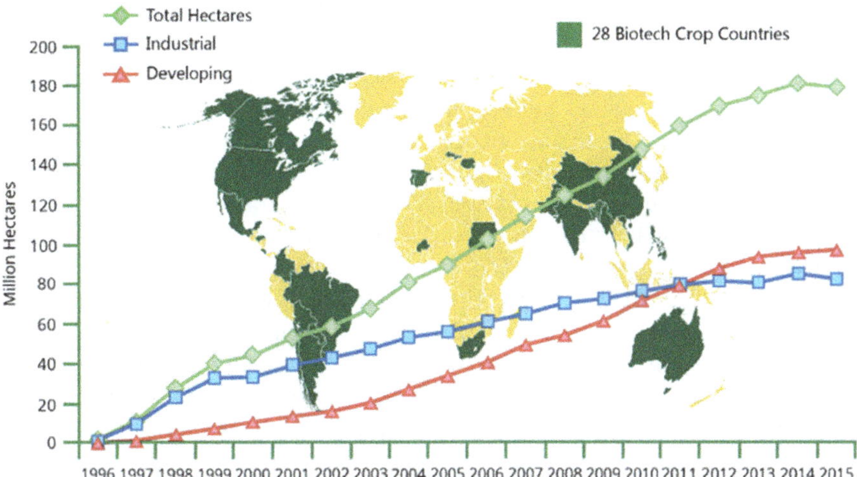

Fig. 1.3 Global area under biotech crops

Fig. 1.4 Group photo with author in picture at the International Conference on plant transformation technologies

recently pteridophytes have also attracted attention, and their tissue culture for biotechnological applications has been elaborately covered in recent book by Fernandez et al. (2011).

Experimental systems based on plant cell and tissue culture are characterized by the use of isolated parts of plants, called explants, obtained from an intact plant body and kept on or in a suitable nutrient medium. This nutrient medium functions as replacement for the cells, tissue, or conductive elements originally neighboring the explant. Such experimental systems are usually maintained under aseptic conditions. Otherwise, due to the fast growth of contaminating microorganisms, the cultured cell material would quickly be overgrown, making a rational evaluation of experimental results impossible.

Some exceptions to this are experiments concerned with problems of phytopathology in which the influence of microorganisms on physiological or biochemical parameters of plant cells or tissue is to be investigated. Other examples are co-cultures of cell material of higher plants with *Rhizobia* to study symbiosis or to improve protection for micro-propagated plantlets to escape transient transplant stresses (Peiter et al. 2003; Waller et al. 2005).

Using cell and tissue cultures, at least in basic studies, aims at a better understanding of biochemical, physiological, and anatomical reactions of selected cell material to specified factors under controlled conditions, with the hope of gaining insight into the life of the intact plant also in its natural environment. Compared to the use of intact plants, the main advantage of these systems is a rather easy control of chemical and physical environmental factors to be kept constant at reasonable costs. Here, the growth and development of various plant parts can be studied without the influence of remote material in the intact plant body. In most cases, however, the original histology of the cultured material will undergo changes and eventually may be lost. In synthetic culture media available in many formulations nowadays, the reaction of a given cell material to selected factors or components can be investigated. As an example, cell and tissue cultures are used as model systems to determine the influences of nutrients or plant hormones on development and metabolism related to tissue growth. These were among the aims of the "fathers" of tissue cultures in the first half of the twentieth century. To which extent and under which conditions this was achieved will be dealt with later in this book.

The advantages of those systems are counterbalanced by some important disadvantages. For one, in heterotrophic and mixotrophic systems, high concentrations of organic ingredients are required in the nutrient medium (particularly sugar at 2% or more), associated with a high risk of microbial contamination. How and to which extent this can be avoided will be dealt with in Chap. 3. Other disadvantages are the difficulties and limitations of extrapolating results based on tissue or cell cultures to interpret phenomena occurring in an intact plant during its development. It has always to be kept in mind that tissue cultures are only model systems, with all positive and negative characteristics inherent of such experimental setups. To be realistic, a direct duplication of in situ conditions in tissue culture systems is still not possible even today in the twenty-first century and probably never will be. The organization of the genetic system and of basic cell structures is, however, essentially the same, and therefore tissue cultures of higher plants should be better suited as model systems than, e.g., cultures of algae, often employed as model systems in physiological or biochemical investigations.

The domain *cell and tissue culture* is rather broad and necessarily unspecific. In terms of practical aspects, basically five areas can be distinguished (see Figs. 1.5 and 1.6), which here shall be briefly surveyed before being discussed later at length. These are callus cultures, cell suspensions, protoplast cultures, anther cultures, and organ or meristem cultures.

Callus Cultures (See Chap. 3)

In this approach, isolated pieces of a selected tissue, so-called explants (only some mg in weight), are obtained aseptically from a plant organ and cultured on or in a suitable nutrient medium. For a primary callus culture, most convenient are tissues with high contents of parenchyma or meristematic cells. In such explants, mostly only a limited number of cell types occur, and so a higher histological homogeneity exists than in the entire organ. However, growth induced after transfer of the explants to the nutrient medium usually results in an unorganized mass or clump of cells—a callus—consisting largely of cells different from those in the original explant.

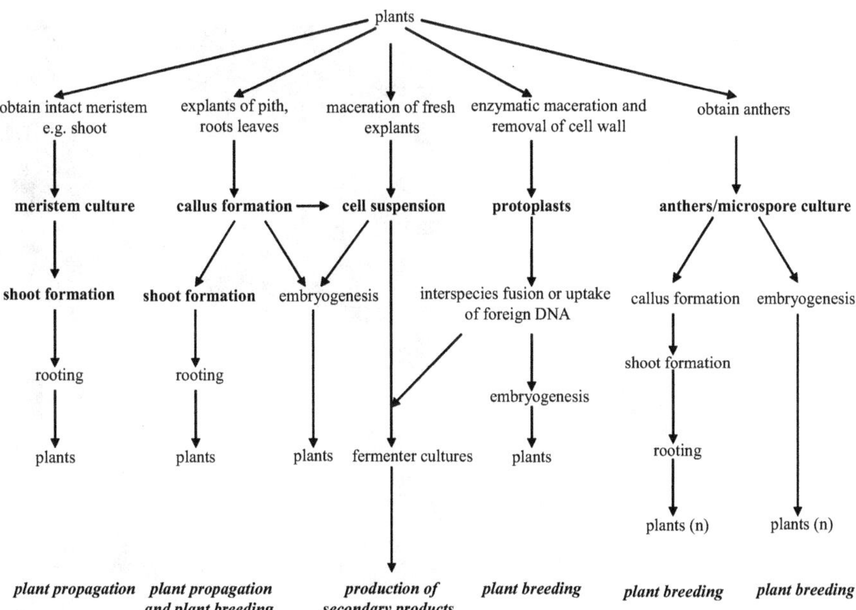

Fig. 1.5 Schematic presentation of the major areas of plant cell and tissue cultures and some fields of application

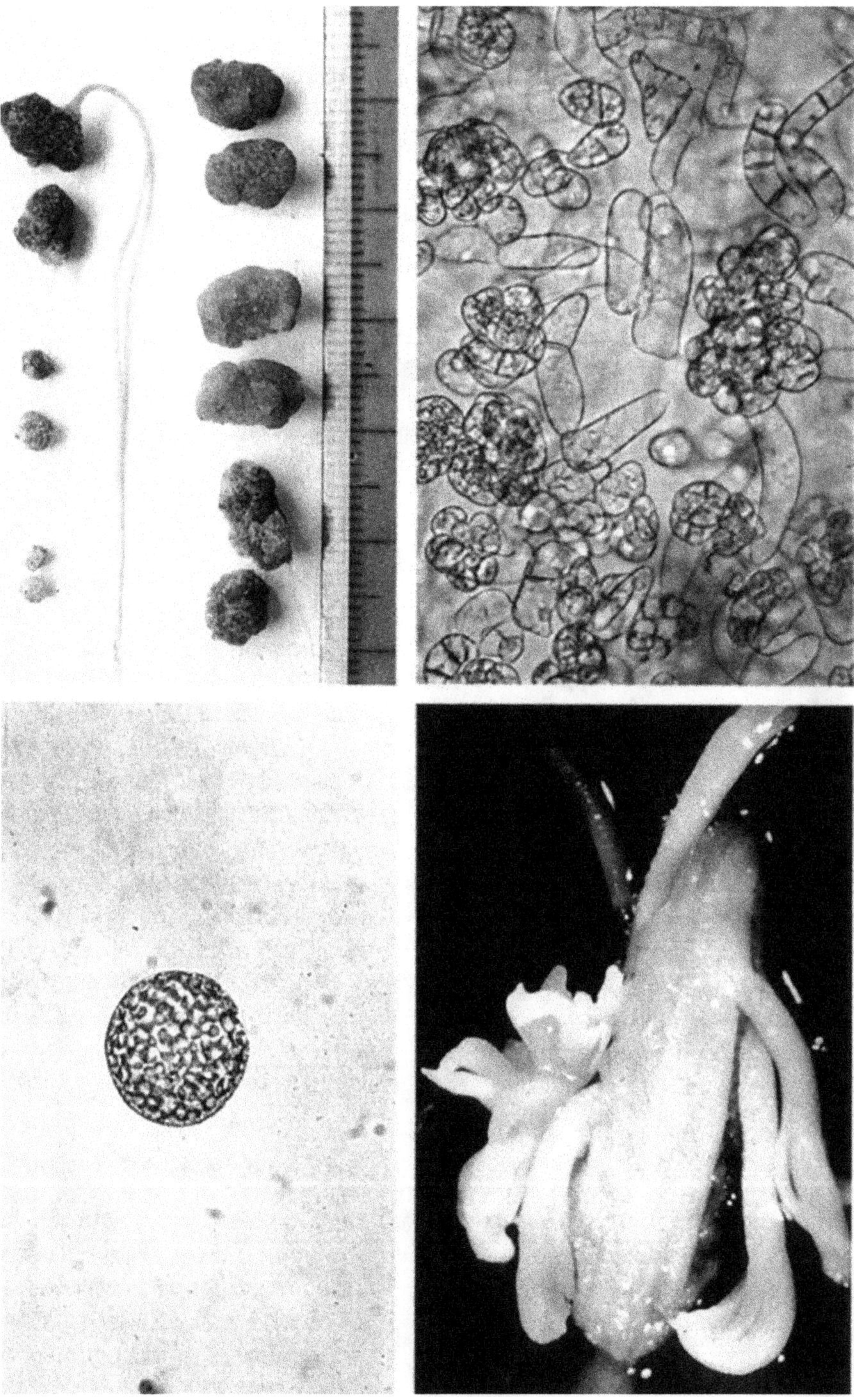

Fig. 1.6 Various techniques of plant cell and tissue cultures, some examples: (top left) callus culture, (top right) cell suspension culture, (bottom left) protoplast culture, (bottom right) anther culture

Cell Suspensions (See Chap. 4)

Whereas in a callus culture, there remain connections among adjacent cells via plasmodesmata; ideally in a cell suspension, all cells are isolated. Under practical conditions, however, also in these cell populations, there is usually a high percentage of cells occurring as multicellular aggregates. A supplement of enzymes able to break down the middle lamella connecting the cells in such clumps or a mechanical maceration will yield single cells. Often, cell suspensions are produced by mechanical sheering of callus material in a stirred liquid medium. These cell suspensions generally consist of a great variety of cell types (Fig. 1.6), and are less homogenous than callus cultures.

Protoplast Cultures (See Chap. 5)

In this approach, initially the cell wall of isolated cells is enzymatically removed, i.e., "naked" cells are obtained (Fig. 1.6), and the explant is transformed into a single-cell culture. To prevent cell lysis, this has to be done under hypertonic conditions. This method has been used to study processes related to the regeneration of the cell wall and to better understand its structure. Also, protoplast cultures have served for investigations on nutrient transport through the plasmalemma but without the confounding influence of the cell wall. The main aim in using this approach in the past, however, has been interspecies hybridizations, not possible by sexual crossing. Nowadays, protoplasts are still essential in many protocols of gene technology. From such protoplast cultures, ideally plants can be regenerated through somatic embryogenesis to be used in breeding programs.

Haploid Techniques (See Chap. 6)

Culturing anthers (Fig. 1.6), or isolated microspores from anthers under suitable conditions, haploid plants can be obtained through somatic embryogenesis. Treating such plant material with, e.g., colchicines, it is possible to produce dihaploids, and if everything works out, within 1 year (this depends on the plant species) a fertile homozygous dihaploid plant can be produced from a heterozygous mother plant. This method is advantageous for hybrid breeding, by substantially reducing the time required to establish inbred lines.

Often, however, initially a callus is produced from microspores, with separate formation of roots and shoots that subsequently join, and in due time haploid plants can be isolated. Here, the production of "ploidy chimeras" may be a problem. Another aim in using anther or microspore cultures is to provoke the expression of recessive genes in haploids to be selected for plant breeding or gene transfer purposes.

Plant Propagation, Meristem Culture, Somatic Embryogenesis (See Chap. 7)

In this approach, mostly isolated primary or secondary shoot meristems (shoot apex, axillary buds) are induced to shoot development under aseptic conditions. Generally, this occurs without an interfering callus phase, and after rooting, the plantlets can be isolated and transplanted into soil. Thereby, highly valuable single plants—e.g., a hybrid—can be propagated. The main application, however, is in horticulture for mass propagation of clones for the commercial market, another being the production of virus-free plants. Thus, this technique has received a broad interest in horticulture and also in silviculture as a major means of propagation.

Some Endogenous and Exogenous Factors in Cell Culture Systems (Chap. 8)
Growth and differentiation of excised tissues are regulated by an interplay of endogenous and exogenous factors in coordination with genetic influences. This chapter deals with growth regulators, nutrient uptake, and role of physical factors in growth and differentiation of cultured plant tissues.

Primary Metabolism (Chap. 9)
Cultured plant cells have the capacity to develop chloroplasts which was initially recorded by Professor Dr Karl-Hermann Neumann in Germany as early as 1964 and brought to the notice of Professor Steward who initially did not believe it as per statements of Professor Dr Neumann. However, subsequently elaborate studies by Professor Dr Neumann lead to not only elucidation of chloroplast development in isolated carrot root explants but also studies on carbon dioxide fixation by such chloroplasts in "sputniks"; the special type of flasks on a rotatory machine under illuminated conditions not only developed chloroplasts but also showed their function with regard to carbon fixation. Professor Ashwani Kumar while working in lab of Professor H.C. Arya during early 1970s also developed autotrophic cultures of *Arachis hypogaea* which were able to grow without carbon supplement at Jaipur. During 1st International Plant Tissue Culture Association (IAPTC) meeting organized by Professor H.E. Street at Leicester, several groups presented their work on primary metabolism from USA, France, Japan Germany, and India. There I (AK) met Professor Dr K-H Neumann who was pioneer on autotrophic callus cultures. After listening to my work on photosynthetic cultures (Kumar 1974) he invited me to join his lab. with support from Alexander von Humboldt Foundation. This enabled me to continue this work and major part of it is presented here. Man had just landed on moon during that period and idea to use such autotrophic cultures to provide food to the astronauts was ripe. Professor Jack Widholm from the USA, Professor Yamada and his team at Japan, and Professor Neumann and Professor Reinert in Germany were some of the pioneer workers in this field. Such autotrophic cultures provided basic understanding of mechanism of photosynthesis in cultured plant cells. The chapter deals with carbon and nitrogen metabolism in cultured plant tissues and their applications.

Secondary Metabolism (Chap. 10)
Plant tissue cultures and organ cultures were thought to be excellent source of production of secondary metabolites under controlled conditions. Leading scientists from Germany, the USA, France, and Japan believed that by regulating the cultural conditions and developing better fermenter systems, the dream of obtaining secondary metabolites from cultured tissues could be realized. Professor Yamada and Sato met initial success, and Japan and Germany with Professor Reinert and Professor Zenk lead this research. Biotechnological innovations and metabolic engineering

further supported the production of secondary metabolites and other biologically active compounds. Metabolic engineering of plant cells and production of heterologous expression systems have been optimized for production of Secondary metabolites.

Rapid development of metabolic engineering and synthetic biology of microorganisms shows many advantages to replace the current extraction of these useful high-price chemicals from plants. Attempts to obtain benzylisoquinoline alkaloids (BIAs) which include many pharmaceutical compounds, such as berberine (antidiarrheal), sanguinarine (antibacterial), morphine (analgesic), and codeine (antitussive) have been reported successfully in microorganisms. Nakagawa (2016) presented total biosynthesis of opiates by stepwise fermentation using engineered *Escherichia coli*. Application of biotechnology for production of biofuels, bioethanol and biodiesel, has gained attention (Chap. 10).

Genetic Problems and Gene Technology (Chap. 13)
Plant tissue culture and its application in biotechnology have attracted industries to this field. Plant tissue culture and biotechnology some of the topic elaborated include somaclonal variations, ploidy stability, evaluation of genetic fidelity during long-term conservation, molecular markers, DNA fingerprinting and characterization of germplasm, and estimation of genetic diversity, gene technology, modes of gene modification, and wild crossing have been covered in addition to mutagenesis, genetic transformations techniques and applications, and gene silencing. The first CRISPR-edited crops presented to the US regulatory system can be cultivated and sold without oversight by the US Department of Agriculture (USDA) (Waltz 2016). This opens new possibilities of achieving goal of feeding increasing world population and meeting their environmental needs with green technologies without causing further pollution. Plant tissue culture and its biotechnological options open new vistas, and this book tries to update recent findings. Zinc finger nucleases (ZFNs) and transcription activator-like effector nucleases (TALENs) have been utilized to allow for site-specific gene mutation, replacement, or integration through non-homologous end joining or homologous recombination (Wilson and Roberts 2014). In contrast to ZFNs and TALENs, the Cas9 protein is by nature a sequence-specific nuclease (Xie et al. 2016. Newer genetic engineering (GE) techniques that don't involve plant pests are quickly supplanting the old ones, and the USDA appears to be saying it does not have the authority to regulate the products of these techniques. The agency ruled similarly on plants transformed with other gene editing techniques, such as zinc finger nuclease and transcription activator-like effector nuclease systems. Invention of new genome engineering strategies, such as TALENs and CRISPR/Cas9 systems, would greatly facilitate researchers to develop more advanced strategies within shorter periods. Recently synthetic biology is perceived as an interdisciplinary field. It acts at the interface between recombinant DNA technology and biotechnology and enables reprograming of living cells to perform novel and improved functions. According to Xie et al. (2016), cell engineering in synthetic biology is based on the concept that biological components from different organisms can be reassembled into genetic networks that operate either in parallel or together with

natural biological systems to improve, restore, or add essential functions to cells. However, synthetic biology is beyond the scope of this publication hence would not be dealt in detail in this edition.

A new chapter has been added to provide practical aspect of plant tissue culture, and its application in biotechnology based on our practical experiences and lab exercises carried out on day to day basis (Chap. 13). This new chapter has been added based on practical being conducted in the Institute a guide is prepared for all those interested in undertaking plant tissue culture in lab, in industry, and in research labs. Exact details of the procedures are given in simple language so that the guide can be followed easily by beginners and experts alike.

Summary of Some Physiological Aspects in the Development of Plant Cell and Tissue Culture (Chap. 14)

Attempts have been made in this chapter to elaborate on recent applications of plant cell and tissue culture with some examples and recent developments in focus. Millions of plants are commercially propagated annually via micropropagation which is labor-intensive and associated with developmental abnormalities; there have been many efforts to develop cost-effective and simple bioreactors with the aim of automating micropropagation. However, designing reactors for plant tissue culture must reconcile environmental factors (shear stress, aeration, RH, nutrient supply) with healthy plant development, cost, and simplicity of use. Numerous studies have applied bioreactors in plant cell and organ culture to obtain specific metabolites. In addition, a considerable number of researchers have cultured plant propagules in bioreactors to produce high-quality seedling.

References

Bender L, Kumar A, Neumann KH (1985) On the photosynthetic system and assimilate metabolism of *Daucus* and *Arachis* cell cultures. In: Neumann KH, Barz W, Reinhard E (eds) Primary and secondary metabolism of plant cell cultures. Springer, Berlin, pp 24–42

Dar ZA, Meena PD, Kumar A (2012) Principles of plant breeding. Pointer Publishers, Jaipur, p 250

Deroles SC, Davies KM (2014) Prospects for the use of plant cell cultures in food biotechnology. Curr Opin Biotechnol 26:133–140

Fernandez H, Kumar A, Revilla BMA (2011) Working with ferns: issues and applications. Springer, Dresden, 350 pp

Kumar A (1974) In vitro growth and chlorophyll formation in mesophyll callus tissues on sugar free medium. Phytomorphology 24:96–101

Kumar A (2013) Biofuels utilisation: an attempt to reduce GHG's and mitigate climate change. In: Nautiyal S, Kaechele H, Rao KS, Schaldach R (eds) Knowledge systems of societies for adaptation and mitigation of impacts of climate change. Springer, Heidelberg, pp 199–224

Kumar A, Roy S (2011) Plant tissue culture and applied plant biotechnology. Avishkar Publishers, Jaipur, 346 pp

Kumar A, Sopory S (eds) (2010) Applications of plant biotechnology: *in vitro* propagation, plant transformation and secondary metabolite production. New Delhi. I.K. International, 606 pp

Kumar A, Bender L, Jeske C, Neumann K-H, Senger H, Strasberger G (1977) The development of photosynthetic system of carrot tissue culture. Proceedings of the International Congress on Photosynthesis, Reading, p 207

Kumar A, Bender L, Neumann K-H (1978) The development of the photosynthetic apparatus of tissue cultures (*Daucus* and *Arachis*) and autotrophic growth. Proceedings of the 4th International Congress for Plant Tissue and Cell Culture, Calgary, p 173

Kumar A, Bender L, Neumann, K-H (1980) Development of photosynthetic apparatus in *Daucus carota* L. Callus cultures. Proceedings of 8th American Society for Photobiology Annual Meeting, Colorado Springs, Colorado, Feb 1980, p 71

Maurino VG, Weber APM (2012) Engineering photosynthesis in plants and synthetic microorganisms. J Exp Bot. https://doi.org/10.1093/jxb/ers263

Nakagawa A, Matsumura E, Koyanagi T, Katayama T, Kawano N, Yoshimatsu K, Yamamoto K, Kumagai H, Sato F, Minami H (2016) Total biosynthesis of opiates by stepwise fermentation using engineered *Escherichia coli*. Nat Commun 7:10390. https://doi.org/10.1038/ncomms10390

Neumann K-H, Bender L, Kumar A, Szeque M (1982) Photosynthesis and pathways of carbon in tissue cultures of *Daucus* and *Arachis*. Proceeding International Congress of Plant tissue and Cell Culture at Tokyo, pp 251–252

Neumann K, Kumar A, Imani J (2009) Plant cell and tissue culture—a tool in biotechnology basics and application. Springer, Dresden, 333 pp

Peiter E, Imani J, Yan F, Schubert S (2003) A novel procedure for gentle isolation and separation of intact infected and uninfected protoplasts from the central tissue of *Vicia faba* L. root nodules. Plant Cell Environ 26:1117–1126

von Caemmerer S, Quick WP, Furbank RT (2012) The development of C4 rice: current progress and future challenges. Science 336:1671–1672

Waller F, Aschatz B, Baltruschat H, Fodor J, Becker K, Fischer M, Heier T, Hückelhoven R, Neumann C, von Wettstein D, Franken P, Kogel KH (2005) The endophytic fungus *Piriformospora indica* reprograms barley to salt-stress tolerance, disease resistance, and higher yield. Proc Natl Acad Sci USA 102:13386–13391

Waltz E (2016) CRISPR-edited crops free to enter market, skip regulation. Nat Biotechnol 34 (6):582–582. https://doi.org/10.1038/nbt0616-582

Weber APM, Brautigam A (2013) The role of membrane transport in metabolic engineering of plant primary metabolism. Curr Opin Biotechnol 24:256–262. https://doi.org/10.1016/j.copbio.2012.09.010

Wilson SA, Roberts SC (2012) Recent advances towards development and commercialization of plant cell culture processes for synthesis of biomolecules. Plant Biotechnol J 10:249–268. https://doi.org/10.1111/j.1467-7652.2011.00664

Wilson SA, Roberts SC (2014) ScienceDirect Metabolic engineering approaches for production of biochemicals in food and medicinal plants. Curr Opin Biotechnol 26:174–182. https://doi.org/10.1016/j.copbio.2014.01.006

Xie M, Haellman V, Fussenegger M (2016) ScienceDirect synthetic biology: application-oriented cell engineering. Curr Opin Biotechnol 40:139–148. https://doi.org/10.1016/j.copbio.2016.04.005

Historical Developments of Cell and Tissue Culture Techniques

2

Possibly the contribution of Haberlandt to the Sitzungsberichte der Wissenschaftlichen Akademie zu Wien more than a century ago (Haberlandt 1902) can be regarded as the first publication of experiments to culture isolated tissue from a plant (*Tradescantia*). To secure nurture requirements, Haberlandt used leaf explants capable of active photosynthesis. Nowadays, we know leaf tissue is rather difficult to culture. With these experiments (and others), Haberlandt wanted to promote a "physiological anatomy" of plants. In his book on the topic, with its 600 odd pages, he only once cited his "tissue culture paper" (page 13), although he was not very modest in doing so. Haberlandt wrote (Fig. 2.1):

> Gewöhnlich ist die Zelle als Elementarorgan zugleich ein Elementarorganismus; mit anderen Worten: sie steht nicht bloß im Dienste der höchsten individuellen Lebenseinheit, der ganzen Pflanze, sondern gibt sich selbst als Lebenseinheit niedrigen Grades zu erkennen. So ist z.B. jede von den chlorophyllführenden Palisadenzellen des Phanerogamenlaubblattes ein elementares Assimilationsorgan, zugleich aber auch ein lebender Organismus: man kann die Zelle mit gehöriger Vorsicht von dem gemeinschaftlichen Zellverbande loslösen, ohne daß sie deshalb sofort aufhören würde zu leben. Es ist mir sogar gelungen, derartige Zellen in geeigneten Nährlösungen mehrere Wochen lang am Leben zu erhalten; sie setzten ihre Assimilationstätigkeit fort und fingen sogar in sehr erheblichem Maße wieder zu wachsen an.

In English, this reads:

Usually, a cell is an elementary organ as well as an elementary organism—it is not only part of an individual living unit, i.e., of the intact plant, but also is itself a living unit at a lower organizational level. As an example, each palisade cell of the phanerogamic leaf blade containing chlorophyll is an elementary unit of assimilation, and concurrently a living organism—careful isolation from the tissue keeps these cells alive. I have even been able to maintain such cells living in a suitable nutrient medium for several weeks; assimilation continued, and considerable growth was possible (Härtel 2003).

© Springer Nature Switzerland AG 2020

K.-H. Neumann et al., *Plant Cell and Tissue Culture – A Tool in Biotechnology*,

https://doi.org/10.1007/978-3-030-49098-0_2

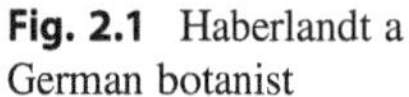
Fig. 2.1 Haberlandt a
German botanist

With this, the theoretical basis of plant and tissue culture systems as practiced nowadays was defined. Apparently, this work was of minor importance to Haberlandt, who viewed it only as evidence of a certain independence of cells from the whole organism. Nevertheless, it has to be kept in mind that at the time Schleiden and Schwann's theory of significance of cells was only about 30 years old (cf. Schwann 1839). Later, Haberlandt abandoned this area of research and turned to studying wound healing in plants. A critical review is given by Krikorian and Berquam (1986).

It was not before the late 1920s–early 1930s that in vitro studies using plant cell cultures were resumed, in particular due to the successful cultivation of animal tissue, mainly by Carrell. In a paper published in 1927, Rehwald reported the formation of callus tissue on cultured explants of carrot and some other species, without the influence of pathogens. Subsequently, Gautheret (1934) described growth by cell division in vitro of cultured explants from the cambium of *Acer pseudoplatanus*. Growth of these cultures came to a halt, however, after about

Fig. 2.2 First meeting of the National Correspondents for IAPTC at the Congress in Leicester, 1974 (Gamborg 2002)

18 months. Meanwhile, the significance of indole acetic acid (IAA) became known, as a hormone influencing cell division and cellular growth. Rehwald did not continue his studies, but based on these, Nobecourt (1937) investigated the significance of this auxin for growth of carrot explants. Successful long-term growth of cambium explants was reported at about the same time by Gautheret (1939) and White (1939)

For Gautheret and Nobecourt, continued growth could be maintained only in the presence of IAA. White, however, was able to achieve this without IAA, by using tissue of a hybrid of *Nicotiana glauca* and *Nicotiana langsdorffii*. Intact plants of this hybrid line are also able to produce cancer-like outgrowth of callus without auxin. Many years later, a comparable observation was made on hybrids of two *Daucus* subspecies produced by protoplast fusion, yielding somatic embryos for intact plants (Sect. 7.3) in an inorganic nutrient medium. *Daucus* and *Nicotiana* have remained model systems for cell culture studies until now but have recently been rivaled by *Arabidopsis thaliana* (Fig. 2.2).

In the investigations discussed so far, the main aim was to unravel the physiological functions of various plant tissues and their contributions to the life of the intact plant. In the original White's basal medium often used, not much fresh weight is produced, and this is mainly by cellular growth. Only a low rate of cell division has been observed.

A new turn of studies was induced in the late 1950s and early 1960s by the work of the research group of F.C. Steward at Cornell University in Ithaca, NY, and of F. Skoog's group in Wisconsin. Steward was interested mainly in relations between nutrient uptake and tissue growth intensity. To this end, he attempted to use fast- and slow-growing tissue cultures of identical origin in the intact plant as model systems. He was aware of the work of van Overbeck et al. (1942), who used coconut milk, i.e., the liquid endosperm of *Cocos nucifera*, to grow immature embryos derived from hybrids of crossings between different *Datura* species. Usually, the development of embryos of such hybrids is very poor, and they eventually die. Following the application of coconut milk, however, their development was accomplished. A supplement of coconut milk to the original medium of P. White induced vigorous growth in quiescent carrot root explants (secondary phloem), compared to that in the original nutrient medium. For Steward, this meant he now had an experimental system in which, by addition or omission of coconut milk, it was possible to evaluate the role played by variations in growth intensity of tissue of identical origin in the plant (Caplin and Steward 1949). The supplement of coconut milk induced growth mainly by cell division that resulted in dedifferentiation of the cultured root explants, and the histological characteristics of the secondary phloem tissue were soon lost. This probably provoked P. White, at a conference in 1961, to ask "What do you need coconut milk for?"

The observation of the induction of somatic embryogenesis in cell suspensions was an unexpected by-product of such experiments (Steward et al. 1958; see Sect. 7.3), a process described at about the same time also by Reinert (1959). Contrary to Steward, who observed somatic embryogenesis in cell suspensions derived from callus cultures, Reinert described this process in callus cultures.

At the beginning of the 1950s, the Steward group initiated investigations to isolate and characterize the chemical components of coconut milk responsible for the vigorous growth of carrot explants, after its supplementation to the nutrient medium. Similar influences on growth became known for liquid endosperms of other plant species, like *Zea* or *Aesculus*, and these were consequently included into the investigations. Some years ago, when already retired, Steward (1985) published a very good summary of these investigations, and therefore no detailed discussion of this work will be attempted here, but some highlights will be recalled.

In summary, using ion exchange columns, three fractions with growth-promoting properties have been isolated from coconut milk. These are an amino acid fraction that, to promote growth, can be replaced by casein hydrolysate or other mixtures of amino acids. Then came the identification of some active components of a neutral fraction. This fraction contains mainly carbohydrates and other chemically neutral compounds. Particularly active in the carrot assay were three hexitols, i.e., myo- and scyllo-inositol and sorbitol. Of these, the strongest growth promotion was obtained with m-inositol: 50 mg/l of this as supplement induced the same amount of growth as did the whole neutral fraction of coconut milk. Actually, earlier also White (1954) recommended an m-inositol supplement to the media as a promoter of growth. Finally, there remains the so-called active fraction of coconut milk to be characterized, the analysis of which is yet not really completed. Still, the occurrence of 2-isopentenyladenine and of and some derivatives of these has been detected, and it seems justifiable to label it as the cytokinin fraction of coconut milk. The occurrence of these cytokinins would be responsible for the strong promotion of cell division activity by coconut milk, as will be described later.

In terms of when they were discovered, cytokinins are a rather "young" group of phytohormones, the detection of which is tightly coupled with cell and tissue culture. The first characterized member of this group was accidentally detected in autoclaved DNA. Its supplementation to cultured tobacco pith explants induced strong growth by cell division, and consequently it was named kinetin (Miller et al. 1955). Chemically, kinetin is a 6-substituted adenine. In plants, this compound has not been detected yet; it should be the product of chemical reactions associated with the process of autoclaving and deviating from enzymatic in situ reactions.

Using tobacco pith explants, Skoog and Miller (1957) carried out by now classic experiments demonstrating the influences of changes in the auxin/cytokinin concentration ratio on organogenesis in cultures. If auxin dominates, then the formation of adventitious roots is promoted; if cytokinins dominate, then the differentiation of shoot parts is observed. At a certain balance between the two hormone groups in the medium, undifferentiated callus growth results (Skoog and Miller 1957). These results are not as distinct in other experimental systems, but the principle derived from these experiments seems to be valid, and to some extent it can be applied also to intact plants.

As mentioned above, the liquid endosperm of *Zea* exerts a similar influence on growth as does coconut milk. Based on the work of the Steward group, Letham (1966) isolated the first native cytokinin, and fittingly it was named zeatin. Shortly after, a second native cytokinin, 2-isopentenyladenine, was identified, which is a precursor of zeatin. Since then, several derivatives have been described, and today more than 20 naturally occurring cytokinins are known, a number that will certainly grow.

In the early 1960s, the way was paved to formulate the composition of synthetic nutrient media able to produce the same results as those obtained with complex, naturally occurring ingredients such as coconut milk or yeast extracts (of unknown composition). Nowadays, mostly the Murashige–Skoog medium (Murashige and Skoog 1962) is used, with a number of adaptations for specific purposes (cf. MS medium; see tables and further information in Chap. 3). In such synthetic media, somatic embryogenesis in carrot cultures was soon also induced (Halperin and Wetherell 1965; Linser and Neumann 1968).

Another line of research was initiated by the National Aeronautics and Space Administration (NASA), which started to support research on plant cell cultures for regenerative life support systems (Krikorian and Levine 1991; Krikorian 2001, 2003). Since the early 1960s, experiments with plants and plant tissue cultures have been performed under various conditions of microgravity in space (cf. one-way spaceships, biosatellites, space shuttles and parabolic flights, and the orbital stations Salyut and Mir), accompanied by ground studies using rotating clinostat vessels.

Table 2.1 Some examples of patent applications in Japan in the 1970s

Ingredient	Plant species
Berberine	*Coptis japonica*
Nicotine	*Nicotiana tabacum*
Hyoscyamine	*Datura stramonium*
Rauwolfia alkaloid	*Rauwolfia serpentine*
Camptothecin	*Camptotheca acuminata*
Ginseng saponins	*Panax ginseng*
Ubiquinone 10	*Nicotiana tabacum, Daucus carota*
Proteinase inhibitor	*Scopolia japonica*
Stevioside	*Stevia rebaudiana*
Tobacco material	*Nicotiana tabacum*
Silkworm diet	*Morus bombycis*

Neumann's (1966) formulation of the NL medium (see tables and further information in Chap. 3) was based on a mineral analysis of coconut milk (NL, Neumann Lösung, or medium). The concentrations of mineral nutrients in this liquid endosperm were applied, in addition to those already used for White's basal medium; moreover, 200 mg casein hydrolysate/l was supplemented, and kinetin, IAA, and m-inositol were applied at the concentrations given in the tables.

Using such synthetic nutrient media, it was possible to investigate the significance of each individual ingredient for the growth and differentiation of cultured cells, or for the biochemistry of the cells, including the production of components of secondary metabolism. This will be dealt with in later chapters of the book.

In the early 1960s appeared the first reports on androgenesis (Guha and Maheshwari 1964) and on the production and culture of protoplasts (Cocking 1960). Concurrently, systematic studies on components of secondary metabolism, mainly of medical interest, were initiated. At that time, cell and tissue cultures were at an initial peak of enthusiasm and popularity, which stretched from the end of the 1960s to the second half of the 1970s. The state of knowledge was such as to stimulate expectations of an imminent practical application of these techniques in many domains, e.g., plant breeding and the production of enzymes and that of drugs for medical purposes. To this end, considerable financial resources were made available from governments, as well as from private companies. Potential applications seemed limitless and included rather exotic ones such as the production of food for silkworms. These high investments were accompanied by first applications for patents (some examples from that time are given in Table 2.1). In the late 1970s, however, reality caught up—promises made by scientists (or at least by some) to sponsors, and expectations raised for an early application of these techniques on a commercial basis were not fulfilled—a "hangover" was the result.

All projects envisaged in that period had aspects related with cellular differentiation and its control. It was realized that without a clear understanding of these fundamental biological processes, enabling scientists to interfere accordingly to reach a given commercial goal, only an empirical trial and error approach was possible. In that pioneer phase in the commercialization of cell and tissue culture, a parallel was often drawn with the early days in the commercial use of microbes, i.e., the production of antibiotics with its originally low yield. It seemed to be necessary only to select high-yielding strains. Compared to microbes, however, the biochemical status of cultured plant cells is less stable, and many initially promising approaches were eventually found to lead to a technological blind alley. Furthermore, it has to be kept in mind that at the advent of antibiotics, no competitor was on the market. By contrast, for substances produced by plant cell cultures, well-established industrial methods and production lines exist. Also, the commercial production of enzymes and other proteins found solely in cells of higher plants would be based on microbes transformed by inserting genes of higher plants. Evidently, of more importance is certainly somatic embryogenesis to raise genetically transformed cell culture strains, and to produce intact plants for breeding—on condition that the transformation be carried out on protoplasts or isolated single cells.

A first system of this kind was reported by Potrykus in 1984 at the Botanical congress in Vienna (see Sect. 13.6.3). Kanamycin resistance was incorporated into tobacco protoplasts, from which kanamycin-resistant tobacco plants were obtained. Here, cell culture techniques were an indispensable, integral part of the experiments. Later, these basic principles were applied in many other systems, and today, after hundreds of genetic transformations, 100,000s hectares are planted with genetically transformed cultivated plants (see Sect. 13.6.3). An initial attempt to introduce commercially useful traits into plants was to prolong the viable storage period of tomatoes (Klee et al. 1991); these tomatoes became known as "Flavr Savr." In spite of being patented (Patent EP240208), commercial success was rather limited, and they were never permitted on the European market.

It was known for a long time that green cultured cells are able to perform photosynthesis (Neumann 1962, 1969; Bergmann 1967; Neumann and Raafat 1973; Kumar 1974a, b; Kumar et al. 1977, 1989, 1990; Neumann et al. 1977; Roy and Kumar 1986, 1990; Kumar and Neumann 1999; see review by Widholm 1992). In the 1980s were published the first papers reporting the prolonged cultivation of green cultures of various species growing at normal atmosphere in an inorganic nutrient medium (Bender et al. 1981; Neumann et al. 1982; Kumar et al. 1983a, b, 1984, 1987, 1989, 1999; Bender et al. 1985). Subsequently, the ability of such cultures to produce somatic embryos was demonstrated (see Chaps. 7 and 9). More recently, methods have been published to raise immature somatic embryos of the cotyledonary stage under autotrophic conditions, yielding intact plants (Chap. 7). It remains to be seen to which extent such material will be useful to obtain plants with special genetic transformations involving photosynthesis. Later, more details on this will be given (see Sect. 13.6.3).

Based on much earlier work in Knudson's laboratory at Cornell University in 1922 (cf. Griesebach 2002), in the early 1960s Morel (1963) reported a method to propagate *Cymbidium* by culturing shoot tips on seed germination medium supplemented with phytohormones in vitro. At Cornell, probably the first experiments with orchid tissue culture were performed, and inflorescence nodes of *Phalaenopsis* could be induced to produce plantlets in vitro cultured aseptically on seed germination media. Indeed, the Knudson C medium (with some variations) is still in use for orchid cultivation in vitro. During the last 40 years, techniques have been found to propagate many plant species, mainly ornamentals, generally employing isolated meristems for in vitro culture (see Chap. 7). These methods were developed empirically by trial and error, and the propagation in vitro of many plant species is used commercially. Up to the 1960s, orchids belonged to the most expensive flowers—the low price nowadays is due to propagation by tissue culture techniques (even students can afford an orchid for their sweetheart at their first date!).

In the following, the various branches of cell and tissue cultures will be described, including methods for practical applications.

Stasolla and Thorpe (2010) reviewed the historical development of plant tissue culture (see also Kumar and Sopory 2010). Genetic transformation has represented a more precise and predictable method for producing plants with new and desirable traits (Vasil (2007). A step heavily targeted in transgenic studies is the fixation of carbon by RuBisCO (see Datta 2007). Compared to C3 plants, C4 plants have evolved mechanisms to minimize RuBisCO oxygenation which include (1) an increase of CO2 concentration around RuBisCO and (2) the conversion of CO2 into C4 acids mediated by phosphoenolpyruvate carboxylase (PEPC). Over the past few years, there have been several attempts to express PEPC in C3 plants in an effort to enhance plant productivity. Gehlen et al. (1996) were the first to constitutively express this enzyme, although without great success. It was only a few years later that Ku et al. (2000) produced transgenic rice able to accumulate PEPC. Such plants showed an increased net photosynthetic rate and represented the first example of how the installation of a single gene from another source is able to trigger a new pathway in the host plant (Datta 2007). Following these initial successes, several labs are currently trying to express in rice the other two enzymes participating in the C4 pathways, pyruvate orthophosphate dikinase and NADPmalic enzyme (Datta 2007).

A large proportion of crop production is generally lost worldwide due to abiotic stresses, such as drought, extreme temperature, and salinity, as well as biotic factors such as bacteria, fungi, and insects. Over the past few years, several plant, bacterial, and animal genes have been assessed in their ability to confer resistance to biotic and abiotic stress, especially in model systems, such as *Arabidopsis* and tobacco (Datta 2007).

Studies with [14C]-glucose and [14C]-acetate during organogenesis in tobacco (Thorpe and Beaudoin-Eagan 1984) and radiata pine (Obata-Sasamoto et al. 1984) showed preferential incorporation of label into various metabolites at higher rates during meristemoid and primordium initiation compared to control tissues, as well as elevated levels of 14CO2 release. Label was found in the lipid, amino acid, organic acid, and sugar fractions. Label from [14C]-glucose went mainly into malate, citrate, glutamate, glutamine, and alanine, but there were no qualitative, only quantitative differences between SF and NSF cotyledons.

Approximately twice as much label went into protein, in SF cotyledons from [14C]-glucose, an observation also confirmed with 35S-methionine (Thompson and

Thorpe 1997). Similar results have been obtained as well with [14C]-acetate feeding (Joy et al. 1994). These studies showed inter alia enhanced metabolism during meristemoid and primordium initiation and metabolism indicative of high energy and amino acid requirements. In contrast, feeding either tissue with [14C]-bicarbonate led to higher incorporation into the NSF tissues (Obata-Sasamoto et al. 1984; Thorpe and Beaudoin-Eagan 1984). In radiata pine a higher level of activity of ribulose bisphosphate carboxylase/oxygenase was found in NSF compared to SF cotyledons, while the reverse was observed for phosphoenolpyruvate (PEP) carboxylase (Kumar et al. 1988). Similarly, enhanced activity of PEP carboxylase has been found in tobacco callus during meristemoid formation (Plumb-Dhindsa et al. 1979). These latter observations indicate the importance of non-autotrophic CO_2 fixation for organogenesis, presumably relative to malate metabolism and production of reducing power.

All of the above findings are consistent with the hypothesis that organized development involves a shift in metabolism that leads to changes in the content and the spectrum of both structural and enzymatic proteins (Thorpe 1980, 1983). These metabolic changes which precede or are coincident with the differentiation process must be considered to be causative (Stasolla and Thorpe 2010).

Production of pathogen-free plants and germplasm storage are related topics in that a major use of pathogen-free plants is for germplasm storage (Thorpe 1990). Pathogen attack and infection are quite common across plant species especially in susceptible species such as strawberries, which are subjected to many viruses and mycoplasmas (Boxus 1976). Although in some cases these pathogens cause visible damages and plant death, in many circumstances the symptoms are not detectable but do have a negative effect on yield and quality (Thorpe and Harry 1997). Virus eradication through the use of virus-free stocks has resulted in increased productivity ranging from 30 to 300% (Bhojwani and Razdan 1983).

References

Bender L, Kumar A, Neumann KH (1981) Photoautotrophe pflanzliche Gewebekulturen in Laborfermentern. In: Lafferty RM (ed) Fermentation. Springer, Wien, pp 193–203

Bender L, Kumar A, Neumann KH (1985) On the photosynthetic system and assimilate metabolism of *Daucus* and *Arachis* cell cultures. In: Neumann KH, Barz W, Reinhard E (eds) Primary and secondary metabolism of plant cell cultures. Springer, Berlin, pp 24–42

Bergmann L (1967) Wachstum grüner Suspensionskulturen von *Nicotiana tabacum* var. Samsun mit CO_2 als Kohlenstoffquelle. Planta 74:243–249

Bhojwani SS, Razdan MK (1983) Plant tissue culture: theory and practice. Developments in crop science, vol 5. Elsevier, Amsterdam

Boxus P (1976) Rapid production of virus-free strawberry by in vitro culture. Acta Hortic 66:35–38

Caplin SM, Steward FC (1949) A technique for the controlled growth of excised plant tissue in liquid media under aseptic conditions. Nature CIXIII:920–924

Cocking EC (1960) A method for the isolation of plant protoplasts and vacuoles. Nature 187:962–968

Datta SK (2007) Impact of plant biotechnology in agriculture. In: Pua EC, Davey MR (eds) Biotechnology in agriculture and forestry, Transgenic crops IV, vol 59. Springer, Berlin, pp 1–31

Gamborg OL (2002) Plant tissue culture. Biotechnology. Milestones. In Vitro Cell Dev Biol-Plant 38:84–92

Gautheret RJ (1934) Culture de tissu cambial. C R Acad Sci Paris 198:2195–2196

Gautheret RJ (1939) Sur la possibilité de réaliser la culture indéfinie de tissu de tubercules de carotte. C R Acad Sci Paris 208:118–120

Gehlen J, Panstruga R, Smets H, Merkelbach S, Kleines M, Porsch P (1996) Effects of altered phosphoenolpyruvate carboxylase activities on transgenic C3 plant *Solanum tuberosum*. Plant Mol Biol 32:831–848

Griesebach RJ (2002) Development of *Phalaenopsis* orchids for the mass-market. In: Janick J, Whipkey A (eds) Trends in new crops and new uses. ASHS Press, Alexandria, VA, pp 458–465

Guha S, Maheshwari S (1964) In vitro production of embryos from Anthers of Datura. Nature 204:497. https://doi.org/10.1038/204497a0

Haberlandt G (1902) Kulturversuche mit isolierten Pflanzenzellen. Sitzungsber Akad Wiss Wien Math-Naturwiss Kl, Abt J 111:69–92

Halperin WS, Wetherell DF (1965) Ammonium requirement for embryogenesis *in vitro*. Nature 205:519–520

Härtel O (2003) Gottlieb Haberlandt (1854–1945): a portrait. In: Laimer M, Rücker W (eds) Plant tissue culture. Springer, Vienna

Joy RW IV, Bender L, Thorpe TA (1994) Nitrogen metabolism in cultured cotyledon explants of *Pinus radiata* during de novo organogenesis. Physiol Plant 92:681–688

Klee HJ, Hayford MB, Kretzmer KA, Barry GF, Kishore GM (1991) Control of ethylene synthesis by expression of a bacterial enzyme in transgenic tomato plants. Plant Cell 3:1187–1193

Krikorian AD (2001) Novel application of plant tissue culture and conventional breeding techniques to space biology research. In: Bender L, Kumar A (eds) From soil to cell—a broad approach to plant life. Giessen Electronic Library, GEB. http://geb.uni-giessen.de/geb/ebooks_ebene2.php

Krikorian AD (2003) Stress and genome shock in developing somatic embryos in space. In: Biotechnology 2002 and beyond. Proceedings of the 10th IAPTC Congress, Orlando, FL. Kluwer, Dordrecht, pp 347–350

Krikorian AD, Berquam DJ (1986) Plant cell and tissue cultures: the role of Haberlandt. Bot Rev 35:59–67

Krikorian AD, Levine HG (1991) Development and growth in space. In: Bidwell RGS, Steward FC (eds) Plant physiology, a treatise, vol X. Academic Press, New York, pp 491–555

Ku, M.S.B Ranade, U. Hsu, T-P Li, X. Jiao, D-M. Ehleringer, J. Miyao M, Matsuoka M. (2000) Photosynthetic performance of transgenic rice plants overexpressing maize C4 photosynthesis enzymes. In: Sheehy JE, Mitchell PL, Hardy B (eds) Redesigning rice photosynthesis to increase yield. Studies in plant science, vol 7. Elsevier Science B.V. 293, pp 193–204

Kumar A (1974a) Effect of iron and magnesium on the growth and chlorophyll development of the tissues grown in culture. Indian J Exp Biol 12:595–596

Kumar A (1974b) In vitro growth and chlorophyll formation in mesophyll callus tissues on sugar free medium. Phytomorphology 24:96–101

Kumar A, Bender L, Jeske C, Neumann K-H, Senger H, Strasberger G (1977) The development of photosynthetic system of carrot tissue culture. Proceedings of Intl. congress on photosynthesis. Reading, p 207

Kumar A, Neumann K-H (1999) Comparative investigations on plastid development of meristematic regions of seedlings and tissue cultures of *Daucus carota* L. J Appl Bot Angew Bot 73:206–210

Kumar A, Sopory S (eds) (2010) Applications of plant biotechnology: *in vitro* propagation, plant transformation and secondary metabolite production. I.K. International, New Delhi, 606 pp

Kumar A, Bender L, Neumann KH (1983a) Photosynthesis in cultured plant cells and their autotrophic growth. Int J Plant Physiol Biochem 10:130–140

Kumar A, Bender L, Pauler B, Neumann KH, Senger H, Jeske C (1983b) Ultrastructural and biochemical development of the photosynthetic apparatus during callus induction in carrot root explants. Plant Cell Tissue Organ Cult 2:161–177

Kumar A, Bender L, Neumann KH (1984) Growth regulation, plastid differentiation and the development of photosynthetic system in cultured carrot root explants as influenced by exogenous sucrose and various phytohormones. Plant Cell Tissue Organ Cult 3:11–28

Kumar A, Bender L, Neumann KH (1987) Some results of the photosynthetic system of mixotrophic carrot cells (*Daucus carota*). In: Biggins J (ed) Progress in photosynthesis research, vol III(4). Martin Nijhoff, Dordrecht, pp 363–366

Kumar PP, Bender L, Thorpe TA (1988) Activities of ribulose bisphosphate carboxylase and phosphoenolpyruvate carboxylase and 14C-bicarbonate fixation during in vitro culture of radiata pine. Plant Physiol 87:675–679

Kumar A, Roy S, Neumann KH (1989) Activities of carbon dioxide fixing enzymes in maize tissue cultures in comparison to young seedlings. Physiol Plant 76:185

Kumar A, Roy S, Neumann K-H (1990) Activities of carbon dioxide fixing enzymes in maize tissue cultures in comparison to young seedlings. In: Baltscheffsky M (ed) Current research in photosynthesis, vol IV. Kluwer, Dordrecht, pp 275–278

Kumar A, Bender L, Neumann K-H (1999) Characterization of the photosynthetic system (ultrastructure of plastid, fluorescence induction profiles, low temperature spectra) of *Arachis hypogaea* L. callus cultures as influenced by sucrose and various hormonal treatments. J Appl Bot Angew Bot 73:211–216

Letham DS (1966) Regulators of cell division in plant tissue. II. A cytokinin in plant extracts: isolation and interaction with other growth regulators. Phytochemistry 5:269

Linser H, Neumann KH (1968) Untersuchungen über Beziehungen zwischen Zellteilung und Morphogenese bei Gewebekulturen von *Daucus carota* L. I. Rhizogenese und Ausbildung ganzer Pflanzen. Physiol Plant 21:487–499

Miller CO, Skoog F, Okomura FS, von Salza MH, Strong FM (1955) Isolation, structure and synthesis of kinetin, a substance promoting cell division. J Am Chem Soc 78:1375–1380

Morel G (1963) La culture in vitro du méristème apical de certaines orchidées. C R Acad Sci Paris 256:4955–4957

Murashige T, Skoog F (1962) A revised medium for rapid growth and bioassays with tobacco tissue. Physiol Plant 15:473–497

Neumann KH (1962) Untersuchungen über den Einfluß essentieller Schwermetalle auf das Wachstum und den Proteinstoffwechsel von Karottengewebekulturen. Dissertation, Justus Liebig Universität, Giessen

Neumann KH (1966) Wurzelbildung und Nukleinsäuregehalt bei Phloem-Gewebekulturen der Karottenwurzel auf synthetischem Nährmedium. Congr Coll Univ Liege 38:96–102

Neumann KH (1969) Der Eintritt der Elemente Kohlenstoff, Wasserstoff und Sauerstoff in den Stoffwechsel. In: Linser H (ed) Handbuch der Pflanzenernährung und Düngung. Springer, Wien, pp 301–374

Neumann K-H, Kumar A, Bender L (1977) Autotrophic growth and photosynthetic system of carrot and peanut tissue culture. Proceedings of Intl. Congress on Photosynthesis. Reading, p 270

Neumann KH, Raafat A (1973) Further studies on the photosynthesis of carrot tissue cultures. Plant Physiol 51:685–690

Neumann KH, Bender L, Kumar A, Szegoe M (1982) Photosynthesis and pathways of carbon in tissue cultures of *Daucus* and *Arachis*. In: Proceedings of the International Congress Plant Tissue and Cell Culture, Tokyo, pp 251–252

Nobecourt PC (1937) Culture en série de tissus végétaux sur le milieu artificiel. C R Acad Sci Paris 205:521–523

Obata-Sasamoto H, Villalobos VM, Thorpe TA (1984) 14C-metabolism in cultured cotyledon explants of radiata pine. Physiol Plant 61:490–496

Plumb-Dhindsa PL, Dhindsa RS, Thorpe TA (1979) Non-autotrophic CO2 fixation during shoot formation in tobacco callus. J Exp Bot 30:759–767

Rehwald L (1927) Zeitschr Pflanzenkrankenkeiten Pflanzenschutz 37:65–86

Reinert J (1959) Über die Kontrolle der Embryogenese und die Induktion von Adventivembryonen an Gewebekulturen aus Karotten. Planta 53:318–333

Roy S, Kumar A (1986) Induction of chlorophyll development in maize callus cultures. In: Proceedings of the 6th International Congress Plant Tissue and Cell Culture, Minneapolis, MN, p 367

Roy S, Kumar A (1990) Development of photosynthetic apparatus in callus cultures derived from a C4 plant. In: Baltscheffsky M (ed) Current research in photosynthesis, vol III. Kluwer, Dordrecht, pp 877–880

Schwann T (1839) Mikroskopische Untersuchungen über die Uebereinstimmung der Struktur und dem Wachstum der Thiere und Pflanzen. Sander, Berlin

Skoog F, Miller CO (1957) Chemical regulation of growth and organ formation in plant tissues cultured in vitro. Symp Soc Exp Biol 11:118–140

Stasolla C, Thorpe TA (2010) Tissue culture: historical perspectives and applications. In: Kumar A, Sopory S (eds) Applications of plant biotechnology: *in vitro* propagation, plant transformation and secondary metabolite production. I.K. International, New Delhi, pp 1–39

Steward FC (1985) From metabolism and osmotic work to totipotency and morphogenesis: a study of limitations versus multiple interaction. In: Neumann KH, Barz W, Reinhard E (eds) Primary and secondary metabolism of plant cell cultures. Springer, Berlin, pp 1–11

Steward FC, Mapes MO, Mears K (1958) Growth and organized development of cultured plant cells. Am J Bot 45:705–708

Thompson MR, Thorpe TA (1997) Analysis of protein patterns during shoot initiation in cultured *Pinus radiata* cotyledons. J Plant Physiol 151:724–734

Thorpe TA (1980) Organogenesis in vitro: structural, physiological, and biochemical aspects. Int Rev Cytol Suppl 11A:71–111

Thorpe TA (1983) Morphogenesis and regeneration in tissue culture. In Owens LD (ed) Genetic engineering: applications to agriculture. Beltsville Symphony No 7. Rowan & Allanheld, Totowa, NJ, pp 258–303

Thorpe TA (1990) The current status of plant tissue culture. In: Bhojwani SS (ed) Plant tissue culture: applications and limitations. Elsevier, Amsterdam, pp 1–33

Thorpe TA, Beaudoin-Eagan LD (1984) 14C-metabolism during growth and shoot formation in tobacco callus. Z Pflanzenphysiol 113:337–346

Thorpe TA, Harry IS (1997) Application of tissue culture to horticulture. Proceedings of the third international ISHS symposium on in vitro culture and horticultural breeding. Jerusalem, pp 39–49

van Overbeck J, Conklin ME, Blakeslee A (1942) Cultivation in vitro of small *Datura* embryos. Am J Bot 29:472–477

Vasil IK (2007) A short history of plant biotechnology. Phytochem Rev (published on line in October 2007)

White PH (1939) Controlled differentiation in a plant tissue culture. Bull Torrey Bot Club 66:507–519

White PH (1954) The cultivation of animal and plant cells. Ronald Press, New York

Widholm JM (1992) Properties and uses of photoautotrophic plant cell cultures. Int Rev Cytol 132:109–175

Callus Cultures

3

After the transfer of freshly cut explants into growth-promoting conditions, usually on the cut surface cell division is initiated, and as a form of wound healing, unorganized growth occurs—a callus will be formed. Following a supplement of growth hormones to the nutrient medium, this initial cell division activity will continue, and this unorganized growth will be maintained without morphological recognizable differentiation. However, under suitable conditions, the differentiation of adventitious roots, shoots, or even embryos can be initiated. Such culture systems can be used to study cytological or biochemical processes of growth related to cell division, cell enlargement, and differentiation. For a description of callus cultures, the culture of carrot root explants here serves as detailed example. Significant deviations from this experimental system will be dealt with later.

Depending on the objectives of the investigations, the culture of the isolated tissue will be either on a solid medium (0.8% agar, 0.4% Gelrite) or in a liquid medium. For both, usually glass vessels are employed, and after transfer of the medium, sterilization by autoclaving follows. As a substitute for glass vessels, sterile "one-way" containers made of plastic material are available on the market (Table 3.1). These are quite costly, however, and it therefore depends on the financial situation of the laboratory which of the two alternatives is favored. To exclude influences of components dissolved from the plastic, control investigations using glass containers are always recommended.

After cooling of the autoclaved vessels containing the nutrient medium, the explants are inoculated. The actual culture is usually carried out in growth rooms at temperatures of 20–30 °C under illumination conditions varying from continuous darkness to 10,000 lux, from fluorescent lamps. The lids on the vessels are closed by aluminum or paraffin foil, and consequently sufficient air humidity is provided for at least 4 weeks of culture.

For agar cultures, besides some shelves and climatization, no other provisions are required. Liquid cultures, however, if submersed, require sufficient continuous aeration. Using Erlenmeyer flasks as culture vessels, rotary shakers with about 100 rpm usually give good results (Fig. 3.1).

© Springer Nature Switzerland AG 2020

K.-H. Neumann et al., *Plant Cell and Tissue Culture – A Tool in Biotechnology*,

https://doi.org/10.1007/978-3-030-49098-0_3

Table 3.1 Autoclavability of some plastics (Thorpe and Kamlesh 1984)

Autoclavable	Not autoclavable
Polypropylene	Polystyrene
Polymethylpentene	Polyvinylchloride (PVC)
Teflon	Styrene acrylonitrile
Acryl	Tefrel
Polycarbonate	Polyethylene
Polysulfone	Polyallomer

An interesting setup for liquid cultures is a device called an auxophyton, developed in the early 1950s by the Steward group at Cornell University (Fig. 3.2). Here, wooden discs with clips are mounted onto a slowly rotating, nearly horizontal metallic shaft. Onto these clips, glass containers of 3 cm diameter closed on both sides (about 70 ml volume) are fixed, to which 15 ml liquid medium is applied. For gas exchange, an opening of about 1.5 cm with a collar of about 1.5 cm is maintained. The shaft rotates at 1 rpm, resulting in the nutrient medium being continuously mixed and aerated. Due to the development of a film of liquid, the cell material is usually fixed to the glass of the container, being alternately exposed to air and to the nutrient medium. With this setup, a better reproduction of data on growth and development is generally observed than is the case with shaker or agar cultures, especially in physiological or biochemical investigations. These "Steward tubes" (or T-tubes; Fig. 3.2, top) in our standard experiments are supplied with three explants each. For many biochemical investigations, however, this is not enough cellular material. Based on the same principle for the production of more material, so-called star flasks (or nipple flasks) were developed (Fig. 3.2, bottom).

The inner volume of these vessels is 1000 ml, usually 250 ml of medium is applied, and 100 explants are inoculated. Due to the nipples in the wall of the container during rotation of the shaft to which they are mounted, the cellular material is fixed, and again as with T-tubes, alternate exposure to air and nutrient medium is achieved. Basically, the same principle of alternating exposure of the cultures to the nutrient medium and the air was applied may years later to develop the RITA system, and similar setups described in Chap. 7.

To prevent microbial contamination, the culture vessels can be closed by cotton wool wrapped in cheesecloth, as a simple method. However, many other materials, such as aluminum foil, or more costly products on the market, can be used instead.

3.1   Establishment of a Primary Culture from Explants of the Secondary Phloem of the Carrot Root

To illustrate the method to obtain a primary culture, in the following a description of the original procedure of the Steward group for callus cultures from carrot roots will be described step by step (Fig. 3.3). This procedure can usually be adapted for use with other tissue types.

Fig. 3.1 Top culture room with stationary cultures; bottom cell suspensions as liquid cultures on a rotating shaker

Fig. 3.2 Auxophyton (Steward et al. 1952): top T-tubes; bottom "star flask"

Preparation

- To obtain discs of the carrot root, a simple cutting platform is used (Fig. 3.3d); beforehand, this is wrapped in aluminum foil or a suitable paper bag and placed for 4 h into a drying oven at 150 °C for sterilization.
- Lids to close apertures in the culture vessels are prepared from aluminum foil by hand, and the vessels are labeled according to the design of the experiment.
- Preparation of the nutrient medium follows (see below) and adjustment of the pH of the medium with 0.1N NaOH and 0.1N HCl.
- The nutrient medium is transferred to the culture vessel (15 ml each) by means of a pipette or more conveniently by using a dispenset. If stationary cultures are to be set up, it is necessary to apply also agar in solid form (e.g., 0.8%).

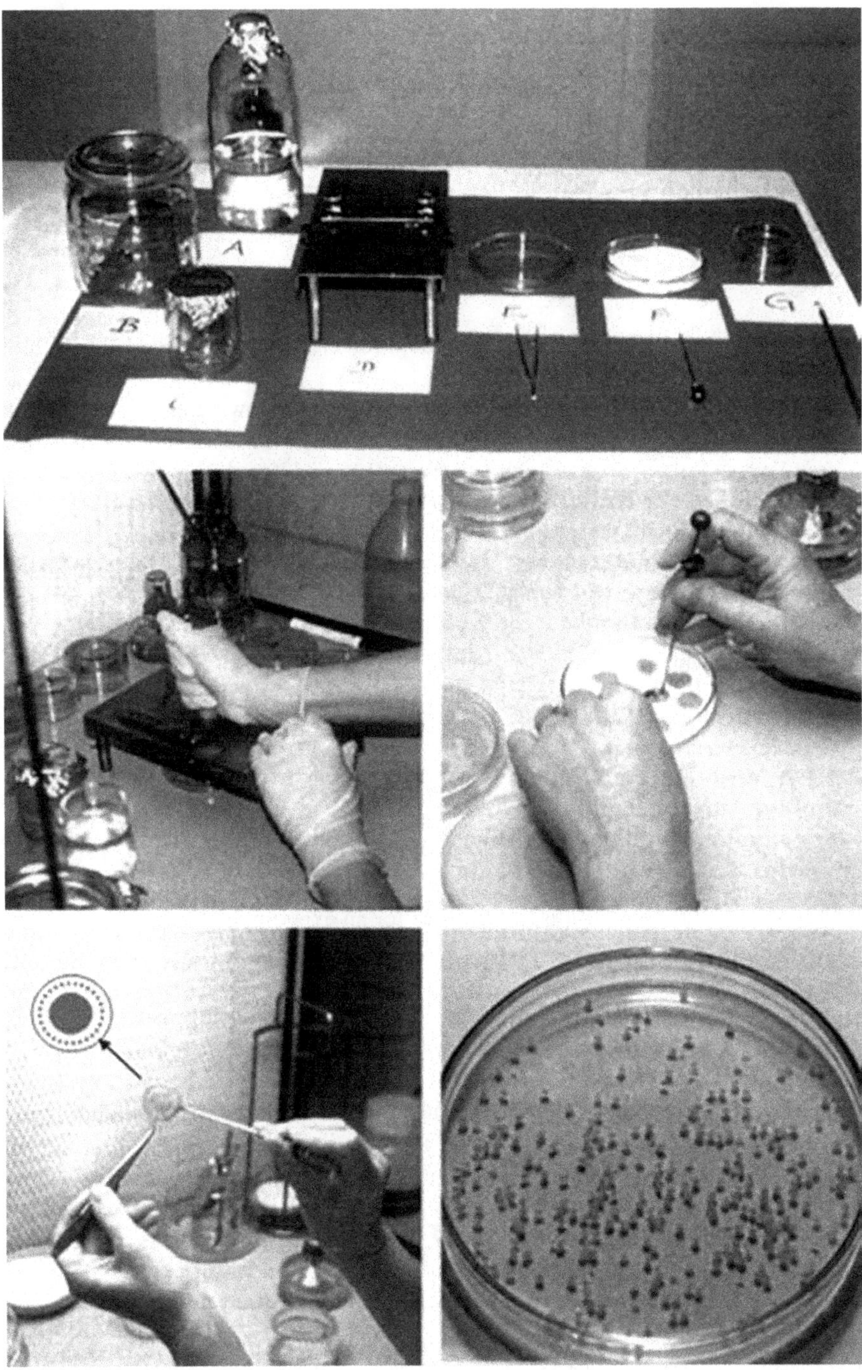

Fig. 3.3 Preparation of explants from a carrot root. Top equipment used for explantation: (*A*) sterilized aqua dest. to wash the tissue, (*B*) jar for surface sterilization of the carrot root, (*C*) jar in which to place the sterilized carrot root, (*D*) cutting platform to obtain root discs, (*E*) Petri dish to receive the root discs and sterilized forceps to handle the root discs, (*F*) Petri dish with filter paper in

Table 3.2 Some disinfectants used in tissue culture experiments and the concentrations applied (Thorpe and Kamlesh 1984)

Disinfectant	%
Sodium hypochlorite (5% active chlorine)	20.0
Calcium hypochlorite	25.0
Bromine water	1.0
Mercury chloride	0.2
Ethanol	70.0
Hydrogen superoxide	10.0
Silver nitrate	1.0

- The culture vessel is closed with aluminum foil caps and sterilized at 1.1 bar and 120 °C for 40 min in an autoclave.
- For each carrot root to be used for explantation, the following equipment should be sterilized (Fig. 3.3): several Petri dishes (diameter 9 cm) furnished with three to four layers of filter paper (autoclaving); one Petri dish for placing forceps, needle, and troquar (Fig. 3.3f); one Petri dish and two 1 l beakers (dry sterilization); 1 l of aqua dest. distributed in several Erlenmeyer flasks (autoclaving); for each carrot to be used in the experiment, two forceps, one troquar, and one needle with a loop made of platinum or stainless steel, wrapped into aluminum foil and dry-sterilized.
- All work to obtain explants for culture is carried out in a sterilized inoculation room or more conveniently on a laminar flow (aseptic working bench). This has to be switched on 30 min before starting the experimental work.

Procedure

To determine the vitality and potential growth performance of the explants before surface sterilization, a disc of the diameter of the carrot root is cut, and with the troquar explants are cut. These are put into a beaker with water, and if the explants swim on the surface, the root is not suitable for an experiment. Explants of healthy carrots sink to the bottom of the container.

- After the selection of a suitable carrot, the root is scraped and washed with aqua dest., dried with a paper towel, and wrapped into three to four layers of paper towel.
- The carrot is placed into a 1 l beaker and covered with a sterilizing solution (e.g., 5% hypochlorite; see Table 3.2) for 15 min. Sterile gloves are needed for further processing. If gloves are not used, then it is necessary to wash one's hands here and then frequently in the following steps, with ethanol or a clinical disinfectant (e.g., Lysafaren).

Fig. 3.3 (continued) which to place the root disc for cutting the explants and troquar (or cork borer) to cut the explants, (*G*) jar in which to place the explants for rinsing and needle (at the tip, with a loop) for explant transfer. (Middle) left cutting discs from the carrot root, (right) cutting of explants from the disc. (Bottom, left) root disc after cutting the explants, right freshly cut carrot root explants

- The forceps are dipped into ethanol (96%), flamed, and placed into a sterile Petri dish.
- Sterilized water is poured into a sterile Petri dish, ready to receive the explants.
- From the cutting platform, the cover is removed and placed in the center of the sterile working bench. One sterile Petri dish (higher rim) is placed directly under the cutting platform, with sterile forceps.
- The carrot is taken out of the sterilization solution and the cover removed, and it is washed carefully with sterilized water. Starting with the root tip, 2 mm discs (knife adjusted accordingly) are cut with the help of the cutting platform, using exact horizontal strokes (Fig. 3.3). Such strokes are required as a prerequisite to later obtain explants from the tissue of the carrot root selected. If a horizontal stroke is missed, then the explants of the secondary phloem (our aim) will often be contaminated by cells of the cambium.
- After having obtained the number of discs desired (from each disc, about 15–20 explants from the secondary phloem can be obtained), two forceps are flamed and put into a sterile Petri dish.
- Cutting the explants (Fig. 3.3): the root discs are transferred (with a sterile forceps) into a Petri dish containing filter paper. With the help of the sterilized troquar, about 20 explants are cut at a distance of about 2 mm from the cambium. The explants are transferred from the troquar to a Petri dish filled with sterilized water. It is practical to cut about 50 explants more than strictly needed.
- To remove contaminating traces of the sterilizing solution used for the roots, the explants should be repeatedly rinsed with sterilized water (five to six times). After the last washing, almost all the water is removed from the dish. Only the liquid required to moisten the surface of the explants remains in the Petri dish.
- The needle, with a loop at the tip used for the transfer of the explants into the culture vessels, is dipped into abs. ethanol and flamed. After cooling of the needle, the explants are transferred into the culture vessel with the nutrient medium. The needle with explants should never touch the opening of the culture vessel (cf. avoid the generation of a "nutrient medium" for microbes). After the transfer of explants to several culture vessels, the needle should be flamed again. Immediately after the inoculation of the explants, the vessels are covered by lids (e.g., aluminum foil). As a further precaution, the opening of the vessel and the lid can be flamed before closing.
- After the work on the laminar flow, the culture vessels with the explants are transferred to the climatized culture room.

If it is difficult to obtain sterile cultures from plant material grown in a non-sterile environment, then explants can be obtained from seedlings derived from sterilized seeds in an aseptic environment. For this, the seeds are first placed into a sterilizing solution for 2–3 h, and it is advisable to use a magnetic stirrer. The duration of sterilization and the type of sterilization solution used usually have to be determined empirically for each tissue and each plant species (Table 3.2). Seeds with an uneven seed coat, or with a cover of hairs, may cause problems. It may be of help to add a few drops of a detergent, e.g., Tween 80. After surface sterilization, the seeds are washed in autoclaved water. For germination, the sterilized seeds are then transferred to either sterilized, moist filters in Petri dishes (or another suitable container) or a sterile agar medium. The greater the chances of contamination, the smaller is the number of seeds recommended per vessel.

The cutting of explants from the seedling is usually done with the help of a scalpel or similar device (e.g., scissors, a razorblade, a cork borer) sterilized in a drying oven; the device should be frequently flamed. More procedures to this end, using embryo tissue, or explants of immature embryos, are described later in other chapters (e.g., Chap. 7).

3.1.1 Sterile (Aseptic) Technique: Microbe Eradication

For the successful establishment and maintenance of plant cell, tissue and organ cultures are aseptic technique absolutely necessary. The in vitro environment in which the plant material is grown is also ideal for the proliferation of microorganisms. In most cases the microorganisms outgrow the plant tissues, resulting in their death. Contamination can also spread from culture to culture. The purpose of aseptic technique is minimizing the possibility that microorganisms remain in or enter the cultures.

Autoclaving is the method most often used for sterilizing heat-resistant items. In order to be sterilized, the item must be held at 121 °C, 15–16 psi (1.1 bar, for calculation see http://www.umzugs.com/psi-bar.htm), for at least 15 min. Therefore time in the autoclave will vary, depending on volume in individual vessels and number of vessels in the autoclave.

Organic compounds such as some growth regulators, amino acids, vitamins, and antibiotics may be degraded during autoclaving. These compounds require filter sterilization through 0.2 or 0.45 μm membranes. Nutrient media that contain thermo labile components are typically prepared in several steps. A solution of the heat-stable components is sterilized in the usual way by autoclaving and then cooled to 45–50 °C under sterile conditions. Solutions of the thermo labile components are filter sterilized. The sterilized solutions are then combined under aseptic conditions to give the complete medium (Table 3.3).

Table 3.3 Some disinfectants used in tissue culture experiments and concentration applied

Sodium hypochlorit (3% active chlorine)	20.0%
Calcium hypochlorit	25.0%
Bromine water	1.0%
Mercury chloride	0.2%
Ethanol	70.0%
Hydrogen superoxide	10.0%
Silver nitrate	1.0%
Plant preservative mixture (biocide)	2%
Vegelys	0.1%

3.1.2 Initial Contamination

Most contamination is introduced with the explant because of inadequate sterilization or just very dirty material. It can be fungal or bacterial. This kind of contamination can be a very difficult problem when the plant explant material is harvested from the field or greenhouse. Initial contamination is obvious within a few days after cultures are initiated. Bacteria produce "ooze" on solid medium and turbidity in liquid cultures. Fungi look "furry" on solid medium and often accumulate in little balls in liquid medium.

3.1.3 Latent Contamination

This kind of contamination is usually bacterial and is often observed long after cultures are initiated. Apparently the bacteria are present endogenously in the initial plant material and are not obviously pathogenic in situ. Once in vitro, however, they increase in titer and overrun the cultures. Latent contamination is particularly dangerous because it can easily be transferred among cultures.

3.1.4 Introduced Contamination

Contamination can also occur as a result of poor sterile technique or dirty lab conditions. This kind of contamination is largely preventable with proper care.

3.1.5 Surface Sterilizing of Plant Material

Prior good care of stock plants may lessen the amount of contamination that is present on explants. Plants grown in the field are typically more "dirty" than those grown in a greenhouse or growth chamber, particularly in humid areas. Explants can be obtained from seedlings derived from sterilized seeds in an aseptic environment. For this the seeds will be put first into the sterilizing solution for 2–3 h, and it is advisable to use a magnetic stirrer. The duration of the sterilization and the sterilization solution used usually has to be determined empirically for each tissue and each plant species (Table 3.3). Seeds with an uneven seed coat or with a cover of hairs may cause problems. It may be a help to add a few drops of a detergent like Tween 20 or 80. After surface sterilization, the seeds shall be washed in autoclaved water. After this for germination the sterilized seeds are either transferred to sterilized moist filters in Petri dishes (or another suitable container) or a sterile agar medium. The greater the chance for contamination, the smaller the number of seeds per vessel is recommended.

Here the cutting of explants usually is done with the help of a scalpel or a similar device (e.g., scissors or a cork borer) sterilized in a drying oven, and the device should be frequently flamed. Some more procedures to this end using embryo tissue or explants of immature embryos are described. *Antibiotics have been used to control this kind of infection.* Misra et al. (2010) reported that Augmentin was selected as the most suitable and effective antibiotic for controlling bacterial infection of *Enterobacter ludwigii* in leaf explants of *Jatropha curcas.* The cultures could be grown for a long term of 2 years without any further contamination (see also Miró-Canturri et al. 2019).

Recently, we have tested the effect of Vegelys (a natural solution, from *Allium* extracts, www.phytoauxilium.com) against pathogenic microorganism in vitro culture. The results show significant elimination of various microbes in plant cells and tissue culture in lower concentration (Imani et al. unpublished data). Additionally, the Vegelys can be used also an effective agent for surface sterilization (e.g., barley, artichoke, bitter gourd), which does not negatively affect the seed germination as with commercially available agents.

3.2   Fermenter Cultures (See Also Chap. 10)

Basically, the same principles as those just described can be applied to fermenter or bioreactor cultures. Although the bioreactor in Fig. 3.4 was originally developed for cultures of algae, this simple equipment (Fa. Braun, Melsungen, volume 5 l) has been successfully used to culture cells of several higher plants (Bender et al. 1981). After applying a "light coat" for illumination (ca. 33 W/m^2), investigations on the photosynthesis of photoautotrophic cultured cells in a sugar-free medium have been carried out with success (see Chap. 9).

The bioreactor in the figure is filled with 4 l of nutrient medium, sterilized in a vertical autoclave; to check the success of autoclaving before the transfer of cells, it is placed in the culture room for 3 days. If IAA is a constituent of the nutrient medium, then the fermenter has to be kept in the dark to prevent its photooxidation, to be observed within a few days. If the bioreactor is still sterile after that time, then the cell material is transferred with a sterile glass funnel and a silicon pipe of 1 cm diameter. The fermenter has to be placed in front of the laminar flow to position the funnel in the sterile air stream of the inoculation cabinet. A sufficient growth of the culture can be achieved with an inoculation of about 30 g fresh weight for the 4 l of medium in the container (see also somatic embryogenesis, Sect. 7.3).

As an alternative, the separate sterilization of the container and the nutrient medium has also been successfully employed. The sterilization of the nutrient medium is the same as that described above, and the empty container was sterilized by autoclaving for 35 min at 1 bar and 130 °C. For harvesting, the content of the bioreactor is simply poured out through some layers of fine cheesecloth.

Basically a bioreactor to culture plant material should provide adequate mixing, while minimizing shearing stress and hydrodynamic pressure. Since the 1970s, much work has been invested in developing airlift bioreactors, which seemed the most promising construction to fulfill these requirements. Still, hardly any damage was observed by using the bioreactor described above to produce somatic embryos of *Daucus,* or *Datura* cell suspensions for the production of scopolamine or atropine (see Chap. 10).

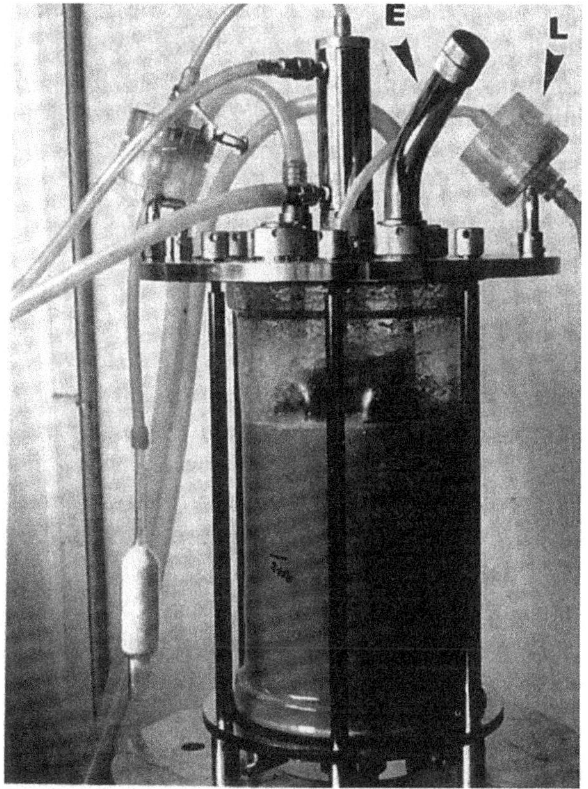

Fig. 3.4 Bioreactor with a carrot cell suspension (*E* inoculation devise, *L* air filter)

As an alternative to reusable glass containers, several devices made of disposable plastic have been developed to reduce operational costs. As an example, the presterilized Life Reactor™ system developed by M. Ziff of the Hebrew University, and R. Levin of Osmototek, a company engaged in the development of "Advanced Products for Plant Tissue Culture," is mentioned. This system is available with a volume of 1.5 or 5 l. Citing from an advertisement for the 1.5 l vessel: "Producing up to 1000 plantlets per litre of liquid medium, this easy to handle system allows research and small commercial laboratories to carry out multiplication on a relatively large scale, in less than a square meter of space, with minimal manpower and at an easily affordable price. The body is a V-shaped bag from a special, heavy duty plastic laminate material. At the bottom of the vessel is a porous bubbler, which is connected to an inlet in the wall. During operation, sterile, humidified air is supplied through this port. Near the top of the vessel is a 1.5 diameter inoculation port, through which the plant material is initially added and later withdrawn. This is closed with an autoclavable cap. One of two ports on the cap is used to exhaust excess air and another is covered by a silicon rubber septum. This can be used to apply additions in aseptic manner."

Development of novel fermentation processes for manufacturing PNPs on the basis of the metabolic engineering of microorganisms has emerged recently as an interesting and commercially attractive approach due to several advantages including the utilization of environ mentally friendly feedstocks, low energy requirements, and low waste emission (Yuan and Alper 2019) (see also Chap. 10).

3.3   Immobilized Cell Cultures

Besides the methods described above, so-called batch cultures, attempts have also been made to establish continuous systems in bioreactors. Here, in analogy to animal cell cultures, the cells are fixed on a stationary carrier. Whereas animal cells have "self-fixing" properties to attach autonomously to a glass surface or on synthetic materials like Sephadex, difficulties arise for plant cells, probably due to the rigid cell wall. A way out of this dilemma is the capture of the cells in the interior of the carrier material.

Originally, calcium alginate was used as carrier; meanwhile, a number of polymers have been tested, such as agar, agarose, polyacrylamide, and gelatin. Pure synthetic materials, like polyurethane, or nylon and polyphenyloxide, have also been examined. All these have advantages and disadvantages, and often polyurethane is preferred. This material possesses a large inner volume (97% w/v), and the capture of the cells is brought about by a passive invasion of the carrier material (see Fig. 3.5). The carrier has to be submersed into the cell suspension, and in the pores of the foam, cells continue to divide and grow until the whole inner volume is invaded. This method requires no additional chemicals to fix the cells to the carrier, and no negative influences on the vitality and metabolism of the cells has been observed to date. Polyurethane is stable in the usual nutrient media, also during prolonged experimental periods. These cells fixed on polyurethane can be transferred to a flatbed container or to a column where they are bathed by a continuous stream of nutrient media (Lindsey et al. 1983; Yuan et al. 1999).

Also here, as for bioreactors with microbes, circular setups with reuse of the medium were successful. Such continuous arrangements serve to produce substances of the primary or secondary metabolism of plant cells (e.g., Yin et al. 2005, 2006), which can be also extruded to the medium. This can be even increased, compared to free cell suspensions, as reported for immobilized cultures of *Juniperus chinensis* (Premjet and Tachibana 2004) using a 3% alginate gel to produce podophyllotoxin. More on this will be given in Chap. 10. This also offers possibilities, at desired culture stages, to change the composition of the nutrient medium to direct cell production.

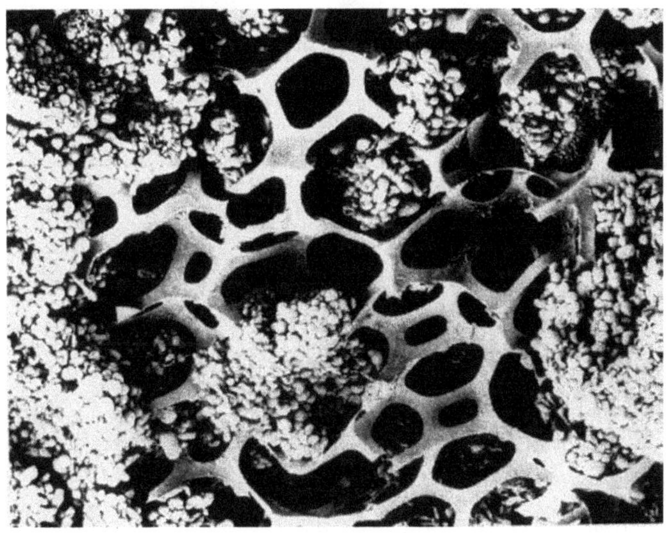

Fig. 3.5 Plant cells in polyurethane foam (photograph by M.M. Yeoman)

3.4   Nutrient Media

Nutrient media occupy a central significance for the success of a cell culture system. Although almost all intact higher plants are able to grow autotrophically in light under normal air conditions and sufficient supply of water and mineral nutrients, this is not the case for all plant organs and tissue. For example, roots or the developing seeds require the import of assimilates from shoot tissue, or phytohormones produced in other remote tissue to stay alive, function, and grow.

This situation is also characteristic for cells of the various cell culture systems being isolated from the intact plant body. The nutrient medium is a substitute for an import of substances, derived in the intact plant from other parts of its body with distinct metabolic properties. Although some cell culture systems have been reported to grow fully photoautotrophically in an inorganic nutrient medium (see Chaps. 9 and 10 for details), by far most culture systems are either heterotrophic or, in the light after the development of chloroplasts, at best mixotrophic. For the cultures, a supplement of carbohydrates to the medium is necessary to fulfill the requirements of energy, as well as carbon, oxygen, and hydrogen as raw material for synthesis. For this purpose, usually mono- or disaccharides are supplied. In most media, sucrose is used at various concentrations, and for most investigations of the growth and development of cultures, it has proved sufficient. Moreover, good growth can be obtained by using monosaccharides, and many other materials, sometimes quite unconventional, are also employed. Generally speaking, a "best" carbohydrate does not really exist for all plant cell cultures—which will be chosen as a supplement to

Table 3.4 Compositions of some nutrient media in use for plant cell and tissue cultures (for 1 l aqua dest.; the compositions of the stock solutions are given in Table 3.5)

Nutrient medium[a]	BM	MS	NL	NN	B5
g sucrose	20.00	30.00	20.00	20.00	50.00
g casein hydrolysate	–	–	0.20	–	0.25
ml mlycine solution	1.0	1.0	–	–	–
ml mineral solution	100.0	100.0	100.0	100.0	100.0
ml Fe solution	1.0	10.0	10.0	10.0	10.0
ml Mg solution	10.0	–	30.0	–	7.0
ml vitamin solution	1.0	1.0	1.0	10.0	1.0
ml folic acid solution	–	–	–	1.0	–
ml biotin solution	–	–	–	10.0	–
ml 2.4D solution	–	1.0	–	–	1.0
ml IAA solution	–	0.5	1.0	0.05	–
ml kinetin solution	–	–	1.0	–	–
ml BAP solution	–	1.0	–	–	–
ml m-inositol solution	–	2.0	10.0	–	10.0
ml coconut milk (CM)	100.00	–	–	–	–
pH	5.6	5.5	5.7	5.5	5.7
g Agar	8.0	8.0	8.0	8.0	8.0

[a]Nutrient medium: BM, White (1954); MS, Murashige and Skoog (1962); NL, Neumann (1966); NN, Nitsch and Nitsch (1969); B5, Gamborg et al. (1968)

the nutrient medium always depends on the tissue, the study aim, and the plant species. These have to be determined in preliminary investigations.

In Tables 3.4, 3.5, 3.6, 3.7, and 3.8, the composition of some nutrient media employed nowadays are given. The concentration of sucrose is usually 2–3%. A second key component of a nutrient medium is the mixture of mineral salts, which has undergone considerable changes since the first publication of a nutrient medium for plant cell cultures by P. White in 1954. An important difference to this in most modern media is an increase in the concentration of phosphorus; this could be increased by the factor of about 10 applying coconut milk to the medium, as practiced by the Steward group at Cornell University in the 1950s and 1960s. Now, also the concentrations of most of the other mineral components are enhanced. The group of micronutrients has been extended by the application of copper and molybdenum. In White's medium, nitrate is the only source of inorganic nitrogen. If grown in the light, a period of 8–10 days is required to develop sufficient functional chloroplasts with the ability to provide an efficient system to reduce nitrite. In White's basal medium, only a supplement of glycine serves as a source of reduced nitrogen. The same function, though more powerful, is associated with the supplement of casein hydrolysate to other nutrient media, consisting of many amino acids (see Table 3.6). Based on the nutrient medium published by Murashige and Skoog (MS medium), ammonia is also supplied as a source of reduced nitrogen in many media, in the form of various salts.

Table 3.5 Compositions of some nutrient media in use for plant cell and tissue cultures: stock solutions

Mineral solution (in 1 l aqua dest.)	BM	MS	NL	NN	B5
Macronutrients (g)					
KNO_3	1.00	19	8.72	9.50	30.00
NH_4NO_3	–	16.50	–	7.20	–
$MgSO_4 \times 7H_2O$	–	3.70	–	1.85	5.00
$CaCl_2 \times H_2O$	–	4.40	–	0.68	–
KH_2PO_4	–	1.70	–	0.68	–
$NaH_2PO_4 \times 2H_2O$	0.21	–	2.34	–	1.50
KCl	0.78	–	0.65	–	–
$Na_2SO_4 \times 10H_2O$	2.50	–	2.45	–	–
$Ca(NO_3)_2 \times 4H_2O$	2.50	–	4.88	–	–
$(NH_4)_2SO_4$	–	–	–	–	1.34
Micronutrients (mg)					
$MnSO_4 \times H_2O$	56.00	170.00	36.00	250.00	100.00
H_3BO_3	15.00	62.00	15.00	100.00	30.00
$ZnSO_4 \times 7H_2O$	12.00	86.00	15.00	100.00	30.00
$Na_2MoO_4 \times 2H_2O$	–	2.50	3.30	2.50	2.50
$CuSO_4 \times 5H_2O$	–	0.25	6.20	0.25	0.25
$CoCl_2 \times 6H_2O$	–	0.25	–	–	–
KI	9.00	8.30	7.00	–	7.50
Mg solution					
$MgSO_4 \times 7H_2O$	36.00	–	36.00	–	36.00
Fe solution (g/l aqua dest.)					
Fe-EDTA	–	4.63	4.63	4.63	4.63
Fe-tartrate	5.0	–	–	–	–
Glycine solution (mg/100 ml aqua dest.)					
1glycine	300.00	200.00	–	20.00	–
Hormones (mg/100 ml aqua dest.)					
M-inositol solution	–	500.00	500.00	–	500.00
2.4D solution	–	22.10	–	–	10.00
IAA solution	–	200.00	200.00	200.0	–
Kinetin solution				10.00	
BAP solution	–	100.00	–	–	–
Coconut milk	10%				

The third major component of a nutrient medium is a mixture of vitamins usually containing thiamine, pyridoxine, and nicotinic acid. The cells isolated from the intact plant body are generally not able to produce enough of these compounds, essential in particular for the metabolism of carbohydrates and of nitrogen. Exceptions to these requirements are again those autotrophic cells mentioned above (see also Sect. 9.1).

As further components of nutrient media in Tables 3.4, 3.5, 3.7, and 3.8, various phytohormones or growth substances are listed. These are able to replace coconut milk as a supplement, used widely in the earlier days of plant cell cultures.

Table 3.6 Concentrations of some amino acids in casein hydrolysate (as mg/l nutrient medium, by an application of 200 ppm per liter nutrient medium)

Amino acid	Concentration	Amino acid	Concentration
Lysine	12.1	Alanine	5.4
Histidine	3.6	Valine	7.9
Arginine	4.3	Methionine	4.4
Aspartic acid	13.3	Isoleucine	5.9
Threonine	6.3	Tyrosine	5.1
Serine	9.3	Phenylalanine	5.4
Glutamic acid	36.1	Leucine	4.3
Proline	19.3	Glycine	3.4

Table 3.7 Concentrations of mineral nutrients in some nutrient media used for cell and tissue culture (final concentration at the beginning of culture, mg/l nutrient medium)

Nutrient medium[a]	BM[b]	MS	NL	NN	B5
Nitrogen[c]	138.00	841.00	179.00	619.00	444.00
Phosphorus	43.00	39.00	47.00	16.00	39.00
Potassium	312.00	783.00	371.00	152.00	116.00
Calcium	60.00	94.00	83.00	35.00	32.00
Magnesium	102.00	53.00	107.00	18.00	97.00
Sulfur	121.00	70.00	165.00	24.00	98.00
Chlorine	31.00[d]	167.00	31.00	63.00	57.00
Boron	0.272	1.10	0.27	1.80	0.54
Manganese	1.41	6.20	1.30	9.10	3.60
Zink	0.39	2.00	0.33	2.30	0.46
Iron	3.002	3.00	3.00	3.00	3.00
Molybdenum	0.005	0.10	0.14	0.10	0.10
Copper	0.04	0.01	0.16	0.01	0.01
Iodine	0.57	0.64	0.57	–	0.57
Cobalt	–	–	0.01	–	–

[a]See Table 3.4
[b]Nutrients in 10% coconut milk in the medium were included
[c]The following organic nitrogen sources were supplied (mg N/l): 35.4 with coconut milk and from glycine in BM; 0.4 as glycine in MS; 31 as casein hydrolysate in NL; 39 as casein hydrolysate in B5
[d]Concentration in coconut milk not available

The requirement for a supply of phytohormones or other growth substances, and its influence on the growth and development of cultured cells, depends primarily on the plant species and variety, the tissue used for explantation, and the aim of the investigation or other use of the cultures. In Table 3.8, some examples are given for influences of various nutrient media with one or the other supply of hormones from our own research program to induce primary callus cultures (cf. Table 3.9).

If primary explants contain also meristematic regions, then considerable growth can be induced already in a hormone-free medium, which usually can be increased by the application of an auxin. By contrast, if so-called quiescent tissue is the origin

Table 3.8 Final concentrations of organic components in some media used for plant cell and tissue culture at the beginning of the experiment (mg/l)

Nutrient medium[a]	BM	MS	NL	NN	B5
Sucrose	20,000.0	30,000.0	20,000.0	20,000.0	20,000.0
Casein hydrolysate	–	–	200.0	–	250.0
Glycine	3.0	2.0	–	2.0	–
Nicotinic acid	0.5	0.5	0.5	0.5	1.0
Pyridoxine	0.1	0.5	0.1	0.5	0.1
Thiamine	0.1	0.1	0.1	0.5	0.1
Biotin	–	–	–	0.5	–
Folic acid	–	–	–	5.0	–
M-inositol	–	100.0	50.0	–	50.0
IAA	–	1.0	2.0	0.1	–
Kinetin	–	–	0.1	–	–
BAP	–	1.0	–	–	–
Coconut milk	100.0				

[a]See Table 3.4

Table 3.9 Growth (mg fresh weight/explant) of explants of the secondary phloem of the carrot root and from the pith of tobacco in some liquid nutrient media (21 days of culture, 21 °C, average of three experiments; original explants: carrot 3 mg, tobacco 7 mg)

Nutrient medium[a]	BM	MS	NL	NN	B5
Carrot	297	37	264	13	6
Tobacco	89	109	231	60	29

[a]See Table 3.4

of explants, such as the secondary phloem of the carrot root, then growth without hormonal stimulation is very poor, consisting mainly of cell enlargement. In a later chapter, the endogenous hormonal system of cultured explants and its interaction with exogenous hormones stemming from the nutrient medium will be discussed in detail (Chap. 12). In describing the various culture systems, the significance of hormones related to specific cell reactions will also be addressed.

Most nutrient media contain an auxin, usually naphthylacetic acid (NAA) or 2.4-dichlorophenoxy acetic acid (2.4D); sometimes, also the natural auxin indole acetic acid (IAA) is used as supplement. These three auxins are distinguished by chemical and metabolic resistance to breakdown or inactivation. IAA is characterized by the highest lability. Mainly through photooxidation, but also due to metabolic breakdown, IAA is soon lost from the system. The stability of NAA is higher, and 2.4D exhibits the highest stability. Differences in stability correlate with the time required to induce rhizogenesis in rapeseed cultures, with earliest appearance of roots in IAA treatments (Table 3.10). The main and most obvious function of auxins is to stimulate cell division. If a high cell division activity is to be maintained for prolonged periods, which usually prevents differentiation of cultures, then the metabolically very stable 2.4D is the auxin of choice. If the experimental aim is to

Table 3.10 Influence of some auxins (2 ppm IAA, 2 ppm NAA, 0.2 ppm 2.4D) on fresh weight, number of cells per explant, and rhizogenesis of explants of rapeseed (cv. Eragi, petiole explants, 21 days of culture, NL liquid culture; Elmshäuser 1977)

	0	IAA	NAA	2.4D
mg F. wt./explant	29	51	134	121
No. of cells $\times$ 10^3/explant	186	296	916	2114
Root formation				
Days after beginning of culture	–	7	14	42
mg roots/explant	–	13	38	n.d.

initiate processes of differentiation that usually require a short period of cell division, then IAA and NAA are more suitable supplements. If, however, the formation of adventitious roots is not desired, as is often the case in primary cultures, then a doubling of the auxin concentration will usually help to prevent this. These are simply some general remarks, however, and as long as more reliable knowledge of the plant hormonal system is not available, the handling of auxin as an ingredient of nutrient media for cell cultures has to be determined empirically for each culture system.

Further promotion of cell division activity, especially by use of the more labile auxins, can be achieved by a simultaneous application of a cytokinin to the nutrient medium. Often, the synthetic cytokinin kinetin is used at very low concentrations (0.1 ppm). Higher concentrations can be quite toxic. For some culture systems, natural occurring cytokinins are also employed, such as zeatin or 2-isopentenyladenine (2-iP) and the synthetic 6-benzyladenine (BA).

In many nutrient media, also m-inositol is used, discovered as a functional component of coconut milk in the early 1960s by the Steward group at Cornell University (Pollard et al. 1961). However, an application of this substance to nutrient media was already suggested by P. White in the early 1950s (White 1954). Inositol is a component of many cellular membranes and plays an important role in cell signaling systems. A short review, though based mainly on results in animal systems, is given by Wetzker (2004). Considering the rather high concentrations of inositol used in many nutrient media, it is actually not a hormone, but often significant responses like those associated with hormones can be observed. Yet, its concentration is too low for it to be considered as a nutrient, e.g., as an energy source.

Still, already during the 1960s, it was observed that, under some circumstances, inositol can functionally replace IAA. Since then, the occurrence of a conjugate of IAA and inositol has been isolated, and one possible function could be the formation of a pool of such IAA–inositol conjugates to protect IAA from breakdown; alternatively, the formation of such conjugates of IAA could be one way to inactivate it. This was reported for the formation of IAA conjugates with glucose or aspartic acid. In most cases, m-inositol increases the action of IAA, as well as that of cytokinins.

If thermolabile components used in nutrient media risk being altered or destroyed by autoclaving for the sterilization of the medium, then sterile filtration is employed at least for these ingredients. In our laboratory, for example, all components with radioactive isotopes are also filter sterilized. Another example for filter sterilization is fructose; during autoclaving, it is transformed into a number of toxic substances that inhibit the growth of cultured cells.

For sterile filtration, the nutrient medium or other compounds are passed through a bacterial filter (pore size 0.2 μm). For this procedure, a broad range of suitable equipment can be found on the market. The simplest and cheapest approach is to use a syringe with a filter adapter, though this is suitable only for small volumes of liquid. For the sterilization of devices and filters, the instructions of the manufacturer should be followed.

3.5   Evaluation of Experiments

In many cases, the evaluation of experimental results is by determination of fresh weight (of air-dried material) and of dry weight (after drying at 105 °C until constant weight). For a rough determination of the growth of cell suspensions, the cells have to be separated and the packed cell volume (PCV) determined. The least destructive and cheapest approach is to use a hand centrifuge at low revolution speed and calibrated centrifuge beakers.

If the aim is to distinguish between growth by cell division and by cell enlargement, then the tissue has to be macerated to determine the number of cells in a defined piece of tissue; in a given cell suspension volume, this would be by counting the individual cells on a grid (a hemocytometer) under a microscope.

The tissue to be macerated is first put in a deepfreeze at -20 °C for 24 h. After thawing, the cell material is placed usually overnight into the maceration solution (0.1N HCl and 10% chromic acid 1:1 v/v). Before cell counting, the macerated material is squashed with a glass rod and several times pumped through a syringe (90 μm). Eventually, an aliquot of the macerate is placed on the counting grid for the counting of cells within a given area. Usually, ten counts are performed per treatment. For maceration, a few hundred mg fresh weight (or less) is generally sufficient (Neumann and Steward 1968). If the maceration fails, or if it proves to be unsatisfactory (cf. too many cells are destroyed and therefore unusable for counting), the composition of the maceration solution has to be empirically adapted to the tissue being investigated. Many callus cultures and most cell suspensions do not require prior freezing. The volume of the maceration solution in μl should be ten times the fresh weight of the tissue to be macerated in mg. For calculation:

$$N = \frac{X * (\mathrm{Mv} + \mathrm{f.wt.})}{\mathrm{VK} * n}$$

where N is the cell number per explant, X the number of cells per count (average of several counts), Mv the volume of maceration solution, f. wt. the fresh weight of macerated tissue in mg, n the number of explants macerated, and VK the volume of the grid in µl (see image of chamber).

3.6   Maintenance of Strains: Cryopreservation

In many instances, it is desirable to maintain certain cell strains viable for prolonged periods, even up to several years. This is especially the case for extensive investigation required to relate several metabolic areas for which the use of the same genome is required.

To maintain such cell strains or cell lines, subcultures have to be regularly set up. This necessity is due to the growth of the cultures, depletion of components of the nutrient medium by the growing cultures, and accumulation of excretions of the cultures, or of dead cells. The frequency of setting up subcultures depends on the extent of these factors at intervals of a few days up to months or even years.

As an example, for *Phalaenopsis* cultures that produced high amounts of polyphenols excreted to the medium, a subculture had to be set up every 2 or 3 days. For slowly, photoautotrophically growing *Arachis* cell cultures, however, a subculture interval of 6–8 weeks was sufficient. Subcultures are usually stationary on agar.

A method usable for subcultures of many plant species consists of an aseptic transfer of vigorously growing callus pieces (10–15 per vessel), with a diameter of about 5 mm, to an Erlenmeyer flask (120 ml) containing 15 ml nutrient medium. Often, the subculture interval is about 4 weeks.

However, this method carries an important disadvantage, i.e., cytological and cytogenetical instability of the cell material after some passages. In terms of the purpose of the procedure, other negative effects include variations in metabolic processes, or even in the organization of the genome (see Chap. 13). Such programs are usually labor-intensive and require much storage space. To some extent, a storage with less frequent subcultures can be achieved by varying the culture conditions, like lowering the temperature of the culture room, limiting the nutrients (mainly sugar), and reducing light intensity. For cold-tolerant species, the temperature can be reduced to nearly 0 °C.

To circumvent such problems, especially during long-term storage, cryopreservation was adapted to plant cell cultures. With this technique, originally developed for animal systems, it is possible to preserve cell suspensions or somatic embryos of cell cultures; after thawing, these can be to be revived with up to 80% success. Protocols for more than 100 plant species are available. These somatic embryos, as well as cell suspensions can be kept in a cryopreserved state for several years. The storage temperature is that of liquid nitrogen (−196 °C), at which cell division and metabolic activities cease (Sminovitch 1979). Otherwise, no modification and variation occur in the cells at this temperature, and maintenance cost and storage space are much reduced, compared to those for strains maintained by multiple subculturing as described above.

The most important aspect of cryopreservation is to prevent the formation of ice crystals, which could destroy cell membranes. The success of the technique, however, differs among cultures of various plant species. Generally, smaller cells are more suitable than bigger ones, as are cells with a broad range of cytoplasm/vacuole ratios. A more recent review of the technique can be found in Engelmann (1997).

In a first step, exponentially growing cell cultures are transferred to the same nutrient medium as that described above but which is supplemented with 6% mannitol for 3–4 days to reduce cell water osmotically under the original conditions. The mixture used for cryopreservation is set up at twice the concentration of components used in the working solution, and it is filter sterilized. This mixture consists of 1M DMSO (dimethylsulfoxide), 2M glycerin, and 2M sucrose. The pH is set at 5.6–5.8. In all, 10 ml of this mixture and 10 ml of the cell suspension are each chilled for 1 h on ice and then combined. This highly viscous solution has to be vigorously shaken. After this, 1 ml of the mixture of cells and cryopreservation solution is placed into sterile polypropylene vials and left on ice a while. Having prepared all the vials, these are transferred to a chilling device, and the temperature is lowered slowly in 1 °C intervals until −35 °C and left at that for 1 h. Finally, the vials are stored in liquid nitrogen.

To revitalize the cell material, the vials should be thawed fast in warm water at 40 °C, and immediately after thawing a transfer of the material to an agar medium is required. After the initiation of growth, a transfer into a liquid medium can be performed (Seitz et al. 1985). The survival rate of *Daucus* or *Digitalis* cells preserved by this technique is between 50 and 75%; for *Panax* cultures, it is less. Evidently, as already mentioned, variations between species exist.

The methods referred to above were developed during the 1970s and 1980s and were modified later especially to store differentiated material like apices or somatic embryos. The basic differences relative to the older methods are a rapid removal of most or all freezable water, followed by very rapid freezing, resulting in so-called vitrification of the cellular solutes. This procedure leads to the formation of an amorphous glassy structure, and the detrimental influences of the formation of ice crystals on cell structure are avoided. Variations of this principle are encapsulation–dehydration, vitrification, encapsulation–vitrification, pre-growth desiccation, and droplet freezing (for summary, see Engelmann 1997). Actually, the methods of encapsulation–dehydration were developed based on earlier investigations on the production of artificial seeds (see Sect. 7.5). Here, cells are first pre-grown in liquid media enriched with high sucrose or some other osmoticum for some days, desiccated to a water content of about 20% (fresh weight), and then rapidly frozen. The samples are encapsulated in alginate beads. Survival rates of cell material are generally high (Engelmann 1997), and the technique has been applied to plants of temperate (apple, pear, grape), as well as of tropical climates (sugarcane, cassava). Encapsulation–vitrification consists of encapsulation in alginate beads and a treatment with vitrification solutions before freezing (Matsumoto et al. 1994).

An interesting variation of these methods consists of preservation of potato apices (Schäfer-Menuhr et al. 1996) in droplets of a cryoprotective medium. Dissected apices are pre-cultured in DMSO for a few hours and then, frozen as droplets, placed on aluminum foil and stored in liquid nitrogen. This method has been successfully applied to about 150 varieties, with an average recovery rate of 40%.

As an example of successful cryopreservation of trees, an extensive review of its application for storage of poplar cells is mentioned (Tsai and Hubscher 2004). This technique seems to be also established practice in somatic embryogenesis to preserve the clonal germplasm of 23 coniferous tree species (Touchell et al. 2002). Here, however, methods using slow cooling dominate and also vitrification methods were applied. For fruit trees, reference is made to Reed (2001). The problem of safe storage will become increasingly important after modified genomes are used for plant breeding; indeed, commercial application for the production of medicinally important germ lines, in particular of recombinant proteins, is envisaged (Imani et al. 2002; Hellwig et al. 2004; Sonderquist and Lee 2008).

3.7   Some Physiological, Biochemical, and Histological Aspects

The carrot root explant system was originally developed by the Steward group at Cornell University, using coconut milk as a source primarily of hormones but also of nutrients (Caplin and Steward 1949). Later, coconut milk, with its unknown and often variable composition (depending on its origin), could be replaced by a mixture of some additives like IAA or other auxins, m-inositol, and kinetin (Neumann 1966; see Tables 3.2, 3.4, and 3.7). Furthermore, the inorganic nutrients in coconut milk were added to the original nutrient medium of White (1954). In this nutrient medium, about the same growth response of cultured carrot root explants could be induced as in White's medium supplemented with coconut milk (NL, see Table 3.8). With such a chemically defined system, it was now possible to characterize the significance of its components for cell division, cell growth, and differentiation and to study the underlying physiological and biochemical processes.

Nowadays, coconut milk has largely lost its originally high significance for tissue culture systems. However, occasional problems arise where it is worthwhile to give this liquid endosperm a second glance. If all other nutrient media fail to induce growth of explants, often a supplement of coconut milk (e.g., 10% v/v) can be successful. To obtain coconut milk, the germination openings of the nut are opened with a borer, and the crude liquid is first cleaned by pouring it through several layers of cheesecloth, followed by autoclaving for sterilization and to remove proteins by precipitation. The hot coconut milk is filtered and then deep-frozen until use, when it is thawed in a water bath at about 60 °C. Coconuts most useful to obtain coconut milk are available from late fall until end December in Europe, at least in Germany.

The fresh weight of cultured carrot root explants in NL3 indicates an initial lag phase of 5–6 days, followed by an exponential phase of 2–3 weeks, and then a stationary phase. As can be seen in Fig. 3.6, the pattern in cell division activity reflects that of fresh weight production. During the exponential growth phase, an increase in cell number, i.e., in cell division activity, obviously dominates in the capacity of the cells to grow.

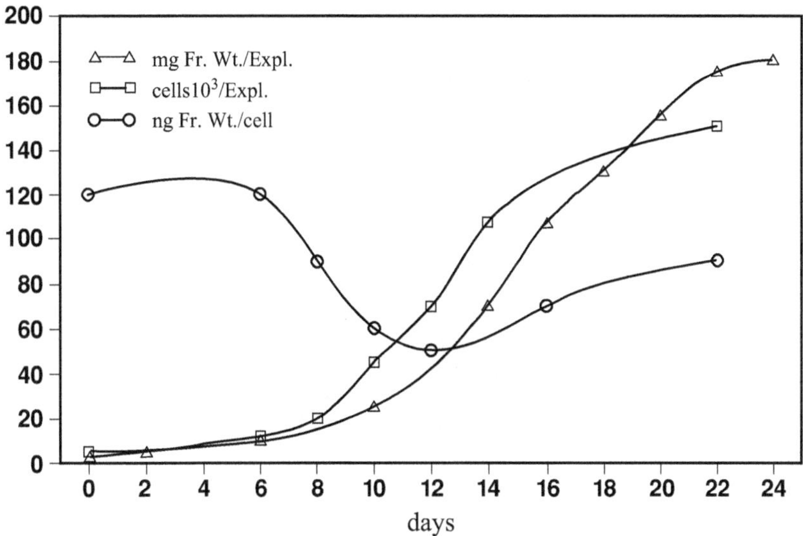

Fig. 3.6 Fresh weight, average cell number, and cell weight of callus cultures from the secondary phloem of the carrot root during culture for 24 days in NL medium (cf. Table 3.4), supplemented with 50 ppm m-inositol, 2 ppm IAA, and 0.1 ppm kinetin (21 °C, continuous illumination at 4500 lux)

Consequently, the average size of the cells is reduced, compared to that in the original explants. Later, due to a slowing down of cell division activity (cf. a reduction in the number of cell divisions per unit time and primarily cellular growth or a decrease in the number of cells engaged in active cell division, or both), the cells of the cultured explant increase in average size (late log phase and transition into the stationary phase). On average, cell size now approaches again that in the original explants. The duration of the various phases varies greatly among explants from different carrot roots, from roots of different varieties, and among explants of different species. Furthermore, marked differences can be observed in cell division activity during the exponential phase of cultures of different origin, despite being grown under identical conditions. This will be dealt with in the description of influences of hormones in the nutrient medium (Chap. 8) and other environmental factors of culture systems.

The carrot callus may seem morphologically unstructured but looking at hand cuttings even with the naked eye shows the presence of anatomical layers. Microscopic inspection of sections of 2–3-week-old callus cultures clearly reveals the existence of several cell layers (Fig. 3.7). On the periphery, a layer consisting of two or three rather large cells (in width) can be seen, followed by a broad layer of smaller cells. Toward the center of the explant, there is a sheath some cells wide, again of bigger cells. In the center itself, remains of the original explant can often be found. In older cultures, sometimes a hole can be seen. On scanning electron microscope

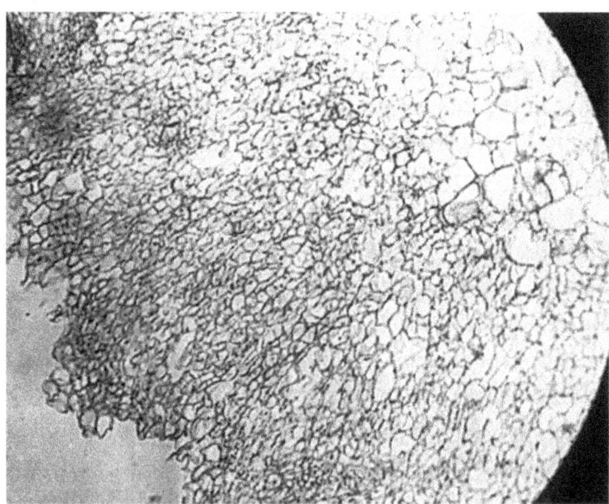

Fig. 3.7 Section of a carrot callus after 3 weeks of culture

graphs from the surface of cultured explants in Fig. 3.8, distinct species-specific differences can be observed.

The surface of the carrot callus looks quite compact, consisting of morphologically quite similar cells. However, the surface of cultures of *Datura* and *Arachis* are only loosely structured, and deep channels penetrating several layers of cells deep into the cultured explant can be observed (Fig. 3.8). Looking at the *Datura* culture, a clear differentiation of cell form is obvious. The significance of organization of the explant surface for the formation of cell suspensions will be discussed later (Chap. 4).

The surface of the carrot root callus is different. Starting from the 6th day onward, emerging new cells from within the explant can be observed pushing aside remains of cells cut when obtaining the original explants (Fig. 3.8). Already this indicates the initiation of cell division two or three cell layers below the surface of the explants.

The inspection of serial cuttings of the middle layer of the explants, consisting of small cells engaged in active cell division, indicates a random occurrence of many so-called meristematic nests (Fig. 3.9).

These are composed of about 100 cells and correspond to small cell clusters common in actively dividing cell suspensions. In the center of these meristematic nests, very small cells can be seen, and cell size increases toward the periphery. Although small, newly formed tracheids occur occasionally, it is safe to conclude that from the center of the meristematic nests toward the periphery, cell division activity is reduced and the age of the individual cell increases. The small layer of larger cells at the periphery of the explants then should be from the periphery of meristematic nests near the surface of the explants. The same holds true for the layers of bigger cells toward the center of the explants. Based on this anatomical description, the propagation of cells in a growing callus with high cell division activity should occur predominantly in the center of these meristematic nests.

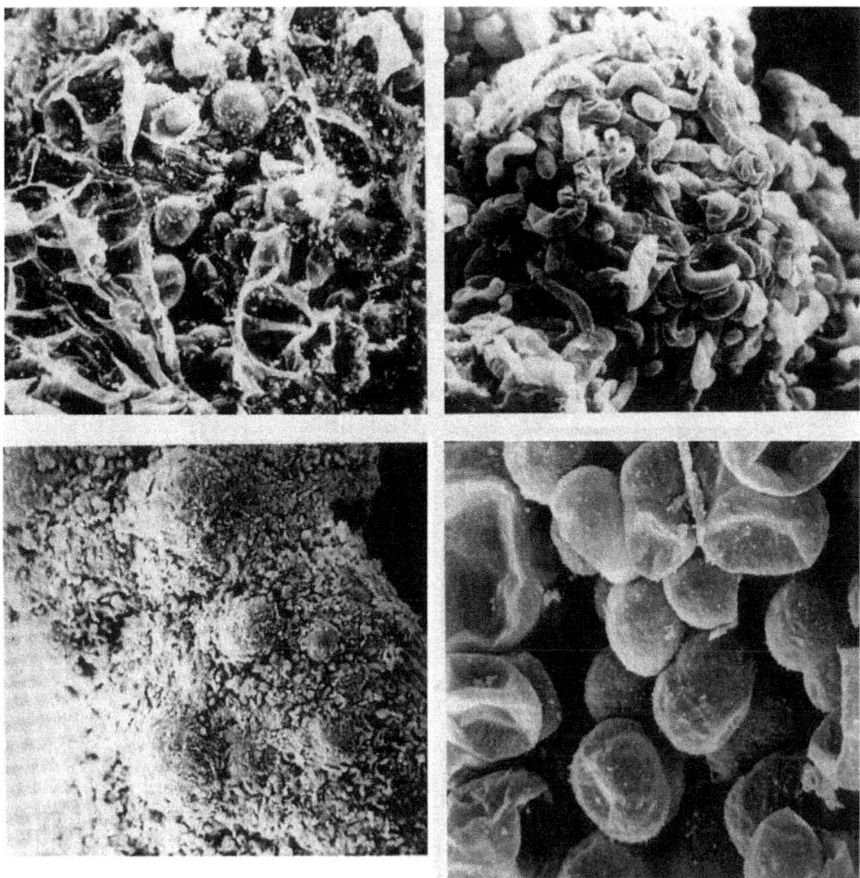

Fig. 3.8 Scanning electromicrographs of the surface of a callus culture. (Top, left) *Datura innoxia*, (right) *Arachis hypogea*. (Bottom, left) *Daucus carota*, primary explants after 6 days of culture, (right) after 21 days of culture

For a better understanding of the growth performance of explants, and the basic factors involved, a short description of the origin and genesis of the meristematic nests shall be given. About 12–15 h after explantation and initiation of culture, one to two cell layers below the cut surface of the explants on the periphery, cell division is initiated, apparently at random (Fig. 3.10).

These cell division initials seem to be the origin of the meristematic nests observed later. This process of initiation of cell division continues for another 1–2 days, and the originally rather large distances between these initials of cell division activity are reduced, i.e., new division centers are induced. Concurrently, cell division in the previously induced cells continues. Consequently, after 4–5 days of culture, such initials exist at various stages of development beside each other.

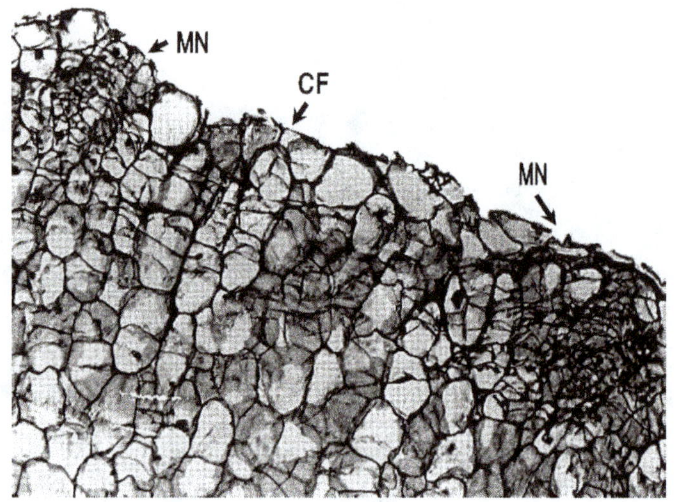

Fig. 3.9 Section of a carrot root explant cultured for 6 days (NL medium). Note the two meristematic nests (*MN*) and cell fragmentation (*CF*)

Histological differences related to the time of initiation of these centers have to date not been observed.

Those first cell divisions result from septation of the large, highly vacuolated parenchymatic cells of the original tissue. A preliminary indication of the initiation of cell division is the formation of long threads of cytoplasm crossing the central vacuole, containing so-called phragmosomes. The nucleus enters these cytoplasmic threads, and the first cell division occurs (Fig. 3.10). As a result of such processes, up to eight daughter cells can be produced from a single parenchymatic cell. This mode of cell division initiation is clearly distinct from that observed in cultured petiole explants at initiation, to produce adventitious roots or somatic embryos (see Sect. 7.3).

A cytophotometric determination of DNA content in cells of these initiation regions of cell division at the periphery of the explants indicates a clear maximum of 4C cells 20 h after the initiation of culture. In cytology, the C-value characterizes the level of DNA of a given cell. A 2C-value in diploid cells indicates that the cells are in the G1-phase of the cell cycle; 4C cells would be in G2 (see Chap. 12). If it is assumed that the majority of cells at explantation are in the G1-phase, then the 4C-value indicates position in the G2-phase after prior passage through the S-phase, with a doubling of the DNA and the initiation of active cell division (24 h). Twelve hours later, again a maximum of cells with 2C is observed, indicating the return to G1 and with this the completion of the first passage of the cell cycle. This first passage is quite synchronized. During this first passage of the cell cycle, hormonal influences have to date not been observed cytologically. In those explants cultured in a hormone-free nutrient medium, after this first cell cycle passage, higher DNA values dominate. By contrast, in those explants growing on media supplemented

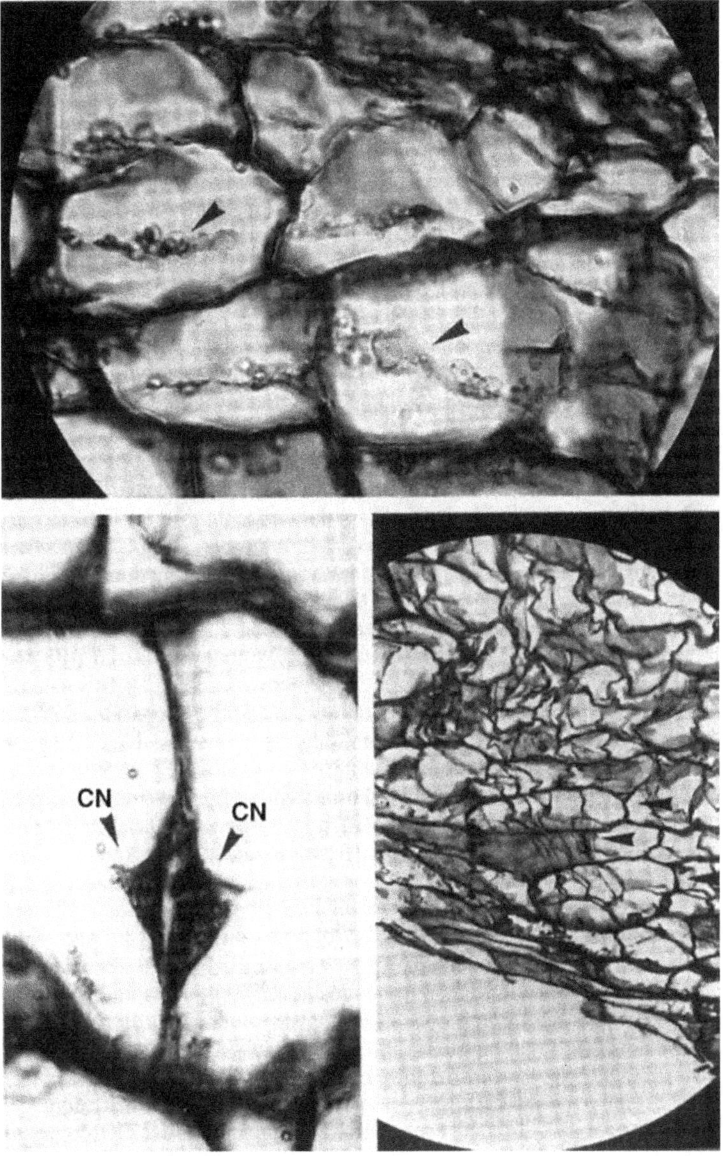

Fig. 3.10 Initiation of cellular fragmentation in cultured carrot root explants (secondary phloem). (Top) appearance of phragmosomes (see arrows); (bottom, left) nuclear division (*CN* cell nuclei), (right) multi-fragmentation in cells

with IAA or inositol, and even more pronounced for those media additionally supplemented with kinetin, a considerable population of cells continues with the passage through the cell cycle and cell division (Chap. 12; Gartenbach-Scharrer et al. 1990).

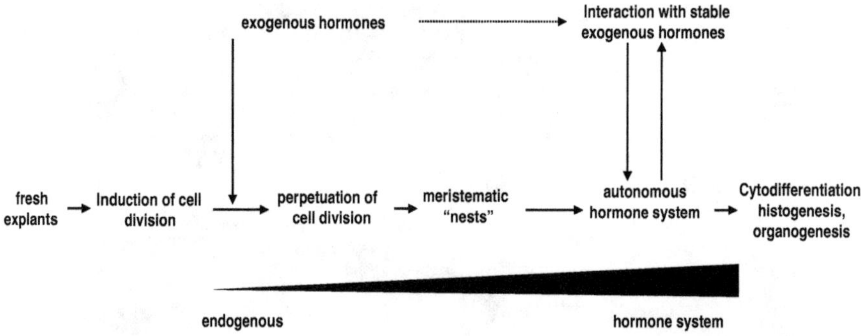

Fig. 3.11 Hypothetic interaction between hormone supply to the nutrient medium and an endogenous hormone system of callus cultures

The function of these phytohormones supplied with the nutrient medium apparently is to induce a continuation of cell division originally evoked as a wound reaction at explantation (Fig. 3.11). This is in agreement with the production of ethylene shortly after the initiation of culture (see Chap. 11).

Without here going into details of an endogenous hormonal system of cultured explants, which will be dealt with later (Chap. 11), it is already pointed out that 24 h after the initiation of culture, IAA synthesis, and somewhat later that of 2iP occur. Apparently, shortly after the initiation of culture, the explants are able to establish an endogenous hormonal system. This native hormonal system is able to determine the course of further development, in interaction with the hormones supplied with the nutrient medium. Presently, it seems that these exogenous phytohormones are responsible for maintaining ongoing cell division activity, originally provoked as a wound reaction at explantation. In explants cultured in a hormone-free medium, cell division stops after about two rounds. Thus, the treatments with hormonal supplements are able to produce meristematic nests with meristematic cells not found in the case of hormone-free treatments, and these probably are able to develop the endogenous hormonal system required for further development of the cultured explants. In Fig. 3.11, a schematic representation of these concepts is given.

In an attempt to interpret results of physiological or biochemical investigations of cell populations, like callus cultures, it has to be kept in mind that these data are an average of reactions of all cells in the setup. Strong variations among individual cells have to be considered, just as was shown for the cytological and morphological variations revealed by microscope inspection. To some extent, this can be circumvented by synchronization, as will be described later for cell suspensions (see Chap. 12). Like in physics, an apparent dichotomy exists between the certainty and stability of the macro-subject and the uncertainty/complexity of the individual cell (or atom). In microbiology, this is referred to as "quantal microbiology." Another aspect is disturbances of cellular performances, if in such investigations changes in the environment occur due to the use of experimental equipment. For example, it is not possible to reliably measure the distribution of carotene in living

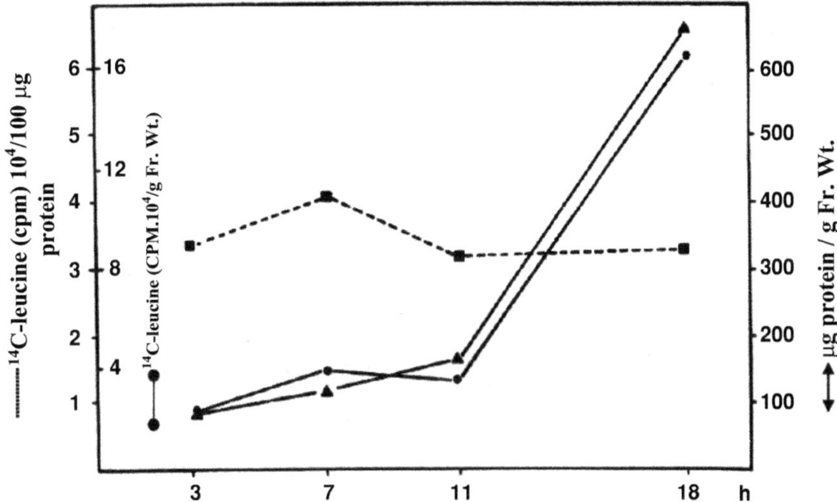

Fig. 3.12 Protein concentration and ^{14}C incorporation from L-(U ^{14}C) leucine in cultured carrot root explants during early stages of culture (Gartenbach-Scharrer et al. 1990)

cells of traverse sections of the carrot root by use of a photometer microscope. Due to the high light intensity of the microscope's lamp, carotene will be destroyed. Actually, here Heisenberg's so-called unschärfe principle has to be applied also to such experimental setups in biology.

As demonstrated in Fig. 3.12, some hours after the beginning of culture, an intensive synthesis of protein is initiated, certainly related to the formation of those threads of cytoplasm going through the central vacuole, as described above. Using ^{14}C-labeled leucine as a tracer to characterize the protein moiety synthesized during the early stages of callus induction, distinctly different protein specimens were synthesized in a hierarchical sequence. Two-dimensional electropherograms of soluble protein at the beginning of culture were stained with Coomassie brilliant blue to visualize all proteins on the electropherogram (Fig. 3.13), and by producing fluorograms, those proteins labeled with radioactive leucine synthesized during a 3-h period were traced (Fig. 3.14) at early stages of culture. By means of Coomassie brilliant blue, all proteins present in the soluble fraction at a given time are visualized and on the fluorograms those synthesized de novo at that time.

These investigations made it possible to subdivide the soluble protein into three classes. One class responds only to Coomassie brilliant blue, another can be made visible only on fluorograms, and a third responds to both. The first class would be proteins characteristic of the original tissue, i.e., the secondary phloem of the carrot root, which are no longer synthesized during culture. The second group would be synthesized only during culture labeled as characteristic for the transformation of cells related to callus induction, and the third group would be proteins of carrots, possibly so-called household proteins, already present at explantation, and that are continuously synthesized after culture initiation (Gartenbach-Scharrer et al. 1990).

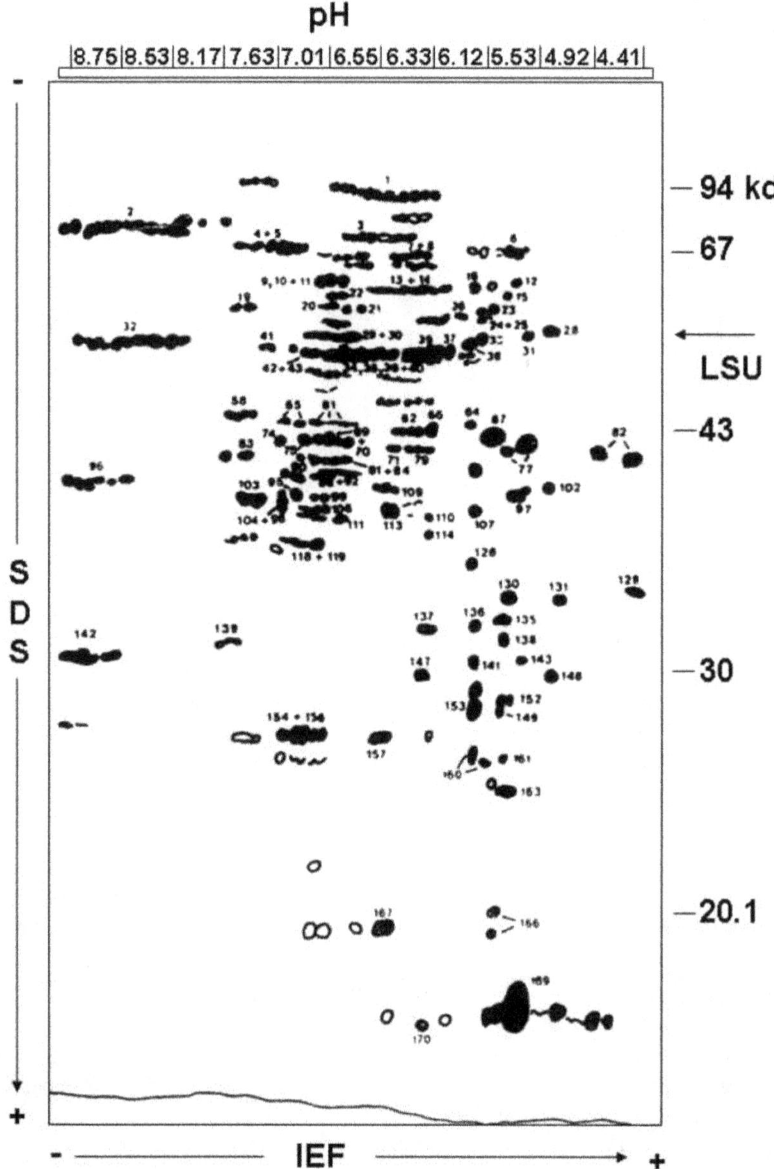

Fig. 3.13 Two-dimensional gel of soluble protein of cultured carrot root explants (secondary phloem): staining with Coomassie brilliant blue, semi-schematic representation extracted from Gartenbach-Scharrer et al. (1990). *LSU* Large subunit of chloroplast protein

At the time of these investigations, proteomics had not yet emerged, and a correlation of the occurrence of a protein and cellular processes could not be attempted. Besides a hierarchical sequence of the synthesis of proteins, however, clear

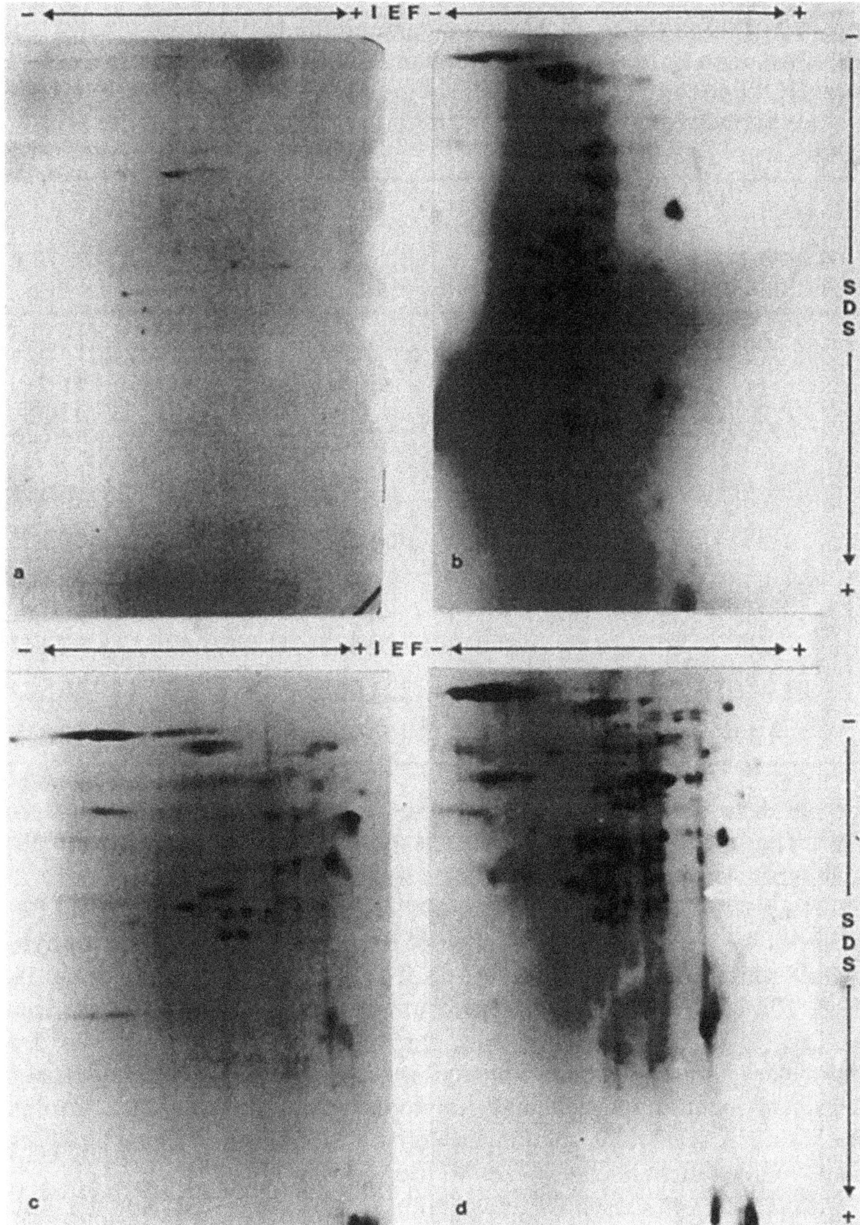

Fig. 3.14 ^{14}C distribution (applied as L-(U ^{14}C) leucine) on two-dimensional electropherograms of soluble protein of cultured carrot root explants, at various stages of initiation of the experiment. Each labeling was for 3 h: (**a**) 0–3 h, (**b**) 4–7 h, (**c**) 8–11 h, (**d**) 15–18 h (Gartenbach-Scharrer et al. 1990)

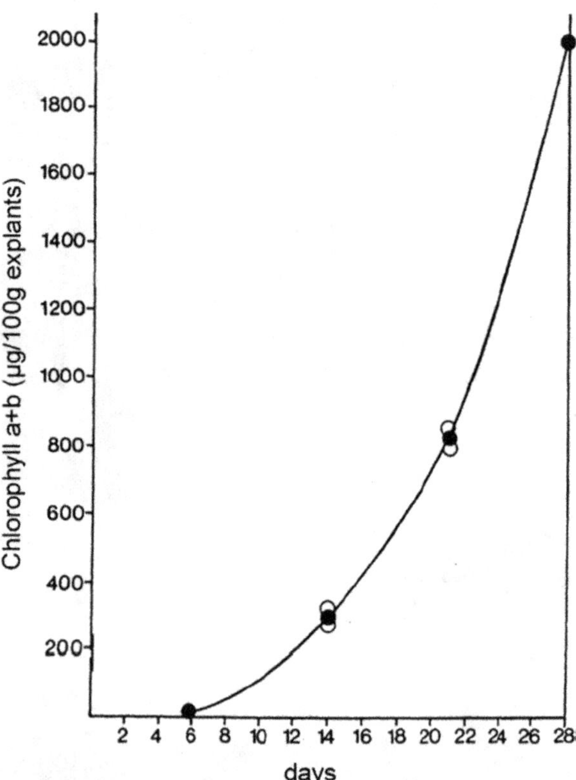

Fig. 3.15 Chlorophyll concentration of explants of the secondary phloem during a 28-day culture period (NL3)

differences in the dynamics of distinct proteins could be seen. As an example, the synthesis of one protein increases continuously, but that of another increases only up to the 11th hour in culture, and thereafter declines. Variations in the kinetics of synthesis and breakdown of individual proteins should bring about fast changes in the composition of the protein moiety of the cells during the first hours of culture and the initiation of cell division (see also Chap. 12).

The specific activity of the protein, however, remains more or less constant (Fig. 3.12). With respect to quantity, the protein fraction concerned seems to be rather small.

Also beyond this early initial period due to the high rate of cell proliferation, the concentration of newly synthesized proteins is strongly increased (Fig. 3.12). On a cellular basis, protein content increases up to the 6th day in culture in the treatments with IAA and inositol and up to the 9th day in those additionally supplied with kinetin. After these peaks, from about the 12th day onward, the protein content per cell decreases, and then a slow increase can be observed again.

The culture of carrot root phloem explants in the light results in the synthesis of chlorophyll, and from about the 5th or 6th day of culture onward, a continuous increase in chlorophyll concentration can be observed (see Fig. 3.15). Ultrastructural investigations from the 3rd day onward revealed a gradual transformation of carotene-containing chromoplasts present in the root explants, initially into amylochloroplasts and eventually into chloroplasts. Proplastids characteristic for young cells of the shoot apex were not observed (Sect. 9.1).

As a result of these changes in differentiation during culture, there occurs also a differentiation of the nutrient system of the cultures. During the first 8–10 days after initiation of the cultures, heterotrophic nutrition dominates, with the sucrose in the medium as a source of carbon and energy. Casein hydrolysate, with its amino acids as component of the nutrient medium, serves as a source of reduced nitrogen. With the establishment of the photosynthetic system, nitrate can be used as nitrogen source. Photosynthesis increasingly contributes carbon and energy, and from about the 10th day onward, a mixotrophic nutrition is established. Following a transfer into a sugar-free medium at this stage, the cultures can continue growing, via photosynthesis (see Sect. 9.1). After about 20–25 days, the sucrose and the casein hydrolysate originally supplied at t0 have been used up by the cultures, and eventually autotrophic nutrition is established. This transition coincides with the initiation of the stationary growth phase of the cultures. A transfer of the explants at this stage into a fresh medium supplemented with sucrose and casein hydrolysate initiates again an intensive cell division activity, and histologically the formation of "annual rings" can be observed. A detailed description of the nutritional system and its differentiation in primary explants will be given in Chap. 9.

If the explants are cultured on IAA and inositol but without kinetin and despite an ample supply with nutrients in the medium, a transition into the stationary phase occurs already on the 10th to 12th day of culture. This indicates the dominance of the hormone regime over the nutritional one, which here should actually have a modifying function in this system. Some deviation from this general statement will be discussed together with the formation of somatic embryos.

At the transition of the kinetin-free cultures to the stationary phase, usually the formation of adventitious roots on the explants can be observed (Fig. 3.16). Using shoot explants of rapeseed, or *Datura* and other plant species at that stage of the cultures, shoots appear later, followed by the formation of adventitious roots. Roots and shoots eventually join, and these structures can be isolated and subsequently transferred into soil for further cultivation. The significance of the process for the propagation of plants and its weaknesses that limit its application will be discussed later in detail. These observations already indicate, however, that in this case there is an influence of the kind of original tissue on the mode of differentiation during culture.

Welwitschia mirabilis is a rare primitive gymnosperm but phylogenetically very important. It is a member of the monogeneric family Welwitschiaceae. It is slow-growing but proved as an excellent experimental material to work on antiaging gene There are very few reports on tissue culture of this plant mainly because of the nonavailability and intractable nature of the plant. The callus was induced from mature leaf-tip explants of 26-year-old plant collected from CSIR–NBRI campus. The induced callus was grown at a faster pace, and induction of organized shiny and smooth, globular structures were observed, but no shoots could develop from the callus (Misra et al. 2015).

Fig. 3.16 Influence of kinetin (0.1 ppm) on the formation of adventitious roots of cultured explants of the secondary phloem of carrot roots (after 3 weeks of culture). (Left) with kinetin; here, roots are sometimes formed after 4–6 weeks of culture; (right) without kinetin

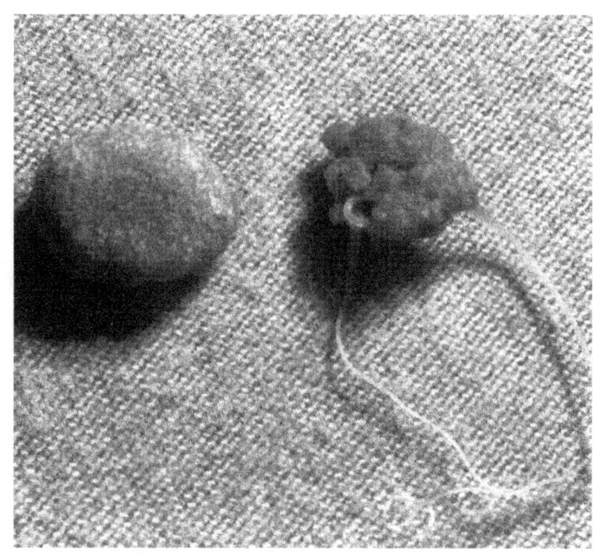

References

Bender L, Kumar A, Neumann KH (1981) Photoautotrophe pflanzliche Gewebekulturen in Laborfermentern. In: Lafferty RM (ed) Fermentation. Springer, Wien, pp 193–203

Caplin SM, Steward FC (1949) A technique for the controlled growth of excised plant tissue in liquid media under aseptic conditions. Nature CIXIII:920–924

Elmshäuser HA (1977) Untersuchungen über die vegetgative Vermehrung von Brassicaceen—insbesondere Raps (*Brassica napus* L. ssp *oleifera* Metzg. Sinsk.). Dissertation, Justus Liebig Universität, Giessen

Engelmann F (1997) In vitro conservation methods. In: Ford-Lloyd BV, Newburry HJ, Callow JA (eds) Biotechnology and plant genetic resources. Conservation and use. CABI, Wallingford, pp 119–162

Gamborg OL, Miller RA, Ojima K (1968) Nutrient requirements of suspension cultures of soybean root cultures. Exp Cell Res 50:151–158

Gartenbach-Scharrer U, Habib S, Neumann KH (1990) Sequential synthesis of some proteins in cultured carrot explant (*Daucus carota* L.) cells during callus induction. Plant Cell Tissue Organ Cult 22:27–35

Hellwig S, Drossart J, Twyman RM, Fischer R (2004) Plant cell cultures for the production of recombinant proteins. Nat Biotechnol 22:1425–1422

Imani J, Berting A, Nitsche S, Schaefer S, Gerlich WH, Neumann KH (2002) The integration of a major hepatitis B virus gene into cell-cycle synchronized carrot cell suspension cultures and its expression in regenerated carrot plants. Plant Cell Tissue Organ Cult 71:157–164

Lindsey K, Yeoman MM, Black GM, Mavituna F (1983) A novel method for the immobilization and culture of plant cells. FEBS Lett 155:143–149

Matsumoto T, Sakai A, Yamada K (1994) Cryopreservation of in vitro grown apical meristems of wasabi (*Wasabia japonica*) by vitrification and subsequent high plant regeneration. Plant Cell Rep 13:442–446

Miró-Canturri A, Ayerbe-Algaba R, Smani Y (2019) Drug repurposing for the treatment of bacterial and fungal infections. Front Microbiol 10:41. https://doi.org/10.3389/fmicb.2019.00041

Misra P, Gupta N, Toppo DD, Pandey M, Mishra MK, Tuli R (2010) Establishment of long term proliferating shoot cultures of elite Jatropha curcas L. by controlling endophytic bacterial contamination. Plant Cell Tissue Org Cult 100:189–197

Misra P, Purshottam DK, Goel AK, Nautiyal CS (2015) *Welwitschia mirabilis*—induction, growth and organization of mature leaf callus. Curr Sci 109(3):567–571

Murashige T, Skoog F (1962) A revised medium for rapid growth and bioassays with tobacco tissue. Physiol Plant 15:473–497

Neumann KH (1966) Wurzelbildung und Nukleinsäuregehalt bei Phloem-Gewebekulturen der Karottenwurzel auf synthetischem Nährmedium. Congr Coll Univ Liege 38:96–102

Neumann KH, Steward FC (1968) Investigations on the growth and metabolism of cultured explants of Daucus carota. Planta 81:333–350. https://doi.org/10.1007/BF00398020

Nitsch JP, Nitsch C (1969) Haploid plants from pollen grains. Science 163:85–87

Pollard JK, Shantz EM, Steward FC (1961) Heitols in coconut milk: their role in nature of dividing cells. Plant Physiol 36:492–501

Premjet D, Tachibana S (2004) Production of podophyllotoxin by immobilized cell cultures of *Juniperus chinensis*. Pakistan J Biol Sci 7:1130–1134

Reed BM (2001) Implementing cryogenic storage of clonally propagated plants. Cryo-Letters 22:97–104

Schäfer-Menuhr A, Müller E, Mix-Wagner G (1996) Cryopreservation. An alternative for the long-term storage of old potato varieties. Potato Res 39:507–513

Seitz U, Reuff I, Reinhard E (1985) Cryopreservation of plant cell cultures. In: Neumann K-H, Barz W, Reinhard E (eds) Primary and secondary metabolism of plant cell cultures. Springer, Berlin Heidelberg, pp 323–333

Sminovitch D (1979) Protoplasts surviving freezing to –196 and osmotic dehydration in 5 molar salt solutions prepared from the bark of winter black locust trees. Plant Physiol 63:722–725

Sonderquist RG, Lee JM (2008) Enhanced production of recombinant proteins from plant cells. In: Kumar A, Sopory S (eds) Recent advances in plant biotechnology and its applications. I.K. International, New Delhi, pp 330–345

Steward PC, Mapes MO, Mears K (1952) Investigation on growth and metabolism of plant cells. I. New techniques for the investigation of metabolism, nutrition and growth in undifferentiated cells. Am J Bot 16:57–77

Thorpe TA, Kamlesh RP (1984) Clonal application: adventitious buds. In: Vasil IK (ed) Cell culture and somatic cell genetics of plants, vol 1. Academic Press, New York, pp 49–60

Touchell DH, Chiang VL, Tsai CJ (2002) Cryopreservation of embryogenic cultures *Picea mariana* (black spruce) using vitrification. Plant Cell Rep 21:118–124

Tsai C-J, Hubscher SL (2004) Cryopreservation in *Populus* functional genomics. New Phytol 164:73–81

Wetzker R (2004) Irrungen und Perspektiven der Phosphoinositid 3-Kinase-Forschung. Biospektrum 5:620–622

White PH (1954) The cultivation of animal and plant cells. Ronald Press, New York

Yin DM, Wu JC, Yuan YJ (2005) Reactive oxygen species, cell growth, and taxol production of *Taxus cuspidata* cells immobilized on polyurethane foam. Appl Biochem Biotechnol 127:173–185

Yin DM, Wu JC, Yuan YJ (2006) Spatio-temporal distribution of metal ions and taxol of *Taxus cuspidate* cells immobilized on polyurethane foam. Biotechnol Lett 28:29–32

Yuan S, Alper HS (2019) Metabolic engineering of microbial cell factories for production of nutraceuticals. Microb Cell Factories 18:46. https://doi.org/10.1186/s12934-019-1096-y

Yuan Q, Xu H, Hu Z (1999) Two-phase culture of enhanced alkaloid synthesis and release in a new airlift reactor by *Catharanthus roseus*. Biotechnol Tech 13:107–109

Cell Suspension Cultures

4

Most cell suspension cultures originate from callus cultures due mainly to mechanical impact in agitated liquid media. In stationary cultures on agar, a suspension can be produced commonly by use of a sterile glass rod or squeezing with a scalpel. In particular with 2.4D in an agar medium, a loosely connected cell population develops on the opposite side of the agar, which can be easily scraped off with a scalpel. Enhanced production of recombinant proteins from plant cells is an important application (Sonderquist and Lee 2008).

An improvement can often be obtained by using ammonia as nitrogen source, probably due to the excretion of protons as exchange for its uptake by the cells.

Callus cultures in a liquid nutrient medium are usually agitated, and after 10–14 days, this mechanical impact results in the development of cell suspensions consisting of cells from the periphery of the explants (Fig. 4.1). Besides healthy cells that continue to grow, such a suspension contains also dead or decaying cell material. If the methods described above fail to succeed, then an enzymatic maceration of callus material should be attempted (0.05% crude macerozyme, 0.05% crude cellulose Onozuka P-1 500, and 8% sorbitol; King et al. 1973). Another possibility to produce a cell suspension is to first obtain protoplasts, as described later (Chap. 5).

The definition of a cell suspension still provokes controversial discussions. The original aim in the 1950s was to establish culture systems in which, similarly to algae cultures, a suspension of cells of higher plants would consist solely of single cells. In practice, this aim was reached only for a few systems, using a hanging drop method. All other attempts failed. Even in experiments starting with a population of single cells in a liquid medium, cell aggregations of various size will develop soon after initiation of growth, coexisting with some free cells (Fig. 4.1b).

4.1 Methods to Establish a Cell Suspension

As done for callus cultures, the description of how to obtain a cell suspension shall be illustrated in a practical example that can be easily adapted to many other systems.

© Springer Nature Switzerland AG 2020

K.-H. Neumann et al., *Plant Cell and Tissue Culture – A Tool in Biotechnology*,

https://doi.org/10.1007/978-3-030-49098-0_4

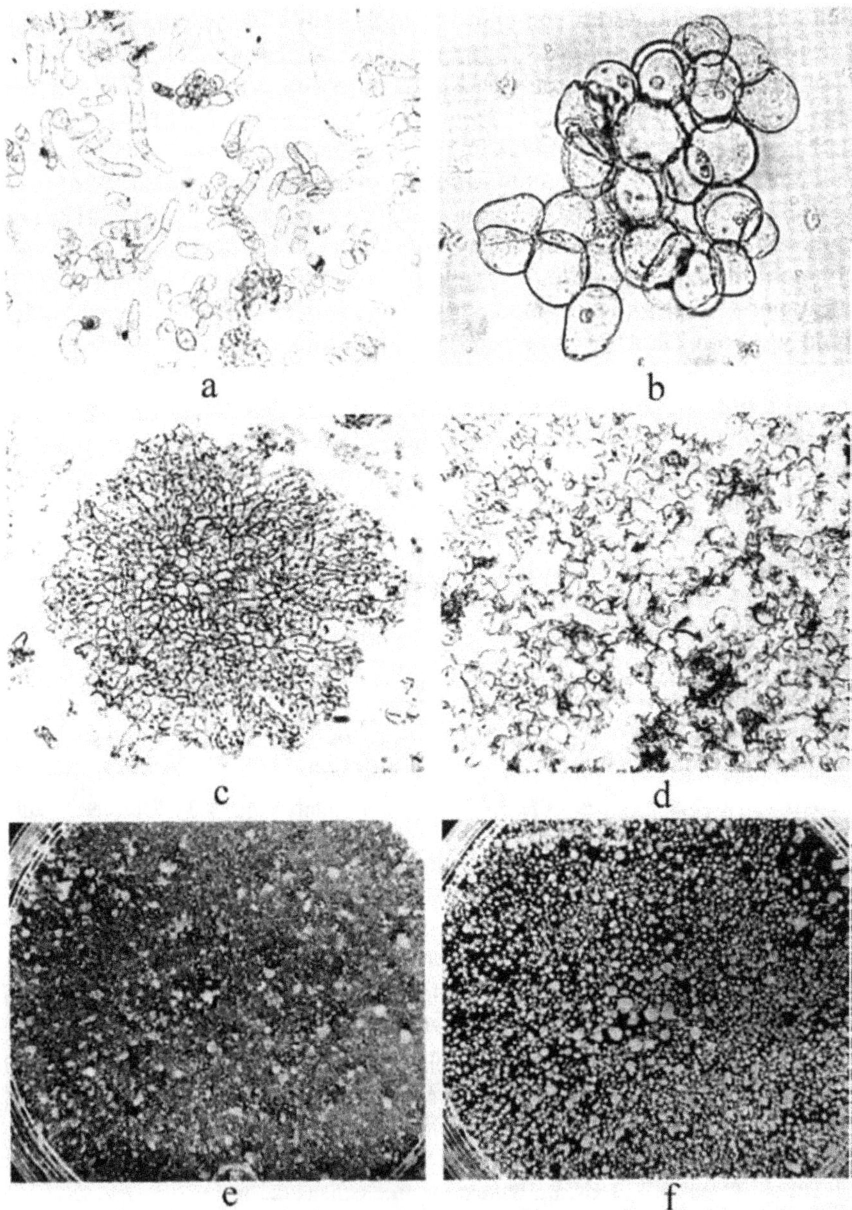

Fig. 4.1 Histomorphological characterization of suspension cultures of *Datura innoxia* (Kibler and Neumann 1980). (**a**) Inoculum (250 μm filtrate, ca. 25×), (**b**) microcluster (ca. 125×), (**c**) histological structure of a secondary cluster cultured with kinetin (ca. 25×), (**d**) histological characterization of cell material cultured without kinetin, (**e**) cell suspensions of *Datura innoxia* without kinetin, and (**f**) with kinetin in the medium (1×)

In this example, shoot explants of *Datura innoxia* were originally used to produce callus cultures to study the synthesis of secondary metabolites. For a better understanding, the establishment of callus cultures, now from shoot tissue, will be briefly described.

Explants of the uppermost (youngest) internode are cut using an extremely sharp scalpel. For sterilization, cut ends are briefly dipped into liquid paraffin to prevent the entrance of the agent used for surface sterilization, for 5–6 min in the hypochlorite solution already described (Sect. 3.1).

Using a laminar flow (aseptic working bench) for all further handling, internode segments 1–2 cm long are rinsed four to five times with sterilized distilled water. After this, the paraffin cover and the epidermis are removed with the help of a sterilized scalpel, and with a second sterilized scalpel, the tissue is cut into segments about 1 mm thick. These discs are cut into halves, which then serve to establish callus cultures. These segments are bigger than those used to establish primary carrot cultures, having a weight of about 7–8 mg and consisting of about 30,000 cells each. If the diameter of the disc is even larger, then even more explants can be obtained. The nutrient medium is given in Table 3.4 (NL medium) and is suitable for both stationary and liquid cultures. Within 3 weeks of culture, a highly proliferate callus develops from which peripheral cells can easily be scraped off with the help of a scalpel. This cell material is transferred to the MS medium with agar (Table 3.4) supplemented with kinetin, for growth at 27 °C at a 12 h light/dark rhythm. After two subcultures at an interval of 3–4 weeks, the subsequent subcultures are initiated every 2 weeks. Also for subculturing, only peripheral cell material from newly developed callus pieces is used.

After the production of sufficient cell material, the rather loosely connected clusters are transferred into a liquid medium of the same composition in Erlenmeyer flasks on a shaker. Within 1 week, a dense cell suspension develops. An inoculation of 5 g fresh weight corresponds to a cell density of 40,000 cells per ml of nutrient medium, sufficient for optimal proliferation of the cell population (Fig. 4.2).

4.2 Cell Population Dynamics

A cell suspension usually consists roughly of three fractions, i.e., free single cells of various shapes, cell aggregates consisting of up to ten cells or more, and finally cell groups with a threadlike morphology. These fractions can be isolated by suitable sieving techniques. Investigations to characterize these three fractions indicated that cell proliferation by division occurs predominantly in cell aggregates, which are comparable to the meristematic nests of callus cultures (Chap. 3). In both, very small cells can be seen in the center, and cell size increases toward the periphery. Highest cell division activity occurs in the center of these structures.

Due to the agitation of the shaker, the outermost cells of the cell aggregates are mechanically removed (see Fig. 4.3) and then represent the fraction of free single cells. These cells should be older and mostly quiescent in terms of cell division activity. However, some of these cells preserve the ability to divide, or this is

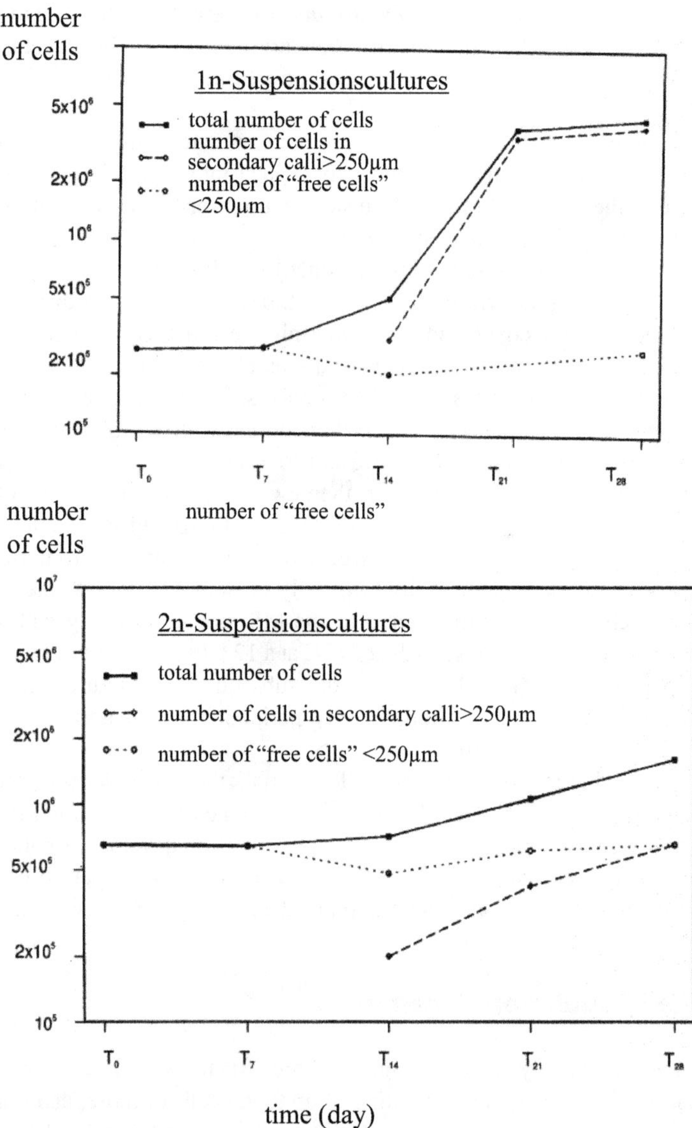

Fig. 4.2 Growth of haploid (top) and diploid (bottom) cell suspensions of *Datura innoxia* (Kibler and Neumann 1980)

re-induced. Such cells are possibly the origin of the third fraction, the cellular threads. A similar organization can be observed in carrot cell suspensions. As an example, the threadlike structure in Fig. 4.4 observed in a carrot suspension seems to be the result of three cell divisions. One terminal cell differentiates into a tracheid-like structure, the other accumulates anthocyanin, and the four central cells showing

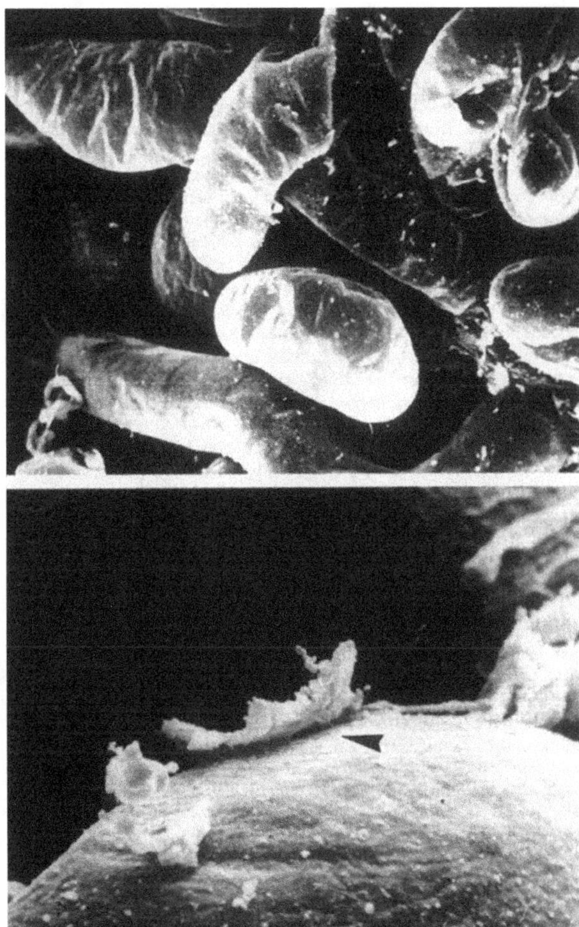

Fig. 4.3 Loosely structured surface of a callus (top) and remains of a mechanical break-off of a cell in a suspension (bottom, see arrow; photographs by A. Kumar)

chlorophyll accumulation would be the youngest cells derived from the last rounds of cell division. The great differences in the structure of the two terminal cells point to an unequal first cell division, with differences in the distribution of cytoplasm. The nutrient medium can be regarded as identical for both cells.

A determination of DNA concentration indicated a near-cytogenetic homogeneity only for cells in the aggregates (secondary calli). In the population of free single cells, a strong inhomogeneity exists, sometimes with very high DNA content per cell (Fig. 4.5). This observation is consistent with results obtained from callus material. Here, also the lowest C-values of a ploidy level can be found in the center of the meristematic nests with high cell division activity.

In both cases, these small cells in haploid cultures were found to have a DNA content essentially identical to that of microspores of the same species (G1-phase cells) or twice that of G2-phase cells. In diploid cultures, the DNA content was either

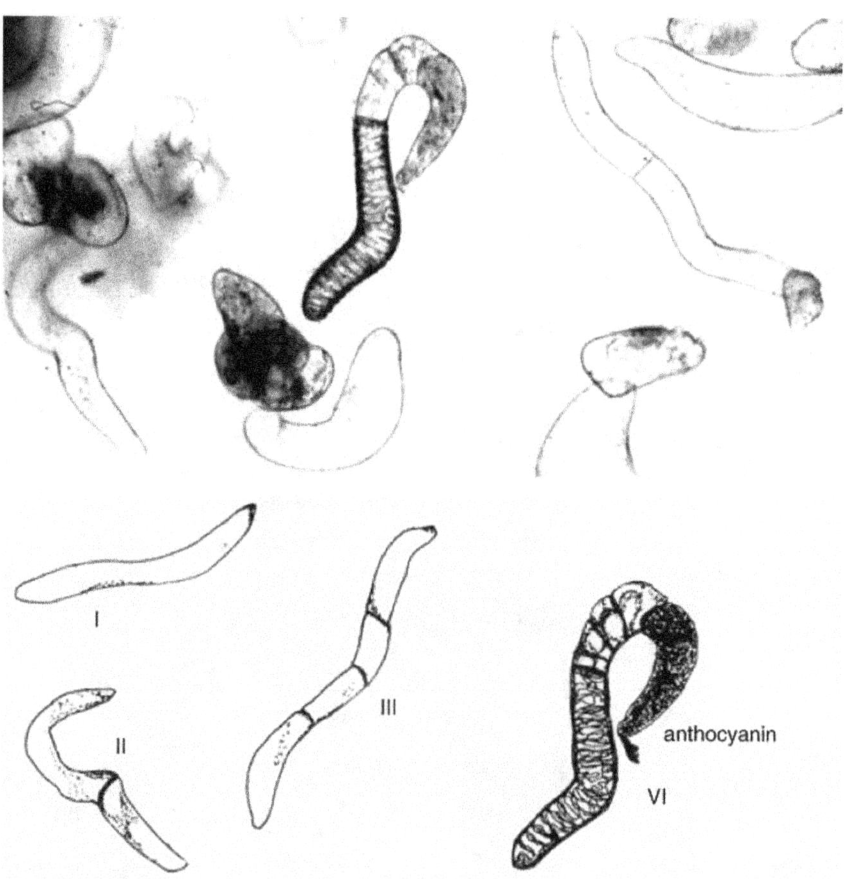

Fig. 4.4 A thread of cells in a cell suspension culture of carrot in White's basal medium containing 10% coconut milk. Top A thread consisting of six cells, resulting from three divisions of a single cell. By the third division, the inner four cells seem to be produced. (Bottom) the right terminal cell contains anthocyanin, and the left terminal is a trachea. The differences of differentiation of the terminal cells would be due to an unequal first cell division of the "mother cell." The higher degree of specialization of the terminal cells, compared to that of the four inner cells, could be due to more time elapsed since division took place relative to the last division

twice that of G1-phase cells of haploids or four times the value of microspores in G2-phase cells. In older cells located between meristematic nests in callus material, which would be comparable to the fraction of free cells in the suspension, a broad variation in C-values was determined. Apparently, cytogenetic stability is linked to the age of the cells, i.e., the length of time elapsed since the last division. In young material with high cell division activity, a high percentage of cells contains DNA characteristic of the ploidy level. A supplement of kinetin, which increases cell division activity, results in a higher cytogenetic stability and homogeneity of the cell population (Sect. 13.1).

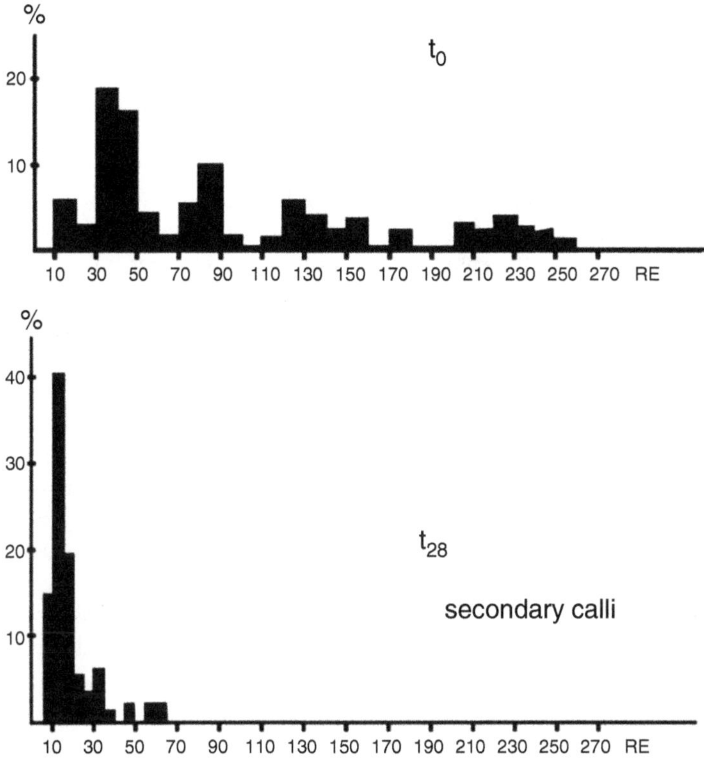

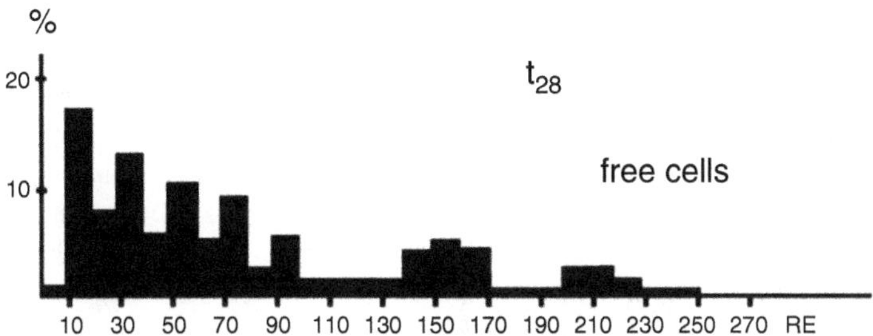

Fig. 4.5 DNA content of nuclei (microfluorometric determination, relative units) of haploid cell suspension cultures of *Datura innoxia* at inoculation (t_0) and after 28 days of culture ($n = 13.5$; after Kibler and Neumann 1980)

In cell suspensions, many cell structures occur that are morphologically difficult to classify. However, some well-defined cell types can also be observed, e.g., tracheids, as described above (Fig. 4.4). In a cell suspension, all free single cells are bathed in the same nutrient solution, and therefore the morphological diversification of its components should be based on the origin of the individual cell. The significance of unequal cell divisions has already been mentioned above—whatever the cause of this phenomenon may be. A direct relation between cell shape and vitality has not been observed.

In cell suspensions of some species like *Daucus* in an IAA-supplemented medium (NL medium, Table 3.4), after some weeks of culture, the formation of early stages of embryo development can be observed, and these can eventually be raised to intact plants (somatic embryogenesis; for details, see Sect. 7.3).

Using the methods described above, only limited amounts of cell material can be produced, usually not sufficient to study physiological or biochemical problems of primary or secondary metabolism or, e.g., somatic embryogenesis. If greater amounts of material are required, fermenter cultures are performed (see also Sects. 3.2 and 10.9). As an example, fermenter cultures of *Datura innoxia* shall be described. Here, within 2 weeks it was possible to produce 1 g of dry weight per day in a liquid nutrient medium of 3.5 l originally inoculated with a cell suspension of 30 g fresh weight. The cell suspension was obtained by a method used to raise cytogenetically stable material, as described later. Pre-culture is carried out in 200 ml nutrient medium (MS+kinetin, see Table 3.4) in a 750 ml Erlenmeyer flask on a shaker (see above). For initiation of the pre-culture, the vessel is inoculated with 1–2 g fresh weight (90–250 μm fraction). The main aim of the pre-culture is to propagate the cells. After 10–14 days of pre-culture, the content of the vessel (cells and nutrient medium) is transferred to the fermenter, as described above (see Sect. 3.2). In the fermenter, cell aggregates as well as free single cells occur.

The principle to distinguish between a propagation phase and a production phase is also applied to fermenter cultures used for biotechnological purposes. Here, fermenters of much larger volume are used; to produce cell suspensions for inocula-tion, however, smaller laboratory fermenters are used initially, as described later. Usually, the cell suspension is transferred with some nutrient medium from the smaller to the next bigger fermenter. For a semi-continuous culture, it is common practice to remove part of the cell material in certain intervals of time for processing and to apply fresh nutrient medium. As described later (Chap. 10), plant cell suspensions are already today cultured in fermenters with a volume of thousands of liters (Mitsu Petrochem. Ind. Ltd.), e.g., to produce shikonin derivatives using cultures of *Lithospermum officinale*. Also propagation via somatic embryogenesis has been carried out in a fermenter (e.g., *Daucus*; see Sect. 7.3).

To maintain cell strains in a healthy condition for prolonged periods, subcultures have to be made frequently, usually at 1- or 2-week intervals, and with a dilution of 1:5 after 1 week and 1:10 after 2 weeks with the fresh nutrient medium. The optimal dilution and the subculture frequency have to be determined for each individual strain. As described above, also cryopreservation is often used to maintain cell suspensions (Sect. 3.6).

References

Kibler R, Neumann KH (1980) On cytogenetic stability of cultured tissues and cell suspensions of haploid and diploid origin. In: Sala F, Parisi B, Cella R, Ciferri O (eds) Plant cell cultures: results and perspectives. Elsevier/North Holland, Amsterdam, pp 59–65

King J, Mansfield KJ, Street HE (1973) Control of growth and cell division in plant cell suspension cultures. Can J Bot 51:1807–1823

Sonderquist RG, Lee JM (2008) Enhanced production of recombinant proteins from plant cells. In: Kumar A, Sopory S (eds) Recent advances in plant biotechnology and its applications. I.K. International, New Delhi, pp 330–345

Protoplast Cultures

5

With suitable enzymes the cell wall of plant cells can be removed through hydrolysis of its macromolecular building material, i.e., "naked" cells called protoplasts are derived. In an isotonic medium, these protoplasts are healthy and can survive. Protoplasts are used to investigate a broad range of physiological problems reaching from the significance of the cell wall for nutrient uptake to mechanisms related to the synthesis of the cell wall. In an early investigation, Bush and Jacobson (1986) show for protoplasts the same kinetics, time course, and pH response, e.g., of potassium uptake as the intact cells of a suspension.

Besides such basic problems since the 1960s in many instances, protoplasts were used to solve problems of practical plant breeding.

It is an old dream of plant breeders to produce hybrids of different plant species not to be obtained by cross pollination to have plant material with properties characteristic of both parents. The probably most prominent example is a hybrid of potatoes and tomatoes as parents with the ability to produce tomatoes as fruits and potatoes growing on subterranean stolons. As can be seen from a reproduction from Strasburger's *Lehrbuch der Botanik* printed at the beginning of last century (Fig. 5.1; cf. it is missing in later editions), this was also the aim of Winkler's occlusion experiments using two solanaceous species. A histological inspection of the shoot apex clearly shows that the hybrids obtained are chimeras with quite interesting morphologies of fruits and leaves. The arrangement of cell layers of both "parents" is probably the result of a mixture of cells of wound callus formation on the cutting edges made for the occlusion procedure.

Based on first successful fusion experiments with protoplasts of different species in the 1970s of last century, hybrids of tomatoes and potatoes produced by fusion of protoplasts of the two species were reported (see Fig. 5.2). Mesophyll protoplasts of *Solanum lycopersicum* and callus protoplasts of *Solanum tuberosum* were fused by Melchers (1978; 80 mM $CaCl_2$, 4.5% PRG, pH 10), but the plants produced were sterile.

Restriction analyses of chloroplast DNA and characterization of RuBisCO of both parents as well as the hybrid by electrophoresis clearly indicated the occurrence

K.-H. Neumann et al., *Plant Cell and Tissue Culture – A Tool in Biotechnology*,
https://doi.org/10.1007/978-3-030-49098-0_5

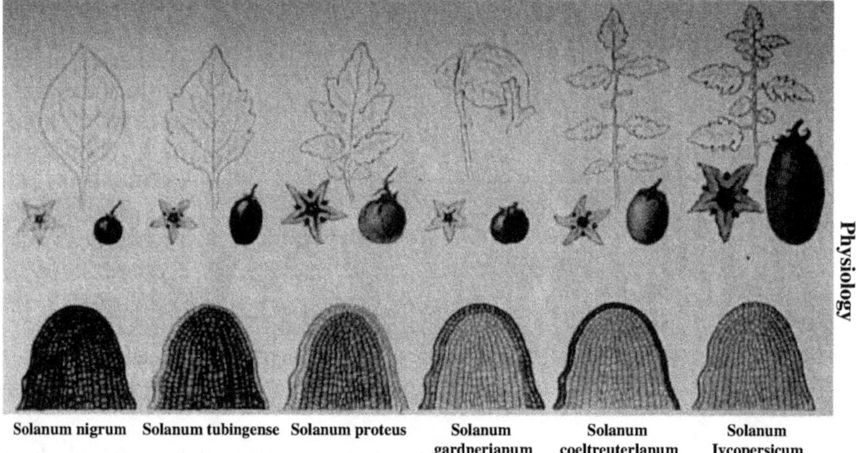

Fig. 5.1 Grafting chimeras of *Solanum nigrum* and *Solanum lycopersicum* and parents (original H. Winkler). A leaf, a flower, the shoot apex, and a fruit are shown for each hybrid. In the apex, the cell layers stemming from *S. nigrum* are dark colored; those from *S. lycopersicum* are light (from Strasburger et al. 1913)

of two types of hybrids, despite fusion of the nuclei. Still, the number of chromosomes was higher than those of either parent. One type of hybrid apparently contained plastids only of the potato (potatoes) and the other those of the tomato (tomatoes). Mixed cases were not found, but only a limited number of individuals were investigated, of which two thirds were potatoes and one third were tomatoes. A successful fusion was identified after microscopically detecting the fusion of color-free (pre-grown in the dark) potato protoplasts, with protoplasts of light-green tomato plants containing a genetically disturbed chlorophyll system. A transfer of the potato cells from darkness to the light resulted in the formation of chlorophyll, and regenerates had leaves with normal chlorophyll concentration. Callus cultures of the tomato parent regenerated only adventitious roots. Regenerates of the fusion experiments with normal chlorophyll concentration were either of potato origin or offspring of a protoplast fusion, i.e., a hybrid. Based on numerous morphological properties, it was possible to distinguish between potatoes and the hybrid. Interestingly, a gas chromatographic analysis of volatile components of undifferentiated callus cultures of hybrids indicated the occurrence of substances absent in those of the parents.

These early experiments proved the possibility of producing crosses between different species, though these hybrids could not be used in practical plant breeding programs.

At about the same time, the fusion of *Arabidopsis thaliana* and *Brassica campestris* was reported by Gleba and Hoffmann (1978) and somewhat later the production of "synthetic" rapeseed plants by in vitro fusion of protoplasts of *Brassica oleracea* and of *Brassica campestris* (Schenck 1982), which are thought to be the parents of rapeseed following a spontaneous hybridization about 1000

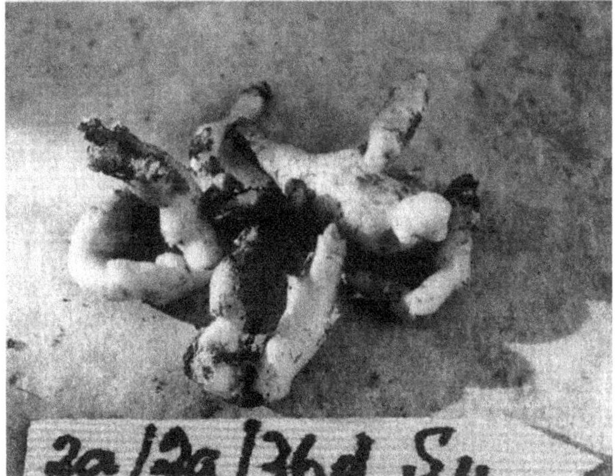

Fig. 5.2 A hybrid produced by protoplast fusion of *Solanum tuberosum* and *Solanum lycopersicum* (potatoes; hybrid nucleus of *S. tuberosum* and *S. lycopersicum*, plastids from *S. lycopersicum*). (Top) fruits developed after pollination with *Solanum stenotomum*. (Bottom) underground storage organs

years ago. In the examples given above, plant species of the same genera or family were used as hybridization partners. Later, fusions were attempted also of species of quite distant systematics, like tobacco and carrot producing so-called NICA plants (Dudit et al. 1987). This fusion was successful, and callus was produced, but the regeneration of plants failed. This could be achieved after the genome of carrot protoplasts was destroyed by irradiation with X-rays. As can be seen in Fig. 5.3, these NICA plants have the habitus of tobacco plants with narrow leaves.

The original aim of protoplast cultures was to produce new genomes with properties exhibited by neither parent. This has now been replaced by many experimental systems able to insert selected foreign genes into a recipient genome—gene technology. The first to successfully use this approach were possibly Potrykus and his research group in Basel (Potrykus et al. 1987). Here, a virus was employed as a vector to transfer the genetic information for resistance to the antibiotic kanamycin to tobacco protoplasts. These transformed protoplasts could be raised to intact plants

Fig. 5.3 The habitus of leaves of NICA plants (left) and that of the tobacco parent (right). NICA plants are the result of protoplast fusions of *Daucus* and *Nicotiana*

carrying this resistance. Gene technology will be dealt with in a later chapter discussing its advantages and shortcomings (Sect. 13.6.3).

Even nowadays, protoplasts are often employed as recipients of foreign genetic material and to produce plants through somatic embryogenesis that can be used in plant breeding programs. As an example, some results on using protoplasts of rice in gene technology shall be mentioned. Of the several methods available to transfer foreign genetic material based on biolistics, or *Agrobacterium*-mediated transfer to competent cells, best results and highest efficiency are achieved by a direct introduction into rice protoplasts. Whereas it is almost routine to obtain protoplasts of japonica varieties, those of indica rice are still recalcitrant to tissue culture procedures. A method to this end was published by Zhang (1995).

5.1   Production of Protoplasts

The methods described in this chapter were originally developed to obtain protoplasts from leaves of various Brassicaceae (Elmshäuser et al. 1979) Later, these could be successfully adapted to other plant species (various orchids, *Datura*, carrots, and others). For sterilization, the tissue used to obtain protoplasts is first exposed to 70% ethanol for 1 min, followed by submergence into a hypochlorite solution (0.6%) for 20 min. After this, the leaf material is washed four times with sterile aqua dest. and then transferred for 15 min to the nutrient medium used

Table 5.1 Culture media used for protoplast culturing

Component	Concentration	Component	Concentration
Macro- and micronutrients (mg/l; Gamborg et al. 1968)			
$NaH_2PO_4 \times H_2O$	1110.0	$MnSO_4 \times H_2O$	10.000
KNO_3	3000.0	H_3BO_3	3.000
$(NH_4)_2SO_4$	134.0	$ZnSO_4 \times 7H_2O$	2.000
$MgSO_4 \times 7H_2O$	250.0	$Na_2MoO_4 \times 2H_2O$	0.250
$CaCl \times 2H_2O$	1025.0[a]	$CuSO_4$	0.025
Fe-EDTA	46.3[a]	KI	0.750
Organic components (mg/l; Kartha et al. 1974)			
Nicotinic acid	1.0	Mannitol	0.5 M[a]
Thiamine	10.0	2.4D	2.3×10^{-6} M
Pyridoxine	1.0	BA	4.4×10^{-6} M
M-inositol	100.0	NAA	1.6×10^{-5} M
Glutamine	200.0[a]		
Casein hydrolysate	250.0[a]		
Glucose	2500.0		
Ribose	125.0		
Enzyme solution to produce protoplasts (pH 6.2)			
Cellulase Onozuka SS1 500	2.0%		
Mazerozyme[b]	1.0%		
Pectinase (Serva)	0.5%		
Potassium dextran sulfate[b]	0.5%		
Mannitol	0.5 M		

Elmshäuser et al. (1979); macro- and microelements as in B5 medium, Table 3.4
[a]Changed from the original
[b]Welding & Co., Hamburg

subsequently for the cultivation of the protoplasts (Table 5.1), without the organic components. Instead, 0.4 M mannitol is supplied to detach the plasmalemma from the cell wall through plasmolysis. After this pre-incubation period, the leaves (still intact) are cut into pieces approximately 0.5 mm in length. Then, 150–200 mg fresh weight of this leaf material is incubated with 10 ml of the enzyme solution in Table 5.1, in 60 × 15 mm plastic Petri dishes. Beforehand, the enzyme solution is passed through a membrane filter (45 μm) for sterilization. The Petri dishes are sealed with Parafilm and covered with aluminum foil to prevent illumination.

The leaf material is left for 6 h at 28 °C in darkness in the enzyme solution. At the end of this incubation, the individual protoplasts are detached by gentle shaking (Fig. 5.4). By passing through a glass filter, or glass wool, the remaining leaf material is removed. After this, the protoplasts are separated from the incubation medium by gentle centrifugation at about 100 g with the help of a hand centrifuge, and the sedimented material containing the protoplasts is washed four times with the nutrient medium to be used for culture of the protoplasts (Table 5.1). Finally, 2 ml of the protoplast suspension is transferred to the same plastic Petri dishes as described above for incubation.

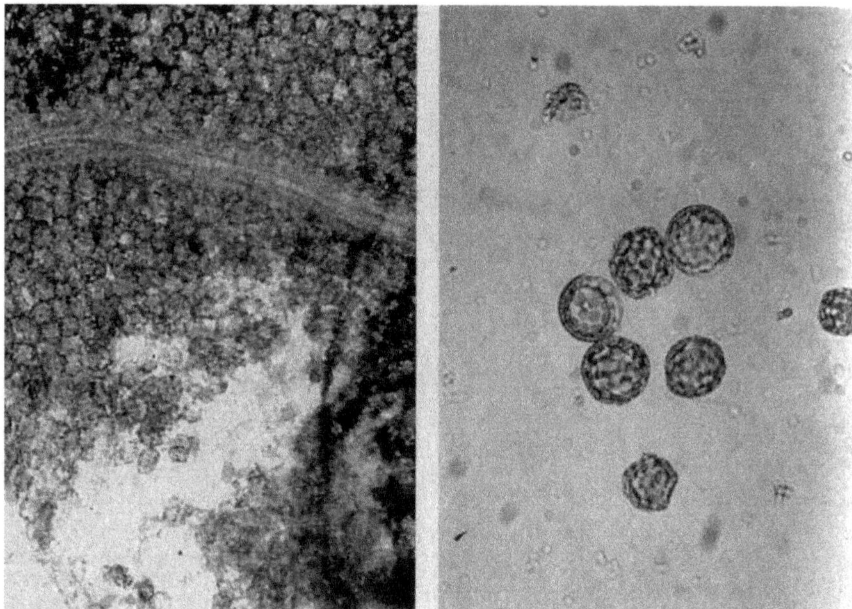

Fig. 5.4 The isolation of protoplasts (*Brassica* ssp.: (left) initiation of disintegration and first free protoplasts, (right) freshly isolated protoplasts

In experiments to produce protoplasts of several Brassicaceae, the highest efficiency in obtaining a healthy population was achieved by using leaves of 6-week-old plants. Important were also the growing conditions of the donor plant immediately before the experiment. An illumination period of 10 h was optimal. Possibly during this illumination period, the cells accumulate enough energy by photosynthesis to withstand the stress of being transformed into protoplasts. An optimal density at the beginning of culture is about 10^5 protoplasts per ml of suspension.

It was possible to successfully replace the enzymes in the solution by the culture supernatant from *Clostridium cellulovorans*, as shown for cultured cells of tobacco and *Arabidopsis thaliana*.

During the first 40 h after the protoplast production, culture is carried out at 500 lux, followed by a period of 5 days at 2000 lux. The temperature is kept at 26–28 °C, under 12/12 h light/dark illumination. Then follows the application of 0.2 ml of fresh medium of the same composition as that originally used but in which mannitol is replaced by sucrose (2%; Table 5.2). After another 7 days, 2 ml of this nutrient medium is supplied per Petri dish, and the total volume is partitioned into two Petri dishes of the same size and volume as that of the former. Cell aggregates produced after another 10 days are transferred onto agar.

About 3 h after the start of incubation in the enzyme solution, a disintegration of tissue and the first free-floating protoplasts can be observed (Fig. 5.4). In the spherical protoplasts, the chloroplasts initially gather at the periphery of the cell, and later, i.e., about 30–35 h after the start of the experiment, an accumulation of

Table 5.2 Nutrient solution to culture cell aggregates developed from protoplasts

Component	Concentration
Sucrose	2.0%
Agar	0.6%
Casein hydrolysate	0.1%
2.4D	0.2 ppm
Kinetin	0.1 ppm

Macro- and microelements following MS medium, vitamins following Gamborg et al. 1968, Table 3.4

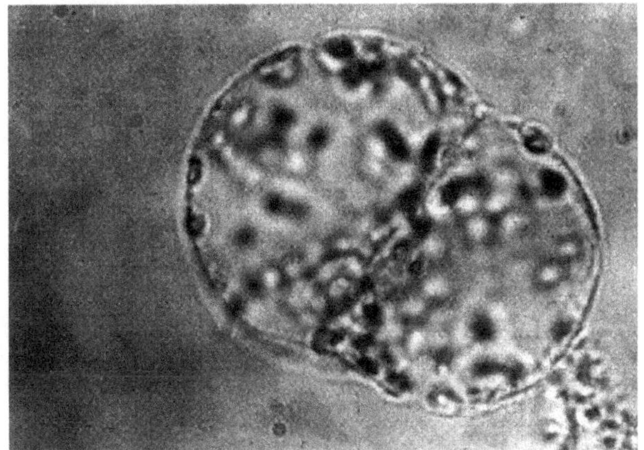
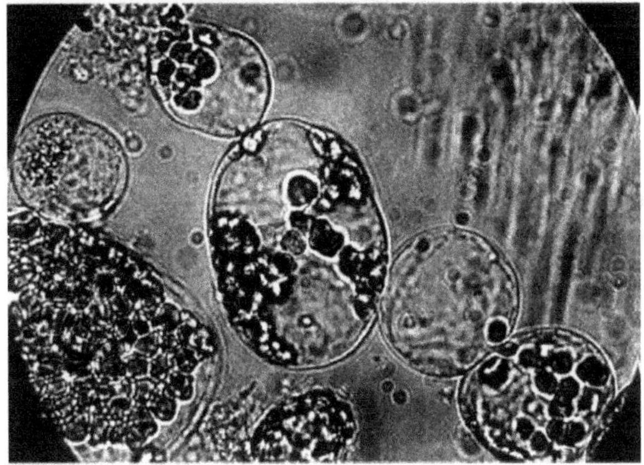

Fig. 5.5 Protoplast isolation: (top) regenerated cell wall (70 h after isolation), (bottom) first cell division (about 100 h after isolation)

chloroplasts occurs around the nucleus. Often these organelles exhibit a brownish color. Using Calcofluor White as a stain specific for cell wall material, the beginning of restructuring of the cell wall can be observed. Concurrently, the originally spherical protoplasts become oblong (oval; Fig. 5.5), and at about 100 h after isolation, the initiation of the first cell divisions can be observed. Apparently, the regeneration of the cell wall is a prerequisite to initiate cell division—a plausible explanation of this phenomenon is to date not possible.

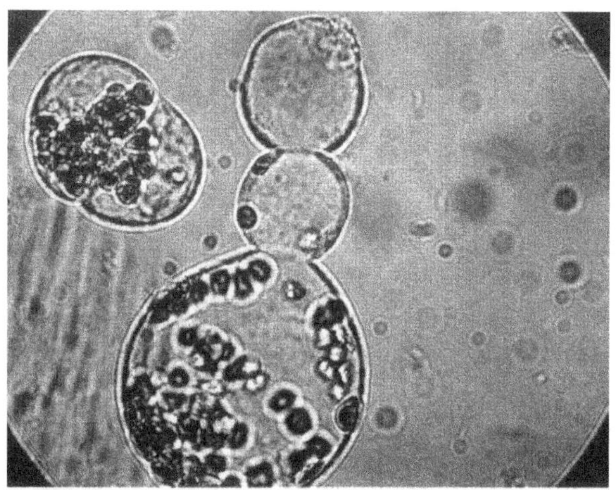

Fig. 5.6 "Budding" of a protoplast (above; turnip rape)

An interesting, though usually negatively viewed phenomenon, is a "budding" of protoplasts (Fig. 5.6) during the regeneration of the cell wall. Apparently during the formation of the new cell wall, parts of the cytoplasm protrude through areas of the cell wall not yet completely regenerated. In these buds, no material belonging to the cell nucleus has been detected, but occasionally some plastids can be seen. Cells with such buds cannot survive, and after 2 weeks at the latest, they die. If the concentrations of the components of medium in Table 5.1 are halved, then budding can be considerable reduced, but not entirely prevented.

Cell division activity of protoplast material is usually limited, accounting for about 2–3% of cells after isolation. These few cells are the origin of cell clusters consisting of 200–300 cells after 2 weeks of culture. An increase of the population of healthy cells able to divide can be observed after a supplement of 0.05% charcoal, to absorb toxic substances produced during the process of protoplast isolation. This increase can be up to tenfold. These clusters can be used to produce somatic embryos and eventually intact plants, as shown for *Daucus* and others.

The basic principle of the many methods described in the literature is the same as that described above; the procedure adopted has to be worked out for each plant species or tissue used.

5.2   Protoplast Fusion

The major aim of protoplast fusion has been to combine the genomes of two species that cannot be combined by pollination. Due to the fast development of gene technology during the last 10–15 years, through which selected genes can be transferred from a donor genome to the genome of any other species, this aim can be achieved more precisely. Still, it may sometimes be desirable to include protoplast fusion in one or the other research program, and so the topic shall be briefly discussed here.

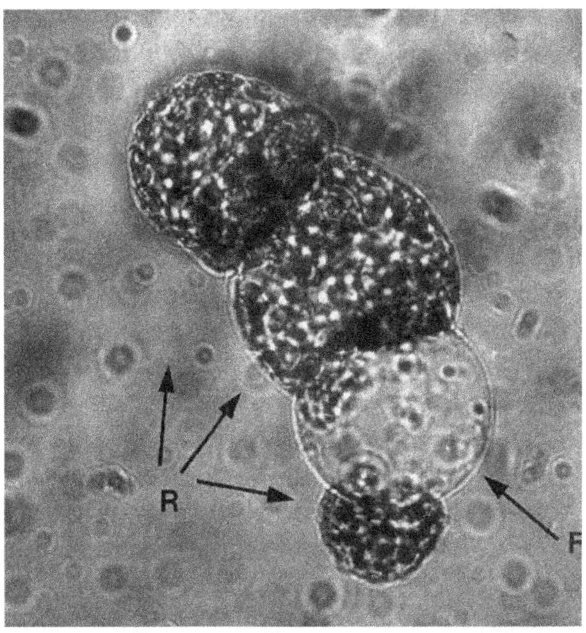

Fig. 5.7 Protoplasts from a leaf (*R* turnip rape) and from roots (*F* fodder kale, *Brassica oleracea* var. viridis) are attached to each other

If protoplasts of different species are mixed, then a high percentage of fusionates are autofusion products of specimens of the same origin. To distinguish these from those of fusion between any two species, a reliable marker is required. The simplest way to this end is the use of different tissues with distinct morphological or other characteristics. A good example is cells with chloroplasts from the leaves of one "parent" and cells free of chloroplasts from another part of the other parent (Fig. 5.7). Isoosmolarity of the two types of cells is a prerequisite. In an early experiment using protoplasts of rapeseed leaves and protoplasts of cells from the carrot root, an "explosion" of the leaf protoplasts was observed due to the high sugar content of the carrot root cells. Still, this was an indication of a successful combination.

Other markers are anthocyanins or other pigments of plants. The use of foreign genetic material as markers introduced into fusion partners will be described in detail in Sect. 13.6.3. Markers are not required if single isolated protoplasts of different origin are fused. Here, using a micromanipulator, these two protoplasts are brought into contact in a mini jar, and the fusion is often accomplished by electroshock.

The plasmalemma, and consequently the protoplast, exhibits an excess of negative charges on its surface. This hinders a spontaneous attachment of two protoplasts. After about 40 years of research on protoplast culture, with many attempts to overcome this problem, only two methods are generally considered as really practical. In the one case, the macromolecular polyethylene glycol is applied at high concentrations (28%), and in the other electroshock is used. Both methods can be employed in various forms, associated with various costs; for example, the common laboratory will suffice for the polyethylene glycol method, but for electrofusion the original self-made equipment is today replaced by expensive, commercially made devices.

The success of fusion experiments is related to the temperature at which the original plant material grew, with higher efficiency at lower (10 °C) than at higher (25 °C) temperatures. Apparently, this is related to the fluidity of membranes, which depends on their composition, particularly for membrane lipids—protoplasts characterized by membranes containing more unsaturated fatty acids exhibit an increased rate of protoplast fusion.

Protoplast fusion of two plant species aims at the production of new genomes with the genetic information of both "parents." Here, one way to create a new genome is to apply X-rays to one "parent" (50 kr). This results in chromosomes being injured and partly eliminated in the following cell divisions (see NICA plants). The still viable chromosomes form a new genome with the untreated cells of the other "parent." Often, haploid material is employed. Using gene technology methodology, it is today possible to insert defined genetic material, i.e., single genes, which will be described later (Sect. 13.6.3).

Besides mixing genetic information of the nuclei of protoplasts of different origin, also plastids and mitochondria can be merged. Some more recent data have become available on the fate of the chondriom. Before first division, the mitochondria apparently elongate and then divide, causing an increase in number. After this, an actin-dependent dispersion of the mitochondria results in a uniform distribution throughout the cytoplasm. This has been observed after fusion during the first 4–8 h of protoplast culture; within 24 h, a near-complete mixing of mitochondria of the fused protoplasts was achieved. The mixed mitochondria population is passed on to daughter cells (Sheahan et al. 2005).

References

Bush DR, Jacobson L (1986) Potassium transport in suspension cells and protoplasts of carrot. Plant Physiol 81:1022–1026

Dudit D, Maroy E, Praznovsky T, Oloh Z, Gyorgyey J, Cella R (1987) Transfer of resistance traits from carrot into tobacco by asymmetric somatic hybridization. Regeneration of fertile plants. Proc Natl Acad Sci U S A 84:8434–8438

Elmshäuser HA, Forche E, Neumann KH (1979) Untersuchungen zur Protoplastengewinnung verschiedener *Brassica*-Arten. Ber Dtsch Bot Ges 91:313–318

Gamborg OL, Miller RA, Ojima K (1968) Nutrient requirements of suspension cultures of soybean root cultures. Exp Cell Res 50:151–158

Gleba YY, Hoffmann F (1978) Hybrid cell lines *Arabidopsis thaliana* and *Brassica campestris*: no evidence for specific chromosome elimination. Mol Gen Genet 165:257–264

Kartha KK, Michayluk MR, Rao KN, Gamborg OL, Constabel F (1974) Callus formation and plant regeneration from mesophyll protoplasts of rape plants (*Brassica napus* L. cv. Zephyr). Plant Sci Lett 3:265–271

Melchers G (1978) Potatoes for combined somatic and sexual breeding methods, plants from protoplasts fusion of protoplasts of potato and tomato. In: Alfermann AW, Reinhard E (eds) Production of natural compounds by cell culture methods, Proc. plant cell cultures. Munchen, pp 306–311

Potrykus I, Paszkowski J, Saul M, Negrutin I, Schilito RD (1987) Direct gene transfer to plants: facts and future. In: Green CE, Somers DA, Hackett WP, Biesboer DD (eds) Plant tissue and cell culture. A.R. Liss, New York, pp 289–362

Schenck HR (1982) *Brassica napus*-successful synthesis by protoplast fusion between *B. oleracea* and *B. campestris*. In: Fujiwara A (ed) Plant tissue culture 1982. Japanese Association for Plant Tissue Culture, Tokyo, pp 639–640

Sheahan MB, McCurdy DW, Rose RJ (2005) Mitochondria as a connected population: ensuring continuity of the mitochondrial genome during plant cell dedifferentiation through massive mitochondrial fusion. Plant J 44:744

Strasburger E, Noll F, Schenck H, Schimpfer AFW (1913) Lehrbuch der Botanik. G. Fischer, Jena

Zhang S (1995) Efficient plant regeneration from Indica (type I) rice protoplasts of one advanced breeding line and three varieties. Plant Cell Rep 15:68–71

Haploid Techniques

6

During a systematic screening of reactions of various parts of flowers of *Datura innoxia* cultured in vitro, Guha and Maheshwari (1964) observed the development of haploid plants from anthers containing immature microspores. Later, especially tobacco anthers were extensively investigated by various research groups, and mostly microspores of this species were used to test the suitability of this technique for hybrid breeding programs. Meanwhile, the production of haploids of several hundreds of plant species has been reported in the literature (see Kumar and Sopory 2010; Saikat et al. 2011). The developmental reprogramming of microspores toward embryogenesis and formation of haploid plant makes isolated microspore culture a rewarding system in the field of plant molecular biology and biotechnology. Transformation of the microspores and derived embryos will yield homozygous lines (double haploid) that could be used further in breeding purposes. (Kumar and Roy 2006, 2011; Kumar and Sopory 2008). Of these, only a few have been used, with limited success, in breeding programs; some reasons for this will be discussed later.

6.1 Application Possibilities

A prerequisite to use the heterosis effect reproducible in hybrid breeding is the availability of homozygous parent lines. The production of such inbred lines requires many backcrossings of heterozygotic parent material. Inbred lines with desired properties are also required for outbreeding plant species. A considerable reduction of the time required to produce such plant material can be achieved by the use of haploids. Haploid higher plants are infertile, and therefore before haploids can be used in breeding programs, a diploidization is required (e.g., using colchicines). With the methods described later, such dihaploid plants can be ideally produced within 1 year. Considering the time necessary for the selection of haploid plants for further use in hybrid breeding, and the propagation of the selected plants (usually by rooting), one needs about 5 years to produce the first hybrid seeds. A time schedule

© Springer Nature Switzerland AG 2020

K.-H. Neumann et al., *Plant Cell and Tissue Culture – A Tool in Biotechnology*,

https://doi.org/10.1007/978-3-030-49098-0_6

to produce a tobacco hybrid, out of the pioneer days of the technique, can be seen in the following summary:

- 1976: anther culture and raising of haploid plants
- 1977: propagation by cuttings and selection of diploid twigs for rooting
- 1978: propagation of dihaploid plants by rooting
- 1979: crosspollination of selected dihaploid parents
- 1980: planting of F1 hybrids

Various methods have been used for the diploidization of haploids, of which only two will be mentioned here. The method of Jensen (1974, 1986), originally described for barley, uses young plants (five-leaf stage) submersed in a solution of 0.1% colchicines (without chloroform) containing 2% DMSO and 0.2–0.5 ml Tween for 5 h in the light at 20–22 °C. For dicots, the method of Ockendon (1986) will be described. Here, a colchicine solution of 0.05% is applied directly to the shoot apex of a young plant, with a microsyringe (10 µl). An alternative is the application with cotton wool soaked with the colchicine solution. For both methods, a success of more than 70% has been reported.

Actually, three experimental approaches are available to produce haploid plants: (1) the anther culture method or that of microspores derived thereof, (2) the embryogenesis of isolated unfertilized egg cells, and (3) production from hybrids of species from which the set of chromosomes of one parent has been eliminated during development. Before a detailed description of the first method will be given, the other two will be briefly summarized.

The now classical example for the utilization of interspecies hybridization is a crosspollination of *Triticum aestivum* and *Hordeum bulbosum* (the bulbosum method). Here, one set of chromosomes (*Triticum*) is organized at metaphase, whereas that of the other parent remains unorganized in the cell and is lost during the following cell divisions. For the bulbosum method, also *Hordeum vulgare* can be used as a suitable parent (Fig. 6.1), and haploids, i.e., dihaploids, obtained by this method were soon used in breeding programs (Kasha and Rheinsberg 1980). Five years after initiation of this breeding program, the first new variety (Mingo) was available. During the last decades, many more examples of interspecies hybrids have been reported, and a summary can be obtained from Gernand et al. (2005). In this paper, experiments are described to also follow the fate of the chromosomes of the "loosing" partner of the hybridization of wheat × pearl millet, by elimination. All pearl millet chromosomes were eliminated between 1 and 3 weeks after pollination. Chromosome elimination involves the formation of nuclear extrusions and the postmitotic formation of micronuclei the chromatin of which is fragmented later.

As will be discussed later, plants produced by such methods exhibit a higher degree of cytogenetic stability, compared to those derived by the anther culture method. Whereas in principle the anther or microspore method is applicable to all plants, the bulbosum method depends on the availability of suitable parent species.

Although basically there exists the possibility of obtaining haploid plant material by culture of immature egg cells, due to the easier handling of anthers, their

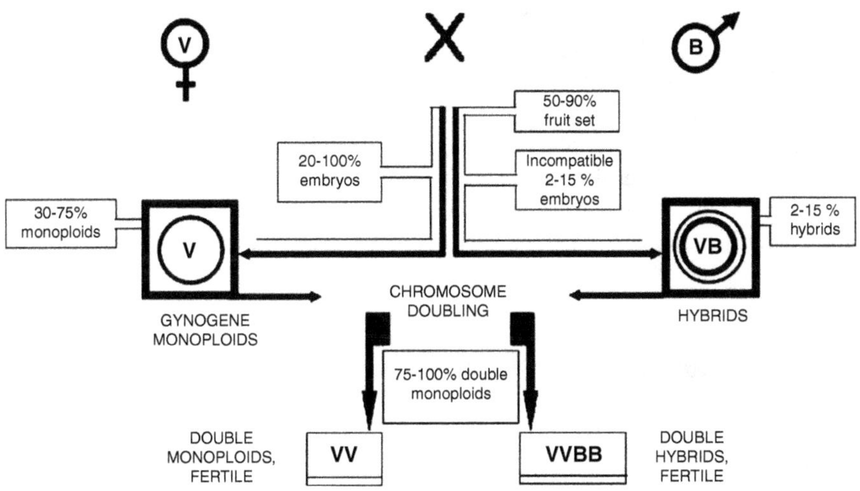

Fig. 6.1 Results of crossbreeding of *Hordeum vulgare* (*V*) and *Hordeum bulbosum* (*B*). Note the high percentage of fruit set (50–90%) with 20–100% of embryo development (after Jensen 1986)

microspores are preferred. A summary on gynogenesis was published some years ago by Keller et al. (1987), in which nine successful example are listed: *Hordeum*, *Triticum*, *Oryza*, *Beta vulgaris*, *Gossypium*, *Ephedra*, *Nicotiana*, *Crepis*, and *Lolium*. During recent years, there has been intensified research to exploit the possibility of gynogenesis—let's wait and see!

A comparison of androgenetic and gynogenetic derived plants will be given below. An example is the use of protoplasts of dissected ovules. Whereas unfertilized protoplasts of barley did not divide, those fertilized developed into microcalli, and if co-cultivated with microspores undergoing embryogenesis, these developed embryonic structures and eventually fertile plants. If cultured alone, these microcalli degenerated (Holm et al. 1994). Another way is to use a floral dip method, as for *Arabidopsis* for genetic transformation with *Agrobacterium tumefaciens* carrying the GUS gene and the 35S promoter, 5 days or more before anthesis. GUS activity was detected only in developing ovules, and not in pollen or pollen tubes. This selectivity could be due to the special developmental path of *Arabidopsis* flowers. Here, the gynoecium develops as an open structure to form closed locules about 3 days before anthesis (Desfeux et al. 2000).

Reports can be found describing superiority for androgenic plants, others for gynogenic, often only in one or the other trait. Androgenesis, however, was generally considered as more efficient than gynogenesis (Foroughi-Wehr and Wenzel 1993); the success of both techniques seems genetically controlled, and broad variations of genotypes can be observed. In a tobacco system, doubled haploids of either origin or their self-progeny were about equal, but the androgenic material exhibited more vigor and was highly variable (Kumashiro and Oinuma 1985). A more recent paper compares androgenic and gynogenic monoploid plants of *Solanum phurea* (Lough et al. 2001). In contrast to gynogenic plants, androgenic plants had an increase in leaf size of 15–20%, and total tuber yield was about doubled to tripled. Plant height, however, was significantly reduced in androgenic lines. Gynogenesis is often employed to surpass the high percentage of albino plants often observed in androgenic systems.

Another possibility to obtain gynogenic plants could be the use of X-ray-irradiated pollen for fertilization. This should make pollen inactive as gene donor but still capable of inducing cellular division of the ovule. Plant material obtained without fertilization and only with the maternal set of genetic material could be produced.

6.2 Physiological and Histological Background

In many publications, a stress requirement is described as a prerequisite to induce androgenic development. The requirement of a stress treatment depends on the plant species, as well as the species genotype. This can be starvation and osmotic stress induced by a mannitol supplement to the culture medium, as for barley, or a combination of starvation and heat shock for tobacco and wheat, or heat shock alone for rapeseed and pepper. Other stress factors can be colchicines, nitrogen starvation (Heberle-Bors 1983), gamma irradiation, or cold shock; a summary is given by Maraschin et al. (2005). In other reports, androgenesis can be induced without an obvious shock treatment. Considering the high concentration of sucrose (2% and more) in most media, already the transfer of isolated tissue invokes some osmotic stress. Furthermore, the confrontation of cells in vitro with often rather high concentrations of phytohormones in embryogenic systems has to be considered as stress factor. Here, positive influences of ABA on embryogenesis match those of osmotic stresses. Also for the induction of somatic embryogenesis, a number of stress factors are under discussion as being necessary, such as wounding, osmotic stress, starvation, and heavy metal ions. Often a separation of explants from their origin in the intact plant and setting a wound at explantation are considered as stress factors and as prerequisites to induce somatic embryogenesis. The relevance of these factors is discussed elsewhere. Actually, compared to the cells in the original mother plant, all in vitro culture systems incur stress conditions for the cells of cultured explants.

To induce the potential for somatic embryogenesis, the dedifferentiation or rather transdifferentiation (or more modern, reprogramming) of microspores, egg cells, or somatic cells is a prerequisite that is brought about by the environment under in vitro conditions—e.g., nutrient media, and temperature.

Following Maraschin et al. (2005), androgenic development, and probably any other somatic embryogenesis, consists of three major phases: acquisition of embryogenic potential, initiation of cell division, and pattern formation of the dividing cells. Initially, a stress (or phytohormones) induces a reprogramming of cellular metabolism, including a repression of gene expression related to starch biosynthesis and the induction of proteolytic genes and stress-related genes. This is followed by the activation of key regulators of embryogenesis—e.g., the so-called BABY BOOM

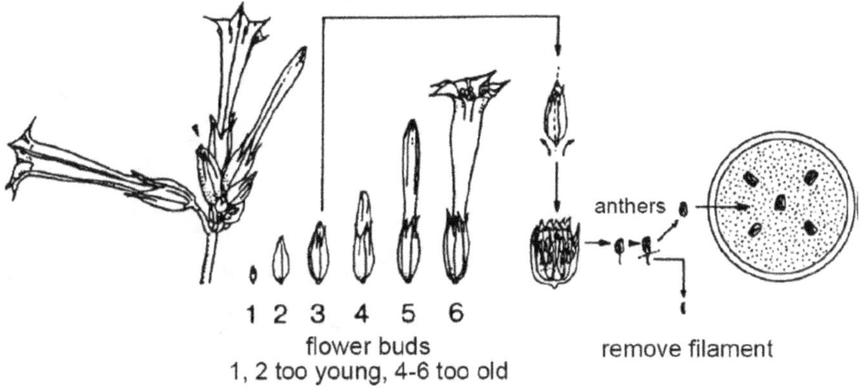

Fig. 6.2 Isolation of anthers (after Reinert and Yeoman 1982)

transcription factor. After induction of cell division during pattern formation of embryos, programmed cell death seems also to play an essential role. This is apparently related to the loss of extracellular ATP (Chivasa et al. 2005), to date of unclear function.

Metabolic rearrangement can be brought about by the ubiquitin 26S proteosomal system, which degrades molecules; autophagy for degrading and recycling organelles in animals via lysosomes and in plants via vacuoles is involved. The breakdown products are eventually metabolized.

Ideally, the induction of androgenesis basically consists of an induction of somatic embryogenesis in microspores after isolation from the mother plant. In many cases, however, first the formation of a callus can be observed, in which later embryos develop. Even more often, root and shoot formation occurs first, and both join later to produce an intact plant.

At the suitable developmental stage (see, e.g., Fig. 6.2), surface-sterilized flower buds are opened under sterile conditions, and the anthers are severed by means of a forceps and placed onto an agar medium (see Fig. 6.3). A suitable medium is given in Table 3.4 (see also NN medium). Two or 3 weeks later, the appearance of the first embryos can be observed (Fig. 6.3). Highly significant for success is the proper developmental stage of the anthers, i.e., their microspores. The optimal stage for microspores of tobacco is the first mitosis that produces the first vegetative and the first generative nuclei. As far as is known, the haploid embryo develops from the vegetative nucleus. Some rough correlation exists between the development of the flower bud and that of the pollen, providing guidelines for the proper developmental stage to obtain the anthers for the experiment. For tobacco, this stage is reached as soon as the corolla can be seen emerging from the calyx (Fig. 6.2). By putting the severed buds overnight into the refrigerator prior to taking the anthers, a 10–15% increase in the generally low number of haploid plants (usually below 0.1%) can be reached.

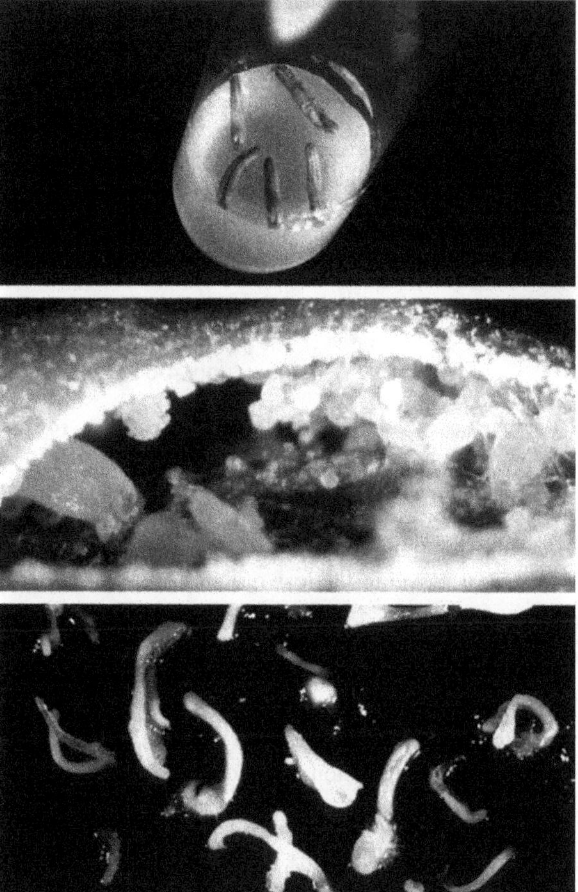

Fig. 6.3 Various stages of anther culture. (Top) anthers of *Datura* on agar, (middle) initiation of embryo development in a ruptured anther, (bottom) embryonic structures isolated from an androgenic anther (photographs by E. Forche)

Heberle-Bors (1983) reports results on some factors that are significant in determining the potential to perform androgenesis and consequently the success per experimental setup. Tobacco was used as model. For a start, a dimorphism of pollen was observed in mature anthers. Unfertile pollen, called P-pollen, is embryogenic and can be separated by density gradient centrifugation from fertile pollen. P-pollen occurs mainly under stress situations promoting male sterility. Examples for this are short day conditions, low temperature, or nitrogen deficiency. Under these conditions, the efficiency of the nutritive tissue of the anther, i.e., the tapetum, is restricted or disturbed, and the pollen cannot develop to maturity. The amount of P-pollen can also be increased by an application of growth substances that promote male sterility (auxins, antigibberellins), and therefore its increase is not solely due to disturbance of the nutritional system of the anther. Furthermore, also plants with a genetically based male sterility usually exhibit a high potential of androgenesis.

Besides influences of variations in the growth conditions of the parent plant on the success to produce haploid plants from its microspores, an experimental system to increase the production of haploid plants from normal and healthy mother plants is described by Moreno et al. (1989). Here, microspores at an early developmental stage (early two-nuclei stage) are cultured in a sugar-free nutrient medium (to preserve isoosmotic conditions, sucrose is replaced by mannose) for 1 week, resulting in a dominance of P-pollen. The abundance of haploid plants seems to be limited only by the survival rate of the pollen following this treatment.

6.3 Methods for Practical Application

Basically, two methods are available, which are practiced with many variations. In the first method, which seems to be the more original, whole anthers are placed on an agar medium, and in the second the anthers are cultured in a liquid medium in which the pollen is liberated by agitation. Using anthers of tobacco, an example for each method will be given.

Agar Culture of Anthers
- Severing the flower buds with a corolla length of 15–25 mm.
- For surface sterilization, the buds are transferred into a hypochlorite solution (0.1% active chlorine) supplemented with some drops of Tween for 10–15 min.
- Washing the buds at the sterile working bench with sterilized water, and transferring onto a sterile Petri dish.
- Removal of the calyx and corolla with a flamed forceps.
- Severing the anthers and transferring into a sterile Petri dish. Gentle removal of the filaments.
- Transfer of the anthers onto an agar nutrient medium (Table 6.1), and see below; 5 ml medium per 50×18 mm Petri dish). The two pollen sacs should touch the surface of the agar, with the furrow oriented toward the air. The dishes are sealed with Parafilm.
- The cultures are transferred to a dark growth cabinet at 28 °C.
- At the appearance of the first embryonic stages after about 2–4 weeks, the cultures are illuminated (16/8 h, 20–25 °C).

$(NH_4)_2SO_4$	463	$MnSO_4 \times 4H_2O$	4.4
KNO_3	2830	$ZnSO_4 \times 7H_2O$	1.5
KH_2PO_4	400	H_3BO_3	1.6
$MgSO_4 \times H_2O$	185	KI	0.8
$CaCl_2 \times 2H_2O$			
Glycine	2.0	Sucrose	20,000–30,000
Thiamine-HCl	0.5		
Nicotinic acid	0.5	Agar	800

Table 6.1 Agar medium for anther cultures (mg/l)

Agar medium	(mg/l)		(mg/l)
$(NH_4)_2SO_4$	463	$MnSO_4 \times 4H_2O$	4.4
KNO_3	2830	$ZnSO_4 \times 7H_2O$	1.5
KH_2PO_4	400	H_3BO_3	1.6
$MgSO_4 \times H_2O$	185	KI	0.8
$CaCl_2 \times 2H_2O$	166	Glycine	2.0
Sucrose	2000	Agar	8000.0
Thiamine-HCl	0.5		
Nicotinic acid	0.5		

Additionally, a freshly prepared EDTA–Fe solution: 27.9 mg $FeSO_4 \times H_2O$ and 37.2 mg EDTA-Na. Of this culture medium, 5 ml is transferred into Petri dishes (50×18 mm) on which, after cooling, the anthers are placed.

This medium was suggested by Sunderland (1984) and originally contained neither phytohormones nor activated charcoal, as used in many systems later. A supplement of 0.5% charcoal or of 1% naphthylacetic acid (NAA), however, often clearly increases the number of haploid plants obtained.

Plant Microspore Techniques
Transfer of transformed immature barley embryos on selection medium
Isolation of wheat/barley microspores
Viability test with FDA
Inoculation of *A. tumefaciens* for agro-infiltration (see also Chap. 13)

6.4 Anther and Ovary Culture

6.4.1 Anther or Microspore Culture

Culturing anthers or isolated microspores from anthers in suitable conditions through somatic embryogenesis haploid plants can be obtained. Treating such plant material with, e.g., colchicines, it is possible to produce diploids, and if everything works out within maximum 1 year (depending on plant species), a fertile homozygous diploid plant can be produced out of a heterozygous mother plant. This method is advantageous for hybrid breeding by reducing substantially the time required to establish inbred lines.

Often, however, first a callus is produced out of microspores with separate formation of roots and shoots which join subsequently and in due time haploid plants can be isolated. Here, the production of chimeras may be sometimes a problem. Another aim by using anther or microspore cultures is to provoke the expression of recessive genes in haploids to be selected for plant breeding or gene transfer.

6.4.1.1 Materials

- Autoclaved blender
- Wheat spikes (6–8)
- Ethanol [75%]
- Mannitol solution [0.2 M]
- 100 and 38 µm stainless steel mesh
- Induction and regeneration medium
- Filter for filter sterilization
- Maltose [20%]
- Parafilm

6.4.1.2 Procedure

Plants Growing Conditions and Selection of Microspore-Containing Spikes of Bobwhite Wheat Variety

About four plants per pot (20–25 cm diameter) are grown in a greenhouse controlled at a temperature of $27 \pm 2\,°C$ (day) and $17 \pm 2°C$ (night) and a light regime (250 µmol/ s/m^2 photon flux density) of 17 h day and 7 h night. Fresh tillers containing microspores enclosed within the anthers in the middle section of a spike at the mid to late uninucleate stage (when wheat heads were about 3–4 cm out of the sheath) of development were cut below the second node and placed in a clean container with distilled water. The stage of development of microspores is established by microscopically examination of crushed anthers. In the uninucleate stage, the microspore should be round and nucleus clearly visible. The nucleus should be located at the periphery of the microspore opposite of the pore (Fig. 6.4). Tillers can be stored in water at 4 °C for up to a week without significant loss in microspore viability.

The following procedures take place under aseptic conditions in a laminar hood (All equipment should be autoclaved).

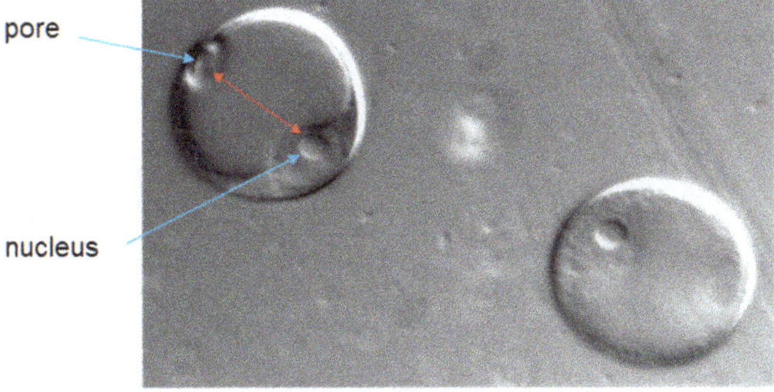

Fig. 6.4 In the uninucleate stage, the microspore is round and nucleus clearly visible and located at the periphery of the microspore opposite of the pore (red arrow)

Treatment of Donor Spikes
- The spikes covered partly by the uppermost leaf sheath of the selected donor plants are disinfected by spraying with 75% ethanol.
- After 15 min or until the alcohol has dried out, the spikes are removed from the boot and awns trimmed not to close from the florets.
- Spikes are sterilized with 20% solution of commercial bleach for 15 min and washed three times with autoclaved distilled water.

Isolation of Microspores
- Florets from six to eight spikes are aseptically cut from the internodes into an autoclaved Waring blender cup.
- 50 ml of 0.2 M mannitol solution is added and the blender cup covered with foil.
- The florets are blended at a low speed for 5 s.
- The contents of the blender filtered through a 100 micron stainless steel mesh into a 150 ml beaker and the residue returned to the blender cup.
- 50 ml of 0.2 M mannitol solution is added for another 5 s run.
- The contents are filtered as before and the cup rinsed three times with 5 ml 0.2 M mannitol.
- The combined filtrate is then poured onto the autoclaved 38 micron stainless steel mesh filter to trap the large viable microspores.
- The microspores retained on the mesh are rinsed gently with 30 ml of 0.2 M mannitol into a 150 ml beaker.
- The resulting solution containing viable microspores is transferred into 15 ml centrifuge tubes and the microspores sedimented at 4000 rpm for 5 min (Fig. 6.5).

Pre-treatment for Conditioning of Microspores
- Resuspend the pellet in 10 ml solution A (Table 6.2).
- Sediment the microspores at 4000 rpm for 3 min.
- Pour off the supernatant, and resuspend the cells resuspended in 5 ml solution A and transfer into a sterile 60 × 15 mm Petri dish.

A   B   C   D

Fig. 6.5 (**a**) Spikes in inducer solution. The leaves and awns were removed; (**b, c**) Caryopses into blender (Warring MCII), (**d**) filtrate into 38 μm mesh filter set (Images provided by Prof. Dr. Diter von Wettstein[†], Pullman, USA)

Table 6.2 Compositions of Solution A

Chemicals	mg/L
KCl	1492
$MgSO_4.7H_2O$	246
$CaCl_2.H_2O$	148
KH_2PO4	136
Na_2EDTA	37.3
$FeSO_4.7H_2O$	56
H_3BO_3	3
KI	0.5
$MnSO_4.4H_2O$	8.0
$ZnSO_4.7H_2O$	0.3
Maltose	90,000
2-HNA	50
Kinetin	0.2

pH 6.5
Filter sterilized

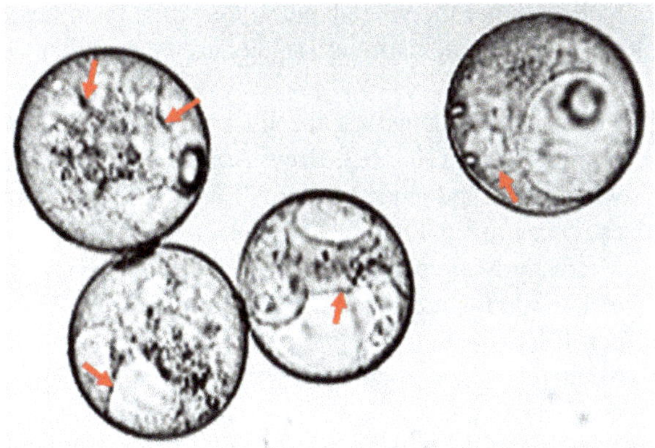

Fig. 6.6 Fibril formed microspores in pretreated medium (red arrow)

- Cover the dish and seal with Parafilm and incubate at 22–25 °C in the dark for 40–70 h or until the fibrillar-appearing cytoplasm of induced microspores is reached. The embryogenic microspores typically have eight or more small vacuoles immediately adjacent to the cell wall. These vacuoles surround the condensed cytoplasm in the center, forming a fibrillar structure (Fig. 6.6).

Induction of Division of Microspores
- Transfer quantitatively the microspores from the Petri dish into a beaker with solution A. Remove solution A by centrifugation at 900 rpm for 3 min and resuspension in 2 ml 0.2 M mannitol.
- The microspores in the 2 ml 0.2 M mannitol solution are layered over 11 ml of 20% maltose solution in a 15 ml centrifuge tube.

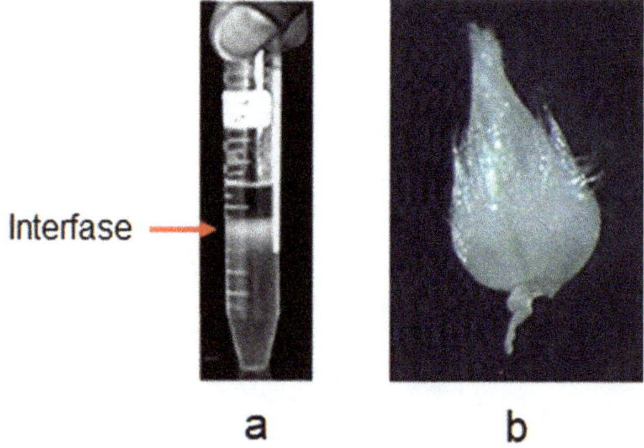

Fig. 6.7 (**a**) Formation of a white band of large microspores at the interphase of the mannitol and maltose solutions (red arrow). (**b**) a wheat ovary

- Centrifugation is at 600 rpm for 3 min. This results in the formation of a white band of large microspores at the interphase of the mannitol and maltose solutions (Fig. 6.7a).

Collect either the white band or the white band together with the upper mannitol phase, and filter them through a 38 micron mesh filter to retain the large microspores.

- Microspores trapped on the filter are washed three times with 2 ml of induction medium (Table 6.3, Brew-Appiah et al. 2013) and transferred into a beaker with 2 ml of induction mrdium (Table 6.3). Concentration of large microspores is estimated using a hemocytometer.
- Microspores are plated in a 60×15 or 35×10 mm petri dish at a minimal density of 1×104/ml and a maximum density of $1 \times 10\,7$ and with a total volume of 3 ml per dish.
- Microspores are co-cultured with live ovaries (Fig. 6.7b) of wheat, three ovaries/ml. Ovaries just fertilized and growing work best. Dishes are covered and sealed with Parafilm and incubated at 27.5 °C in the dark without shaking for embryoid development (Fig. 6.8). For plant regeneration, after 30 days culture in induction medium, the embryogenic calli (~2 mm in diameter) transferred to regenerations medium C-17 (Table 6.4, Pauk et al. 1991). The cultures were kept at 26 °C in dim light [~100 μmol m^{-2} s^{-1} μmol/s photon flux density) with a 16/8 h (day/night) photoperiod. Plantlets are ready for potting 4 weeks after embryogenic calli are transferred in for regeneration (Fig. 6.9).

Advantages
- Producing homozygous doubled haploid (acceleration the breeding progress, 2–4 years reduced)
- Use of homozygous plant for gene mapping
- Production of transgenic plants without chimeras
- Transformation with T-DNA complex to overcome genotypes depending transformation

Table 6.3 Composition of induction medium (mg/L)

KNO_3	1400	Amino acids cocktail (mg/L)	
NH_4NO_3	300	Glutamine	1950
KH_2PO_4	170	Asparagine	120
$MgSO_4 \cdot 7H_2O$	370	Arginine	60
$CaCl_2 \cdot 2H_2O$	440	g-aminobutyric acid	160
$FeSO_4 \cdot 7H_2O$	37.3	Serine	110
$Na^2EDTA \cdot 2H_2O$	22.3	Alanine	60
$MnSO_4 \cdot 4H_2O$	8.6	Casein amino hydrolysate	40
$ZnSO_4 \cdot 7H_2O$	6.2	Cysteine	20
H_3BO_3	37.3	Leucine	20
KI	0.83	Isoleucine	20
$NaMoO_4 \cdot 2H_2O$	0.25	Proline	20
$CuSO_4 \cdot 5H_2O$	0.025	Lysine	20
$CoCL_2 \cdot 6H_2O$	0.025	Phenylalanine	10
Myo-Inositol	300	Tryptophane	10
Thiamine-HCL	0.4	Methionine	10
Pyridoxine-HCL	0.5	Valine	10
Nicotinic acid	0.5	Glycine	5
Ascorbic acid	0.75	Histidine	5
PAA	2.0	Threonine	5
Kinetin	0.5	Tyrosine	10
L-Glutamine	975		
Amino acids, Cocktail	355 mg		
Maltose	90 g		
Phytagel	3 g		
pH	6		

Application of Haploids and Double Haploids

The developmental reprogramming of microspores toward embryogenesis and formation of haploid plant makes isolated microspore culture a rewarding system in the field of plant molecular biology and biotechnology. Transformation of the microspores and derived embryos will yield homozygous lines/doubled haploids (DH) that could be used for further breeding. Since the first haploid *Nicotiana tabacum* plant was obtained in 1967 (Bourgin and Nitsch 1967), the large number of haploid plants has been induced from various species of both herbaceous and ligneous plants, such as *Oryza sativa*, *Setaria italica*, *Zea mays*, and *Aesculus hippocastanum* (Guha et al. 1970; Ban et al. 1971; Murakami et al. 1972; Radojevic 1978).

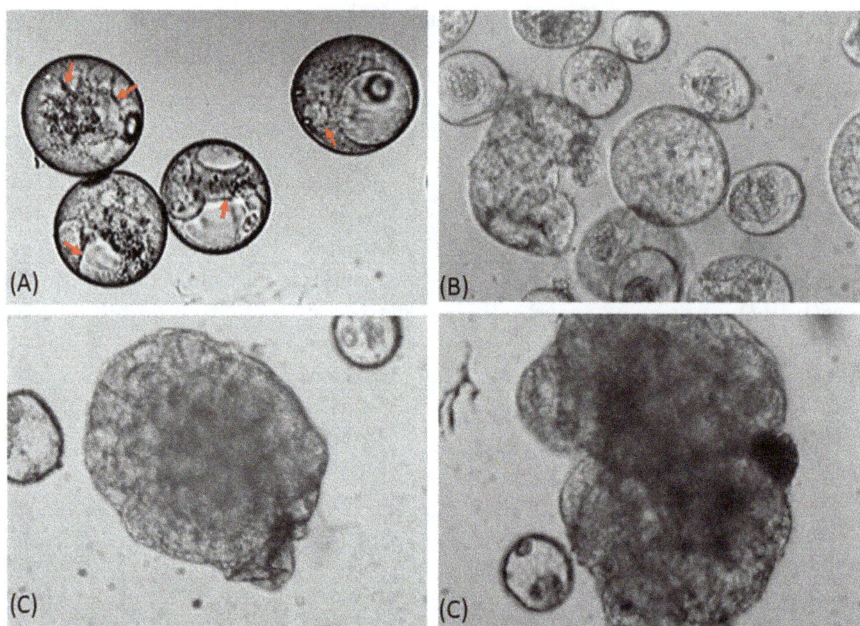

Fig. 6.8 Development process of microspores in induction medium. (**a**) Fibril formed 5 days after induction (red arrow). Cell multiplication and growth at 7 days (**b**); 14 days (**c**) and 21 days (**d**)

Cucurbits

The Cucurbitaceae plant family includes a number of economical and essential garden crops such as cucumber (*Cucumis sativus* var. *sativus* L.), melon (*Cucumis melo* L.), pumpkin (*Cucurbita moschata* Duchesne ex Poir.), squash (*Cucurbita pepo* L.) and summer squash (*Cucurbita pepo* L.), watermelon (*Citrullus lanatus* (Thunb.) Matsum. & Nakai), winter squash (*Cucurbita maxima* Duch. ex Lam.), and other species. All are cultivated throughout the world, across cultures, and have made an indelible mark on modern civilization.

As reported by Dong et al. (2016) haploids and doubled haploids are critical components of plant breeding. They discussed the following four aspects of this issue in detail: (1) the most popular methods of obtaining haploid and DHs in cucurbits and the influencing factors of each method, where the most important results are summarized in Tables 6.1, 6.5 and 6.6. (2) Comparison between methods of chromosome doubling and ploidy identification in cucurbits and among media for efficient DH production. (3) DH application in plant breeding and in basic genetic research, including molecular studies, and (4) important conclusions regarding methodologies and protocols for androgenesis, gynogenesis, and parthenogenesis haploids, basic steps and aspects of each method, and cucumber haploid protocols, specifically, as shown in Fig. 6.10.

Table 6.4 Compositions of C-17 medium (mg/L)

KNO$_3$	1400
NH$_4$NO$_3$	300
MgSO$_4$·7H$_2$O	150
KH$_2$PO$_4$	400
CaCU·2H$_2$O	150
MnSO$_4$·4H$_2$O	11.2
ZnSO$_4$·7H$_2$O	8.6
CUSO$_4$·5H$_2$O	0.025
H$_3$BO$_3$	6.2
Kl	0.83
COCl$_2$·6H$_2$O	0.025
Na$_2$EDTA	37.25
Fe$_2$SO$_4$·7H$_2$O	27.85
Nicotmic acid	0.5
Thiamine HCl	1.0
Pyridoxine HCl	0.5
Clycince	2.0
D-Biotin	1.5
Folic acid	0.5
Myo-inositol	100.0
2,4-D	2.0
Kinetin	0.5
Sucrose (9%)	90,000
Agar	5500
pH	5.8

(A)   (B)

Fig. 6.9 Development of plantlets in regeneration medium. (**a**) embryogenic calli; (**b**) plantlets generated from embryogenic calli (Images provided by Prof. Dr. Diter von Wettstein[†], Pullman, USA)

Table 6.5 Some morphological characteristics of haploid, diploid, and dihaploid plants of *Nicotiana tabacum* var. Xanthi (Zeppernick 1988)

Plant[a]	Height (cm)	Number of leaves	Number of side buds	Number of buds/ flower	Stem	Leaves	Side buds	Fresh weight (g/plant) Buds/ flower	Roots	Yield/ plant	LAI[b]
n	76	28	13	146	84	86	5	31	119	325	0.47
$2n$	112	35	5	52	173	185	2	8	99	467	0.63
$2 \times n$	77	28	17	66	99	134	3	9	138	383	0.62

[a]Harvest: n, 13 July; $2n$, 12 July; $2 \times n$, 12 July
[b]Leaf area index (LAI) = leaf width (in cm)/leaf length (in cm)

Table 6.6 Gibberellin activity of haploid and dihaploid strains and an F1 hybrid (8×4) determined by a dwarf rice method (Tanginbozu) after separation of the ethyl acetate extract of leaves of a defined size by thin layer chromatography (expressed as gibberellin equivalents, ng/g fresh weight; average of four replicates per strain; harvest 1988)

Strain	2/21, n	2/22, $2 \times n$	4/31, n	4/32, $2 \times n$	8/11, n	8/12, $2 \times n$	8×4, $2n$	8×4, $2n$
Area for thin l. chr.	Generative plants							Vegetative plants
Polar	3.1	0.9	0.7	0.2	0.2	0.3	4.1	0.5
Unpolar	0.4	0.4	0.8	0.5	0.3	0.2	0.1	0.1
Total	3.5	1.3	1.5	0.7	0.5	0.5	4.2	0.6

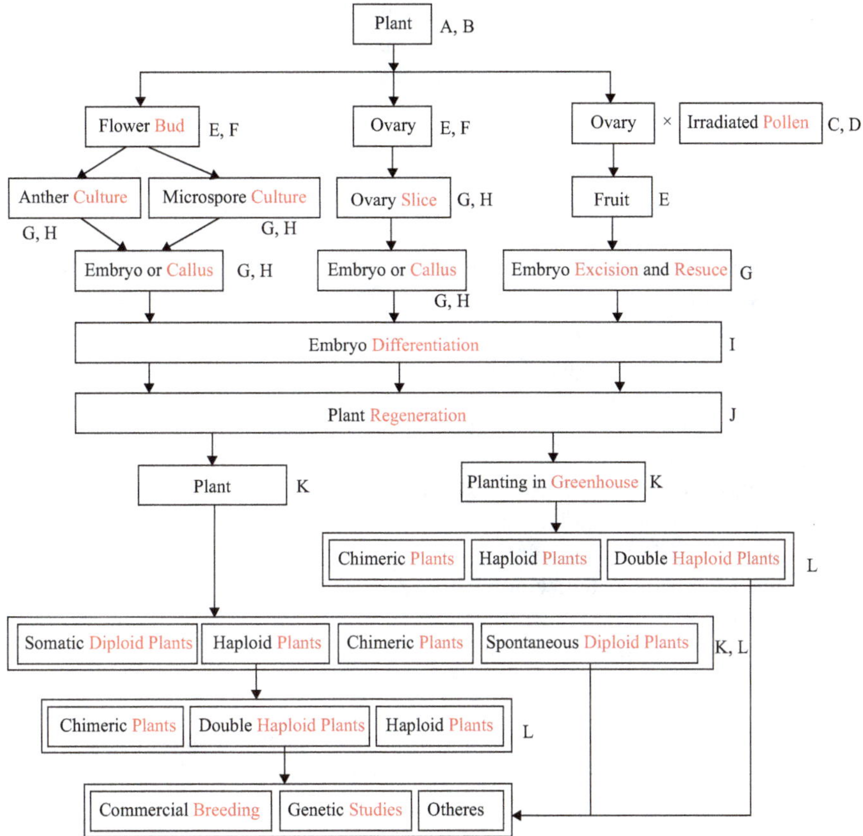

Fig. 6.10 The basic steps and main factors in androgenesis, gynogenesis, and parthenogenesis haploids in cucurbits (Source Dong et al. 2016)

Liquid Culture of Anthers

This method includes a pre-treatment of the anthers before culture to stimulate the development of the microspores into plantlets. This pre-treatment consists of a "cold stress":

- Transfer of the freshly isolated flower buds into a sealable container (Petri dish, polyethylene bag).
- Placement of the container for 3 weeks in the dark into the refrigerator (7–9 °C). Following Sunderland, this "cold" treatment increases the number of haploid plants 10- to 15-fold.
- To obtain the anthers, the method described for agar cultures is followed.
- Transfer of up to 50 anthers into a Petri dish (50 × 18 mm) containing 5 ml of the nutrient medium given above without agar. The anthers float on the surface of the liquid medium.

After a few days, a sufficient number of pollen falls out of the anthers, which can then be used again to start a new culture by transfer to another Petri dish.

After about 2 weeks, the first embryonic structures can be observed; until transplantable young plants are available, the nutrient medium has to be renewed several times.

Still not completely understood is the significance of activated charcoal. This supplement is somehow associated with the inactivation (via absorption) of some agar components that can inhibit androgenesis, and the release of other, water-soluble substances that can promote it (Forche et al. 1981). As mentioned before, another key factor is temperature. A number of plant species are reported to require a storage of the anthers at low temperature (2–4 °C) for at least 24–48 h prior to cultivation. Temperature is important also during culture, and its requirement seems to depend on genetic factors. For example, whereas the tobacco variety "Wisconsin" requires 22 °C to initiate androgenesis, this can be achieved for "Xanthi" only at 28 °C. The requirements discussed here can be regarded only as tendencies, and the exact conditions to induce androgenesis have to be determined for each species or variety. Many examples can be found in the literature on the Internet.

The success of microspore cultures depends also on the growing conditions of the donor plant. For example, temperature again seems to be of significance here. Higher yields of embryos were obtained if microspores originated from plants of rapeseed (*Brassica napus*) grown under a light/dark cycle of 16/8 h at 15/12 °C, compared to 23/18 °C (Lo and Pauls 1992). The authors relate this to a reduction in cytoplasmic granularity and/or exine density.

Finally, a description of haploid plants will be given. The tissue surrounding the haploid microspores in the anther (tapetum connective) is diploid. Thus, reliable methods are needed to distinguish between haploid plants derived from the microspores and diploids derived from the adjacent diploid tissue. Most reliable, of course, is to count the chromosomes in the dividing cells, e.g., in the cells of the root tip. However, usually only one root tip is available per plant, which one does not wish to "sacrifice." As an alternative, the method described now has been used successfully in our investigations for many years, and it requires only a small piece of a leaf or callus. If the plant originates from microspores, then cells with 1C nuclei can be observed. An absence of such cells indicates that the plants are derived from diploid material of the anther. The 1C-value of the species can be determined using its microspores. Problems related to ploidy stability will be dealt with later (Chap. 13).

Microfluorometric determination of the ploidy level (Blaschke 1977; Blaschke et al. 1978). For the standardization of the method, the DNA was measured in the early tetrad stage of the nuclei in the developing microspore. The intensity of fluorescence of these structures is set equal to the DNA content of the haploid nucleus (1n). This old method is particularly suitable for using tissue with no or only very few cell divisions available for chromosome counts.

The plant material to be used for the investigation is first fixed in ethanol/glacial acetic acid (3:1) for 12 h, followed by an ascending alcohol series for dehydration. Then, the cells are embedded in paraffin, and with the help of a microtome, sections of 12–15 µm are cut. The sections are fixed on a slide with a gelatine/glycerine solution (1 g pure gelatine dissolved in 100 ml aqua dest., supplemented with 15 ml glycerine and few crystals of thymol) and dried for 12 h at 40 °C.

After removal of the paraffin on a heated plate, a descending alcohol series is applied, similar to other common histological methods. This is followed by staining with a 0.005% bisbenzimide solution (Dye Höchst 33258) for 10 min (pH 4.4–4.6) and then rinsing for 5 min in running tap water and differentiation for 15 min in 70% ethanol. After embedding the sections in "glycerine for fluorescence microscopy" (Merck), a cover glass is emplaced, and the next day the measurements can be made.

Measurements are made at a wavelength of >490 nm using a Xenon high-pressure lamp for illumination at 500-fold magnification. To determine fluorescence intensity, the lens of the microscope is adapted to the size and shape of the nuclei. For the measurements, one has to be careful to select nuclei that do not touch others nearby. The intensity of fluorescence is stable for at least 6 months. The reading can be compared to values obtained from tetrad-state microspores. After enzymatic maceration of cell suspensions, an automation of C-value determination can be performed by flow cytophotometry. Here, protoplasts are more suitable than intact cells.

Other Methods

Recently, a method was published for DNA determination using genetically transformed *Arabidopsis* material (Zhang et al. 2005), by coupling GFP with the histone 2A. This construct (HTA6) complexed within chromatin is therefore linearly related to the DNA content of the nucleus. This material could also be used for flow cytometry.

Sari et al. (1999) compared four different methods to determine the ploidy level, i.e., chromosome counting, flow cytometry, size and chloroplast number of the guard cells, and some morphological observations, using two cultivars of watermelon. All four were successful and equally reliable. The most easy to perform was the method based on stomata measurements. The length of stomata of haploids was 17–18 µm, the diameter 10–12 µm, and the number of chloroplasts in the guard cells was 6–7. For diploids, the corresponding values were 23–24 µm, 18 µm, and 11–12. With suitable equipment, a rough determination of the ploidy level could also be carried out on intact plants, in the greenhouse or in the field.

Here, questions concerning relations of the ploidy level and the plastome should be followed in detail.

6.5 Haploid Plants

The most obvious morphological characteristics of haploid plants are a reduced height, smaller leaves with a reduced diameter of the leaf lamina, and an excessive number of small fruits (Table 6.5, Fig. 6.11). In Fig. 6.12, it can be seen that also haploid plants produce fruits. Seed formation, however, is absent, and since fruit size often depends on the development of seeds, the fruits remain small.

Fig. 6.11 Leaves, flowers, and fruits of diploid (left) and haploid (right) *Datura innoxia* plants (photographs by E. Forche)

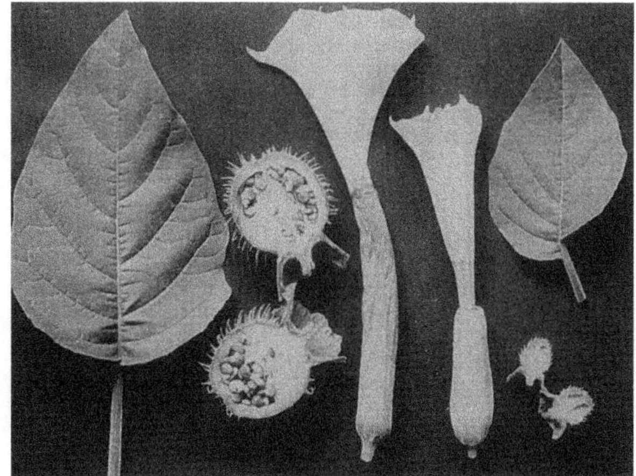

Fig. 6.12 Shape of leaves of three strains of dihaploid (*2n*) tobacco plants derived by androgenesis

Furthermore, deformations of leaves and flowers often occur in haploid plants. In *Datura* plants, flowers with three petals, rather than five, have been observed, and in tobacco flowers, a fused gynoecium and seven to eight anthers could be seen. It seems that two flowers were fused into one; such fusions have been observed also for the leaves. Such deformations have not been observed in normal diploid plants growing under the same conditions. Possibly, the population of diploid plants observed was too small to give a definite result on this problem.

Besides a reduction of time required to establish inbred lines, another benefit of haploid plants is the possibility of bringing recessive genes to realization hidden in the parent generation. To consider the "gene dosage effect" in judging the genomes of plants derived by androgenesis, dihaploid plants are required. Such plants can be produced by applying, e.g., colchicines to the buds (see above). In some plant

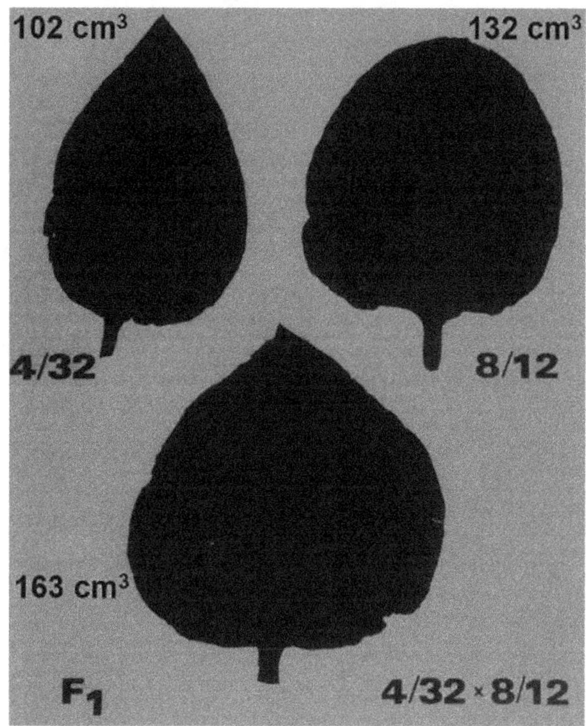

Fig. 6.13 Shapes of leaves of two tobacco strains derived by androgenesis and an F1 hybrid

species (e.g., tobacco), dihaploids are spontaneously produced from leaf buds. If these dihaploids are propagated by cuttings, then the properties of these genomes can be preserved. With such material, China was the first to breed new varieties of tobacco, corn, wheat, and other plants, now since many years in use by farmers.

A clear characteristic of haploid plants is a reduced leaf diameter, at more or less normal leaf length (Table 6.3; cf. Figs. 6.12 and 6.13). Furthermore, haploids enter the flowering stage about 1 week earlier than dihaploids, consistent with the higher concentration of c former (Table 6.6). Because spraying dihaploids with a gibberellin solution (GA_3) induces the same leaf shape, it is reasonable to predict a strong influence of the ploidy level on the native gibberellin system of plants. Nevertheless, it is difficult to see in which way the gibberellin system and the ploidy level are related. Although the synthesis and the general metabolism of this group of phytohormones is quite well understood, its action as a signal to promote morphogenic responses such as leaf shape still requires investigations (for a summary, see Thomas and Sun 2004).

It is difficult to decide to which extent the variations in leaf shape of the three dihaploid strains in Fig. 6.13 are due to the expression of recessive genes. It should be pointed out that these strains were derived from anthers of the same diploid mother plant. These dihaploid plants are fertile and can be used for crossings. The

Table 6.7 Nicotine concentrations of two dihaploid strains of *Nicotiana tabacum* and a hybrid thereof (F1; pot experiments, average of 3 consecutive years; Zeppernick 1988)

Strain	mg nicotine/g dry weight leaves	mg nicotine/plant ("Hauptgut")
4/32	2.2	8.4
8/12	4.1	22.9
F1	6.3	37.2

figure shows clear differences that were preserved through several generations of propagation by seeds. As can be seen from the images of comparable leaves of the two parents and a hybrid, a clear heterosis can be observed for leaf size (Fig. 6.13). The plant height of the hybrid, however, is between those of the parents. Heterosis can be seen again for the concentration of nicotine in the leaves (Table 6.7).

References

Ban Y, Kokubu T, Miyaji Y (1971) Production of haploid plants by anther-culture of *Setaria italica*. Bull Fac Agr Kagoshima Univ 21:77–81

Blaschke JR (1977) Histologische, cytochemische und biochemische Untersuchungen zur Charakterisierung des Kinetineinflusses auf die Entwicklung haploider und diploider Kalluskulturen von *Datura innoxia* Mill. PhD Thesis, Justus Liebig University, Giessen

Blaschke JR, Forche E, Neumann KH (1978) Investigations on the cell cycle of haploid and diploid tissue cultures of *Datura innoxia* Mill. and its synchronisation. Planta 144:7–12

Bourgin JP, Nitsch JP (1967) Obtention de Nicotiana Haploïdes à Partir D'étamines Cultivés In Vitro. Ann Physiol Vég 9:377–382

Brew-Appiah RA, Ankrah N, Liu W, Konzak CF, von Wettstein D, Rustgi S (2013) Generation of doubled haploid transgenic wheat lines by microspore transformation. PLoS One 8(11):e80155. https://doi.org/10.1371/journal.pone.0080155

Chivasa S, Ndimba BK, Simon WJ, Lidsey K, Slabas AR (2005) Extracellular ATP functions as an endogenous external metabolite regulating plant cell viability. Plant Cell 17:3019–3034

Desfeux C, Clough SJ, Bent AF (2000) Female reproductive tissues are the primary target of *Agrobacterium*-mediated transformation by the *Arabidopsis* floral-dip method. Plant Physiol 123:895–904

Dong Y-Q, Zhao W-X, Li X-H, Liu X-C, Gao N-N, Huang J-H, Wang WY, Xu XL, Tang Z-H (2016) Androgenesis, gynogenesis, and parthenogenesis haploids in cucurbit species. Plant Cell Rep 518. https://doi.org/10.1007/s00299-016-2018-7

Forche E, Kibler R, Neumann KH (1981) The influence of developmental stages of haploid and diploid callus cultures of *Datura innoxia* on shoot initiation. Z Pflanzenphysiol 101:257–262

Foroughi-Wehr B, Wenzel G (1993) Andro- and parthenogenesis. In: Hayward MD, Bosemark NO, Romagosa I (eds) Plant breeding: principles and prospects. Chapman and Hall, London, pp 261–277

Gernand D, Rutten T, Varshney A, Rubsova M, Prodanovic S, Brüß C, Kumlehn J, Matzk F, Houben A (2005) Uniparental chromosomal elimination at mitosis and interphase in wheat and pearl millet crosses involves micronucleus formation, progressive heterochromatinization and DNA fragmentation. Plant Cell 17:2431–2438

Guha S, Maheshwari SC (1964) In vitro production of embryos from anthers of *Datura*. Nature 204:497

Guha S, Iyer RD, Guapta N (1970) Totipotency of gametic cells and the production of haploids in rice. Curr Sci 39:174–176

Heberle-Bors E (1983) Induction of embryogenic pollen grain in situ and subsequent in vitro pollen embryogenesis in *Nicotiana tabacum* by treatments of the pollen donor plants with feminizing agents. Physiol Plant 59:67–72

Holm PB, Knudsen S, Mouritzen P, Negri D, Olsen FL, Roue C (1994) Regeneration of fertile barley plants from mechanically isolated protoplasts of the fertilized egg cell. Plant Cell 9:531–543

Jensen CJ (1974) Chromosome doubling techniques in haploids. In: Kasha KJ (ed) Haploids in higher plants: advances and potential. University of Guelph Press, Guelph, Ontario, pp 153–193

Jensen CJ (1986) Haploid induction and production in crop plants. In: Horn W et al (eds) Genetics manipulation in plant breeding. W. de Gruiter, Berlin, pp 231–257

Kasha KS, Rheinsberg E (1980) Achievements with haploids in barley research and breeding. In: Davies DR, Hopwood DA (eds) The plant genome. John Innes Charity, Norwich, pp 215–230

Keller WA, Arnison PG, Cary BJ (1987) Haploids form gametophytic cells—recent developments and future prospects. In: Green CE, Somers DA, Hackett WP, Biesboer DD (eds) Plant tissue and cell culture. A.R. Liss, New York, pp 223–241

Kumar A, Roy S (2006) Plant biotechnology and its applications in tissue culture. I.K. International, Delhi, 307 pp

Kumar A, Roy S (2011) Plant tissue culture and applied plant biotechnology. Avishkar Publishers, Jaipur, 346 pp

Kumar A, Sopory S (2008) Recent advances in plant biotechnology and its applications. I.K. International, New Delhi, 718 pp

Kumar A, Sopory S (eds) (2010) Applications of plant biotechnology: *in vitro* propagation, plant transformation and secondary metabolite production. I.K. International, New Delhi, 606 pp

Kumashiro T, Oinuma T (1985) Comparison of genetic variability among anther-derived and ovule-derived doubled-haploid lines of tobacco. Japan J Breed 35:301–310

Lo KH, Pauls KP (1992) Plant growth environment effects on rapeseed microspores development and culture. Plant Physiol 99:468–472

Lough RC, Varrieur JM, Veilleux RE (2001) Selection inherent in monoploid derivation mechanisms for potato. Theor Appl Genet 103:178–184

Maraschin SF, de Priester W, Spaink HP, Wand M (2005) Androgenic switch: an example of plant embryogenesis from the male gametophyte perspective. J Exp Bot 56:1711–1726

Moreno RMB, Macke F, Hauser MT, Alwen A, Heberle-Bors E (1989) Sporophytes and male gametophytes from in vitro cultured, immature tobacco pollen. In: Cresti M, Gori P, Pacini E (eds) Sexual reproduction in higher plants. Springer, Berlin, pp 137–142

Murakami M, Takahashi N, Harada K (1972) Induction of haploid plants by anther culture in maize: 1. on the callus formation and root differentiation. Kyoto Prefect Univ Fac Agric Sci Rep 24:1–8

Ockendon DJ (1986) Utilization of anther culture in breeding Brussels sprouts. In: Horn W (ed) Genetic manipulation in plant breeding. Verlag W.de Gruyter, Berlin, pp 265–272

Pauk J, Manninen O, Mattila I, Salo Y, Pullii S (1991) Androgenesis in hexaploid spring wheat F2 populations and their parents using a multiple-step regeneration system. Plant Breed 107:18–27

Radojevic L (1978) In vitro induction of androgenic plantlets in *Aesculus hippocastanum*. Protoplasma 96:369–374

Reinert J, Yeoman MM (1982) Plant cell and tissue culture, a laboratory manual. Springer, Berlin

Saikat BK, Datta M, Sharma M, Kumar A (2011) Haploid production technology in wheat and some selected higher plants. Aust J Crop Sci 5:1087–1093

Sari N, Abak K, Pitrat M (1999) Comparison of ploidy level screening methods in watermelon: *Citrullus lanatus* (Thunb.) Matsum, Nakai. Sci Hortic 82:265–277

Thomas SG, Sun T-P (2004) Update on gibberellin signalling: a tale of the tall and the short. Plant Physiol 135:668–676

Zeppernick B (1988) Untersuchungen zur Bedeutung des Ploidieniveaus (n, 2xn) für den Entwicklungsverlauf, das Gibberellinsystem und die Nikotinkonzentration von Blättern und Zellkulturen von *Nicotiana tabacum*, var. Xanthi. Dissertation, Justus Liebig Universität, Giessen

Zhang C, Gong FC, Lambert GM, Galbraith DW (2005) Cell type-specific characterization of nuclear DNA contents within complex tissues and organs. Plant Methods 1:7

Plant Propagation: Meristem Cultures, Somatic Embryogenesis Micropropagation, and Transformation of Somatic Embryos in Bioreactors

7.1 General Remarks and Meristem Cultures

At present, the major commercial application of cell and tissue cultures is plant propagation. Although in some instances also field crops in breeding programs have been propagated in vitro (e.g., cereals, potatoes, sugar beet), the overwhelming practical use is made in propagations of ornamentals. In subtropical and tropical agriculture, however, this field is gaining significance for cash crops. A major factor for its general acceptance is certainly often the vegetative propagation practiced already in conventional farming methods. Here, propagation in vitro is simply an extension of the methods used for centuries. Also, the higher price of individual plants is certainly an important factor to apply in vitro methods for propagation.

In Germany, the economic value of ornamentals (though with a big gap) ranks second to that of cereals, well ahead of sugar beets. As an example, some details will be given for Germany for the years 1991–2000 (compiled by Prof. W. Preil, Ahrensburg). Statistics are available from 1985 onward, and it is interesting to note that the strongest increase in the number of plants occurred from 1985 to 1990. During this period, the increase was more than threefold, from 4943 to 16,407 (all values in thousands). After this "starting period," the increase slowed considerably to about 30%, indicating some kind of saturation (Table 7.1). This is due mainly to the reduction in the number of ornamentals produced, which, however, has been compensated by the rise in orchids, mainly *Phalaenopsis* (from 519 in 1991, to 9150 in 2000).

These plants were produced by 25–29 laboratories (cf. this varies from year to year), of which two were producing >3 million plants annually. Of the orchids, *Phalaenopsis* dominates by far, followed by *Miltonia* and *Cymbidium*. *Anthurium andraeanum* and *Anthurium scherzerianum* lead in the production of ornamentals.

The head runner for berries is *Fragaria*, and for shrubs it is *Rhododendron*, followed by *Rosa* and *Syringa*. Interestingly, potatoes are leading in the group of agricultural crops and vegetables, followed by *Asparagus* (albeit by much less in terms of volume). A complete listing of all plants propagated in Germany can be

© Springer Nature Switzerland AG 2020

K.-H. Neumann et al., *Plant Cell and Tissue Culture – A Tool in Biotechnology*,

https://doi.org/10.1007/978-3-030-49098-0_7

Table 7.1 In vitro propagation in Germany from 1991 to 2000 (in thousands)

	1991	1992	1993	1994	1995	1996	1997	1998	1999	2000
Orchids	2.575	3.951	3.300	3.032	2.797	3.160	3.820	5.740	9.433	12.106
Ornamentals	9.501	7.427	5.756	6.071	5.900	6.327	5.925	5.247	4.260	2.922
Berries	4.780	5.505	4.660	5.226	4.561	5.462	5.090	5.079	3.580	4.038
Woody plants	1.373	1.834	2.132	2.859	3.157	3.609	3.760	3.731	3.894	3.894
Shrubs, aquat. plants	725	991	1.998	1.610	1.729	2.020	1.861	2.836	2.902	1.276
Agricult. plants, vegetables	313	286	490	299	371	306	317	324	448	364
Total	18.837	19.533	18.038	18.370	18.217	20.432	20.622	22.986	24.354	26.670

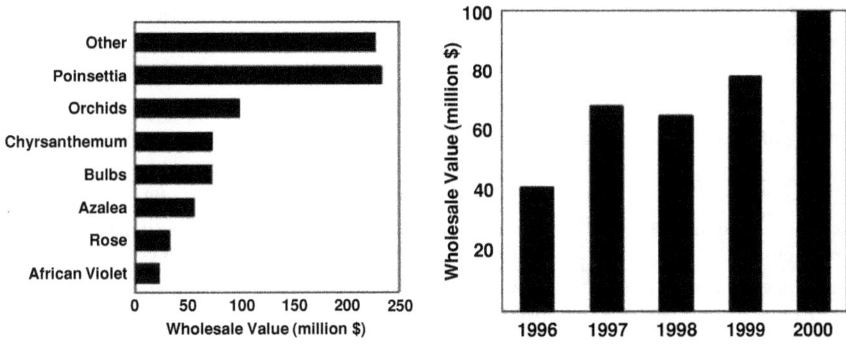

Fig. 7.1   Some commercial aspects of plant propagation in vitro in the USA

obtained from ADIVK on the Internet. The number of plants produced in vitro per species, as well as in total, varies from year to year. After a long stagnation period lasting up to 1995, at about 18,000 plants, a continuous increase has been observed up to 2000 (at least). This is likely due mostly to the accentuated increase in orchids, dominated by *Phalaenopsis*. Here, as well as for ornamentals, a general "in fashion" trend plays a role. Other species, like *Gerbera*, have dramatically reduced in number, from 500 in 1992 to 50 in 2000 or *Saintpaulia* from 1500 to zero in the same period. Especially interesting is the volume of in vitro plants produced for *Cymbidium*. After a peak in production in 1989, there came a decrease, in some years even to zero. In the period 1994–1999, however, the level was of 2000–5000 specimens, and in 2000 the production increased to 26,000 specimens per year.

A similar, though more pronounced, tendency can be observed in the USA (Fig. 7.1). The total sale for orchids was US$100,000,000, with about 75% of the orchid market for *Phalaenopsis* (Griesebach 2002). In a massive cooperative venture, production is today concentrated in the USA for cultivar development, in Japan for initiation of tissue culture, in China for mass propagation, and in The Netherlands for growth to maturity of tissue cultured plants. Orchid growing is nowadays an international business.

Presently, in vitro propagation is performed applying three major techniques, i.e., meristem culture, embryogenesis in vitro, and shoot induction in callus cultures, followed by rooting. The procedure for propagation of ornamentals but also for banana and others is meristem culture, usually initiated by culture of an isolated shoot apex (of various size) by an induction of growth of leave buds. Besides the shoot apex, for some plants like *Cymbidium*, also axillary buds are suitable. Primary explants are obtained aseptically using a sharp scalpel, and after differentiating of shoot organs, these are either rooted on the primary explant on a special rooting medium, and eventually separated, or they are rooted in isolation after severing and continue growing separately. Plants like roses or fruit trees grow then on their own root system without grafting. For banana, 25–30 plants can be obtained from one shoot apex. Using these again for meristem culture, a great number of plants can rapidly be obtained from a single original, cultured apical meristem. This is of

Table 7.2 In vitro procedure, and conventional propagation from seeds of celery plants (adapted from Rowe 1986)

Conventional propagation by seeds	Plant production by somatic embryogenesis in vitro
Sowing of seeds in greenhouse	Obtainment of petiole explants from selected hybrids and induction of callus formation (30–60 days)
80% germination	Formation of cell suspension in liquid medium
Selection and thinning in greenhouse (60–70 days)	Propagation of cell suspension (+auxin) until a packed cell volume (PCV) of ca. 5000 ml (100–150 days)
Hardening in open air (7 days)	Division into portions of about 700 ml and induction of somatic embryogenesis in an auxin-free medium (30 days)
Mechanic transplantation in field	Fractionation of embryos according to size and to obtain uniform populations (about 1.5×10^6 embryos
Harvest after ca. 120 days	Transfer to solid medium in greenhouse (85–90 days), hardening, and mechanical planting in field
Total about 200 days	Total about 400 days

importance for banana plantations susceptible to Panama disease. This disease is due to an infection from the soil, which prevents the conventional propagation by ratoons. Culturing meristems, however, requires a lot of manual labor and is consequently quite costly.

Although the production of somatic embryos has been described in the literature for several hundred plant species, its commercial application to propagation in vitro is still rather rare. One basic reason could be that it is often difficult to definitely predict the success of the setup. The publication of a protocol to produce somatic embryos is not automatically associated with a method that functions reliably day after day. Still, this method of propagation offers the possibility to use a fermenter culture system to install a semiautomatic system. Later, some examples will be discussed in detail. Here, a fully automatic program to control the growth and development of cultures by changing several parameters like the pH of the medium, the CO_2 concentration, and others can be installed. After embryo production, in an appropriate developmental stage, the culture can be transferred to a solid medium to promote the development to young plants. After thinning, these can be transplanted into soil or another solid substrate to grow in open air.

In terms of financial costs, the highest expenditure for in vitro propagation is manual labor. An automation of somatic embryogenesis in bioreactor or fermenter cultures, after its optimization, should reduce this factor considerably (see, however, somaclonal variation, genome stability, and others). In a later chapter, this will be taken up again. Some years ago in a feasibility study of in vitro production, a price of about 20 cents was calculated per transplantable young seedling for meristem cultures, and of about 5 cents for plants produced by somatic embryogenesis in a bioreactor. More recent information can be found in de Fossard (2007). In Table 7.2, there is a procedure to produce celery hybrids commercially for transplanting 20 ha within 1 week. Ever more protocols are becoming available to propagate tropical cash crops like coffee, cacao, and date palms.

During recent years, cell material of immature or mature embryos is being increasingly used for in vitro propagation, in particular of so-called recalcitrant species like the many cereals or woody plants like conifers. As will be discussed elsewhere, in this tissue apparently the competence to somatic embryogenesis is better preserved than in other parts of the plants.

The aims of in vitro propagation vary broadly, and they are usually different for plants with conventional vegetative propagation and those with propagation by seeds. In the former, the aim is a fast propagation of a great number of genetically identical plants from a selected mother plant or to obtain virus-free specimens or to eliminate other pathological organisms commonly transmitted by vegetative propagation. Another aim can be the long preservation of valuable mother plants or the production of plants growing on their own root systems. For seed-propagated plants, the aim often is a reduction of the time required to produce offspring for breeding and other purposes or the cloning of highly valuable hybrids. As described elsewhere, genetic instability can often be observed in callus cultures, and to bypass this meristem cultures (if possible) is preferable.

One of the most important applications of in vitro propagation of vegetatively propagated ornamentals lies in obtaining healthy plants from infected "mother plants." In conventional gardening, viruses and other pests would be transmitted when using rooted cuttings. In these mother plants, however, usually the uppermost shoot apex (<0.5 mm), generally without conductive elements, is free of infections, and its use for in vitro propagation can yield pathogen-free offspring. This is enforced by a heat shock treatment before explantation (up to 40 °C). Still, often a stepwise treatment of several subcultures is necessary. For this work, a very sensitive control method is required to screen the plant material, often available only in tiny amounts. To control virus infection, immunological methods are commonly applied (e.g., an ELISA test). In Fig. 7.2, a schematic summary of the production of healthy offspring from an infected "mother plant" is given.

Despite the many advantages of in vitro propagation discussed above, also some weaknesses of such methods have to be addressed. As already described for rapeseed cultures, clear differences of reactions can be observed between varieties of a given species. Another example of this kind can be seen for the reaction of some *Pelargonium* varieties to various concentrations of IAA and BAP in the nutrient medium (Fig. 7.3). Consequently, it is recommendable to determine optimal conditions to obtain a method suitable for each origin. This is even more obvious for various species. Besides genetic instabilities of callus cultures used for propagation, already discussed several times, another disadvantage is the small number of plants obtained by employing this method. If the mother plant is a chimera, then a segregation of offspring usually has to be expected.

Malformations of plants produced via tissue culture should also be mentioned. These are caused largely by the conditions of tissue culture and can be also observed for plants derived from embryogenesis in vitro. Often, more than one shoot tip appears, as shown in Fig. 7.4 for a carrot plant produced by somatic embryogenesis. Also for mature taproots of this species, twisting and other malformations can be observed, as they are sometimes in seed-derived plants growing in unsuitable soil.

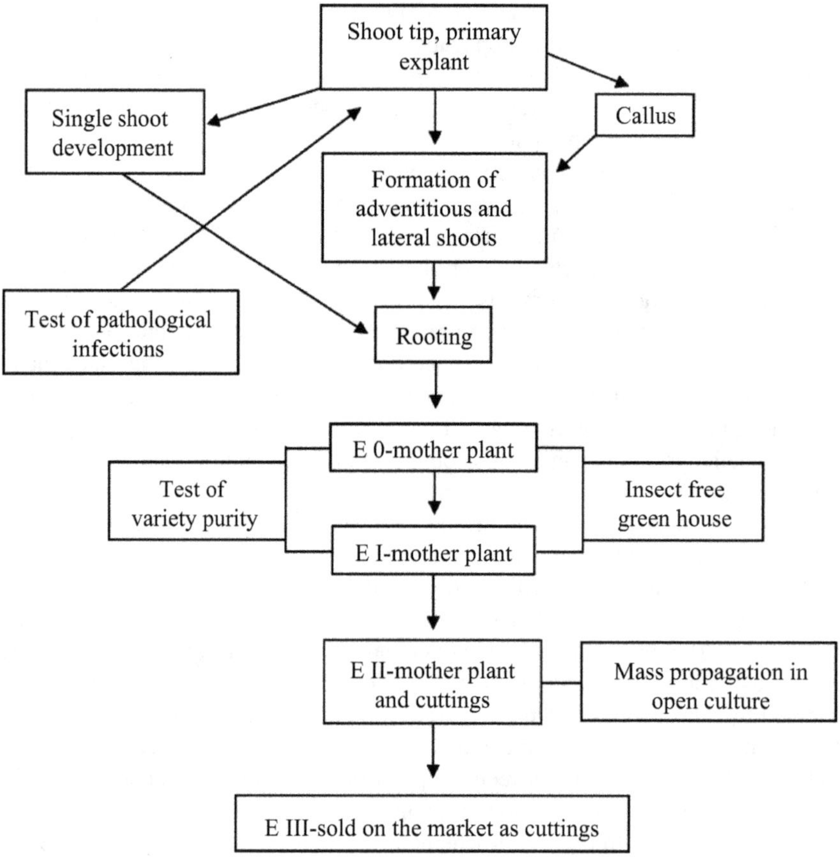

Fig. 7.2 Scheme for the production of *Pelargonium* mother plants and cuttings by means of tissue culture methods and use in gardening (Reuther 1990)

Plants derived from seeds of such plants, however, are perfectly normal, indicating no heritability of such malformations. Thus, such plant material is still usable in plant breeding programs, but is not suitable for sale on the market.

One of the major problems is the adaptation of young plantlets, produced in vitro, to greenhouse conditions and eventually to transplantation into soil. The problems are due to incomplete formation of the cuticle, missing or nonfunctional stomata, and a disturbed conductive system. These shortcomings are related mainly to the water system of the plants but also to the endogenous distributions of assimilates and other substances.

All this would result from the culture in vitro with about 100% air humidity in the vessels used for propagation. Thus, the gradient of air moisture from the ambient air to the developing leaves, needed to induce the normal development of leaves, is absent. The use of water-absorbing pearls or cooling of the shelves used for the culture vessels could help.

% shoot formation

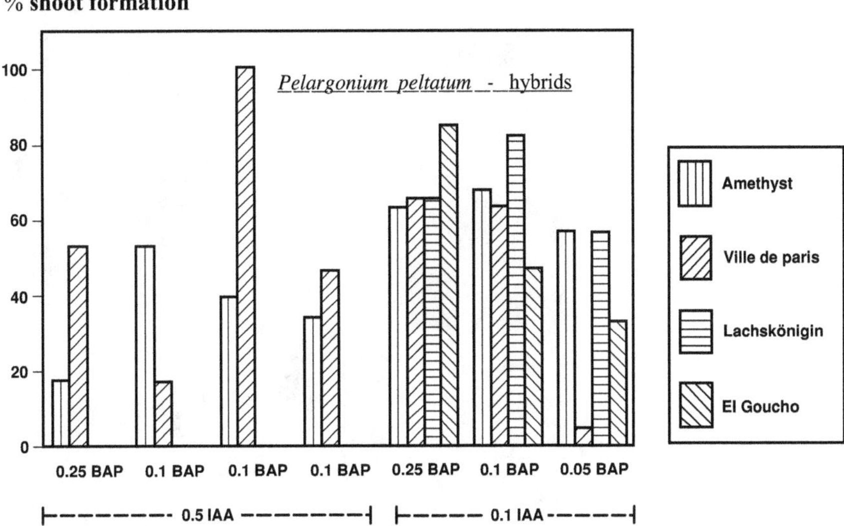

Fig. 7.3 Influences of various concentrations of IAA and BAP (ppm) on shoot formation of some varieties of *Pelargonium peltatum*. The *columns* represent the number of shoots as percentage of the number of shoot tips used

An improvement of transplantation success can often be achieved by removing all older leaves before transplanting into soil. Leaves produced anew are normally developed, with an intact cuticle and a fully developed stomata system.

Especially in commercial propagation, a disadvantage could be the dependence of a successful in vitro propagation on a special developmental stage of the "mother plant" or even of a plant organ used for explantation. Here, the experience of skilled personnel certainly plays a decisive role. Often, dormancy of the origin of explants has to be considered, or an application of growth regulators, especially cytokinins, to the "mother plant" can be of help. As an example, a BAP solution of 100 ppm or more can be sprayed onto the "mother plant," or the plant can be injected with a solution containing 500 ppm BAP.

Before presenting some practical examples, a summary of the major advantages and disadvantages of in vitro propagation will be attempted (see also Hartmann 2002). In vitro propagation will generally be preferred if:

- Propagation is desired year-round, independent of weather conditions.
- Diseases during propagation are to be avoided.
- Plants with difficult genetics are to be maintained and propagated (cf. aneuploidy, polyploidy, sexually sterile strains).
- The plants are to be free of pathogens, especially of virus infections.
- Fast propagation of valuable single plants is desired.

Fig. 7.4 A carrot plant derived by somatic embryogenesis with several shoot tips

In many cases, the best procedure to achieve this should be automated or at least semiautomated somatic embryogenesis. This will be described later in detail. Certainly, some considerable capital investment is required for installation of the equipment required for the technique (see later). Another negative factor for commercial companies in fulfilling contracts for the delivery of plants at fixed dates and costs is the high risk of microbial infection.

Usually for each species, and often for each variety of a given species, optimal conditions for the technique have to be determined. Still, the general principles described here should apply to most cases, although the exact conditions would have to be worked out. As a help, some protocols will be described in more detail. Those for propagation of *Cymbidium* and carrot have been used extensively in our own laboratory; those on raspberry (*Rubus idaeus* L.) have been elaborated in thorough discussions with Geier (e.g., 1986) and Mrs. S. Merkle (Fa. Hummel, Stuttgart, pers. comm.).

7.2   Protocols of Some Propagation Systems

7.2.1   In Vitro Propagation of *Cymbidium*

Orchid seeds are very tiny, contain neither cotyledons nor an endosperm, and usually consist only of some 100 cells. In a natural environment, a mycorrhiza is generally required for germination. Here, within about 1 month a small green, egg-shaped, seedling-like structure develops. This stage is designated as a protocorm. A shoot and a root primordial develop on this structure, and within several months, the first root and the first leaf can be observed. The development of *Cymbidium* starts with an immature embryo in the seed, followed by the intermediate stage as protocorm, and ending with the fully developed plant. As will be shown below, the protocorms can be divided, and from these parts individual plants can be raised from protocorm formation on the cut surface. Basically, the same applies also for the propagation of other orchids, e.g., *Phalaenopsis* Fig. Here, in particular hybrids play an important role. For a summary of *Phalaenopsis* production and commercialization, the reader is referred to Griesebach (2002).

As origin for in vitro propagation, apical shoot meristems (2 mm from the shoot tip), or axillary buds of shoots 3–10 cm in length are used, from which the explants are obtained after four vertical and one horizontal cut with a sharp razorblade. Sterilization is carried out by dipping the explant for 15 min in a hypochlorite solution, followed by a short submergence into 96% ethanol and thorough washing in sterilized water. After this, the scale leaves are removed to expose the apex (see Fig. 7.5). Then, the explant is transferred to the nutrient solution (NL in Table 3.4, supplemented with kinetin, m-inositol, and IAA—NL3), and after 4–7 days, sprouting of buds can be observed. After about 4 weeks, on the cut surfaces created by explantation, some outgrowth can be observed that develops eventually into "protocorms" (primary protocorms).

These can yield plantlets after transfer into Knudson C medium (Table 7.3), or they can be used as basis for the production of additional "protocorms" (secondary protocorms). Such a protocorm contains several vegetative tips, and consequently several plants can develop that can be further increased in number by producing cuttings of the original "protocorm" from which again secondary protocorms arise and so forth. For the formation of such secondary "protocorms," the cuttings are again transferred into NL3 medium. Instead of NL3, Knudson C medium is also suitable, but the time required to produce primary protocorms is reduced by about 2 weeks, and the number of secondary "protocorms" is increased. In Fig. 7.6, some histological sections are given to illustrate protocorm development.

If these protocorms are transferred to the Knudson C nutrient medium, development proceeds, and after about 3 weeks, young plantlets are available. Protocorm propagation and the development of the protocorms into young plants are possible without a supply of growth regulators. As indicated by the data in Table 7.4, however, an influence of such substances is clearly recognizable.

For a production of four secondary protocorms, each of which produces two more protocorms, the reproduction rate per strain and per month amounts to factor

Fig. 7.5 In vitro propagation of *Cymbidium*. (Top) schematic illustration of the preparation of *Cymbidium* explants; for improved visualization, protruding side buds used for isolation are exaggerated (adapted from Reinert and Yeoman 1982). (Middle left) shoot apex of *Cymbidium* after 4 days of culture. (Middle right) *Cymbidium* protocorm cut for new protocorm production. (Bottom left) initial protocorm formation on the cut surface resulting from explantation. (Bottom right) *Cymbidium* protocorm with formation of secondary protocorms (S. Sakr, unpublished images of our institute)

Table 7.3 Composition of the nutrient solution (g/l) for in vitro culture of *Cymbidium* explants (from Knudson 1946)

Component	Concentration
Ca(NO$_3$)$_2$ × 4H$_2$O	1.000
MgSO$_4$ × 7H$_2$O	0.250
KH$_2$PO$_4$	0.250
Fe-EDTA	0.025
(NH$_4$)$_2$SO$_4$	0.500
MnSO$_4$	0.008
Sucrose	30.000

8. Following a calculation by Morel (1974), within a period of 9 months, about one billion plants can be derived from one shoot tip explant. As already mentioned, the introduction of in vitro methods in commercial production has resulted in a dramatic decrease in the prices of individual plants and of cut flowers of orchids. *Sphicolae liocattleya*, Falcon "Alexandri," shall serve as example. According to Griesebach (1986), the price for conventionally propagated plants amounted to US\$500; after the introduction of in vitro propagation, this dropped to about US\$5.

Recently, the production of viable artificial seeds of *Cymbidium* has been reported. Details will be given in the chapter on artificial seeds (cf. Sect. 7.5).

7.2.2 Meristem Cultures of Raspberries

Most methods for in vitro propagation are based on protocols employing meristem cultures. These are also commonly used in commercial applications in horticulture. Under suitable conditions, an apical meristem can be induced to initially produce shoot organs, followed by the formation of a root system. In meristem cultures, an induction of shoot regeneration can be rather easily promoted. For many plant species, including cultivated plants, after rhizogenesis such structures can be raised into intact plants ready to be transplanted into soil. The differentiation of these very young leaf primordials is quite similar to the differentiation status of the cells of apical meristems, with limited commitment to further developmental lines. The leaf primordials are determined to become shoot organs, but not necessarily only leaves. Older (base positioned) buds produce leaves under comparable cultural conditions. Apparently, some control point exists for those cells of the apical meristem to determine further development out of several possibilities. These shoot regenerates can be used again for apical meristem cultures, and soon a great number of offspring from a "mother plant" can be obtained. These plants should be genetically identical. At least plants derived by meristem culture methods show a much lower percentage of somaclonal variation, compared to those derived from callus cultures (see Sects. 13.1 and 13.6.3). However, attempts of molecular characterization of the genome of plants derived by meristem cultures are rare. The uppermost shoot tip without conductive elements is usually free of virus infections, and for the regeneration of species highly susceptible to virus infections, meristem culture is often the only method to produce virus-free plant material. Raspberries (*Rubus idaeus* L.) are such

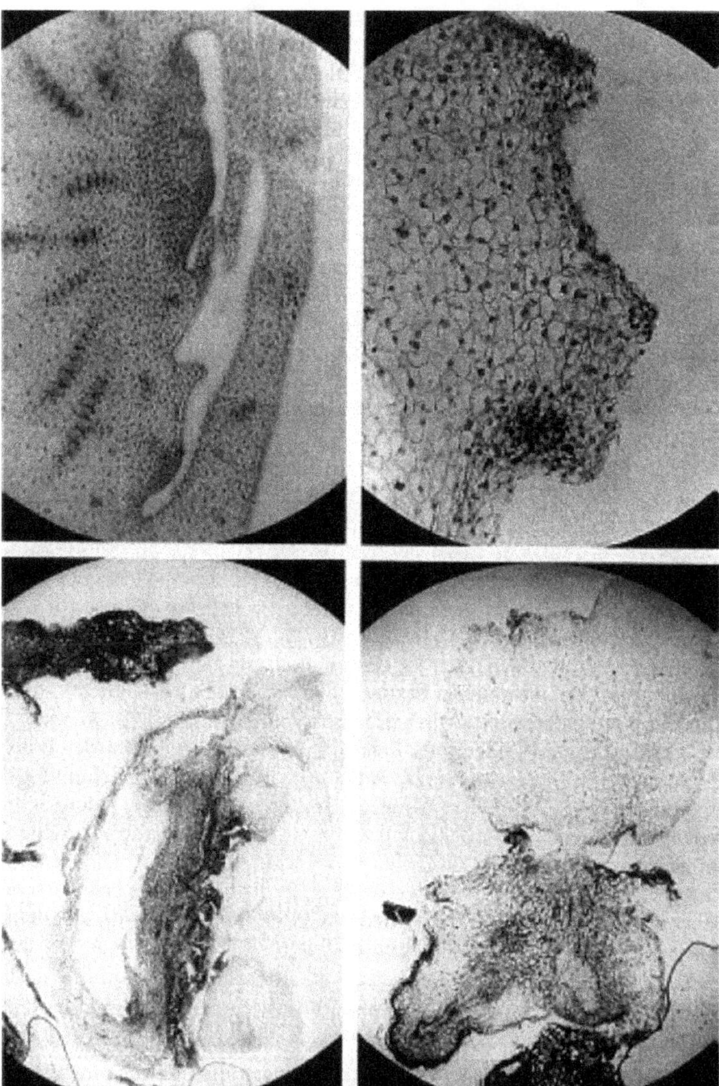

Fig. 7.6 Some histological sections demonstrating some steps in *Cymbidium* culture (S. Sakr, unpublished images of our institute). (Top left) apical shoot meristem at the initiation of culture. (Top right) initial cell division on the cut surface produced at explantation. (Bottom left) primary protocorms on the cut surface of an axillary bud explant. (Bottom right) formation of secondary protocorms

a species, which will be used as an example for propagation by meristem cultures. Here, axillary buds have been used as origin for meristem cultures for many years on a commercial scale by Fa. Hummel and Stuttgart, and the description of the protocol

Table 7.4 Influence of various IAA and kinetin concentrations on protocorm propagation of *Cymbidium* cultures (NL3, see Table 3.4, with 50 ppm m-inositol; 6 weeks of culture, protocorms per culture vessel)

	4 ppm IAA	8 ppm IAA	16 ppm IAA
0.2 ppm kinetin	15	18	18
0.4 ppm kinetin	22	24	28
0.8 ppm kinetin	33	40	45

below was compiled together with Mrs. S. Merkle of this company (see also Merkle 1994).

The conventional method consists of root cuttings 5–20 cm in length obtained in winter to grow in a sand/peat mixture in a greenhouse. The plants developed by this method are commonly infected by viruses, resulting in loss of yield. An alternative is the use of virus-free "mother plants" found at locations where virus infection via transmission by insects from older, free-growing raspberry populations can be prevented. This is possible only if a distance of at least 200 m from these native stands is available.

To employ meristem propagation, first a careful selection of "mother plants" is required, which preferably should be virus-free. If this is not the case, then also root cuttings can be obtained from "mother plants" selected according to yield and other criteria. To produce virus-free plants, some temperature treatment in a room with a temperature up to 36–39 °C is applied to the young plants regenerated from the root cuttings for 4–6 weeks. At this elevated temperature, the propagation of the virus is more inhibited than is the rate of apical cell division in these young plantlets.

After this treatment, the plantlets are surface sterilized with hypochlorite and ethanol, as described before, and the uppermost shoot tip (0.2–0.4 mm) consisting of the shoot apex and one leaf primordium is severed. Most chances to obtain virus-free explants are associated with explants free of conducting elements. Here, however, chances of survival and regeneration are strongly reduced, and the aim should therefore be to find some compromise.

The explants are transferred to an agar medium (0.9%) in test tubes with half the concentration of macronutrients of the MS medium, full concentration of the micro-nutrient and vitamin mixtures of MS, 2% sucrose, 1 g casein hydrolysate/l, and 0.5 ppm 6-BA (Chap. 3). The cultures are in a room with 16 h illumination per day at about 4000 lux. The temperature in the light is set at 24 °C and in the dark at 22 °C. After about 6 weeks, shoot development can be observed, with one shoot per explant. All manipulations and culturing are under aseptic conditions.

At a shoot length of about 1 cm, the explant is transferred to a rooting medium of the same composition as given above, in which 6-BA is replaced by 0.5 ppm IAA. About 2–3 weeks later, root development can be observed; another 2–3 weeks later, the root system is further developed, and the structure can be successfully transferred into a peat substrate. After 3 weeks of "hardening" in the greenhouse, the young plants are transferred to the open air, protected against insect contact. After this

Fig. 7.7 Raspberry cultures in a glass container immediately before separation and transfer into the rooting medium (photograph by S. Merkle)

period, a virus test is performed, and virus-free plants are either used as "mother plants" to obtain root cuttings or sold on the market.

To use the plant material produced by meristem culture for another passage of in vitro propagation, this is transferred to a propagation medium with the same composition as given above, but now containing 1 ppm 6-BA. Due to the higher BA concentration, leaf axillary buds sprout. Within 6 weeks, a propagation factor of about 5 can be obtained. For this, cultivation is carried out in conventional glass vessels (Fig. 7.7). To each vessel, 10 shoots are transferred, and after 6 weeks 50 new shoots are available for further propagation. For this, the shoots are isolated on a sterile bench and then transferred to a fresh nutrient medium. Essentially, this procedure can be repeated unlimitedly, but not more than 30 subcultures are advisable because of genetic instability (see Chap. 13). After producing enough shoots, rooting is initiated as described above, and eventually development continues on the peat substrate. A summary of the whole method is given in Fig. 7.8. The plants produced by this method are at least phenologically homogenous (Fig. 7.9). From the initial propagation cycles, some shoots of each strain are rooted as described above and checked for virus contamination (ELISA test). If virus infection is indicated, then the whole strain has to be discarded. In general, by using the method described above, most viruses can be eliminated.

7.2.3   In Vitro Propagation of *Anthurium* (following Geier 1986)

The origin of explants are young leaves (50–70% final length) that are sterilized by submergence for 5 s into 70% ethanol, followed by a transfer for about 20 min into a

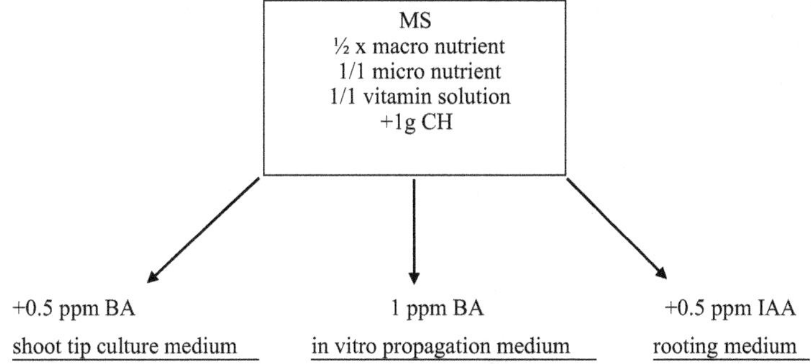

Fig. 7.8 Various nutrient media used for raspberry meristem cultures (S. Merkle, pers. comm.)

Fig. 7.9 Raspberry plants produced by in vitro propagation (photograph by S. Merkle)

hypochlorite solution (1.5% active chlorine, 0.5 ml Tween 20 per liter). To remove the remains of the sterilization solutions, washing with sterilized water first for 10 min, then for 30, and eventually for 60 min is performed. To obtain explants for cultivation, the leaf is cut into squares of 1–1.5 cm, avoiding the midribs. To prevent potentially heavy losses by microbial contaminations, per experiment usually only one explant is transferred to a small vessel. The culture of the explants on medium A (see Table 7.5) results in the formation of callus on the cut surface (continuous darkness). A few days later, some shoot primordials can be observed. If these are isolated and transferred to new media, then callus formation is again initiated, followed by differentiation of shoot primordials.

Table 7.5 Compositions of some nutrient media (mg/l) for in vitro propagation of *Anthurium scherzerianum* from leaf segments (Geier 1986)

Nutrient medium[a]	A	B	C
KNO_3	950	950	950
NH_4NO_3	720	720	720
$MgSO_4 \times 7H_2O$	185	185	185
$CaCl_2 \times 2H_2O$	220	220	220
KH_2PO_4	68	68	68
$FeSO_4 \times 7H_2O$	27.8	27.8	27.8
$Na_2EDTA \times 2H_2O$	37.8	37.8	27.8
$MnSO_4 \times H_2O$	19	19	19
H_3BO_4	10	10	10
$ZnSO_4 \times 7H_2O$	10	10	10
$Na_2MoO_4 \times 2H_2O$	0.250	0.250	0.250
$CuSO_4 \times 5H_2O$	0.025	0.025	0.025
M-inositol	100	100	100
Nicotinic acid	5	5	5
Glycine	2	2	2
Pyridoxine HC	0.5	0.5	0.5
Thiamine HCl	0.5	0.5	0.5
Folic acid	0.5	0.5	0.5
Biotin			
Sucrose	20,000	20,000	20,000
2.4D	0.1	–	–
Benzyladenine	1	0.2–0.5	–
Agar	8000	8000	8000
pH	5.8	5.8	5.8

[a]A, medium for induction and propagation caulogenic callus material; B, medium for shoot propagation by shoot proliferation; C, medium for rooting

Such subculturing can be repeated every 2–3 months (Fig. 7.10). Interestingly, the induction of primary callus formation is possible only in the dark, whereas these subcultures are successful in either light or darkness.

For in vitro propagation, shoots of 1 cm length are used after 4 weeks of illumination (2500 lux for 14 h per day). For rooting, medium C is employed. About 6–8 weeks later, the plants are transferred into soil. A reduction of light intensity to about 300 lux, and lowering of temperature from 25 to 10 °C strongly reduce growth—indeed, even after 2 years, transplantation into soil is successful. By cultivation on medium B (Table 7.5), shoot formation can be induced by the development of adventitious shoots and that of axillary buds into shoots, followed by rooting on medium C. Economic considerations are discussed by Geier (1986).

Fig. 7.10 In vitro propagation of *Anthurium scherzerianum* (Geier 1986). (Top left) plants propagated by in vitro culture methods. (Top right) on medium A in darkness, subcultured leaf callus with excessive shoot formation. (Middle left) leaf explants with callus and shoot formation after 20 weeks of culture in darkness with different cytokinins, otherwise medium A. (Middle right) propagation by shoot proliferation on medium B. (Bottom) rooted shoots on medium C

7.3   Somatic Embryogenesis

The term somatic embryogenesis describes a developmental process of somatic cells that results in morphological structures very similar in appearance to zygotic embryos. These somatic embryos can develop into intact plants, producing flowers and seeds. In an earlier chapter of this book, somatic embryogenesis in cell suspensions or callus cultures obtained from explants of the carrot root have already been briefly described.

Somatic embryogenesis refers to a process that results in obtaining bipolar somatic embryos or non-sexual cells through a series of developmental stages similar to those occurring during in vivo embryogenesis. In this method of in vitro vegetative propagation first developed for *Daucus carota*, embryos are obtained from cells other than zygotic cells (Steward et al. 1958; Reinert 1958). Somatic embryogenesis (SE) has been achieved from hypocotyl-derived callus culture in *Pterocarpus marsupium* (Husain et al. 2010), *Bixa orellana* L. (Parimalan et al. 2011), *Coffea canephora* P ex Fr. (Ramakrishna et al. 2012), *Coffea dewevrei* (Sridevi and Giridhar 2014), *Tylophora indica* (Sahai et al. 2010), and *Solanum sarrachoides* Sendt. (Banerjee et al. 1994). Husain et al. (2010) reported somatic embryogenesis (SE) from hypocotyl-derived callus culture in *Pterocarpus marsupium*. An efficient and reproducible protocol has been developed for the in vitro production of *Tylophora indica* via leaf explants. Regenerated plantlets were hardened, acclimatized, and established in soil with 90% survival rate (Husain et al. 2010). Clonal propagation has been achieved in *Stevia rebaudiana* (Giridhar et al. 2010), *Bixa orellana* L. (Parimalan et al. 2011), *Moringa oleifera* (Saini et al. 2012), *Glycine max* L. (Merr.) Akitha Devi (2012), and *Rivina humilis* L. (Harsha et al. 2012) genetic transformation *Capsicum frutescens* (Gururaj et al. 2012).

Due to further development of biotechnology, somatic embryogenesis in angiosperms and gymnosperms was studied. A complete deciduous plant was obtained in 1970 from a leaf of *Populus tremula* (Winton 1970). Initial reports on somatic embryogenesis in gymnosperms were published in the late 1970s and early 1980s and described embryo-like structures that were incapable of further development in cultured cells of *Pinus banksiana* (Durzan and Chalupa 1976), *Picea glauca*, and *Pseudotsuga menziesii* (Durzan 1980). Subsequently mature somatic embryos, able to convert into plantlets, were generated from immature embryos of *Picea abies* (Chalupa 1985) and from mature embryos of *Picea chihuahuana* into adventitious buds (Lopez-Escamilla et al. 2000). Micropropagation of an *Abies* species was conducted through the callus stage with the use of somatic embryogenesis (Nawrot-Chorabik 2016). Techniques for propagation of clones, through somatic embryogenesis, have contributed to the production of good-quality planting material of *Abies nordmanniana* with the desired phenotypic characteristics.

In Fig. 7.11, this process is summarized by means of a hand drawing for explants of the carrot root, the original source to produce somatic embryos, and in Fig. 7.12 photographs are given for some of the stages. A similar development can be induced in explants of the hypocotyl, petiole, leaf lamina, or other plant parts. Actually, this is a developmental program in which it can be demonstrated that quite a number of

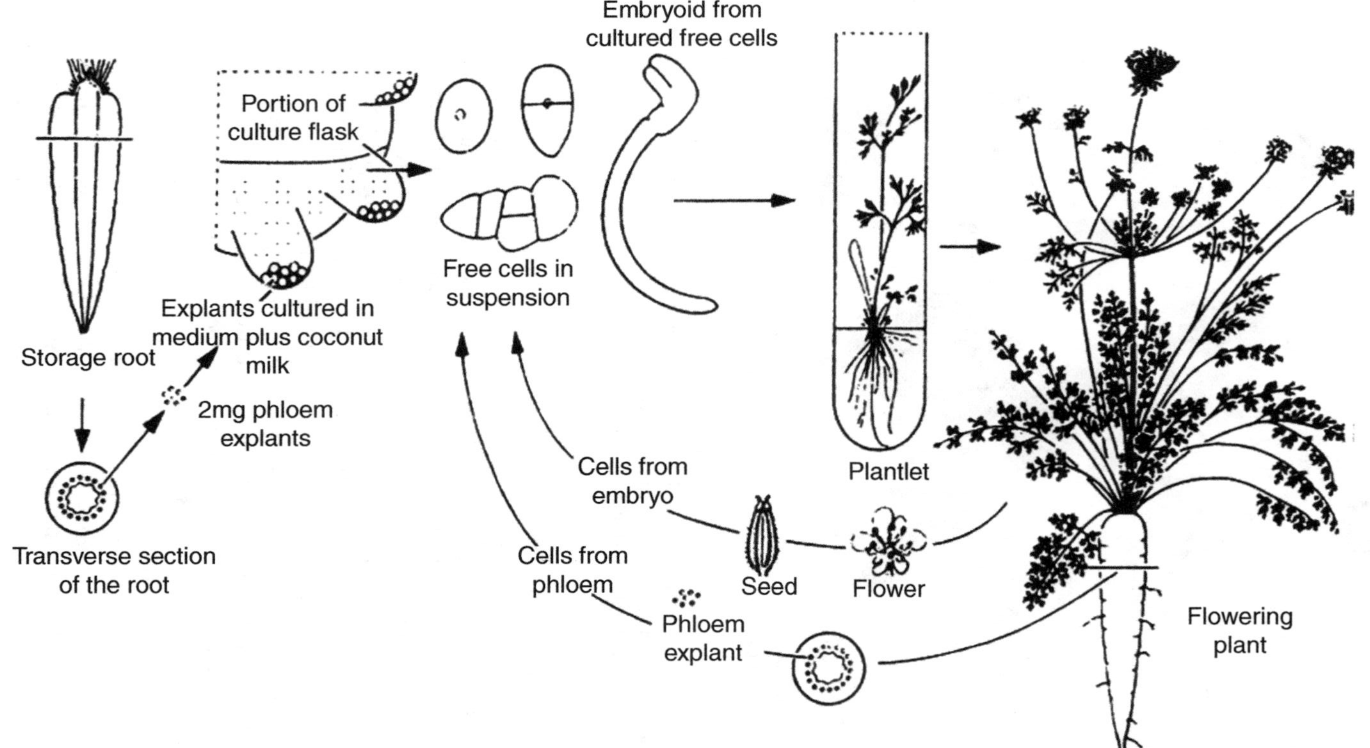

Fig. 7.11 The growth of carrot plants continuously maintained by means of cultured cells (drawing by M.O. Mapes, Steward et al. 1964)

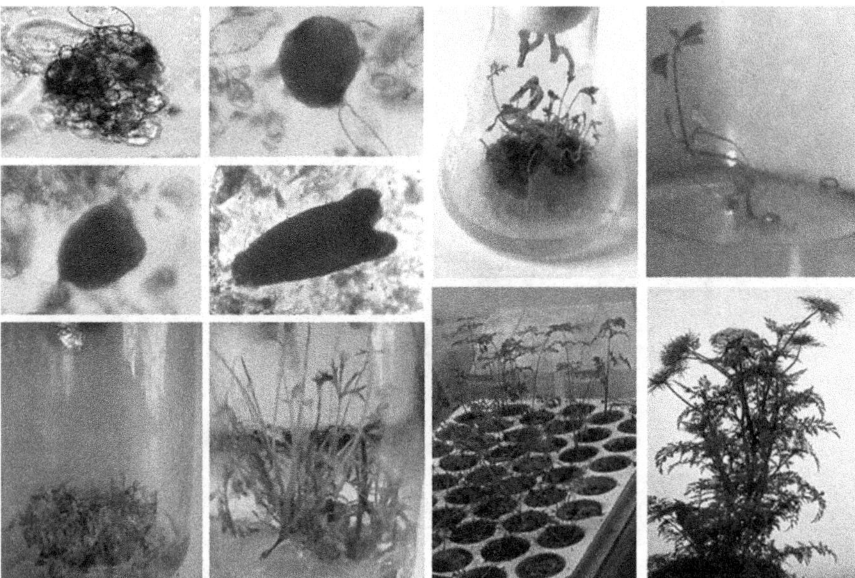

Fig. 7.12 Development of somatic embryos of a carrot cell suspension or a callus culture (Neumann 1995). (Left panel, top) pro-embryogenic cell cluster (left) and globular stage (right); middle heart stage (left) and torpedo stage (right); (bottom) development of plantlets from an embryogenic cell suspension poured onto an agar plate (left) and young plants isolated and transplanted onto agar (right). (Right panel, top) young plants emerging from a carrot callus (left) and plantlet isolated out of callus (observe the cotyledons, right); (bottom) young plants a few days after transfer into soil (left) and flowering plant of callus origin (right)

somatic cells (although not all) can be induced in some ways to react like the fertilized egg, i.e., the zygote, to produce a complete plant to flower and to set seeds. In fact, it is a reversal of the time elapsed since the original mixture of the male and female gametes of the parents was performed at fertilization, and from which the cells of the cultures were obtained. In a way, the "arrow of time" is reversed. Highly differentiated cells are transformed back into the original status they came from. From a certain viewpoint, our embryogenic cell cultures can be compared to embryonic stem cells of mammals presently discussed by many.

The plants produced by somatic embryogenesis, and in particular their offspring obtained from seeds, are hardly distinguishable from those produced by zygotic embryogenesis. Still, some differences exist at the cytological as well as at the molecular level, for which some examples will be given. For anatomy, in Fig. 7.13 the results of early studies by Street and Withers (1974) are summarized.

As an example, for the molecular level in horse chestnut (*Aesculus hippocastanum* L.) during zygotic as well as somatic embryogenesis, the activity of catalase (CAT) and of superoxide dismutase (SOD) increases, but differences in isozyme pattern were detected. In zygotic embryogenesis, a transition from a fast-migrating CAT form on PAGE, to a slowly migrating form occurred in July, i.e.,

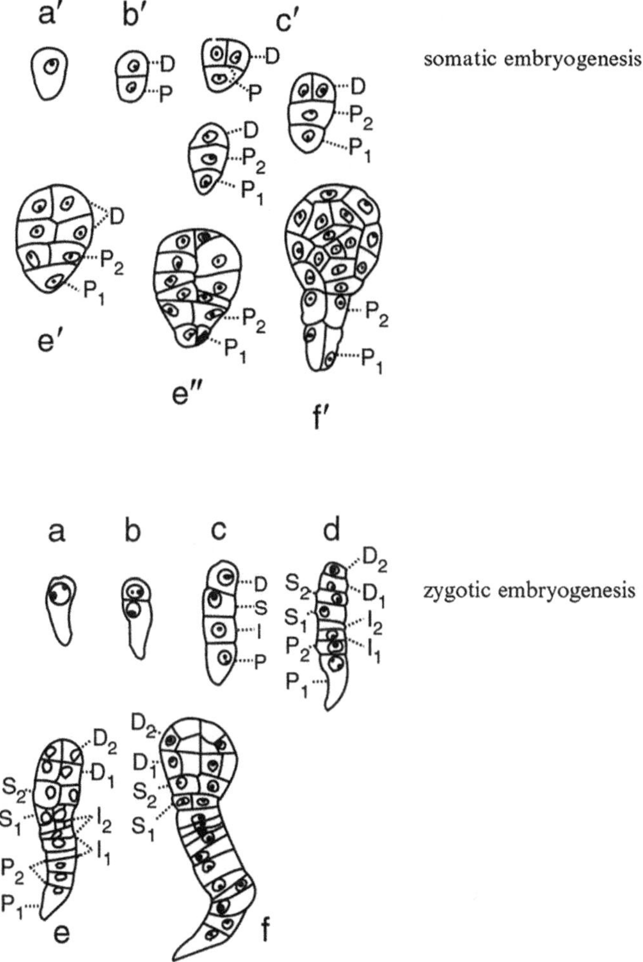

Fig. 7.13 Embryogenesis of *Daucus carota* (Street and Withers 1974) and zygotic embryogenesis (after Borthwick 1931); somatic embryogenesis is based on observations in suspension cultures (a–f and a ¢–f ¢ are comparable stages; D, S, I, and P are cells of the four-cell stage embryos and derivatives of these, e.g., P1, P2)

about 2 months after pollination—in somatic embryo development, these two isoforms of CAT were continuously detectable. For SOD, one Mn-dependent form and five Cu/Zn-dependent forms were detected at all stages of zygotic embryogenesis, whereas during somatic embryogenesis one Mn/SOD and one Fe/SOD were found (Bagnoli et al. 1998).

Also comparing zygotic embryogenesis and somatic embryogenesis in silver fir, SDS/PAGE profiles were very similar, but not identical. For zygotic embryogenesis, six storage proteins were found and for somatic embryogenesis 11. At certain stages, peroxidase activity was lower in somatic than in zygotic embryogenesis, and esterase activity was higher in the former (Kormutak et al. 2003).

The examples given above could indicate the existence of different pathways to produce zygotic and somatic embryos of the same species. It is difficult to evaluate such data. During both forms of embryogenesis, enzyme concentration or activities continuously change, and therefore a direct comparison of developmental stages of zygotic and somatic embryos is difficult to make. Before one can postulate the existence of alternative developmental programs to produce embryos in the same species, more data should evidently be available.

Somatic embryogenesis was discovered in basic studies to understand the differentiation and development of higher plants (Steward et al. 1958; Reinert 1959). Soon, however, also practical applications were envisaged, mainly by plant breeders and in horticulture. Despite intensive work to this end, success was rather limited, mainly due to the high costs of the manual labor required. Consequently, for a long time somatic embryogenesis was largely a system of basic research. With the advent of gene technology, and the requirement to produce plants from transformed cells for further handling, somatic embryogenesis has become an indispensable part of the technique. In this domain, it today serves as a tool, and the basics are of less interest. Most suited for the purpose are cells of embryo origin for which protocols to produce somatic embryos are available for many plant species and can easily be found on the Internet.

Often, also cells of immature embryos are used as origin. The production of somatic embryos in liquid media offers a chance for automation that should substantially reduce production costs, compared to those incurred in propagation by meristem cultures or adventive organogenesis (see above). Some time ago, it was calculated that to produce somatic embryos at a price of 0.01 cent apiece would be profitable (Walker and Sluis 1983), although this mostly by far exceeded the cost of natural propagation by seeds in the field. Exception could be hybrids to be produced by manual crossing, like coffee. Such embryos can be used to produce artificial seeds, as will be described later. For propagation of coffee, a method was published to successfully transfer the embryos even directly into soil, using the RITA system described in the section on liquid cultures (and see below at the end of this chapter).

Based on different aims for work on somatic embryogenesis, below the discussion will be divided into:

- The basics of somatic embryogenesis
- The application of somatic embryogenesis.

Compared to plants produced by embryogenic callus material, or adventive organogenesis, i.e., indirect somatic embryogenesis, a usually higher cytogenetic stability (see Sect. 13.1) can be observed for plants produced by direct somatic embryogenesis. Here, the embryos develop from one cell, as e.g., described for carrots by Li and Neumann (1985), whereas in indirect somatic embryogenesis, a prior callus phase is required. Somatic embryogenesis has been described in the literature for more than 100 species, also on the internet. In most cases, however, this is for indirect somatic embryogenesis.

Again, the description of somatic embryogenesis for a given species in a research article is not identical with a reliable method that works at will every day. To this end, more understanding of the process is required. After all, as mentioned before, somatic embryogenesis is a key process not only for propagation of plants but also to raise transgenic plants after artificial changes in the genome, and it is a model system to understand growth and differentiation in basic research. The aim of further research should be to understand this system to be able to apply it to any plant species at will.

7.3.1 Basics of Somatic Embryogenesis

As an introduction, first some remarks are given on embryogenesis in higher plants in general (see also Neumann and Grieb 1992). Besides zygotic embryogenesis, which occurs in all higher plant species, some detours to this basic process occur as substitutes for propagation by seeds without fertilization, and in some species both are possible. These detours are usually summarized as apomixes (Fig. 7.14). In the case of adventive embryogenesis, or nucellar embryogenesis, generation changes and the formation of the embryo sac are bypassed, and embryos develop from somatic diploid cells of the nucellus or the integuments. This sporophytic form of apomixes often results in polyembryony (e.g., in dandelion, citrus). In gametophytic apomixes, the diploid embryo sac develops from a vegetative cell of the nucellus (somatic apospory) or from the embryo sac mother cell, due to incomplete meiosis (generative apospory). In these cases, the embryo ($2n$) develops parthenogenically

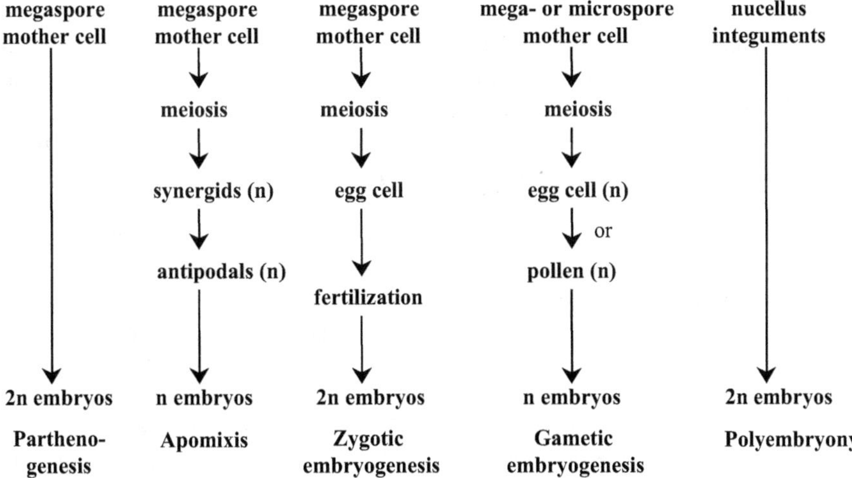

Fig. 7.14 Some alternatives of embryo development of higher plants

Embryo > Hypocotyl > Petiole > Leaf lamina > Root

Fig. 7.15 Competence to somatic embryogenesis in carrot (*Daucus carota* L.)

from the egg cell or from synergids or antipodals (diploid apogamic). Haploid parthenogenic and haploid apogamy occur as nonrecurrent apomixes (Maheswari 1979). Apparently, in addition to the embryo sac mother cell, other cells of the generative apex possess an embryogenic competence, and finally embryo development can be initiated in micro- and macrospores (androgenesis, gynogenesis; see Chap. 6).

The induction of embryo development from these cells requires a stimulus—in zygotic embryogenesis, this is fertilization. Apomictic embryogenesis can be initiated by a number of factors, such as fertilization, pollination, and environmental factors including temperature shocks, and the photoperiod (Nogler 1984). The chemical nature of the actual stimulus is not yet known.

These detours, however, are confined to the generative apex. Still, this natural competence of cells to somatic embryogenesis seems to exist more generally, though usually camouflaged in the intact plant. Here, somatic embryogenesis, as practiced in many tissue culture laboratories, nevertheless has to be induced by specific conditions, as can be provided in vitro by a suitable environment consisting of a nutrient medium containing a stimulus, usually an auxin, sometimes also a cytokinin, or both, and some requirements for appropriate temperature and illumination. In some protocols, also an ABA supplement is beneficial. In most plant species, however, this competence is lost during ontogenesis, and somatic embryos cannot be produced from explants of other parts of the plant. Thus, these species are heuristically defined as recalcitrant. Still, using cells of embryonic origin has often proved to be successful. This is the case, e.g., for some economically important cereals. Here, mature seeds are germinated in a medium containing high concentrations of 2.4D (e.g., 10–15 ppm 2.4D), resulting in excessive callus formation from which a great number of embryos can be derived after a transfer to an auxin-free nutrient medium (Imani 1999; unpublished results of our laboratory).

Generally, some hierarchical order within the plant seems to exist for many species, with highest success using embryonic cell material, followed by that of the hypocotyl, shoot buds, petiole, young leaves, and finally the root (Fig. 7.15).

There are some exceptions to this loss of embryogenic competence during ontogenesis, however, one being *Daucus carota*, the common carrot. Here, it is possible to culture intact 6- to 8-week-old plants aseptically and partly submersed in an appropriate medium (with enhanced salt concentrations) containing an auxin, and within about 4 weeks, somatic embryos appear from all parts of the shoot. If IAA is used as the auxin, then adventitious roots emerge about 2 weeks earlier (Schäfer et al. 1988). In this system, the competence to somatic embryogenesis is apparently preserved beyond the embryo stage, and it can be activated rather easily by a suitable environment. Under these conditions, embryo development can be also observed

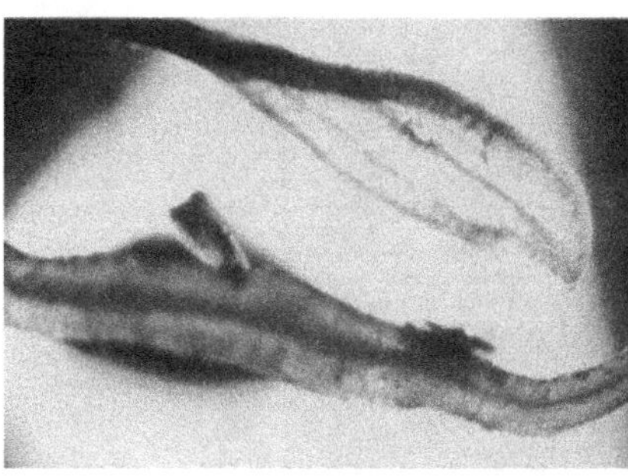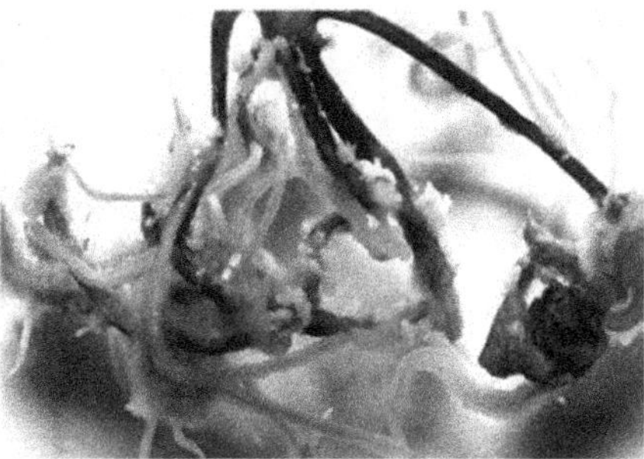

Fig. 7.16 Development of somatic embryos on cultured intact carrot plantlets in an auxin-supplemented medium

directly without prior callus formation, as shown for the emergence of embryos from the hypocotyl of a young plantlet in Fig. 7.16.

Such favorable conditions to initiate somatic embryogenesis apparently can be also invoked under particular genetic circumstances in intact plants on an inorganic agar medium. This was observed for about 6- to 8-week-old plantlets derived by somatic embryogenesis from fusion products of protoplasts obtained from transgenic plants of the wild carrot, and transgenic plants of a domestic variety of *D. carota*. Hygromycin resistance was introduced into the wild carrot strain and 5-methyl-tryptophan resistance into the strain of the domestic variety (Rotin). From these fusion products, plants were raised through somatic embryogenesis, and on the petioles or also on roots of these hybrids, a small callus developed on which later somatic embryos appeared. These could be used to obtain plantlets on which this process was observed again—this was repeated for three "generations" (Fig. 7.17; Chinachit 1991; De Klerk et al. 1997). This kind of development was not observed for sexual crosses of the "parents" of the hybrids or for protoplast fusion products of the genetically unaltered parent genomes. In some as yet unknown way, the introduction of foreign DNA would have changed the developmental control system of the hybrids, possibly related to the hormonal system; no further explanation of these observations can be given at present.

For further studies to understand this developmental process, it is not suitable to use cultured entire plants in vitro but rather more practical to use explants from various organs. In our laboratory, we mainly use explants of petioles of ca. 6-week-old plants. Here, two culture systems are practiced (Fig. 7.18). If the rather stable 2.4D is used as the auxin, then the cultured explants have to be transferred to an auxin-free medium 2 weeks after the beginning of culture, and about 2 weeks later, the various stages of embryo development can be observed. Though rhizogenic centers can be observed that are histologically similar to those of the IAA treatment, as described below, rarely some adventitious roots also develop. In the other system using IAA as an auxin, after about 10 days, adventitious roots appear, and again embryo development occurs about 4 weeks after the isolation of the explants. Since

Fig. 7.17 Development of embryogenic callus structures on petioles of fusion products of two transgenic carrot lines (wild carrot, hygromycin resistant; domestic carrot, 5-methyl-tryptophan resistant on an inorganic agar medium)

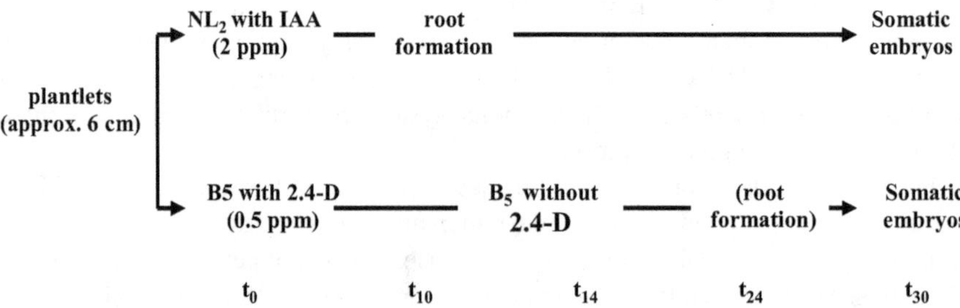

Fig. 7.18 Developmental processes in cultured carrot petiole explants in two different nutrient media

the IAA of the nutrient medium is destroyed after 2–3 days in the light (Bender and Neumann 1978), a transfer to an auxin-free medium is not required.

In the former system, the number of embryos developed is usually higher than in the latter system, and for many biochemical investigations, often the inclusion of root tissue is not desirable. Therefore, in many studies the former system is preferred in others the latter (Neumann 1995).

A culture of 48 h in the 2.4D medium, however, is sufficient to initiate embryo development at t12–t14 after initiation of the culture (Grieb 1991/1992). Apparently, the switch to the embryogenic developmental pathway occurs very early after culture initiation. The auxin 2.4D is quite stable, and an alternative explanation could be a preservation of molecules within the explants for some time. No investigations have yet been made on the fate of this auxin in the cells after its uptake.

Below, some essential requirements for somatic embryogenesis are given; generally, these apply also to zygotic embryogenesis, and in fact essentially to all differentiation pathways.

Requirements to induce somatic embryogenesis:

- Competent cells
- A suitable environment
- A stimulus

Based on these requirements, the following questions and aims have been formulated for further investigations:

- What constitutes embryogenic competence at the cytological and the molecular level?
- How are embryogenic competent cells produced during ontogenesis of plants?
- Molecular organization of the program of somatic embryogenesis, and its realization.
- What is the stimulus to induce the program of embryogenesis in competent cells?

It is not possible to fully discuss these aspects here. Rather, a few examples are given, mainly from our own research program.

Competent cells

The various non-zygotic pathways of embryogenesis originate from cells of the nucellus or the integuments, and the first sign of embryo development after the reception of the stimulus is an intensive growth of cytoplasm in some originally vacuolated cells (Fig. 7.19). The cells reacting to the stimulus are apparently distributed at random in the nucellus or the integuments. This indicates that at a given stage only some, and not all cells of these structures are competent to receive the stimulus and transform it into the initiation of embryo development. Also the megaspore mother cell develops out of a vacuolated cell by an increase of cytoplasm (Fig. 7.20).

In the carrot petiole, originally vacuolated subepidermal cells are embryogenic, and also here the first sign of the induction of somatic embryogenesis is an increase of cytoplasm in these cells. In the following, a detailed description of the carrot petiole system will be given.

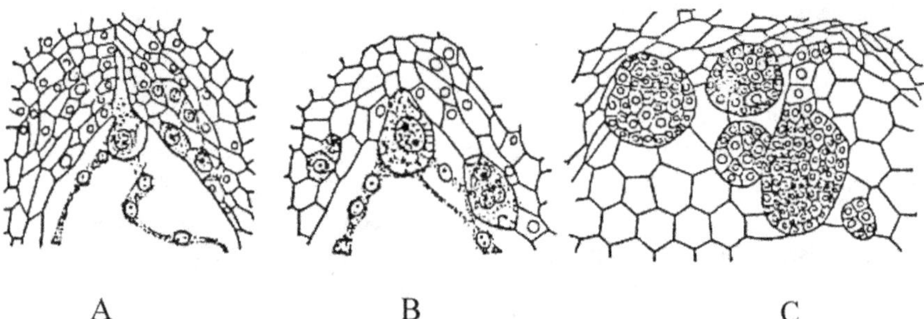

Fig. 7.19 Development of adventitious embryos of *Citrus trifoliate* (after Maheswari 1979). (**a**) Micropylary section of the embryo sac with fertilized egg cell and pollen tube endosperm nuclei. Some cells of the nucellus are enlarged with a big nucleus and dense cytoplasm. (**b**) A more developed stage of (**a**). (**c**) Upper part of the embryo sac with some early stages of embryo development; only the zygotic embryo has a suspensor

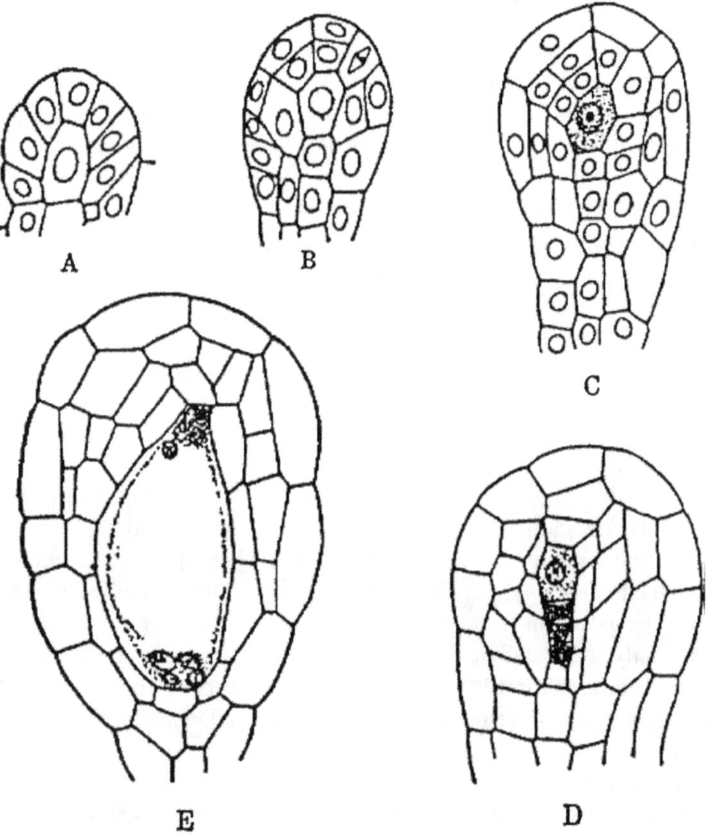

Fig. 7.20 Development of the egg cell and the embryo sac of *Leiphaimos spectabilis* (after Maheswari 1979)

As mentioned before, somatic embryogenesis can be observed in two forms, i.e., as a direct somatic embryogenesis and an indirect form. In direct somatic embryogenesis, the embryo develops from a single cell of, e.g., the hypocotyl or the petiole, without prior callus formation. Here, apparently a somatic cell can be directly transformed into an embryogenic cell. In indirect somatic embryogenesis, first the formation of a callus occurs, and subsequently embryo development can be observed in some of these callus cells. The indirect form has been described for many more plant species than has the direct one. Examples for direct somatic embryogenesis are *Daucus* (Li and Neumann 1985), *Trifolium rubens* (Cui et al. 1988), and *Dactylis glomerata* (Conge et al. 1983). For some plant species like *Daucus*, both forms are possible.

The various developmental processes are confined to specific areas in the petiole, and only some subepidermal cells (sometimes in close proximity to a glandular canal) are truly totipotent and competent to produce somatic embryos in a direct way, without a prerequisite for a callus phase (cf. direct somatic embryogenesis; Li and Neumann 1985; Neumann and Grieb 1992; Neumann 1995; De Klerk et al. 1997). Interestingly, the cells forming the glandular canal, characteristic for the species, contain high concentrations of auxins, as shown by using transgenic plants containing the auxin-sensitive MAS promoter coupled to the GUS gene, although the significance of this aspect is as yet not known (Grieb et al. 1997).

Rhizogenic centers develop near vascular bundles prior to these embryogenic centers. In both cases, some originally vacuolated cells start to produce new cytoplasm before the initiation of cell division. This contrasts with the initiation of cell division in cultured explants of the carrot taproot. Here, the first responses are associated with the formation of phragmosomes localized in strings of cytoplasm traversing the vacuoles (Neumann 1995, see also Chap. 3). In these strings of cytoplasm, nuclear division also takes place.

The cytological events described in Fig. 7.21 are summarized from an extensive histological investigation, and not all stages could be observed in all sections obtained (cf. Fig. 7.22). Looking at longitudinal cuttings of cultured petioles, regenerative centers are irregularly distributed along the axis. Apparently, the regenerative cells in these petiole explants contain at least a second competence to that brought about in the original petiole hidden during its development on the intact plant, i.e., rhizogenesis in the cells near the vascular bundles or embryogenesis in these subepidermal cells. A rough summary for the 2.4D system is given in Fig. 7.23.

For a technical description of direct somatic embryogenesis, the carrot petiole system as used in our laboratory will be summarized as example, based on an earlier description (Neumann and Grieb 1992). The original explants are obtained from petioles of 6- to 8-week-old plants. The explants of about 1 cm length can be severed from petioles of either aseptically germinated or (after surface sterilization) of soil germinated plants, by means of sterilized scissors or a scalpel. Also, young leaves of older plants can serve as origin.

In histological investigations, not only a different morphogenetic capacity can clearly be observed in various parenchyma tissues of the petiole but also a timescale for morphogenetic programs for the tissues described in detail above (Figs. 7.21 and

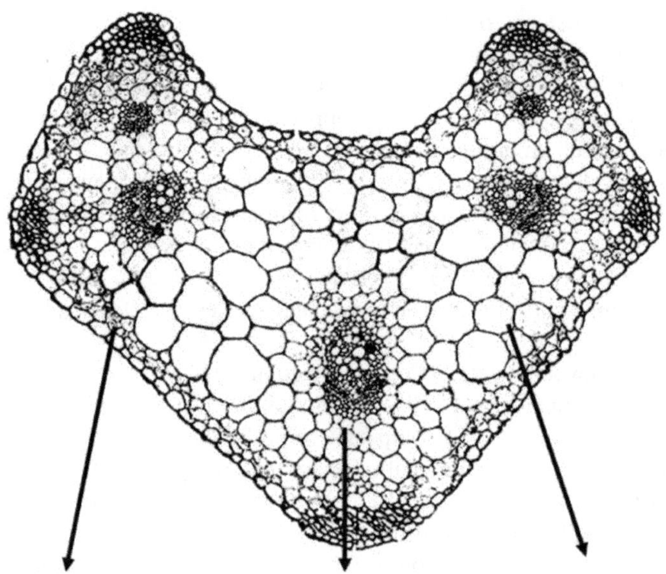

EMBRYOGENIC AREA	RHIZOGENIC AREA	CAULOGENIC AREA
a) growth of cytoplasm and cell division (12 days)	a) growth of cytoplasm and cell division (2 days)	a) growth of cytoplasm and cell division (5 days)
b) tetraoidal stage (14 days)	b) root primordia (5 days)	b) shoot primordia (12 days)
c) globular stage (18 days)	c) root emergence (7 to 10 days)	
d) heart-shaped stage (24 days)		
e) torpedo-shaped stage (28 days)		

Fig. 7.21 Regenerated areas in a carrot petiole and morphogenic reactions during 4 weeks of culture in NL with IAA and m-inositol (time given in days after start of the culture; photograph by J. Imani; Schäfer et al. 1985)

7.22). In nutrient media supplemented with IAA as the auxin (NL) and in the medium with 2.4D (B5), 2–4 days after initiation of culture, the cytoplasm increases in some vacuolated parenchyma cells adjacent to the conductive elements. These cells are usually in close proximity to glandular canals and develop eventually into rhizogenic centers. Further development differs depending on the auxin supplied. In the IAA medium, after 4–6 days in culture root primordia and later adventitious roots can be observed. If 2.4D is used as the auxin, then the rhizogenic centers often show some oriented growth, but root primordia and later roots are not differentiated. Nevertheless, if these cultures are transferred into a hormone-free medium, such root primordia and roots are eventually produced, though the latter in much lower numbers.

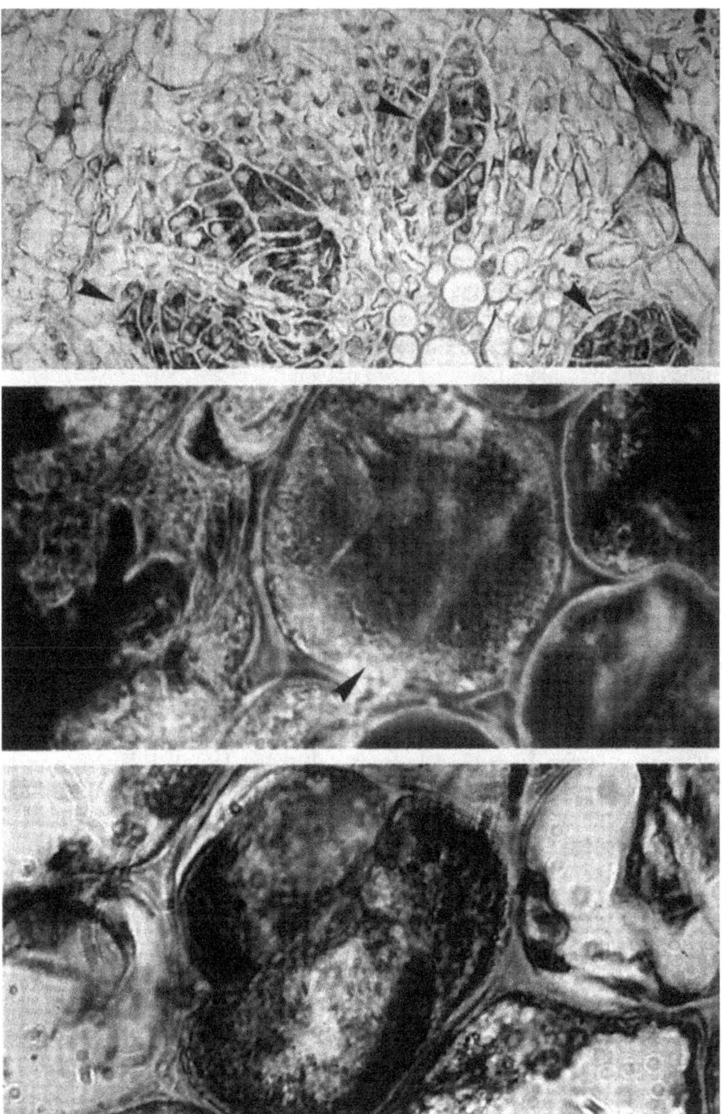

Fig. 7.22 Some histological sections of cultured petiole explants of carrot. (Top) rhizogenic centers near vascular bundles; (middle) cytoplasm-rich cell in the subepidermal area; (bottom) four-cell stage in embryogenic area

Evidently, the competence to rhizogenesis exists also in the 2.4D cultured material, but its realization is prevented by 2.4D in the medium. Some days later, the formation of caulogenic centers can sometimes be observed in the area of big parenchyma cells of the petiole (F. Schäfer, pers. comm.; see Fig. 7.21). After

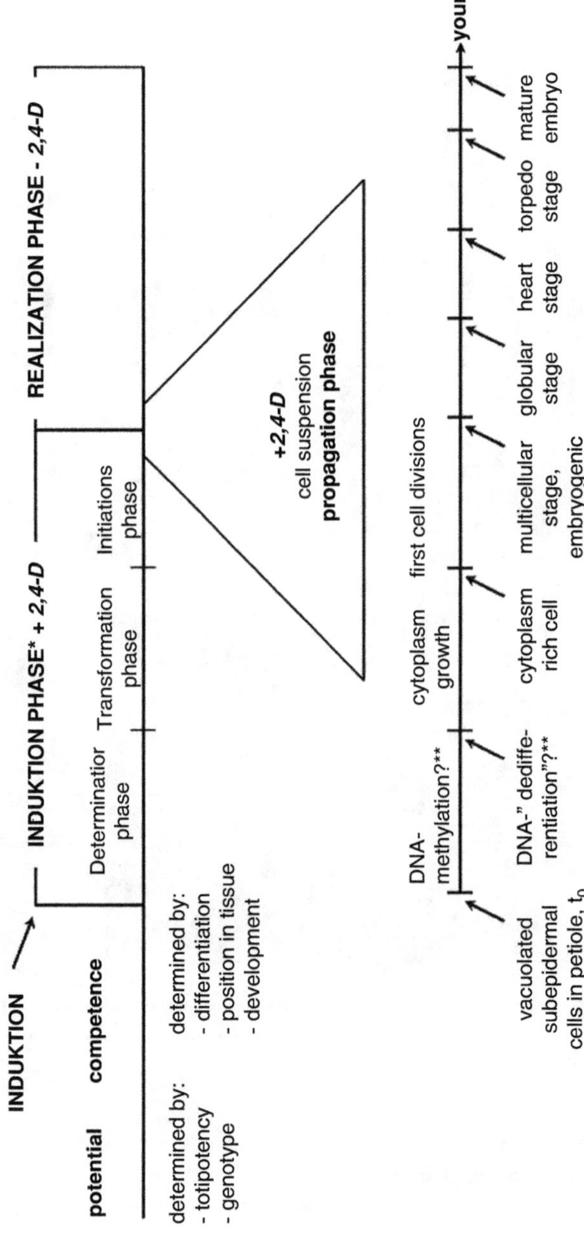

Fig. 7.23 Flow sheet on somatic embryogenesis

10 days of culture, the petioles are transferred from the 2.4D medium into a B5 medium devoid of auxin.

In both systems, after 12–14 days in culture, some vacuolated cells of a subepidermal cell layer again show an increase in cytoplasm without prior cell division, and these cells are transformed into embryogenic cells (Fig. 7.22). Here, it seems to be basically a histological process similar to those occurring in the floral apex during the various stages of development of the egg cell and the embryo sac. At the initiation of these processes, again the cells that will become embryogenic are often adjacent to glandular canals, whatever the significance of this is. Again as already observed for rhizogenesis, further development of the two systems differs. In the cultures in the IAA-supplemented medium, mostly a direct embryo development is initiated, and these cytoplasm-rich cells differentiate first into a four-cell structure, followed by the globular stage, the torpedo stage, and the cotyledonary stage, eventually developing into the young plantlets to be transferred onto a solid medium. If cultured in a 2.4D medium, it is only after transfer from the 2.4D-supplemented nutrient solution into an auxin-free medium that the development of these initiated cells through multicellular stages can be observed. Often, initiated areas break off, and are found freely floating in the medium.

Some cells on the surface of these structures develop into the various stages of embryo development. Again from such clusters, new embryogenic structures can split off, and embryogenesis is initiated again. In both systems, however, often embryos with an abnormal morphology occur (Fig. 7.24). No convincing explanations for these are available yet. After the stimulus to somatic embryogenesis is received, embryogenic competence is preserved through many cell divisions (propagation phase) for years. Apparently, a realization of the embryogenic program can be blocked by an auxin as long as propagation is envisaged in a medium containing 2.4D, like the B5 medium. In Fig. 7.23, a scheme is given as a summary of somatic embryogenesis.

For the initiation of indirect somatic embryogenesis, callus growth resulting from high cell division activity is first required. Early examples are explants of the carrot root (Linser and Neumann 1968) or of *Digitalis* (Luckner and Diettrich 1985, 1987). Apparently, cells of these original explants are not competent to develop somatic embryos. During high cell division activity, some processes of transdifferentiation ("reprogramming") seem to proceed and then result in competent cells. Such systems show that in some cases, competent cells can be generated if absent in the original explant. As already shown for petiole explants and the generative apex, also here only a few cells are competent to develop into embryos. An important function in the induction of somatic embryogenesis could possibly be attributed to adjacent cells, although no convincing evidence is yet available about this aspect. Before this can be clarified, a clear identification of "target cells" to become embryogenic is required. To this end, cytoimmunological and histochemical methods should be employed. On the basis of similarities of somatic embryogenesis and the various apomictic embryogenesis programs, somatic embryogenesis should essentially be simply another example of asexual embryogenesis in higher plants.

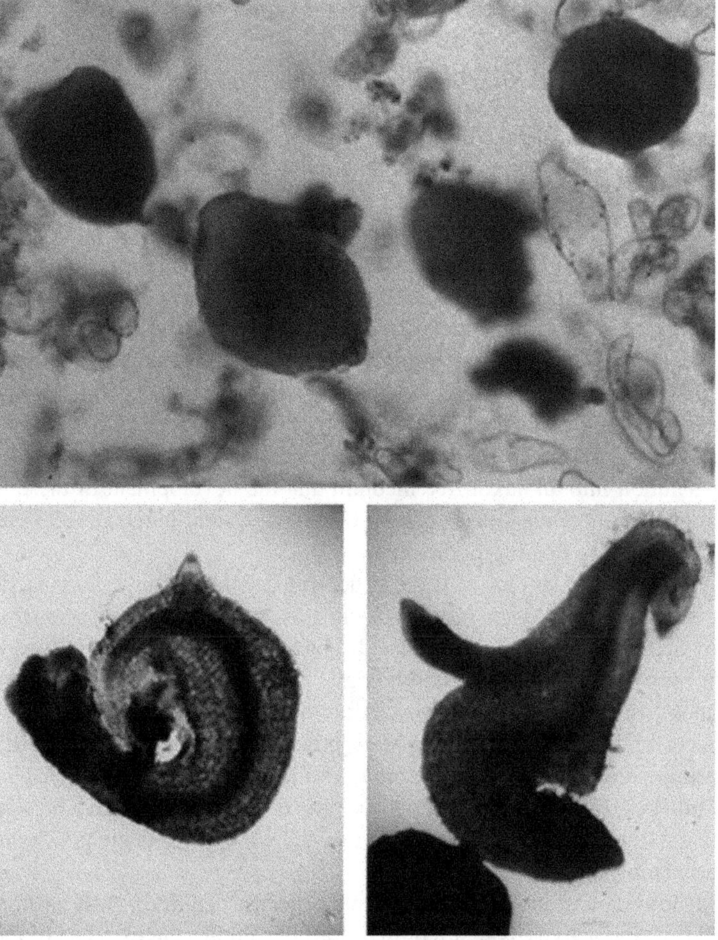

Fig. 7.24 Somatic embryogenesis: (top) various stages of somatic embryogenesis of an embryogenic cell cluster; (bottom) abnormally developed embryos

The influence of the hormonal supplement on morphogenesis varies between species. Whereas in the carrot system, cytokinin application inhibits embryo development, in cultures of *Medicago truncatula*, and in addition to NAA as auxin, BAP as a cytokinin has to be supplied to initiate embryogenesis. With only NAA in the medium, only rhizogenesis is initiated (Nolan et al. 2003).

7.3.2 Ontogenesis of Competent Cells

How and where do competent cells develop to produce somatic embryos? Currently, this question cannot be satisfactorily answered. Cutting an explant from a tissue represents a wound, and as a reaction in a suitable medium like the MS or NL medium, cell division is initiated and a wound callus is produced. This is accompanied by a dedifferentiation of the cells involved. Since embryo development can be induced also in cultured intact plantlets, as described above, this wounding would play no role in the induction of competence. This competence would be induced while the explant is still a part of the intact plant during its ontogenesis. As discussed above, such competent cells can occur more or less in any organ, though indistinguishable from adjacent cells in the same tissue under the microscope, from the embryo to the floral apex. These cells would be able to receive and react to the external stimulus, which can be a plant hormone like an auxin. As shown for transgenic material at initiation of culture, endogenous auxin, i.e., IAA, is evenly distributed within the petiole of carrots (see below). After a few days of culture, the IAA is polarized around vascular tissue, which then develops rhizogenic centers. Some days later, as described above, some subepidermal cells become embryogenic. If TIBA (an inhibitor of auxin transport) is applied, then the accumulation of IAA near vascular tissue does not occur, and somatic embryogenesis is not initiated (J. Imani, unpublished results of our institute). Possibly these rhizogenic centers produce signals that initiate this process. As discussed at length later, some glycoproteins are candidates for such a signal. Still, the question remains what distinguishes these receptive subepidermal cells to become embryogenic, among all other cells? The answer to this should come from a better anatomical and molecular understanding of the ontogenesis of leaves. In our experience, young leaves are better suited, being generally independent of the age of the plant from where the leaf to be used is obtained. Again the petiole is better suited to somatic embryogenesis than is the leaf lamina. The question then arises in which way is the ontogenesis of the subepidermal cells of the petiole different from that of other leaf cells?

In the ontogenesis of petioles, like in the development of the lamina of many dicots, intercalary growth by meristems located between differentiated tissues plays an important role. This could also be of significance for the development of cells (subepidermal cells) between the epidermis and the vascular bundles. Also the small size of the cells of subepidermal tissue, as the origin of embryogenic cells, points to the descendents from intercalary activities. These rather young cells would be receptive to exogenous stimuli from nearby cells, like those of rhizogenic centers or others. Not all cells of the embryogenic subepidermal cell layer are able to produce somatic embryos. In the subepidermal cell layer, as well as those along the petiole axis, the embryogenic cells are distributed at random. Therefore, it is difficult to conceive the occurrence of competition due to diffusion gradients in a given cell group to select only certain cells to initiate embryo development. Maybe differences in the time elapsed since the last cell division, i.e., cell age, play a decisive role that would vary among individual cells of this cell layer. A consequence of this would be cytological and biochemical variation. Further ideas to this end are discussed in Chap. 12. These considerations are based only on the somatic competence of cells of carrot petioles; the situation will certainly be different for other systems.

	Species	Origin
1	*D. halophilus* (e)	Mediterranean
2	*D. capillifolius* (e)	North Africa
3	*D. montevidensis* L. (n)	Mediterranean / South America
4	*D. commutatus* (e)	Mediterranean
5	*D. azoricus ssp* . (e)	Azores, Iran
6	*D. gadacei ssp.* (e)	France
7	*D. pusillus* Michx. (n)	North and South america
8	*D. muricatus* L. (n)	Mediterranean
9	*D. glochidiatus* (n)	Australia
10	*D. maritimus* (e)	Mediterranean
11	*D. maximus ssp.* (e)	Mediterranean
12	*D. carota* (wild carrot) (e)	Germany
13	*D. carota sativus* var. (Rotin) (e)	Germany

(e) = embryogenic; (n) = non embryogenic

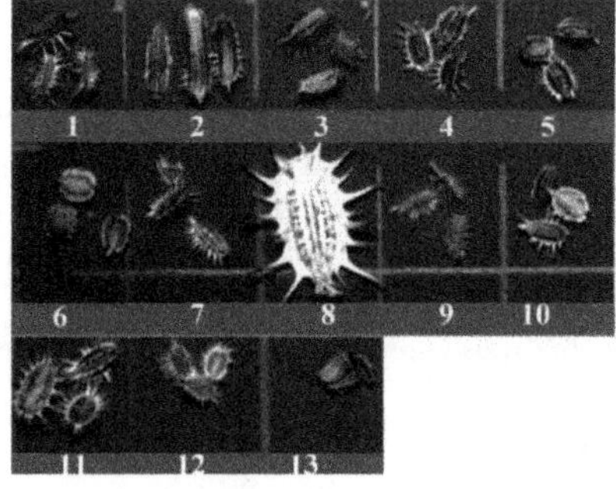

Fig. 7.25 Some embryogenic and non-embryogenic *Daucus* genomes and their seeds

7.3.3   Genetic Aspects: DNA Organization

A rough estimation (see von Arnold et al. 2002) amounts to 3×10^4 genes to be expressed in embryos and seedlings, and 3500 genes seem to be required to complete embryo development. In *Arabidopsis*, about 40 genes seem to direct the formation of all body pattern elements in the *Arabidopsis thaliana* embryo (von Arnold et al. 2002). Genetic factors also play a central role in inducing somatic embryos, i.e., in providing the competence of the species for the process. Here, using petiole explants, strong variations can be found even within a single genus such as *Daucus*. Eight of 12 *Daucus* species or subspecies cultured under identical conditions produced somatic embryos (*D. halophilus, D. capillifolius, D. commutatus, D. azoricus, D. gadacei, D. maritimus, D. maximus, D. carota*), whereas four (*D. montevidensis, D. pussillus, D. muricatus, D. glochidiatus*) were not competent to do so in similar conditions (Fig. 7.25). Still, the non-competent genomes may be embryogenic, using explants of embryonic origin.

Since two of the recalcitrant species/subspecies are native to the Mediterranean, as well as three of the competent ones, by and large the geographic location of origin seems to be of secondary importance (Le Tran Thi, unpublished results of our laboratory).

We used several molecular approaches to characterize the DNA of competent and recalcitrant genomes; one was the RAPD technique, which gives random amplified polymorphic DNA sequences employing PCR, i.e., the polymerase chain reaction. Here, selected primers with special sequences of about ten nucleotides are applied as starters for replication, and the number of nucleotides between the primer sequences yields DNA stretches of various lengths that can be separated on a gel by

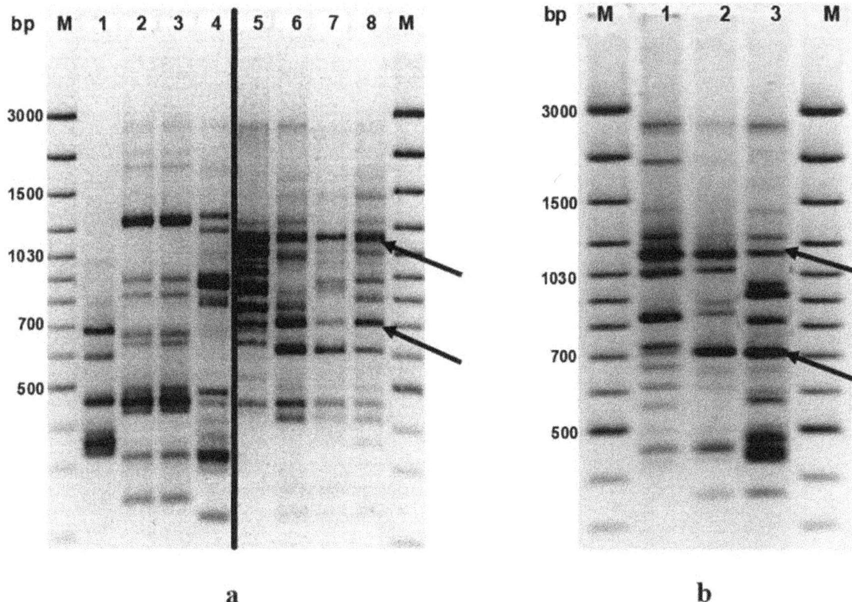

a   b

Fig. 7.26   RAPD analysis (primer 5 ¢-d(GTAGACCCGT)-3 ¢) of some embryogenic (lanes 5–8) and non-embryogenic (lanes 1–4) *Daucus* species and subspecies (arrows indicate fragments of 1.1 and 0.68 kbp). (**a**) Lanes: *M* marker, (*1*) *D. muricatus* L. (n), (*2*) *D. pusillus* Michx. (n), (*3*) *D. montevidensis* Link ex Sprengel (n), (*4*) *D. glochidiatus*(Labill.) Fischer et al. (n), (*5*) *D. carota* ssp. *carota* L. (wild carrot, e), (*6*) *D. carota* ssp. *maritimus* (Lam) Batt. (e), (*7*) *D. carota* ssp. *halophilus* Brot. (e), (*8*) *D. carota* ssp. *maximus* (Desf.) Ball (e). *e* Embryogenic, *n* non-embryogenic. (**b**) See (**a**) for lanes (Imani et al. 2001)

electrophoresis. The location of the primer sequences on the DNA is genetically fixed and characteristic of the species.

The RAPDs of these *Daucus* genomes in Fig. 7.26 were compared, using some 30 primers. For one of these, two areas were identical in the embryogenic species, and absent in the recalcitrant species, i.e., the areas with RAPDs at about 1100 bp and at about 650 of primer 3.

We do not yet know what the function of these "marker DNA" sequences for an embryogenic potential could be and whether they have anything to do with somatic embryogenesis at all. The two conspicuous bands were isolated and sequenced. Without going into all previously published details of this study (Imani et al. 2001), both bands were quite similar in the embryogenic species, with an identity of 70–95% in the nucleotide sequence (Table 7.6).

Neither indicated an open reading frame, and it is safe to conclude that no sequences of genes occur in these stretches of DNA. A search in databanks did not help in further characterization; these DNA stretches had apparently not been described before. Further investigations are required to see whether these bands can be regarded as markers for the ability to produce somatic embryos in cultured

Table 7.6 Degree of identity (%) in the nucleotide sequence of some DNA stretches of various *Daucus* genomes

	Daucus carota halophilus, 1	*Daucus carota maritimus*, 2	*Daucus carota maximus*, 3	*Daucus carota* (wild carrot), 4
1.1 kbp				
1	–	80	84	92
2	80	–	74	82
3	84	74	–	87
4	92	82	87	–
0.68 kbp				
1	–	97	97	92
2	97	–	97	97
3	97	97	–	97
4	92	97	97	–

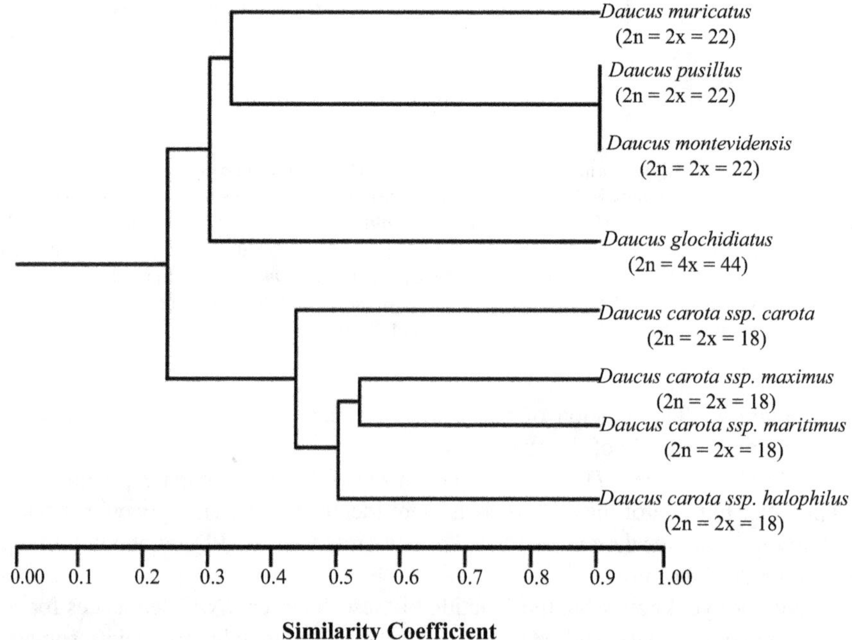

Fig. 7.27 Dendrograms of eight *Daucus* accession construct by RAPD analysis of genomic DNA (primer 5 ¢-d(GTAGACCCGT)-3 ¢) (Imani et al. 2001)

explants from mature plants, at least for the genus *Daucus*. The dendrogram in Fig. 7.27 indicates the genetic relations of these *Daucus* genomes. The genomes with no potential to produce somatic embryos form a separate group with chromosome numbers of more than 18, as found in *Daucus carota*.

Table 7.7 Somatic embryogenesis (s. e.) in cultured petiole explants of some carrot varieties (B5 system, 32 days of culture)

Wild carrot	Vosgeses	Lobbericher	Rote Riesen	Rotin
+[a]	+	−	+	++

[a]Symbols: −, no s. e; +, <50 s. e./15 ml; ++, >50 s. e./15 ml

The group of De Vries used a different approach to find marker genes for somatic embryogenesis. Starting also with carrot cultures, in hypocotyl explants, and later also in established cell suspensions derived thereof, a gene was identified in a population of embryogenic cells. Based on the nucleotide sequence, a protein could be derived that is a leucine-rich repeat receptor-like kinase (SERK). Using a construct consisting of the SERK promoter, and the luciferase gene as reporter, it could be shown that SERK expression occurs only up to the globular stage of embryo development. SERK mRNA was also found in zygotic embryogenesis up to the globular stage, but not in non-pollinated flowers or other tissues. Apparently, SERK is quite specific for the early stages of somatic and zygotic embryogenesis, indicating some similarity of both (Schmidt et al. 1997).

SERK was also detected in *Arabidopsis thaliana*, and here it was shown that SERK is localized in cell membranes from where it can be transported in intracellular vesicles in the cytoplasm. In *A. thaliana*, a gene family of SERK, now called AtSERK, apparently exists of which AtSERK1 is best characterized. Homologs to SERK have been described also for other plant species. In sugarcane cultures (*Saccharum* sp.), for example, the genes of the SERK family are called SoSERK, and this family has five members. SoSERK1 has a 72% identity with AtSERK1. Some members of the SoSERK family have a higher percentage identity with those of other monocots. Also for *Zea mays*, two SERK genes have been described (ZmSERK1 and ZmSERK2). These single copy genes have about 80% identity in nucleotide sequence, and they share similar intron/exon structures to those of the SERK genes. Whereas the expression of SERK2 occurs more or less in all tissues investigated to date, expression of SERK1 dominates in reproductive tissue, especially in microspores. These SERK genes, however, are expressed in embryogenic and non-embryogenic callus cultures. Furthermore, in callus cultures of sugarcane, SoSERKs are also expressed (Engelmann 1997). All this casts some doubt on the specificity of expression during the early stages of embryogenesis, as described above. Doubts on the specificity were already expressed by Nolan et al. (2003) in an extensive investigation using a culture system of *Medicago truncatula*, a legume. Here, a Serk1 gene (MtSERK1) orthologous to AtSERK1 (92% identity) was characterized, and the results suggested that this gene would have a broader role in morphogenesis in cultured tissue, and not only in somatic embryogenesis.

Let us go from species to varieties. Also here differences exist with respect to competence to produce somatic embryos, as reported some years ago (Table 7.7).

Whereas petioles from wild carrots, a French variety (Vosgeses), and the old German variety Rote Riesen are moderately competent under the conditions employed, the variety Lobbericher is not embryogenic under identical conditions;

Table 7.8 DNA density gradient profiles of some carrot varieties (+, present; −, absent)

	Density (g/cm^3)$^{-3}$						
	I	II	III	IV	V	VI	VII
	Main band	1.422	1.448	1.498	1.502	1.520	1.539
Wild carrots	1.485	+	+	+	+	+	+
Lobbericher	1.482	−	−	−	+	+	+
Rote Riesen	1.478	+	−	−	+	+	+
Italian	1.484	−	+	−	−	+	+

the more recent German variety Rotin, however, is highly competent. The DNA of these varieties has been compared by density gradient centrifugation (Table 7.8; Dührssen and Neumann 1980) already years ago. The GC content of DNA sequences obtained by mechanical sheering of total DNA varies, and the density increases with GC content. GC-rich sequences appear as heavy satellites of the main band DNA. The highest number of GC-rich satellites (Cs_2SO_4/Ag^+ density gradient centrifugation) was found for wild carrots; in the domestic varieties, always one or the other of the satellites is missing. If the wild carrot is considered as an ancestor of domestic carrot varieties, then this indicates that during domestication some DNA sequences were lost or altered in concentration (Dührssen et al. 1984).

The wild carrot is embryogenic and also the variety Rote Riesen, whereas Lobbericher and an Italian variety are not. The common denominator of wild carrots and Rote Riesen is the satellite with a density of 1.422 g/cm^3, absent in the non-embryogenic. Here again we do not yet know whether the DNA of this satellite has anything to do with somatic embryogenesis, or if it could be regarded only as a "marker" for the potential.

These GC-rich satellites are due to either highly or moderately repeated sequences generally not coding for proteins, and are nowadays often defined as so-called junk DNA. This probably holds true also for the RAPDs discussed above. The function of this "junk DNA" is to date largely unknown, but one has to keep in mind that it can represent more than 90% of the DNA of an organism. The genes as such could be similar or even identical in these varieties and even in species with different embryogenic potential. What distinguishes them could be the organization of the genetic system, i.e., the ways and sequences in which they are activated to produce proteins and enzymes. Here possibly junk DNA could play a crucial role. A high heritability of embryogenic potential has also been demonstrated in extensive investigations using sunflower cultures for two experimental traits, i.e., the number of embryogenic explants and the number of embryos produced (Flores Berrios et al. 2000).

As mentioned above, apparently not all cells of the petiole explant are able to be induced to use the stimulus transmitted at explantation to develop into root primordia and somatic embryos. Using the carrot system, also cells of other tissues besides those in the petiole can be induced to produce somatic embryos, i.e., these cells also possess embryogenic competence. To trigger this competence, the chemical environment is of importance, i.e., in vitro it is the nutrient medium, and for zygotic embryo development and apomixes, it is the endosperm.

More information became available following the application of proteomics and its methodology, as recently published by Imin et al. (2005), and serving as example. Here two lines of *Medicago truncatula* were studied, one recalcitrant and the other highly embryogenic. More than 2000 proteins were detected, of which 54 were significantly changed in expression during 8 weeks of culture with embryo development of the embryogenic line. Of these, more than 60% had differences between the two lines in the pattern of gene expression. Sixteen could also be identified; still, post-translational modification should be considered.

Of what consists the stimulus to induce the program of embryogenesis in competent cells? Besides the question of what consists the competence of those embryogenic cells in the various tissues of the carrot plant to receive a stimulus, what can we say on the nature of that stimulus to induce somatic cells to produce embryos? Again, basically two possibilities arise:

- The shock of isolation and in general the isolation of the explant from the mother plant.
- The auxin (or any other growth substance) in the nutrient medium
- Or both

As described above, it is possible to induce the production of somatic embryos in intact young plants cultured partly submersed under otherwise identical conditions as petiole explants, and embryos appear on the petioles, leaf lamina, and other parts. Consequently, and although often discussed in the literature, the first argument regarding shock and/or isolation as initiator can be discarded. The other possibility is the auxin as stimulus. This agrees with data on the IAA concentration in an embryonic and a non-embryonic carrot variety. In explants of the embryogenic variety, during induction a steep increase of IAA occurs, and in the non-embryogenic variety, this was not the case (Li and Neumann 1985). Possibly, an increase in endogenous auxin occurs also parallel to the induction of embryo development in zygotic and apomictic embryogenesis. The petiole program as described above can be initiated in light and in darkness.

7.3.4 The Phytohormone System

Competent cells have to be induced to somatic embryogenesis by a trigger, which in the carrot system is an auxin in a suitable nutrient medium such as a modified B5 medium or the NL medium (Tables 3.4 and 3.5) developed in our laboratory (Neumann 1966, 1995). If these induced cells continue to grow in an auxin-free medium, resulting either from photooxidative destruction of IAA or from a transfer into an auxin-free medium, as in the case of the 2.4D medium, then embryo development will proceed. If, however, these induced cells are continuously subcultured in the 2.4D medium, then this commitment will be preserved for

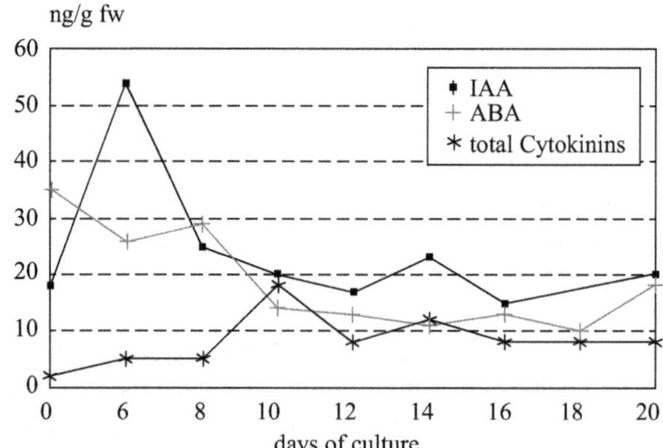

Fig. 7.28 Concentrations of some endogenous phytohormones in cultured carrot petioles (NL system) at several stages during the induction and realization of somatic embryogenesis

many years, and the realization of the embryogenic program will be prevented until these cells are transferred into an auxin-free medium. Many investigations on somatic embryogenesis use cell suspensions isolated often a long time beforehand and of unknown history. These cultures are hardly suitable to investigate somatic competence; at best, they are useful to study the release of the program of embryo development of cells initiated earlier. As in the case of our 2.4D medium, embryo development possibly was inhibited in such cultures by some component of the chemical or physical environment.

Let us first turn to the induction process. As shown for many cultures, the hormonal system plays a key role in the induction of somatic embryogenesis. It is known from earlier investigations that cultured cells develop an endogenous hormonal system different from that of the original explants (Bender and Neumann 1978; Stiebeling and Neumann 1987), and therefore as a first approach, the concentration of some phytohormones was determined at various stages during the induction of somatic embryogenesis in cultured petiole explants (Fig. 7.28; Grieb et al. 1997). Of those phytohormones determined, ABA dominates in the original petiole explants, followed by much lower concentrations of IAA and of cytokinins. Whereas the ABA concentration more or less continuously decreases during culture, the IAA concentration reaches a maximum 6 days after the start of the experiment, concurrently with the start of the development of adventitious roots. After this, the concentration of this native auxin continuously decreases until, on the 14th day, again a small peak appears. On the 10th or 11th day, there is also a small maximum in the cytokinins concentration (though at a lower level), dominated by 2iP and its riboside. This coincides with the onset of cytoplasmic growth in small, originally vacuolated subepidermal cells, to become embryogenically induced as described above. In summary, the cultured petiole explant produces its own hormonal system, with continuous changes in the ratios between concentrations of the phytohormones investigated to date. The significance of such changes to plant development was recognized many years ago by Skoog and Miller (1957), and this could possibly play a decisive role also in cultured petiole explants to become embryogenic. As mentioned before, this is to some extent confirmed by comparing these data, obtained

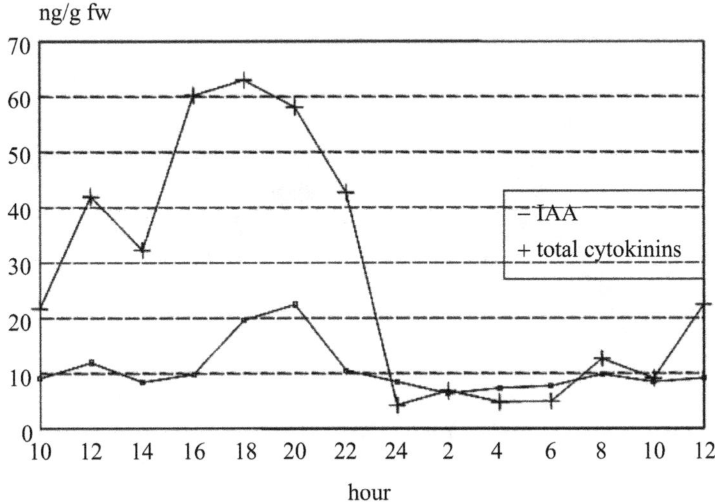

Fig. 7.29 Concentrations of IAA and of cytokinins in cultured carrot root tissue (constant environment: continuous illumination of ca. 5000 lux, 22 °C) during a 26-h experimental period. The samples were taken at 2-h intervals (Nessiem, unpublished results from our institute)

from a highly embryogenic variety, with those from a recalcitrant variety characterized by completely different patterns for the concentrations of IAA and of cytokinins, as published earlier (Li and Neumann 1985). An increase in the endogenous IAA concentration in cultured carrot cells competent to perform somatic embryogenesis was also reported by Michalczuk et al. (1992a, b) and by Pasternak et al. (2002) for an alfalfa culture system. The concentration of phytohormones follows a circadian rhythm also in cultured cells, which has to be considered in interpreting results on the phytohormone system in vitro. This has been demonstrated at least for carrot callus cultures of root origin (Fig. 7.29).

Using petiole explants from transgenic plants containing the auxin-responsive MAS promoter linked to the GUS reporter gene (Figs. 7.30 and 7.31), the distribution of auxin within the cultured petiole could be followed during the induction phase of somatic embryogenesis. As mentioned before, whereas in the original petiole explant at explantation the auxin is more or less evenly distributed throughout the petiole, after 5–6 days in culture, and concurrently with the formation of root primordia near vascular bundles, IAA is now accumulated in this area of the petiole (Fig. 7.32). After 9 days of culture, immediately before root development can be observed, the auxin concentration is substantially reduced in this area, and IAA now accumulates in the emerging embryogenic areas. Apparently, in the cultured petioles not only changes in the total concentrations of IAA occur, but also distinct changes in its distribution related to histogenic events can be observed (Grieb et al. 1997; Imani 1999).

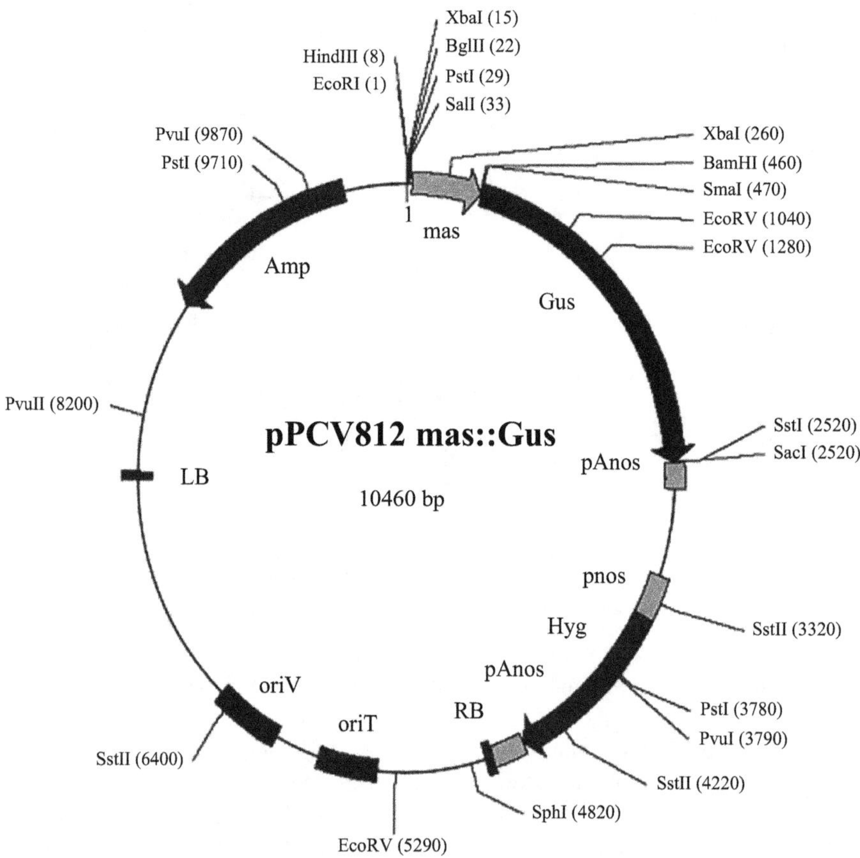

Fig. 7.30 Plasmid pPCV812 with the MAS promoter and the GUS reporter gene: Hyg denotes hygromycin resistance, and Amp denotes ampicillin resistance (courtesy of Dr. Z. Koncz, Max-Planck-Institut Cologne, Germany, who provided the plasmid)

Such transgenic plants will also be excellent tools in general to study the concentration trends of hormones during the development of intact plants. Some examples are given in Fig. 7.33.

Although not required for the *Daucus carota* system to be supplied to the medium to induce somatic embryogenesis, cultures of other plant species often require an ABA or a cytokinin supplement for the process. In *D. carota* cultures, an ABA supplement actually seems to be slightly inhibitory. If ABA plays a role in the process, then it should do so during the first week of culture when its concentration in the petiole explants is high. Of all the *Daucus* genomes investigated to date (see above), *D. carota* exhibits the highest embryogenic potential. To investigate the role of ABA in somatic embryogenesis, first its concentration in the original petiole explants of the various embryogenic and non-embryogenic *Daucus* species was determined. The highest concentration of free ABA was found in the highly

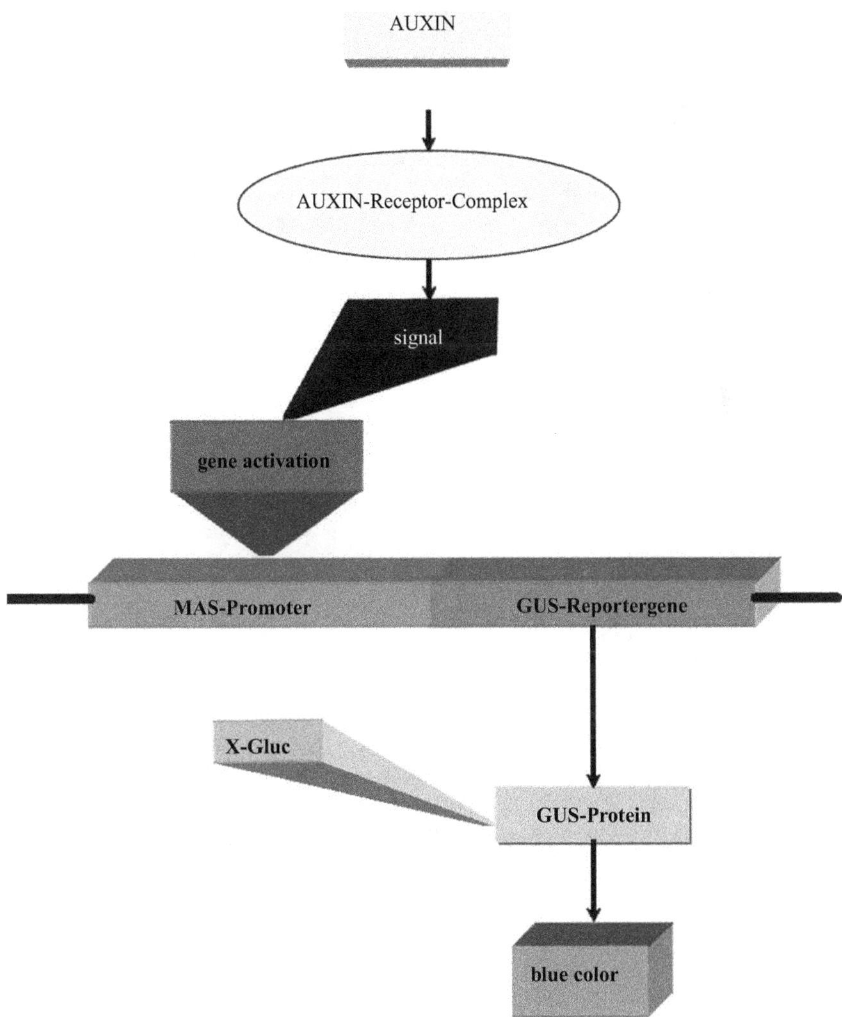

Fig. 7.31 Reaction system of the MAS promoter coupled to the GUS reporter gene

embryogenic domestic carrot variety Rotin. The other embryogenic species, with a much lower potential for the process, had considerably lower concentrations of this phytohormone. Here the number of embryos produced was greatly increased, and the time required for the initiation of embryogenesis was clearly reduced by application of ABA to the medium used for culture of these genomes (Le 2001; and unpublished results of our laboratory). In still unknown ways, ABA seems to be as highly involved in embryogenesis as are auxins. Possibly, the high potential of *D. carota* for somatic embryogenesis is related to the high concentration of free ABA in the original explants of this species at explantation.

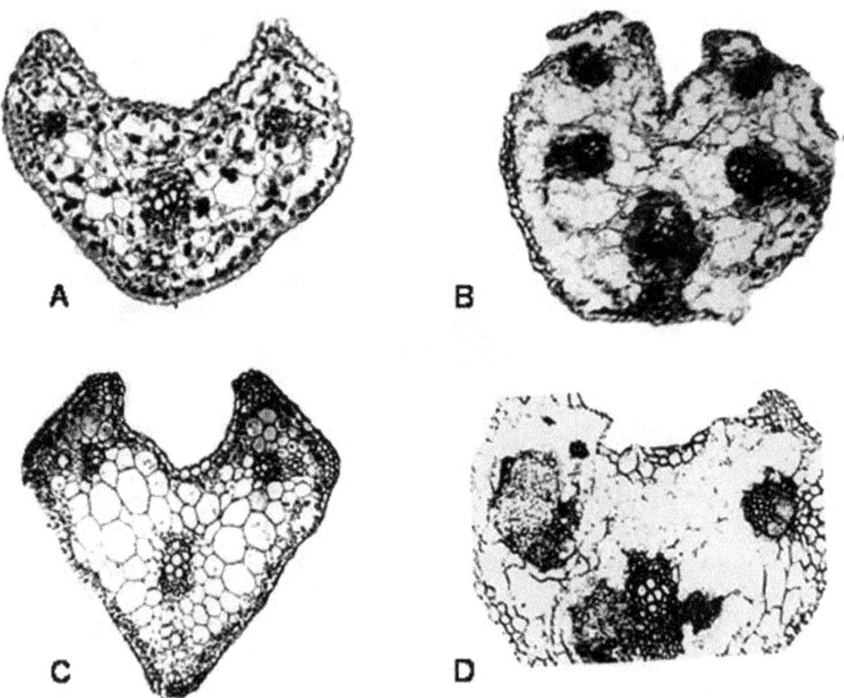

Fig. 7.32 Sections of petiole explants of transgenic carrots. The explants were obtained from plantlets containing the MAS promoter (auxin-sensitive) coupled to the GUS reporter gene. GUS activity was detected after application of X-Gluc to indicate the occurrence of auxin (dark spots in (**a**) and (**b**)). (**a**) At explantation, (**b**) after 5 days of culture (note the strong response of cells forming the glandular canals; dark), (**c**) historadioautogram at d0 after labeling for 3 h with 14 C leucine, (**d**) the same labeling duration as in (**c**) after 7 days of culture

With respect to ABA, Hays et al. (1999) studied its interaction with jasmonic acid (JA) in microspore-derived embryos of *Brassica napus*. The experimental system was the expression of napin and oleosin genes. Treatments with ABA plus JA gave an additive accumulation of mRNA of napin and oleosin. After treatment with JA, however, endogenous ABA levels were markedly reduced. It was concluded that possibly these cultures use endogenous JA to modulate ABA effects on the transcription of these genes. Furthermore, JA could have an effect on ABA, which would be reduced during later stages of seed development.

Besides these small molecules described to date, for *Cryptomeria japonica* a hormone-like plant growth factor involved in somatic embryogenesis was characterized, called phytosulfokine (PSK). This molecule is a sulfate peptide of 102 amino acids, with an aminoterminal hydrophobic signal sequence of 28 amino acids. PSK can be found in monocots as well as dicots, including *Arabidopsis*. This compound is involved in first steps in cellular proliferation, dedifferentiation, and redifferentiation, and a gene coding for its precursor has been identified. As

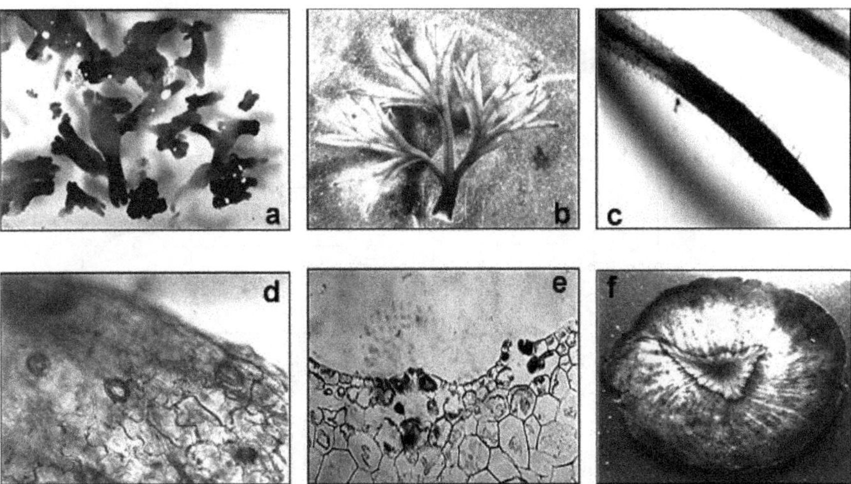

Fig. 7.33 IAA distribution in various parts of transgenic carrot plants after introduction of the system MAS promoter and GUS reporter gene (J. Imani, unpublished results of our institute; see Imani et al. 1999): (**a**) somatic embryos, (**b**) leaf lamina, after chlorophyll extraction, (**c**) root tip, (**d**) leaf with open stomata, (**e**) mechanical wound on petiole, (**f**) cross section of storage root (dark spots indicate GUS activity)

discussed elsewhere, some osmotic stress often has a positive effect on somatic embryogenesis, and the high molecular osmoticum polyethylene glycol (PEG) stimulates the formation in somatic embryos of *Cryptomeria* cultures. If a supplement of PEG and PSK is combined, then a positive interaction of the two can be observed. It is interesting that in this culture system, also GA3 (10 μM) seems to be important, which is an unusual requirement for somatic embryogenesis (Igasaki et al. 2003).

7.3.5   The Protein System

For some of these stages of the petiole system used for hormonal investigations, also the protein synthesis pattern was determined (Grieb 1991/1992; Neumann and Grieb 1992; Grieb et al. 1997). For the localization of protein synthesis activities within the cultured petioles, ^{14}C leucine was applied, and its distribution was followed by histoautoradiography and electrophoresis after protein extraction. Most of the labeling was concentrated in those parts of the petiole that indicated also an accumulation of auxin, i.e., the protein synthesis at one stage or the other would occur in cells engaged in differentiation. For these elaborate studies, the cultured petiole explants in the induction medium, i.e., the B5 medium supplemented with 0.5 ppm 2.4D, were labeled for 3 h each with ^{14}C leucine, starting at 0 h, 5 h, then at day 7, and the last labeling at day 14 of culture. The soluble protein was extracted as described

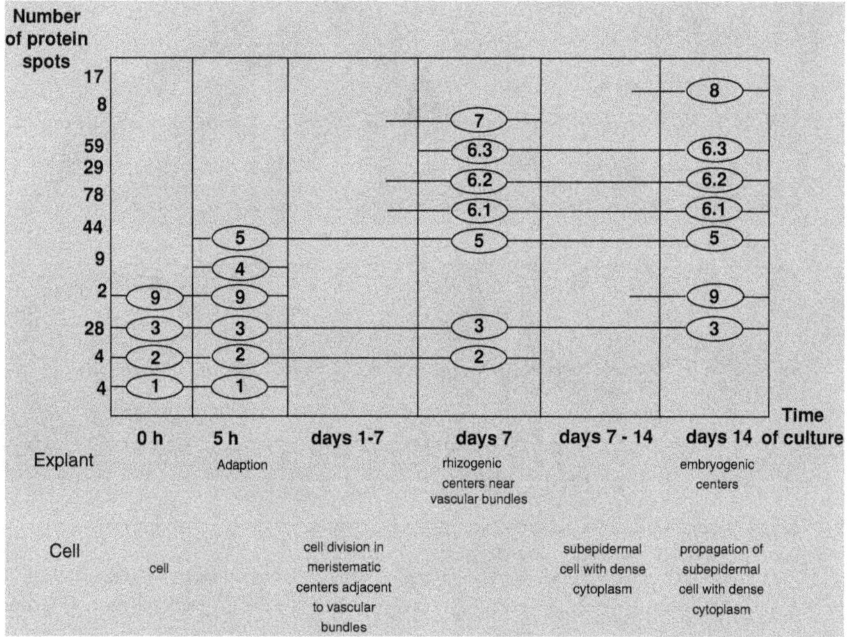

Fig. 7.34 Occurrence of various protein groups in cultured carrot petioles during specific periods of the culture cycle 133 (B5 medium with 0.5 ppm 2.4D; Neumann and Grieb 1992; Neumann 1995)

earlier (Gartenbach-Scharrer et al. 1990), and the extracts were separated by two-dimensional gel electrophoresis, followed by either staining with Coomassie brilliant blue to visualize proteins on the gels or by fluorography to detect the distribution of ^{14}C in the various proteins. In all, ca. 280 proteins were detected on the gels in these investigations, by either one or both detection method. According to the staining, the labeling pattern, and the occurrence during the various labeling periods, the proteins were arranged in nine groups and were related to cytological events during the induction of somatic embryogenesis. A continuous change in the composition of the protein moiety occurs with the initiation or termination of the synthesis of proteins in one or the other group, in a sequential and hierarchical pattern during the induction of somatic embryogenesis. Some proteins, however, were detected throughout the whole experimental period and would represent so-called household proteins (Fig. 7.34). Of special interest are the 17 proteins that could be detected only on day 14, and that would be somehow related to the occurrence of the cytoplasm-rich subepidermal cells destined to be the origin of somatic embryos and/or to the initiation of embryo development (Grieb 1991/1992).

According to the technical standard at the time of the investigation, as a first approach to understand the physiological significance of these proteins for somatic embryogenesis based on the isoelectric point and the molecular weight, a search was undertaken for analogs published in databases like Swiss-Prot or TrEMBL (e.g.,

Table 7.9 Detection of Coomassie brilliant blue-stained and ^{14}C-labeled analogs of proteins of cultured petiole explants of carrot after feeding ^{14}C-leucine (93.3 × 10^4 Bq L-[U[14]C-leucine]) for 3 h in each case

Analogs of proteins during induction phase	IP[a]	MW (kd)	t7		t14	
			CBB	[14]C	CBB	[14]C
Pyruvate decarboxylase (EC 4.1.1)	5.7	60.93	X	X	X	X
A-amylase (EC 3.2.1.1)	5.8	53.67	X			X
LC-RuBisCO (EC 4.1.1.39)	6.2	52.87	X	X	X	X
RuBisCO (EC 4.1.1.39)	5.5	49.89	X	X	X	X
Alcohol dehydrogenase (EC 1.1.1.1)	6.0	40.97	X		X	X
Phosphofructokinase (EC 2.7.1.11	6.6	34.12			X	X
Acetaldehyde dehydrogenase	6.6	32.62	X	X		
DNA-binding protein (homeobox containing genes of carrot)	5.9	34.84		X		X

[a]IP, isoelectric point; MW, molecular weight; t7, t14, days of culture; CBB, Coomassie brilliant blue; ^{14}C, ^{14}C-leucine

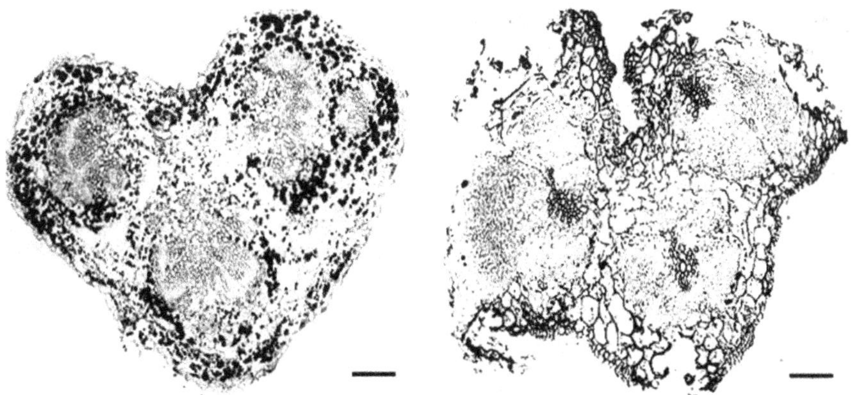

Fig. 7.35 Distribution of starch in cultured carrot petiole explants: (left) 12 days of culture, (right) 14 days of culture

Table 7.9). Of those 17 proteins synthesized only on day 14 after the start of the cultures, analogs could be found for only three. All three are related to carbohydrate metabolism, namely, alpha-amylase, phosphofructokinase, and alcohol dehydrogenase (Mashayekhi-Nezamabadi 2001; Imani et al. 2001; and unpublished results of our laboratory).

A protein described by the Komamine group as characteristic for an embryogenic status was already synthesized on the 7th day of culture, prior to the cytoplasmic growth of subepidermal cells destined to become embryogenic, usually observed from the 10th or 12th day of culture onward. This indicates that the initiation of the embryogenic program starts quite some time before its histological evidence.

Table 7.10 Starch concentration in cultured petiole explants of *Daucus carota* (cv. Rotin) during the induction of somatic embryogenesis

Days of culture	Starch concentration (mg/g fresh weight)
0	0.79 ± 0.48
6	1.10 ± 0.23
9	1.51 ± 0.48
12	0.93 ± 0.35
14	0.66 ± 0.53

As reported for other systems, also in cultured carrot petiole explants, a starch accumulation can be observed during the induction phase of somatic embryogenesis (unpublished results of our laboratory, Pleschka et al.). Histochemical studies on starch distribution indicate starch accumulation near vascular bundles of these cultured explants, and at the end of the induction phase, this disappears almost completely (Fig. 7.35; Pleschka et al.). In Table 7.9 are given some data on the dynamics of a few enzymes involved in carbohydrate metabolism of the petiole explants at two stages of somatic embryogenesis, as derived from two-dimensional electropherograms of ^{14}C-labeled soluble proteins.

Most of these proteins (selected from about 50) could be detected by CBB at t7 (here, appearance of rhizogenic centers), and also de novo synthesis occurs at both stages, as indicated by ^{14}C labeling. Alpha-amylase, the enzyme catalyzing starch breakdown, occurs at t7 as CBB spot, but not labeled, i.e., it is not synthesized. At t14, its concentration is below detection by CBB, but now it is labeled, i.e., it is synthesized de novo, coinciding with substantial starch breakdown in the cultured explants at the end of the induction phase and the development of embryogenic centers. The same pattern can be observed also for alcohol dehydrogenase. As these few examples in Table 7.9 indicate, different protein synthesis programs are in operation during the two stages of culture. It remains to be seen whether these changes in protein synthesis pattern are related to somatic embryogenesis. Such studies were first attempts to understand the coordinated activities of the information–transformation chain (DNA, RNA, protein). New methods developed in the recent fields of proteomics and genomics, like those applied to somatic embryogenesis of rice (Koller et al. 2002) and of *Medicago truncatula* (Imin et al. 2005) should soon be able to yield new approaches to this long-standing problem of understanding the regulation of growth and development—probably, however, this will again be a step that opens many new questions. At the time of these experiments, the terms, methods, and philosophy of proteomics and metabolics were not yet available, and such experiments were first attempts in that direction. Nevertheless, the genomics of *Daucus carota* is still not available today.

A relation of the occurrence of proteins to development and differentiation has been published recently for rice (Koller et al. 2002). Here, 2528 unique proteins have been detected and identified; enzymes involved in central metabolic pathways occur in all tissues investigated, i.e., leaf, root, and seed. Metabolic specialization was associated with tissue-specific enzyme complements, and the majority of metabolic enzymes belong to this group. It has to be kept in mind, however, that such

Table 7.11 Influence of various carbohydrates on the development of somatic embryos in a photoautotrophic cell suspension (*Daucus carota* L., var. Vosgeses, hormone-free medium, ambient CO_2): +, embryogenesis; −, no embryos produced

0.06M		0.003M	
Sucrose	+	Sucrose	+
Glucose	+	3-OM-glucose	−
Ribose	−		
Xylose	−		
Arabinose	−		
Pyruvate	−		

proteomic studies are performed with soluble proteins and that proteins associated with membranes are usually not considered.

As mentioned above for the characterization of the function of alpha-amylase, its synthesis, starch content, and starch distribution were compared by means of histochemistry on the 12th and 14th day. Whereas up to the 12th day, high starch accumulation was observed in the parenchyma cells of the petiole, starch concentration was reduced on the 14th day, presumably the result of the action of the newly synthesized alpha-amylase molecules. This tendency was confirmed by the enzymatic determination of starch (Table 7.10). The glucose resulting from starch breakdown should be phosphorylated by hexokinase, in preparation for further metabolic processing.

Apparently for further processing of starch breakdown products, hexokinase activation with free hexose or glucose-6-phosphate as substrates, as described for animal systems, seems to be required for embryo development. To substantiate these ideas, the occurrence and the metabolic activity of hexokinase should be investigated further. Especially important would be investigations on the histological and cytological distribution of the enzyme. The results of some preliminary investigations to this end have been published recently by Pleschka et al. (2001).

As described for other plant species (Widholm 1992), also cultured cells of carrot can perform photosynthesis (see Sect. 9.1; Neumann 1962, 1995; Neumann and Raafat 1973; Bender et al. 1981). Further studies on the function of free glucose have employed a photoautotrophic cell culture strain that was also embryogenic. The cells of this strain are able to grow slowly at ambient CO_2 in the light without differentiating somatic embryos. This, however, is the case after a supply of sucrose, of glucose, and (less pronounced) of fructose or mannose at low concentrations (Grieb et al. 1994; Pleschka 1995; Table 7.11).

To distinguish between nutritive and regulatory effects of these sugars on somatic embryogenesis, cell suspensions of this autotrophic strain were cultured at an elevated CO_2 concentration of 2.3%. Here, growth is comparable to that after a supplement of 2% sucrose, but also here somatic embryogenesis could not be observed (Grieb et al. 1994; Pleschka 1995; Table 7.12). This, however, is the case after an additional supply of 0.1% sucrose (or less?) to the nutrient medium.

Table 7.12 Influence of sucrose and an elevated CO_2 concentration on the fresh weight, cell number/g f. wt., and somatic embryogenesis (s. e.) of a photoautotrophic carrot cell suspension culture (*D. carota*, var. Vosgeses, 42 days of culture)

Ambient CO_2				2.34% CO_2		
Sucrose concentration (%) in the medium	mg F. wt. increment per 100 mg inoculum	Cell number per g f. wt. $\times 10^6$	S. e.	mg F. wt. increment per 100 mg inoculum	Cell number per g f. wt. $\times 10^6$	S. e.
0	288.6	2.90 ± 0.5	−	753.9	10.49 ± 0.30	−
0.1	436.8	3.48 ± 0.44	+	644.3	7.49 ± 2.28	+
0.5	1282.1	5.62 ± 038	+	1151.8	9.68 ± 1.71	+
1.0	2109.4	6.39 ± 1.4	+	1174.6	8.14 ± 1.61	+
2.0	2341.1	7.28 ± 0.8	+	1875.4	10.85 ± 2.67	+

Table 7.13 Influence of various carbohydrates on the somatic embryogenesis of petiole explants of carrot

Carbohydrates	Somatic embryogenesis
15 mM sucrose (control)	+++
15 mM glucose	+++
15 mM mannose	++
15 mM glucose-1-P	−
6 mM sucrose (control)	+++
6 mM glucose-6-P	+
6 mM fructose-1,6-P	−
3 mM sucrose (control)	++
3 mM 3-OM-glucose	−

These results indicate that the requirement for sugar should not be its only function as a nutrient but also as a regulator in the development of embryos (Grieb et al. 1994; Pleschka 1995; see also above). Glucose can be substituted by mannose or glucose-6-phosphate (Table 7.13; Pleschka et al. 2001), though with reduced efficiency, but not by the same concentration of glucose-1-phosphate in petiole cultures, which is the result of starch breakdown by starch phosphorylase and bypasses hexokinase for further processing.

For induction of somatic embryogenesis, petiole explants were cultured in a modified B5 medium with 2.26×10^{-6}M 2.4D for 14 days, and for realization the petiole explants were transferred into an auxin-free, modified B5 medium for 14 days. The carbohydrates indicated above were each reapplied (as at initiation of culture) to the media employed.

Somatic embryos of *Coffea arabusta* (a hybrid of *Coffea arabica* and *Coffea canephora*) can perform autotrophic growth and development from the torpedo stage onward. However, on a commercial scale, embryo development in autotrophic conditions is initiated at the cotyledonary stage. The main advantage would be a substantial reduction in the risk of microbial contamination (Afreen et al. 2002). An attempt at automation will be described later. Indeed, embryo development in photoautotrophic conditions is quite prominent.

Whereas the investigations of the carrot system were concerned with the induction of somatic embryogenesis, a paper by Dong and Dustan (1996) followed the development of the embryos up to the cotyledonary stage of *Picea glauca*. Here, by differential screening against non-embryogenic tissue, 28 cDNAs with temporal expression were detected. For this development, 2.4D and N6-benzyladenine had to be replaced by ABA.

Of other reports on the synthesis of proteins associated with somatic embryogenesis, only a few shall be discussed here; more can be found on the Internet. From cucumbers, two genes coding xyloglucan endotransglycosilases that were differentially expressed were detected after the induction of somatic embryogenesis (Malinowski et al. 2004). These enzymes seem to be involved in cell wall synthesis during cell growth and differentiation. Some sequence motifs in the promoter region were characterized (responsible for embryo-specific expression, auxin-inducible expression, ethylene-inducible expression). Glycosylated acidic endochitinase excreted to the medium promotes somatic embryogenesis in embryogenic suspensions of *Daucus carota* (von Arnold et al. 2002). Interestingly, an endochitinase from sugar beet stimulates the development of somatic embryos of *Picea abies* at early developmental stages (Egertsdotter and von Arnold 1998).

Another example of proteins expressed during somatic embryogenesis is the arabinogalactan proteins, often found in culture media. These proteins are a heterogeneous group, distinguished by a high content of carbohydrates and some lipids localized in the cell walls and plasma membranes (Majewska-Sawka and Nothnagel 2000). Application of an inhibitor (a synthetic phenyl glycoside) to the medium that binds specifically to arabinogalactans inhibits the somatic embryogenesis of *Daucus carota* and a *Cichorium* (von Arnold et al. 2002). These and other compounds could be the active components of so-called conditioned media obtained after prior cultivation of embryogenic material in the nutrient solution. Such media can often promote somatic embryogenesis in follow-up cultures. In addition to stimulatory molecules, however, this group of glycoproteins includes components inhibitory to somatic embryogenesis (Kreuger and van Holst 1996).

Possibly these glycoproteins are precursors of some signaling molecules like lipochitooligosaccharides (von Arnold et al. 2002), stimulatory for the cell division of plant cells. These substances were originally described to be significant for the formation of nodules of legumes, produced by *Rhizobium* (von Arnold et al. 2002). These compounds promote embryo development up to a late globular stage in carrot cultures (De Jong et al. 1993). Furthermore, in carrot and *Picea alba* embryogenic systems, the effect of chitinase on somatic embryogenesis can be substituted (De Jong et al. 1993; Egertsdotter and von Arnold 1998). As shown in vitro, chitinase is able to cleave arabinogalactan proteins, and both could be found in developing seeds; if arabinogalactans are incubated with chitinase, then somatic embryogenesis is enhanced (van Hengel et al. 2001). Although all this information is still very vague, a hypothesis was put forward that endogenous lipochitooligosaccharides are released from arabinogalactans by endochitinase, as signals to stimulate somatic embryogenesis (von Arnold et al. 2002). Many questions remain, like where do these arabinogalactans come from, what stimulates

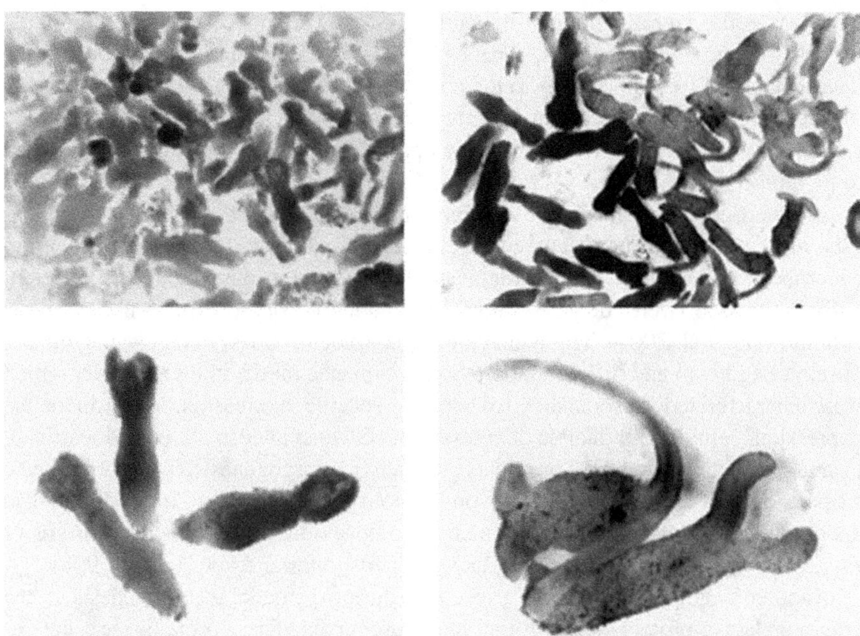

Fig. 7.36 Somatic embryos cultured with casein hydrolysate as only nitrogen source (top left), and after transfer into nitrate-containing medium (top right); higher magnification of embryos in the casein hydrolysate medium (bottom left) and after transfer from the casein hydrolysate medium to the nitrate medium (lower right)

the synthesis, what triggers the synthesis and the activity of chitinases, where is the primary site of action of these signals to initiate the program of somatic embryogenesis, and finally what is the relation with the influence of phytohormones such as auxins.

Another example of this kind of dual function, as that described for hexoses above, is the role of the form of nitrogen applied to the nutrient medium. In experiments dealing with this aspect, the same nitrogen concentrations of ammonia, nitrate, and casein hydrolysate were applied to carrot cell suspensions. In the ammonia treatment, the pH of the medium was decreased, and growth was strongly inhibited. Using casein hydrolysate as the only nitrogen source, growth was vigorously stimulated, but embryo development did not proceed beyond the torpedo stage, not even after a prolonged culture of 8–10 weeks. Here, a rough synchronization of embryo development (often desirable) was observed up to the torpedo stage. In the treatment with only nitrate as nitrogen supply, growth was visually less than in the former treatment, and less embryonic structures were observed, which, however, developed into plantlets, as in the control with a mixture of all three nitrogen compounds. In other experiments with successive application of first casein hydrolysate, followed by nitrate, there was a strong increase in the number of fully developed cotyledonary embryos being synchronously promoted (see Fig. 7.36; Mashayekhi-Nezamabadi 2001).

Working on the influence of mineral nutrients on gene activity, of 1280 mineral nutrition-related cDNAs, 115 were upregulated following nitrate supply after several weeks of nitrogen starvation. Besides genes for nitrate and nitrite reductase and other metabolic enzymes, some were also potentially involved in transcriptional regulation, and two in the regulation of methyltransferases. Some genes were also suppressed (Wang 2000, 2001). Apparently, nitrate not only serves as a nitrogen source but also induces diverse responses at the mRNA level. However, no such information is available for the system of somatic embryogenesis described above. Still, it has to be investigated whether the activity of these genes is induced directly by the nitrate molecule, or rather as an expression of a general stimulation of metabolism as a response to increased availability of the macronutrient nitrogen of which nitrate is a source (Wang et al. 2000, 2001). Genomic analysis of a nutrient response in *Arabidopsis* reveals divers expression patterns, and novel metabolic and potential regulatory genes are induced by nitrate.

The concentration of other mineral nutrients in the culture medium is also important for embryo development. As an example, at zero boron or at very low concentrations of this micronutrient, shoot development of the somatic embryos is almost negligible, and root development by far dominates (Fig. 7.37; Mashayekhi-Nezamabadi 2001; Mashayekhi and Neumann 2006). This is reversed at higher boron concentration.

Here, some relations with the ratios of concentrations of endogenous IAA/cytokinins at certain boron concentrations were determined (see Table 7.14). In the treatment with no boron supplement and with a preference for root development, a high ratio of IAA to cytokinins was determined, whereas at higher boron concentrations and shoot dominance, this ratio is more pronounced for cytokinins. An exception is the treatment with 1 ppm boron, which showed very low cytokinin concentrations—to date, no explanation can be given for this finding. Still, the investigation was repeated several times, with similar results. The tendency is clearly a promotion of root development at lower boron concentrations accompanied by a high IAA/cytokinin ratio, and a preference for shoot development at the expense of root development at a low IAA/cytokinin ratio. This is in agreement with generally accepted ideas based on the earlier reports by Skoog and Miller (1957) mentioned above, dealing with the influences of such changes in the auxin/cytokinin ratio on morphogenesis.

Influences of the concentration of boron on embryo development have already been described by Behrendt and Zoglauer (1996) for *Larix*. The significance of boron was also reported for the development of somatic embryos of loblolly pine (*Pinus taeda* L.), based on extensive investigations of the mineral composition of zygotic and somatic embryos at various developmental stages (Pullman et al. 2003). These examples again indicate a significance of mineral nutrients in the development of somatic embryos, an aspect greatly neglected to date. Indeed, no systematic study is yet available on the function of the mineral nutrition of culture systems in producing somatic embryos. Such a study, including relations to the endogenous hormonal system of the cultures, could be quite rewarding.

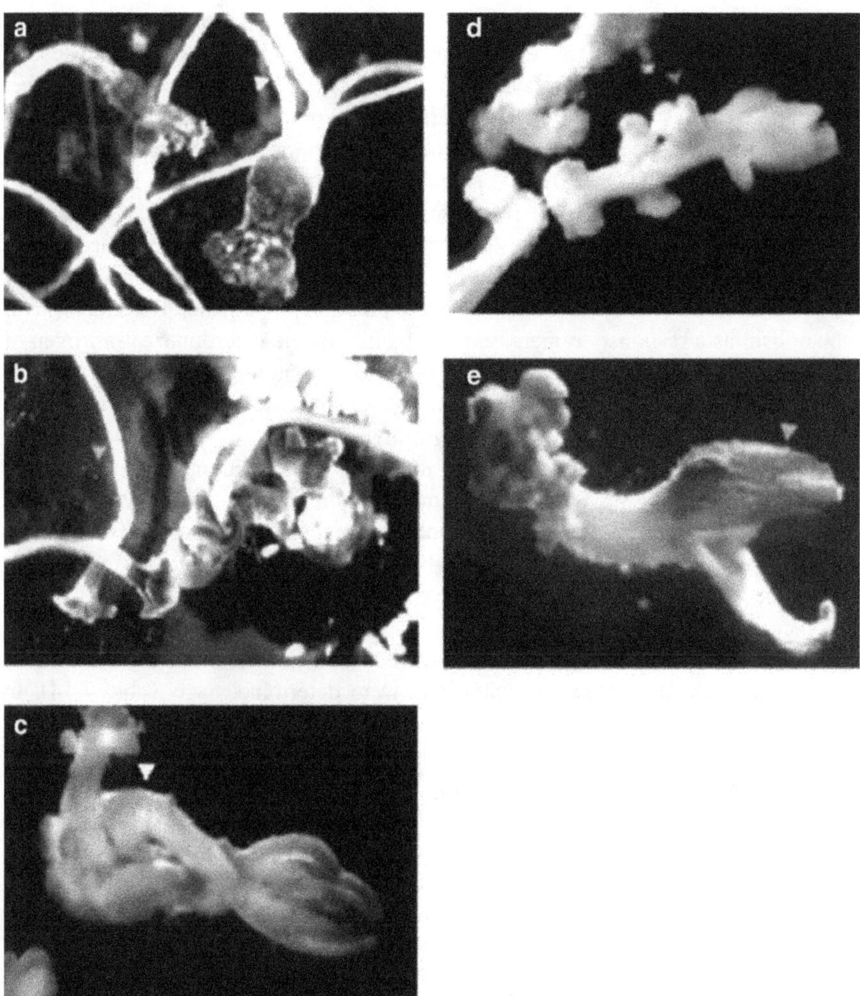

Fig. 7.37 Influences of various boron concentrations on the development of somatic embryos in carrot cell suspensions after 20 days of culture in B5 medium without 2.4D. (**a**) No boron: root development is promoted, and the development of cotyledons is retarded. (**b**) boron 0.1 mg/l: root and shoot development seems to be more or less balanced. (**c**) boron 1 mg/l: root development is inhibited, and shoot growth is promoted. (**d**) boron 2 mg/l: differentiation of adventitious shoot buds is initiated (arrows), and root development is inhibited. (**e**) boron 8 mg/l: there is strong promotion of the development of the cotyledons, and root development is almost completely arrested; adventitious shoot buds can be recognized in the root area (see arrow)

7.3.6   Cell Cycle Studies

Although it is generally assumed that the position of foreign DNA in the receiving genome will have an influence on its realization, evidence on this topic is rather scarce. Within this context, some results will be presented of experiments using cell

Table 7.14 Influence of various boron levels on the concentrations of some endogenous phytohormones, zeatin (Z), zeatin riboside (ZR), dihydrozeatin (DHZ), isopentenyladenine (IP), isopentenyladenosine (IPR), indole-3-acetic acid (IAA), and abscisic acid (ABA) in somatic embryo cultures of *Daucus carota* L., after 21 days[a]

Boron (mg/l)	IAA	ABA	IP	IPR	DHZ	Z	ZR	Total cytokinins	IAA/cytokinins
0.00	3.01d	87.5a	0.19	0.21	0.3	0.05	n.d.	0.74e	4.06
0.10	100.5a	12.15c	0.07	44.68	1.7	0.28	n.d.	46.72a	2.15
0.50	31.75b	5.1d	0.04	30.84	0.89	0.18	n.d.	31.96c	0.99
1.00	12c	23b	0.04	0.3	0.81	0.07	n.d.	1.21e	9.91
4.00	11c	3.58d	0.5	12.85	0.07	0.03	n.d.	13.45d	0.82
8.00	9c	6.9d	0.03	39.84	1.15	0.09	n.d.	41.11b	0.22

[a]Duncan test (significance level 95%), a, b, c indicate significant differences; n.d., not detectable

cycle synchronized carrot cell suspensions. The experimental outline is based on two assumptions: (1) as known from the literature, foreign DNA is preferentially inserted into replicating DNA, and (2) some hierarchical sequence exists in the replication of DNA during the S-phase of the cell cycle, e.g., euchromatin before heterochromatin. The synchronization of the cell cycle was induced (as described elsewhere) using the FDU/thymidine system, and after release of the blockage by FDU after applying thymidine, the S-phase was initiated again. The DNA of rol genes A, B, and C was applied either at 30-min (up to 2 h) or 60-min (2–6 h) intervals. The duration of co-culture was 8 h for each application. To raise embryos and plantlets, the same procedure as described above was used, but in this material in most treatments, embryo development could be observed only after application of 10% coconut milk, for some as yet unknown reason. Based on mainly morphological characteristics, six "morphotypes" could be distinguished, examples of which are given in Fig. 7.38.

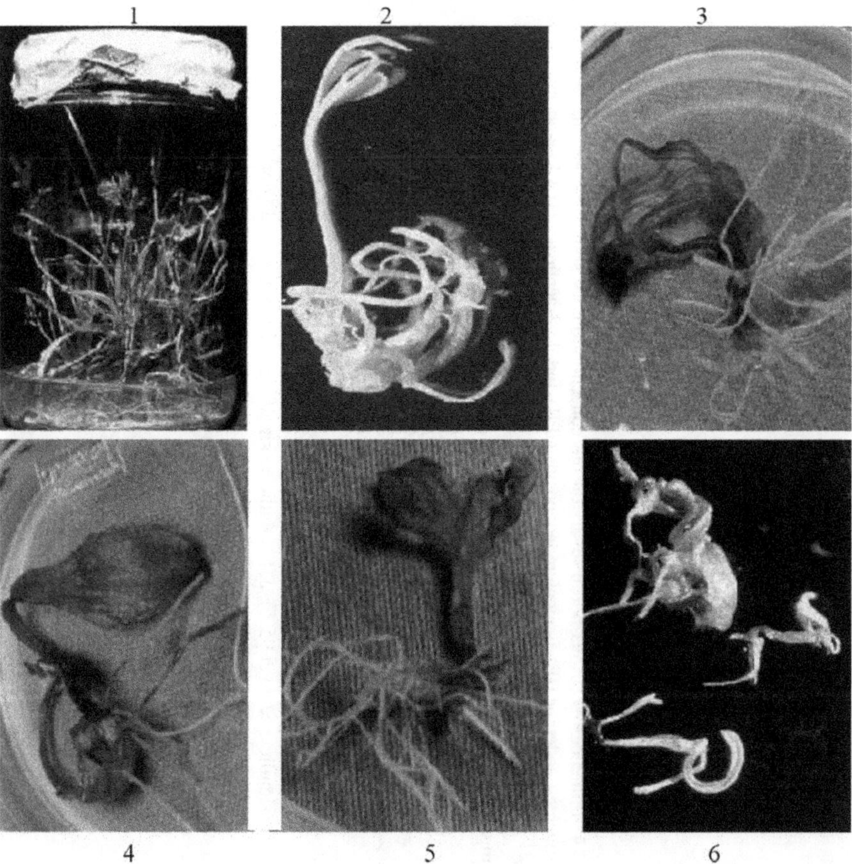

Fig. 7.38 "Morphotypes" observed in carrot cultures after insertion of "rol ABC" genes (Geisler 2001)

Table 7.15 Formation of some morphotypes due to foreign DNA following the application of thymidine (in hours)

Morphotype	1	2	3	4	5	6
Transformation II (8 h co-culture)						
Control (no synchr. and no rol gene appl.)	+	−	−	−	−	−
0 h	+	−	+	−	+	−
0.5 h	+	−	+	+	+	+
1 h	+	−	+	−	+	−
1.5 h	−	−	+	−	+	−
2 h	−	−	+	−	+	−
3 h	+	−	+	−	−	−
4 h	+	+	−	−	−	−
6 h	+	−	+	−	−	+

The occurrence of these morphotypes in the various treatments is summarized in Table 7.15. These are preliminary data, and based on the results already obtained, the experimental setup should be changed in some ways. Still, some relation between the status of the replication system at application of the foreign DNA and morphogenesis can be recognized, indicating the significance of the position of foreign DNA in the receiving genome. This is substantiated by southern blots of some treatments, given as examples in Fig. 7.39. Here, the foreign DNA is integrated at different positions according to the time of application in the cell cycle of the receiving genome, at one treatment (rol genes applied 60 min after initiation of the S-phase) also at two positions. This can be also observed for the 6-h treatment in the form of an additional band at 25 kbp, though here some doubts arise. This band could be due to incomplete digestion of DNA. An additional aspect is the integration of the three rol genes, which, although applied as a mixture, indicate individual positions of insertion into the DNA. At present, no unequivocal interpretation of this is possible.

A detailed description of the whole investigation is given by Geisler (2001) in her Ph.D. Thesis. There also the data for other durations of co-culture can be found.

These investigations were an early attempt to gain some information on possibilities to insert foreign genetic material at selected targets within the receiving genome. Meanwhile, many efforts can be seen in the literature to obtain targeted mutations and also gene replacements by site-specific induction of double-strand breaks of DNA (cf. Pabo et al. 2001).

7.4   Practical Application of Somatic Embryogenesis

In the previous section, investigations on somatic embryogenesis dealt with understanding more of the development of embryos for which somatic embryogenesis served as a surrogate. Furthermore, this was also used as a model system for basic studies related to growth, differentiation, and cell development of higher plants. Somatic embryogenesis, however, has also great practical significance for plant

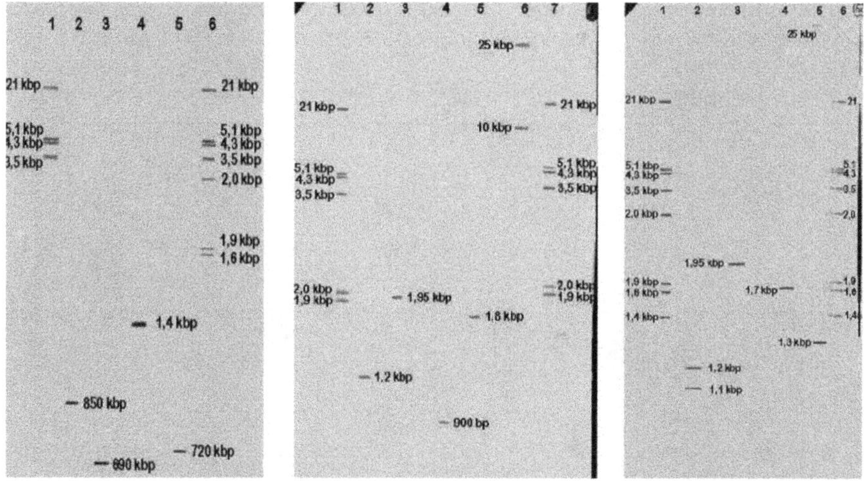

Fig. 7.39 Southern blots of cell cultures transformed with rol genes. Right slots: *1* marker, *2* t0, *3* t0.5, *4* t1 to t5, *5* t2, *6* marker: examples of southern blots of DNA after co-culture with "rol genes" for various durations (Geisler 2001). Middle slots: *1* marker, *2* appl. at t0, *3* appl. at t1, *4* appl. at t2, *5* appl. at t3, *6* appl. at t0, *7* marker: occurrence of high molecular DNA (25 kbp), incomplete digestion of DNA, results are not valid. Left slots: *1* marker, *2* appl. at t1, *3* appl. at t2, *4* appl. at t3, *5* appl. at t4, *6* marker

propagation, including the production of artificial seeds, plant breeding, and gene technology. In the following, some examples will be given. As already published by the Steward group at Cornell University decades ago, somatic embryogenesis can be also induced in explants of embryos. Only rather recently, this was applied to so-called recalcitrant species after it was clear that in immature embryos, mature embryos, and sometimes also in early seedling stages, the competence to become embryogenic was preserved, to be lost later in development.

Here, often cell suspensions are used that are suitable, first, to integrate foreign genetic material, and, second, to raise these new genomes into intact plants via somatic embryogenesis for selection, breeding, and, finally, propagation by seeds. In this domain in recent years, an explosion in the number of papers describing protocols for dicots, monocots, and several conifers has been observed in the literature. Here, somatic embryogenesis is used as a "tool," with less attention being paid to the basics of the process. These protocols can be easily found on the Internet.

As an example, a method used to produce somatic embryos of *Arabidopsis thaliana* will be described (Hecht et al. 2001) and for monocots that for wheat. The protocol to produce somatic embryos in carrot suspensions originally obtained from petiole explants has already been described. In principle, the methods are quite similar. It is surprising to note that a method using non-embryonic somatic cells of *Arabidopsis thaliana* to induce somatic embryos was developed only rather recently (Ikeda-Iwai et al. 2003). Here, stress treatments like heavy metal ions (CdCl), osmotica like mannitol or sorbitol, and dehydration were applied to apical shoot tip explants of 5- or 6-day-old seedlings.

The use of embryonic cells to induce somatic embryogenesis starts with the placement of surface sterilized seeds of *Arabidopsis* into the MS medium containing 2% sucrose (w/v), 4.5 μM 2.4D, 10 mM MES (2-(N-morpholino)-ethanesulfonic acid) at pH 5.8. After a treatment at 4 °C for 2 days, cultures were transferred onto a rotary shaker at 25 °C in the light at 3000 lux (16 h light/8 h dark), and the germinating seedlings developed callus aggregates. After 3 weeks, and some subcultures with fresh MS medium, green embryogenic clusters with a smooth surface were transferred into a 2.4D-free medium for 1 week for embryo development. Non-embryogenic callus material had a yellowish appearance.

For somatic embryogenesis of monocots, an example of our own research group will be given (Imani 1999). Here, somatic embryos are produced from callus material derived from wheat or barley seeds. The seeds are sterilized first for 1 min in 70% ethanol, followed by gentle shaking in a diluted sodium hypochlorite (1:1.5) solution (ca. 7% active chlorine) for about 30–45 min, which contains also a drop of Tween 80. After this, the seeds are washed several times with sterilized water under sterile conditions and then placed on B5 medium supplied with 10 ppm 2.4D-containing agar plates. Culture is for 4–6 weeks in the light at 28 °C. As shown in Fig. 7.40, first some callus is produced from which, after transfer onto a hormone-free B5 medium, embryos, and finally seedlings develop.

The same method was used also to produce somatic embryos of barley, carrots, *Hypericum*, and other dicots.

Some differences can be observed between the pathways of angiosperms and gymnosperms. In gymnosperm cultures starting with aggregates of a few cells in an auxin- and cytokinin-containing medium, three types of PEMs are successively produced within 25 days. The embryo develops in the hormone-free medium first by producing some kind of suspensor, which then degenerates within 3 weeks, and the development of mature embryos requires a supplement of ABA to the medium (Fig. 7.41; von Arnold et al. 2002).

Somatic embryos can be also produced in a bioreactor. In the experiment in Fig. 7.42, 3 g of an embryogenic carrot cell suspension was passed through a sieve of 90 μm to remove the biggest aggregates, which were discarded. The smaller clumps were cultivated for 1 week in the B5 medium with 2.4D in an Erlenmeyer flask on a rotary shaker. After this pre-culture, the cells were incubated in a bioreactor with 4 l of B5 without the auxin. Immediately before this, the cells were transferred for 24 h in an auxin-free medium to remove 2.4D on the surface of the cell material. After a culture period of 2 weeks, a great number of young plants could be observed in the bioreactor (Fig. 7.42), a rough estimate amounting to about 100,000 plantlets. In addition, many different stages of embryo development occurred in the suspension, which in a prolonged experimental period could have developed into plantlets (Imani 1999).

As mentioned above, based on the principles of the Steward auxophyton method, a technique called "RITA bioreactor" was developed some years ago.

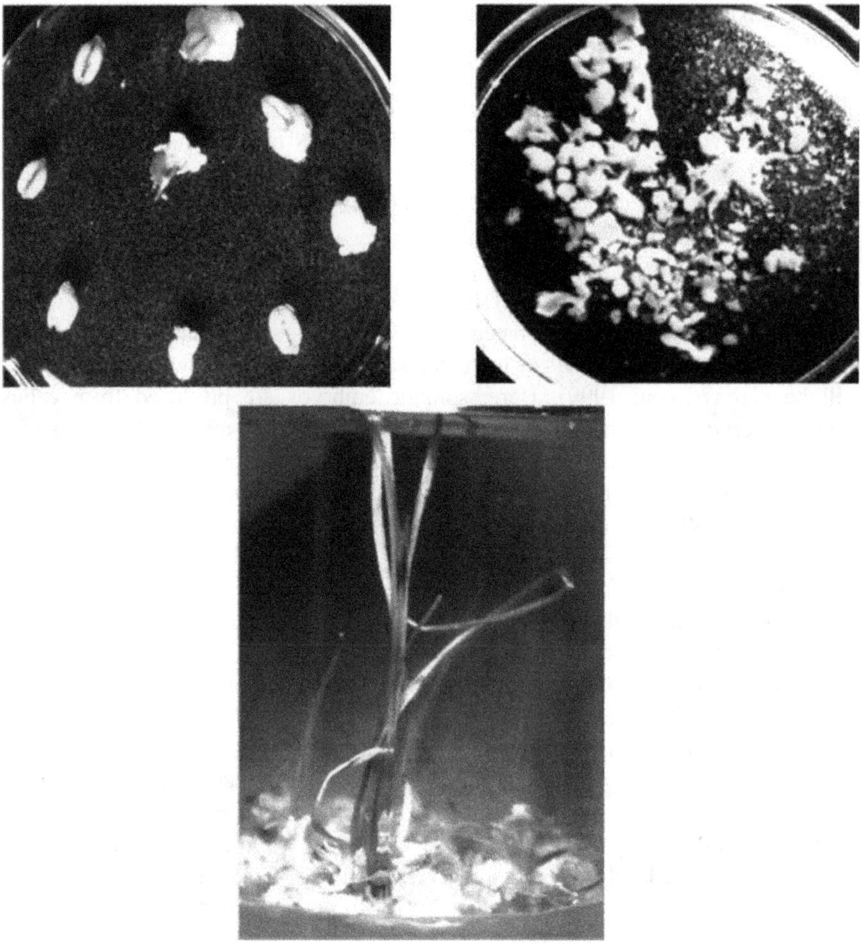

Fig. 7.40 Somatic embryogenesis in wheat seeds

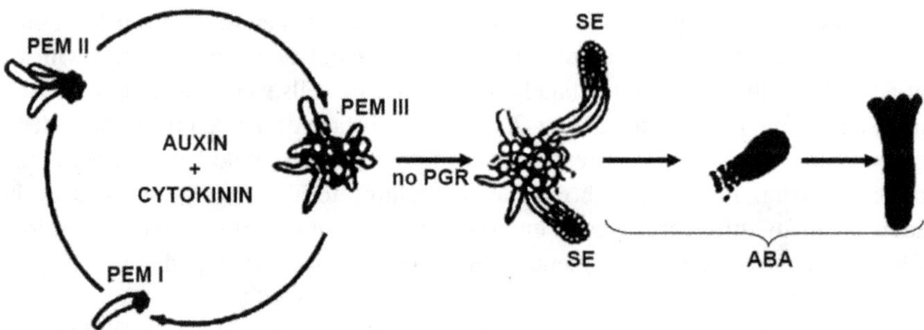

Fig. 7.41 Schematic presentation of somatic embryogenesis of gymnosperms. The graph was adapted from von Arnold et al. (2002)

Fig. 7.42 Somatic embryogenesis and development of plantlet of a carrot suspension in a bioreactor

The unit consists of two containers, one mounted on top of the other (Fig. 7.43). The plant material is placed in the upper container and the nutrient solution in the lower. By use of a pneumatic system, the plant material is bathed by the nutrient solution from the lower container for various durations, from twice for 1 min per day upward, according to the plant species, the developmental stage, and other criteria. A review of the technique is given by Etienne and Berthouly (2002). With this method, the production costs per plant can be dramatically reduced, depending on the plant species—e.g., for sugarcane, this amounts to about 50(46)%, compared to the standard procedure on semisolid medium, mainly due to the reduction in labor and space.

This system has been used to produce coffee plants in vitro using F1 hybrids. Complete embryo development was achieved in this system within 4 months. For development of the embryos, the immersion frequency was set at two times for 5 min per day. The embryos were eventually transferred to a mixture of soil (2 parts), sand (1 part), and coffee pulp (1 part), after chemical sterilization of the substrate. As can be seen in Fig. 7.43, the plants produced by this technique appear absolutely normal (Etienne-Barry et al. 1999).

Fig. 7.43 Production of coffee plants by means of somatic embryogenesis using the RITA bioreactor system (Etienne-Barry et al. 1999)

Another, similar system is the TRI bioreactor reported by Afreen et al. (2002), in which only the root zone of cotyledonary stage somatic embryos is in contact with the nutrient solution under autotrophic conditions. Here, the roots are immersed for 15 min every 6 h. The selection of suitable embryos and the transfer into the system, however, have to be performed by hand (see Fig. 7.44). The conversion into plantlets was 84%.

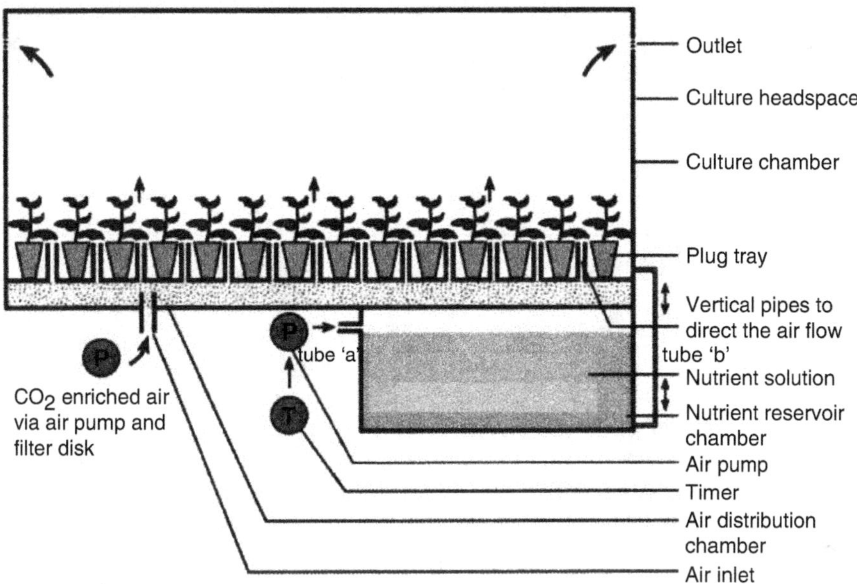

Fig. 7.44 Schematic diagram of the temporary root zone immersion (TRI) bioreactor with forced ventilation system (Afreen et al. 2002)

7.5   Artificial Seeds

Despite the successful sowing of in vitro produced somatic embryos directly into soil, as described for coffee, work is pursuing to develop methods to produce artificial seeds. Here as example, mature somatic embryos are covered by alginate, which contains some nutrients as a replacement for the endosperm, to which also some fungicide can be mixed (Redenbaugh et al. 1987). These artificial seeds are sown like conventional seeds into soil. Protocols are available for a number of plant species, like carrots, cotton, lettuce, celery, and alfalfa.

This has also been employed using cultures of explants of mature *Pinus patula* trees (Malabadi and Staden 2005). Somatic embryos derived from vegetative shoot apices were encapsulated into sodium alginate. Such synthetic seeds could still be germinated after 120 days of storage at 2 °C, and normal plants developed. More information and descriptions of methods for individual species can be found on the Internet. The aim is to produce homogenous plant material for practical applications, e.g., of outbreeders, or of F1 hybrids. A method for automatic production of such seeds is available. The costs of 100 "artificial seeds" amount to nearly 2.5 cents, which is considerably higher than the costs for conventional seeds. Artificial seeds, however, could be of interest to obtain hybrids, e.g., of cauliflower, broccoli, or *Geranium*.

Recently, the production of viable artificial seeds was reported also for *Cymbidium*, an orchid. Here, 3-month-old protocorm-like bodies (PLBs) were encapsulated in a sodium alginate solution. The survival rate of these seeds was 100%, and this was not affected after 1 year of storage in a sucrose-free liquid medium. After coating with chitosan solution, these seeds were successfully transferred to a non-sterilized substrate in the greenhouse (Nhut et al. 2005).

7.6   Embryo Rescue

For products of crossings of species of a genus, the development of embryos is often abnormal, and a technique called embryo rescue is employed to obtain viable plants. In such crossings, the unripe embryo is aborted in the F1 generation. To promote further development of embryos, these are cultured in a suitable medium in vitro. Some years ago, this method was used to raise plants of hybrids of an early ripening strain of cherries and of a cherry strain with a reduced stem height. Embryos of the hybrids were prematurely aborted, and in order to raise seedlings, those embryos were obtained 35–40 days after full bloom and cultured in vitro. The development of shoots was unproblematic; however, the development of the root system required the application of extracts of cotyledons of stratified cherry seeds. Later, these extracts could be replaced by GA3 and inositol (Abou-Zeid and Neumann 1973).

In interploid sexual hybridization of a number of citrus strains to obtain improved seedless triploid acid fruit hybrids, embryo rescue was employed to avoid embryo abortion caused by endosperm failure. More developed embryos, i.e., embryos with fully developed cotyledons, were more suitable than those in earlier developmental stages. Up to 65% developed into normal plants, with differences between the crosses (Viloria et al. 2005).

As another example, this technique was also used successfully to raise hybrids of Thompson seedless grapes and a number of other *Vitis* species. Thompson seedless grapes are highly susceptible to downy mildew (*Plasmospara viticola* Beri. and de Toni), and the aim of the investigation was to introgress resistance to this disease into this high-quality subspecies (*Euvitis*, $2n = 38$) by pollination with a resistant subspecies (*Muscadinia*, $2n = 40$). A summary is given in Fig. 7.45 (Bharathy et al. 2005). Here, also embryo rescue was employed.

In Vitro Propagation and Genetic Transformation (See Also Chap. 13):
As reported by Singh and Khawale (2008), a large number of tropical and subtropical fruit trees have been propagated in vitro. They include banana, mango, citrus, papaya, grape, pomegranate, pineapple, guava, litchi, jackfruit, bael, tamarind, jamun, passion fruit, carambola, loquat, and sapota (see also Kumar and Sopory 2008, 2010).

Kumar and Vijay (2008) developed in vitro propagation method for medicinal plant *Asparagus racemosus* through shoot bud differentiation from nodal segments. Micropropagation of ringal bamboo *Drepanostachyum falcatum* (Nees) Keng F. has been developed through nodal segments (Saini et al. 2016). Sahai et al. (2010)

Fig. 7.45 Use of an embryo rescue program to obtain hybrid plants in a *Vitis* breeding program. (**a**) Immature berries, (**b**) ovules, (**c**) germination of embryos, (**d**) early stages of embryos, (**e**) hybrid plantlet in test tube, (**f**) hybrid plants (Bharathy et al. 2005)

reported efficient and reproducible protocol for the in vitro production of *Tylophora indica* via leaf explants. Regenerated plantlets were hardened, acclimatized, and established in soil with 90% survival rate.

Javed and Anis (2015) reported an efficient in vitro micropropagation system for *Erythrina variegata*, a multipurpose tree legume. Genetic improvement of castor (*Ricinus communis* L.) through tissue culture and biotechnological tools has been achieved (Sujatha 2008). Chandler (2013) reported that global production value of ornamentals and value added products are US$250–400 billion.

Song et al. (2010) induced embryogenesis in celery, *Apium graveolens* L., (Apiaceae family) transformed with *Agrobacterium tumefaciens*. According to Pinto et al. (2013) in *Eucalyptus*, mature somatic embryos usually do not develop in the presence of auxin and plant regeneration has been usually achieved in auxin free medium or, occasionally, in media containing cytokinins and/or gibberellic acid (GA3). In *E. citriodora*, mature embryos germinated easily when transferred to an auxin-free medium (Muralidharan and Mascarenhas 1995). Potts and Dungey (2004) reported that for pulp production and increasingly for solid wood *E. grandis*, *E. urophylla* and their hybrids are the most favored in tropical and subtropical regions, while *E. globulus* is favored in temperate regions. Pinto et al. (2013) reviewed the most relevant recent advances on the SE process in *Eucalyptus*, from the somatic embryo induction to the plant acclimatization. Tiwari and Trivedi (2010) reported morphogenesis and somatic embryogenesis along with production of edible quality guava (*Psidium guajava*) fruits.

Plant Resistance to Insects and Flower Color

Another important aspect evolved by the convergence of tissue culture and transformation approaches is the development of insect pest resistant lines (Kumar and Roy 2011). As reported by Datta (2007), the introduction and commercialization of Bt varieties has reduced chemical pesticide application by 40%. Genetic modification for the introduction of resistance to insects and diseases, longevity, and novel flower ornamental crops has improved their value (Chandler and Brugliera 2011). Recently several genes which have important role in host pathogen interaction have been isolated and characterized. They included vital genes, namely, ornithine decarboxylase, argininosuccinate lyase, MAP kinases, chorismate synthase, peroxisomal biogenesis factor 6, and β-1,3-glucanosyltransferse in fungal pathogens (*Fusarium oxysporum* and *Colletotrichum gloeosporioides*), and acetylcholinesterase, chitinase, chitin synthase, ecdysone receptor, intestinal mucins, and sericotropin in insect pests (*Helicoverpa armigera* and *Plutella xylostella*) for their control in tomato/chili, and tomato/brinjal/cauliflower, respectively, through host-induced RNAi (Natarajaswamy et al. 2013; Mamta and Rajam 2016).

Though transformation of a number of agriculturally important plant species has been reported, such efforts on medicinally important plants have been very few (Gómez-Galera et al. 2007). However, *Agrobacterium tumefaciens* has been used in genetic transformation of *Artemisia annua* (Vergauwe et al. 1998; Wang et al. 2016b), *Scrophularia buergeriana* Miq. (Park et al. 2003), *Commiphora wightii* (Arnott) Bhandari (Kumar and Nadgauda 2014), *Cerasus humilis* (Wang et al. 2016a), and *Bambusa balcooa* (Ghosh et al. 2013). Currently, Taxus plant cell fermentation (PCF) is used to mass produce the antimitotic paclitaxel (Taxol) and related precursors (Mountford 2010). Pandey et al. (2010) described *Agrobacterium tumefaciens*- mediated transformation of *Withania somnifera*—an important Indian medicinal plant.

Ovary/Ovule/Embryo Culture for Crop Improvement: Brassica

The in vitro ovary culture, ovule culture, and sequential embryo culture have been the most effectively utilized techniques to overcome post-fertilization barriers. To overcome the difficulties in crossing diverse varieties, an in vitro sequential embryo rescue technique—involving ovary, ovule, and embryo culture—was established that allows for introgression of desired genes/traits across compatibility barrier (Agnihotri 1993, 2010). The half-seed technique was employed on in vitro grown seedlings/transgressive segregants obtained from in vitro cultures and selected for desired nutritional quality traits in both *B. napus* and *B. juncea* (Agnihotri and Kaushik 1998). The ovary/ovule culture has also been successfully utilized to transfer tolerance to the two most important fungal diseases, *Albugo candida* and *Alternaria brassicae*, in *B. juncea* (Gupta et al. 2006, 2010).

Standardization and Utilization of Doubled Haploids in Crop Improvement: Indian Mustard, *Brassica juncea* (L.) Czern and Coss

Agnihotri et al. (1996) studied the role various factors play in inducing embryogenic development of isolated microspores in *B. juncea*. Further work focused on enhancing the yield of haploid embryos so that a high throughput method can be developed to aid plant breeding methods. These embryos also had a high frequency for conversion into haploid and spontaneous doubled haploid plantlets (Agnihotri et al. 1996). The method established for induction of microspore embryogenesis and conversion of plantlets into flowering DH plants was then utilized in conjugation with chemical mutagenesis at three levels, namely, isolated microspores, haploid embryos, and donor plant level to generate genetic variability for various agronomic and economically superior traits (Prem et al. 2011). This also provided new source to tolerance for fungal diseases *Albugo candida* and *Alternaria brassicae* (Kaur et al. 2011; Prem et al. 2012).

7.7 Bioreactors for Micropropagation and Genetic Transformations

Millions of plants are commercially propagated annually via micropropagation which is labor-intensive and associated with developmental abnormalities. There have been many efforts to develop cost-effective and simple bioreactors with the aim of automating micropropagation, but designing reactors for plant tissue culture must reconcile environmental factors (shear stress, aeration, RH, nutrient supply) with healthy plant development, cost, and simplicity of use.

Bioreactors have been used for enhancing the embryogenic tissue proliferation. As reported by Kong et al. (2013) airlift bioreactors (ALBs) enhanced chestnut embryogenic tissue proliferation for genetic transformation and mass propagation. Compared with flasks, ALBs generated higher yields of tissue mass and larger fractions of small cell clumps (\1 mm in diameter) that were optimum target material for transformation. Bioreactor-produced tissue demonstrated high amenability to transformation with reporter genes and chestnut blight resistance candidate genes (CGs) via *Agrobacterium* co-cultivation (Kong et al. 2013).

Numerous studies have applied bioreactors in plant cell (Huang and McDonald 2012) and organ (Srivastava and Srivastava 2012) culture to obtain specific metabolites. In addition, a considerable number of researchers have cultured plant propagules in bioreactors to produce high-quality seedling (Zhao et al. 2012). It is clear from these studies that temporary immersion bioreactor culture systems are appropriate for shoot multiplication and regeneration (Watt 2012) and the continuous immersion system is suitable for the proliferation of propagules (without leaves) such as bublets (Kim et al. 2004), PLBs (Park and Facchini 2000; Yang et al. 2010) and rhizomes (Jin et al. 2007). Gao et al. (2014) reported micropropagation of *Cymbidium* sp. using continuous and temporary air lift bioreactor system.

References

Abou-Zeid A, Neumann KH (1973) Preliminary investigations on the influence of cotyledons on the development of cherry embryos (*Prunus avium* L.). Z Pflanzenphysiol 69:299–305

Afreen F, Zobayed SMA, Kozai T (2002) Photoautotrophic culture of *Coffea arabusta* somatic embryos: photosynthetic ability and growth of different stage embryos. Ann Bot 90:11–19

Agnihotri A (1993) Hybrid embryo rescue. In: Lindsey K (ed) Plant tissue culture manual; fundamentals and applications. Kluwer Academic Publishers, Dordrecht, pp E4: 1–8

Agnihotri A (2010) Synergy of biotechnological approaches with conventional breeding to improve quality of rapeseed-mustard oil and meal. Anim Nutr Feed Technol 10 S:169–177

Agnihotri A, Kaushik N (1998) Transgressive segregation and selection of zero erucic acid strains from intergeneric crosses of *Brassica*. Ind J Plant Genet Res 11(2):251–255

Agnihotri A, Swartz GS, Downey RK (1996) Microspore embryogenesis in *B. juncea*. In: Pareek LK (ed) Trends in plant tissue culture and biotechnology. Agrobotanical Publishers, Rajasthan, pp 218–221

Akitha Devi MK, Sakthivelu G, Giridhar P, Ravishankar GA (2012) Protocol for augmented shoot organogenesis in selected variety of [*Glycine max* L. (Merr.)]. Indian J Exp Biol 50 (10):729–734

Bagnoli F, Capuana M, Racchi ML (1998) Developmental changes of catalase and superoxide dismutase in zygotic and somatic embryos of horse chestnut. Aust J Plant Physiol 25:909–913

Banerjee S, Ahuja PS, Pal A (1994) Somatic embryogenesis from callus culture of *Solanum sarrachoides* Sendt. J Plant Physiol 143:750–752

Behrendt U, Zoglauer K (1996) Boron controls suspensor development in embryogenic cultures of *Larix decidua*. Physiol Plant 97(2):321–326

Bender L, Neumann KH (1978) Investigations on indole-3-acetic acid metabolism of carrot tissue cultures. Z Pflanzenphysiol 88:209–217

Bender L, Kumar A, Neumann KH (1981) Photoautotrophe pflanzliche Gewebekulturen in Laborfermentern. In: Lafferty RM (ed) Fermentation. Springer, Wien, pp 193–203

Bharathy PV, Kulkarni DD, Biradar AB, Karibasappa GS, Solanke AU, Patil SG, Agrawal DC (2005) Production of hybrid plants in Thompson seedless grapes (*Vitis vinifera*) through breeding and in ovule embryo rescue methods. In: Kumar A, Roy S (eds) Plant biotechnology and its application in tissue culture. I.K. International, New Delhi, pp 62–69

Borthwick HA (1931) Developmental of the macrogametophyte and embryo of Daucus carota. Bot Gaz 92:23–44

Chalupa W (1985) Somatic embryogenesis and plantlet regeneration from cultured immature and mature embryos of Picea abies (L.) Karst. Commun Inst For Czech Repub 14:57–63

Chandler S (2013) Genetically engineered ornamental plants: regulatory hurdles to commercialization. ISB News Rep (August):12–14

Chandler S, Brugliera F (2011) Biotechnology in floriculture. Biotechnol Lett 33:207–214

Chinachit W (1991) Die somatische Hybridisierung von Wildkarotten und Kulturkarotten (*Daucus carota* L.) und die Charakterisierung der Hybriden. Dissertation, Justus Liebig Universität, Giessen

Conge BV, Hanning GE, Gray DJ, McDaniel JK (1983) Direct embryogenesis from mesophyll cells of orchard grass. Science 212:850–851

Cui D, Myers JR, Collins GB, Lazzeri PA (1988) In vitro regeneration in *Trifolium*. 1. Direct somatic embryogenesis in *T. rubens* (L.). Plant Cell Tissue Organ Cult 15:33–45

Datta SK (2007) Impact of plant biotechnology in agriculture. In: Pua EC, Davey MR (eds) Biotechnology in agriculture and forestry, Transgenic Crops IV, vol 59. Springer, Berlin, pp 1–31

de Fossard R (2007) Nursery costs. CD January 2007, with interactive spreadsheets (Excel). Agritech Consultants, Shrub Oak, NY

De Jong AJ, Schmidt EDL, De Vries SC (1993) Early events in higher-plant embryogenesis. Plant Mol Biol 22(2):367–377

De Klerk GJ, Arnholdt-Schmitt B, Lieberei R, Neumann KH (1997) Regeneration of roots, shoots and embryos: physiological, biochemical and molecular aspects. Biol Plant 39:53–66

Dong JZ, Dunstan DI (1996) Expression of abundant mRNAs during somatic embryogenesis of white spruce [*Picea glauca* (Moench) Voss]. Planta 199:459–466

Dührssen E, Neumann KH (1980) Characterisation of satellite-DNA of *Daucus carota* L. Z Pflanzenphysiol 100:447–454

Dührssen E, Lanzendörfer M, Neumann KH (1984) Comparative investigations on DNA organization of some varieties of *Daucus carota* L. Z Pflanzenphysiol 113:223–229

Durzan DJ (1980) Progress and promise in forest genetics. In: Proceedings of the 50th anniversary conference. Paper science and technology, the cutting edge, May 8–10, 1979. The Institute of Paper Chemistry, Appleton, WI, pp 31–60

Durzan DJ, Chalupa V (1976) Growth and metabolism of cells and tissue of jack pine (Pinus banksiana). 3. Growth of cells in liquid suspension cultures in light and darkness. Can J Bot 54:456–467

Egertsdotter U, von Arnold S (1998) Development of somatic embryos in Norway spruce. J Exp Bot 49:155–162

Engelmann F (1997) In vitro conservation methods. In: Ford-Lloyd BV, Newburry HJ, Callow JA (eds) Biotechnology and plant genetic resources. Conservation and use. CABI, Wallingford, pp 119–162

Etienne H, Berthouly M (2002) Temporary immersion systems in plant micro propagation. Plant Cell Tissue Organ Cult 69:215–231

Etienne-Barry D, Bertrand B, Vasquez N, Etienne H (1999) Direct sowing of *Coffea arabica* somatic embryos mass-produced in a bioreactor and regeneration of plants. Plant Cell Rep 19:111–117

Flores Berrios E, Sarrafi A, Fabre F, Alibert G (2000) Genotypic variation and chromosomal location of QTLs for somatic embryogenesis revealed by epidermal layers culture of recombinant inbred lines in the sunflower (*Helianthus annuus* L). Theor Appl Genet 101:1307–1312

Gao R, Wu S, Piao X et al (2014) Micropropagation of *Cymbidium sinense* using continuous and temporary airlift bioreactor systems. Acta Physiol Plant 36:117–124. https://doi.org/10.1007/s11738-013-1392-9

Gartenbach-Scharrer U, Habib S, Neumann KH (1990) Sequential synthesis of some proteins in cultured carrot explant (*Daucus carota* L.) cells during callus induction. Plant Cell Tissue Organ Cult 22:27–35

Geier T (1986) *Anthurium scherzerianum* und Gewebekultur. Dtsch Gartenbau 43:2030–2033

Geisler C (2001) Insertion von rol-Genen in zellzyklussynchronisierte Karottenzellkulturen. Dissertation, Justus Liebig Universität, Giessen

Ghosh JS, Chaudhuri S, Dey N, Pal A (2013) Functional characterization of a serine-threonine protein kinase from *Bambusa balcooa* that implicates in cellulose overproduction and superior quality fiber formation. BMC Plant Biol 13:128. https://doi.org/10.1186/1471-2229-13-128

Giridhar P, Sowmya KS, Ramakrishna A, Ravishankar GA (2010) Rapid clonal propagation and stevioside profiles of *Stevia rebaudiana* Bertoni. Int J Plant Dev Biol 4(1):47–52

Gómez-Galera S, Pelacho AM, Gené A et al (2007) The genetic manipulation of medicinal and aromatic plants. Plant Cell Rep 26:1689–1715. https://doi.org/10.1007/s00299-007-0384-x

Grieb B (1991/1992) Untersuchungen zur Induktion der Kompetenz zur somatischen Embryogenese in Karottenpetiolenexplantaten (*Daucus carota* L.)—Histologie und Proteinsynthesemuster. Dissertation, Justus Liebig Universität, Giessen

Grieb B, Groß U, Pleschka E, Arnholdt-Schmitt B, Neumann KH (1994) Embryogenesis of photoautotrophic cell cultures of *Daucus carota* L. Plant Cell Tissue Organ Cult 38:115–122

Grieb B, Schäfer F, Imani J, Nezamabadi Mashayekhi K, Arnholdt-Schmitt B, Neumann KH (1997) Changes in soluble proteins and phytohormone concentration of cultured carrot petiole explants during induction of somatic embryogenesis (*Daucus carota* L.). Angew Bot 71:94–103

Griesebach RJ (1986) Orchid tissue culture. In: Zimmermann RJ (ed) Tissue culture as a plant production system for horticultural crops. Martinus Nijhoff, Dordrecht, pp 343–349

Griesebach RJ (2002) Development of *Phalaenopsis* orchids for the mass-market. In: Janick J, Whipkey A (eds) Trends in new crops and new uses. ASHS Press, Alexandria, VA, pp 458–465

Gupta K, Prem D, Nashaat N, Agnihotri A (2006) Response of interspecific *Brassica juncea* x *Brassica rapa* hybrids and their advanced progenies to *Albugo candida* (Pers.) Kunze. Plant Pathol 55:679–668

Gupta K, Prem D, Agnihotri A (2010) Pyramiding white rust and Alternaria blight resistance in low erucic Brassica juncea [(L.) Czern & Coss] using Brassica carinata. J Oilseed Brassica 1 (2):55–65

Gururaj HB, Padma MN, Giridhar P, Ravishankar GA (2012) Functional validation of *Capsicum frutescens* aminotransferase gene involved in vanillylamine biosynthesis using *Agrobacterium* mediated genetic transformation studies in *Nicotiana tabacum* and *Capsicum frutescens* calli cultures. Plant Sci 195:96–105

Harsha PSC, Khan MI, Giridhar P, Ravishankar GA (2012) *In vitro* propagation of *Rivina humilis* L. through proliferation of axillary shoots and shoot tips of mature plants. Ind J Biotechnol 11:481–485

Hartmann HT, Kester DE, Davies FT Jr, Geneve RL (2002) Plant propagation: principles and practices, 7th edn. Prentice Hall, Englewood Cliffs, NJ

Hays DB, Wile RW, Sheen WC, Pharis RP (1999) Embryo-specific gene expression in microspore-derived embryos of *Brassica napus*. An interaction between abscisic acid and jasmonic acid. Plant Physiol 119:1065–1072

Hecht V, Vielle-Calcada JP, Hartog M, Schmidt EDL, Boutilier K, Grossniklaus U, de Vries SC (2001) The *Arabidopsis* somatic embryogenesis receptor kinase gene is expressed in developing ovules and embryos and enhances embryogenic competence in culture. Plant Physiol 127:803–816

Huang TK, McDonald KA (2012) Bioreactor systems for in vitro production of foreign proteins using plant cell cultures. Biotechnol Adv 30:398–409

Husain MK, Anis M, Shahzad A (2010) Somatic embryogenesis and plant regeneration in *Pterocarpus marsupium* Roxb. Trees 24:781–787

Igasaki T, Akashi N, Ujino-Ihara T, Matsubayashi Y, Sakagami Y, Shinohara K (2003) Phytosulfokine stimulates somatic embryogenesis in *Cryptomeria japonica*. Plant Cell Physiol 44:1412–1416

Ikeda-Iwai M, Umehara M, Satoh S, Kamada H (2003) Stress-induced somatic embryogenesis in vegetative tissue of *Arabidopsis thaliana*. Plant J 34:107–114

Imani J (1999) In situ-Nachweis der Auxinvertcilung in kultivierten Petiolen explantaten von transgenen Pflanzen wahrend der Induktion der somatischen Embryogenese bei *Daucus carota* L. PhD Thesis, Justus Liebig University, Giessen

Imani J, Nezamabadi Mashayekhi K, Neumann KH (1999) Gentechnische Verfahren als Werkzeuge in der pflanzlichen Hormonforschung. Justus Liebig University, Giessen

Imani J, Tran Thi L, Langen G, Arnholdt-Schmitt B, Roy S, Lein C, Kumar A, Neumann KH (2001) Somatic embryogenesis and DNA organization of genomes from selected *Daucus* species. Plant Cell Rep 20:537–541

Imin N, Nizamidin M, Daniher D, Nolan KE, Rose RJ, Rolfe BG (2005) Proteomic analysis of somatic embryogenesis in *Medicago truncatula*. Explant cultures under 6-benzylaminopurin and 1-naphthaleneacetic acid treatments. Plant Physiol 137:1250–1260

Javed SB, Anis M (2015) Cobalt induced augmentation of *in vitro* morphogenic potential in *Erythrina variegata* L.: a multipurpose tree legume. Plant Cell Tissue Organ Cult 120:463–474

Jin H, Piao XC, Sun D, Xiu JR, Lian ML (2007) Mass production of rhizome and shoot of Cymbidium niveo-marginatum using simple bioreactor. J Northeast Forest Univ 35:44–48

Kaur P, Jost R, Sivasithamparam K, Barbetti MJ (2011) Proteome analysis of the Albugo candida-Brassica juncea pathosystem reveals that the timing of the expression of defence-related genes is a crucial determinant of pathogenesis. J Exp Bot 62(3):1285–1298. https://doi.org/10.1093/jxb/erq365

Kim EK, Hahn EJ, Murthy HN, Paek KY (2004) Enhanced shoot and bulblet proliferation of garlic (Album sativum L.) in bioreactor systems. J Hortic Sci Biotechnol 79:818–822

Knudson L (1946) A new nutrient solution for the germination of orchid seed. Am Orchid Soc Bull 15:214–217

Koller A, Washburn MP, Lange BM, Andon NL, Deciu C, Haynes PA, Hays L, Schieltz D, Ulaszek R, Wei J, Wolters D, Yates JR III (2002) Proteomic survey of metabolic pathways in rice. Proc Natl Acad Sci U S A 99:11969–11974

Kong L, Holtz CT, Nairn CJ, Houke H, Powell W a, Baier K, Merkle S a (2013) Application of airlift bioreactors to accelerate genetic transformation in American chestnut. Plant Cell Tissue Organ Cult 117(1):39–50

Kormutak A, Salaj T, Matusova R, Vookova B (2003) Biochemistry of zygotic and somatic embryogenesis in silver fir (*Abies alba* Mill.). Acta Biol Cracov Ser Bot 45:159–162

Kreuger M, van Holst GJ (1996) Arabinogalactan proteins and plant differentiation. Plant Mol Biol 30(6):1077–1086

Kumar S, Nadgauda R (2014). Control of morphological aberrations in somatic embryogenesis of *Commiphora wightii* (Arnott) bhandari (Family: Burseraceaea) through secondary somatic embryogenesis. Proc Natl Acad Sci India Sect B Biol Sci:1–10. https://doi.org/10.1007/s40011-014-0347-2

Kumar A, Roy S (2011) Plant tissue culture and applied plant biotechnology. Avishkar Publishers, Jaipur, 346 pp

Kumar A, Sopory S (eds) (2008) Recent advances in plant biotechnology and its applications. I.K. International, New Delhi

Kumar A, Sopory S (eds) (2010) Applications of plant biotechnology: *in vitro* propagation, plant transformation and secondary metabolite production. New Delhi, I.K. International, 606 pp

Kumar A, Vijay N (2008) *In vitro* plantlet regeneration in *Asparagus racemosus* through shoot bud differentiation on nodal segments. In: Kumar A, Spory S (eds) Recent advances in plant biotechnology. I.K. International, New Delhi, pp 185–197

Le TT (2001) Untersuchungen zur Bedeutung der Abscisnsäure (ABA) für die somatische Embryogenese verschiedener Daucusarten unter Einbeziehung transgener Stämme. Dissertation, Justus Liebig Universität, Giessen

Li T, Neumann KH (1985) Embryogenesis and endogenous hormone content of cell cultures of some carrot varieties (*Daucus carota* L.). Ber Dtsch Bot Ges 98:227–235

Linser H, Neumann K-H (1968) Uentersuchungen ueber Beziehungen zwischen Zellteilung und Morphogenese bei Gewekulturen von Daucus carota L. Physiol Plant 21:487–499

Lopez-Escamilla AL, Olguin-Santos LP, Marquez J, Chavez VM, Bye R (2000) Adventitious bud formation for mature embryos of Picea chihuahuana Martinez, an endangered Mexican spruce tree. Ann Bot 86:921–927

Luckner M, Diettrich B (1985) Formation of cardenolides in cell and organ cultures of *Digitalis lanata*. In: Neumann KH, Barz W, Reinhard W (eds) Primary and secondary metabolism of plant cell cultures. Springer, Berlin, pp 154–163

Luckner M, Diettrich B (1987) Biosynthesis of cardenolides in cell cultures of *Digitalis lanata*—the results of a new strategy. In: Green CE, Somers DA, Hackett WP, Biesboer DD (eds) Plant tissue and cell culture. A.R. Liss, New York, pp 187–197

Maheswari P (1979) An introduction to embryology of angiosperms. McGraw Hill, New Delhi

Majewska-Sawka A, Nothnagel EA (2000) The multiple roles of arabinogalactan proteins in plant development. Plant Physiol 122:3–9

Malabadi R, Staden J (2005) Storability and germination of sodium alginate encapsulated somatic embryos derived from the vegetative shoot apices of mature *Pinus patula* trees. Plant Cell Tissue Organ Cult 82:259–265

Malinowski R, Filipecki M, Tagashira N, Wiśniewska A, Gaj P, Plader W, Malepszy S (2004) Xyloglucan endotransglucosylase/hydrolase genes in cucumber (*Cucumis sativus*)—differential expression during somatic embryogenesis. Physiol Plant 120:678–685

Mamta RKRK, Rajam MV (2016) Targeting chitinase gene of Helicoverpa armigera by host-induced RNA interference confers insect resistance in tobacco and tomato. Plant Mol Biol 90:281–292. https://doi.org/10.1007/s11103-015-0414-y. Impact Factor: 4.257

Mashayekhi K, Neumann KH (2006) Effects of boron on somatic embryogenesis of *Daucus carota*. Plant Cell Tissue Organ Cult 84:279–283

Mashayekhi-Nezamabadi K (2001) Untersuchungen zum Einfluss von Bor auf die somatische Embryogenese bei *Daucus carota* L. Dissertation, Justus Liebig University, Giessen

Merkle S (1994) In vitro-Vermehrung: heute und morgen. Gärtnerbörse 6:282–286

Michalczuk L, Cooke TJ, Cohen JD (1992a) Auxin levels at different stages of carrot somatic embryogenesis. Phytochemistry 31:1097–1103

Michalczuk L, Ribnick DM, Cooke TJ, Cohen JD (1992b) Regulation of indole-3-acetic acid biosynthetic pathways in carrot cell cultures. Plant Physiol 100:1346–1353

Morel G (1974) Clonal multiplication of orchids. In: Withers E (ed) The orchids scientific studies. J. Wiley and Sons, New York, pp 169–222

Mountford PG (2010) The Taxol story—development of a green synthesis via plant cell fermentation. In: Dunn PJ, Wells AS, Williams MT (eds) Green chemistry in the pharmaceutical industry. Weinheim, Wiley-VCH Verlag GmbH & Co. KGaA, pp 145–160

Muralidharan EM, Mascarenhas AF (1995) Somatic embryogenesis in Eucalyptus. In: Jain SM, Gupta PK, Newton RJ (eds) Somatic embryogenesis in woody plants, Vol. 2. Angiosperms. Kluwer Academic Publishers, Dordrecht, pp 23–40

Natarajaswamy K, Naorem A, Rajam MV (2013) Targeting fungal genes by diced siRNAs: A rapid tool to decipher gene function in *Aspergillus nidulans*. PLoS One 8(10):e75443. Impact Factor: 3.534

Nawrot-Chorabik K (2016) Plantlet regeneration through somatic embryogenesis in Nordmann's fir (Abies nordmanniana). J For Res. https://doi.org/10.1007/s11676-016-0265-7

Neumann K-H (1962) Uentersuchungen uber der Einfluss essentialler Schwermetalle auf das Wachstum und den Proteinstoffwechsel von Karottengewebekulturen. Dissertation. Justus Liebig Universitaet Giessen

Neumann KH (1966) Wurzelbildung und Nukleinsäuregehalt bei Phloem-Gewebekulturen der Karottenwurzel auf synthetischem Nährmedium. Congr Coll Univ Liege 38:96–102

Neumann KH (1995) Pflanzliche Zell- und Gewebekulturen. Eugen Ulmer, Stuttgart

Neumann KH, Grieb B (1992) Somatische Embryogenese bei höheren Pflanzen: Grundlagen und praktische Anwendung. Wiss Z Humbold-Universität Berlin Math/Naturwiss 41:63–80

Neumann KH, Raafat A (1973) Further studies on the photosynthesis of carrot tissue cultures. Plant Physiol 51:685–690

Nhut DT, Tien TNT, Huong MTN, Hien NTT, Huyen PX, Luan VQ, da Teixeira S (2005) Artificial seeds for propagation and preservation of *Cymbidium* spp. Propag Ornam Plant 5:67–73

Nogler GA (1984) Gametophytic apomixis. In: Johri BM (ed) Embryology of angiosperms. Springer, Berlin, pp 475–518

Nolan EK, Irwanto RR, Rose RJ (2003) Auxin up-regulates MtSERK1 expression in both *Medicago truncatula* root forming and embryogenic cultures. Plant Physiol 1033:218–230

Pabo CO, Peisach E, Grant RA (2001) Design and selection of novel Cys2His2 zinc finger proteins. Annu Rev Biochem 70:313–340

Pandey V et al (2010) Agrobacterium tumefaciens-mediated transformation of Withania somnifera (L.) Dunal: an important medicinal plant. Plant Cell Rep 29(2):133–141

Parimalan R, Akshatha V, Giridhar P, Ravishankar GA (2011) Somatic embryogenesis and *Agrobacterium*-mediated transformation in (*Bixa orellana* L.). Plant Cell Tissue Organ Cult 105:317–328. https://doi.org/10.1007/s11240-010-9870-x

Park SU, Facchini PJ (2000) *Agrobacterium rhizogenes*-mediated transformation of opium poppy, *Papaver somniferum* L., and California poppy, *Eschscholzia californica* Cham., root cultures. J Exp Bot 347:1005–1016

Park S, Chae Y, Facchini PJ (2003) Genetic transformation of the figwort, Scrophularia buergeriana Miq., an oriental medicinal plant. Plant Cell Rep 21:1194–1198. https://doi.org/10.1007/s00299-003-0639-0

Pasternak TP, Prinsen E, Ayaydin F, Miscolczi P, Potters G, Asard H, Van Onckelen HA, Dudits D, Feher A (2002) The role of auxin, pH and stress in the activation of embryogenic cell division in leaf protoplast-derived cells of alfalfa. Plant Physiol 129:1807–1819

Pinto G, Araújo C, Santos C, Neves L (2013) Plant regeneration by somatic embryogenesis in Eucalyptus spp.: current status and future perspectives. South Forests: J For Sci 75(2):59–69. https://doi.org/10.2989/20702620.2013.785115

Pleschka E (1995) Untersuchungen zum Einfluß von Saccharose und einigen weiteren Kohlenhydraten auf die somatische Embryogenese von mixotrophen und photoautotrophen Gewebekulturen von *Daucus carota* L. Dissertation, Justus Liebig Universität, Giessen

Pleschka E, Mashayekhi-Nesambadi K, Imani J, Lein C, Neumann KH (2001) The significance of carbohydrates in somatic embryogenesis of carrot tissue cultures. J Appl Bot 75:188–194

Potts BM, Dungey HS (2004) Interspecific hybridization of eucalypts: key issues for breeders and geneticists. New For 27:115–138

Prem D, Gupta K, Agnihotri A (2011) Can we predict mutagen induced damage in plant systems mathematically? Insights from zygotic embryo and haploid mutagenesis in Indian mustard (B. juncea). Bot Serb 35(2):139–145

Prem D, Gupta K, Agnihotri A (2012) Harnessing haploid mutations using mutant donor plants for microspore culture in Indian mustard [*Brassica juncea* (L.) Czern and Coss]. Euphytica 184 (2):207–222

Pullman GS, Montello P, Cairnay J, Xu N, Feng X (2003) Loblolly pine (*Pinus taeda* L.) somatic embryogenesis: maturation improvements by metal analysis of cygotic and somatic embryos. Plant Sci 164:955–969

Ramakrishna A, Giridhar P, Jobin M, Paulose CS, Ravishankar GA (2012) Indoleamines and calcium enhance somatic embryogenesis in *Coffea canephora* P ex Fr. Plant Cell Tissue Organ Cult 108:267–278

Redenbaugh K, Slade D, Viss P, Fujii J (1987) Encapsulation of somatic embryos in synthetic seed coats. Hort Sci 22:803–809

Reinert J (1958) Morphogenese und ihre Kontrolle an Gewebekul- turen aus Carotten. Naturwissenschaft 45:344–345

Reinert J (1959) Über die Kontrolle der Embryogenese und die Induktion von Adventivembryonen an Gewebekulturen aus Karotten. Planta 53:318–333

Reinert J, Yeoman MM (1982) Plant cell and tissue culture, a laboratory manual. Springer Verlag, Heidelberg, p 81

Reuther G (1990) Current status and future prospects of large scale micropropagation in commercial plant production. Food Biotechnol 4(1):445–459. https://doi.org/10.1080/08905439009549757

Rowe WJ (1986) New technologies in plant tissue culture. In: Zimmermann RH (ed) Tissue culture as a plant production system for horticultural crops. Martin Nijhoff, Dordrecht, pp 35–52

Sahai A, Shahzad A, Anis M (2010) High frequency plant production via shoot and somatic embryogenesis from callus in *Tylophora indica*, an endangered plant species. Turk J Bot 34:11–20

Saini RK, Shetty N, Giridhar P, Gokare R (2012) Rapid in vitro regeneration method for Moringa oleifera and performance evaluation of field grown nutritionally enriched tissue cultured plants. 3 Biotech 2:187–192

Saini H, Arya ID, Arya S, Sharma R (2016) *In vitro* micropropagation of Himalayan weeping bamboo, *Drepanostachyum falcatum*. Am J Plant Sci 7:1317–1324

Schäfer F, Groskurt E, Neumann KH (1985) Organogenesis and embryogenesis in cultured petioles of carrots (*Daucus carota* L). In: Terzi M, Pitto L, Sung ZR (eds) Proc worksh somatic embryogenesis. San Miniato, Italy, pp 159–171

Schäfer F, Grieb B, Neumann KH (1988) Morphogenetic and histological events during somatic embryogenesis in intact carrot plantlets (*Daucus carota* L.) in various nutrient media. Bot Acta 101:362–365

Schmidt ED, Guzzo F, Toonen MA, deVries SC (1997) A leucine-rich repeat containing receptor-like kinase marks somatic plant cells competent to form embryos. Development 124:2049–2062

Singh SK, Khawale RK (2008) In vitro propagation of tropical and sub-tropical fruit crops. In: Kumar A, Spory S (eds) Recent advances in plant biotechnology. I.K. International, New Delhi, pp 157–184

Skoog F, Miller CO (1957) Chemical regulation of growth and organ formation in plant tissues cultured in vitro. Symp Soc Exp Biol 11:118–140

Song G-Q, Sink KC, Hancock JF (2010) Flux in morphology and GUS expression in somatic embryogenic-derived populations of transgenic celery. In: Kumar A, Sopory S (eds) Applications of plant biotechnology: *in vitro* propagation, plant transformation and secondary metabolite production. I.K. International, New Delhi, 606 pp

Sridevi V, Giridhar P (2014) *In vitro* shoot growth, direct organogenesis and somatic embryogenesis promoted by silver nitrate in *Coffea dewevrei*. J Plant Biochem Biotechnol 23(1):112–118

Srivastava S, Srivastava AK (2012) In vitro azadirachtin production by hairy root cultivation of Azadirachta indica in nutrient mist bioreactor. Appl Biochem Biotechnol 166:365–378

Steward F, Mapes M, Mears K (1958) Growth and organized development of cultured cells. II. Organization in cultures grown from freely suspended cells. Am J Bot 45:705–708

Steward FC, Mapes MO, Kent AE, Holsten RD (1964) Growth and development of cultured plant cells. Science 143:20–27

Stiebeling B, Neumann KH (1987) Identification and concentration of endogenous cytokinins in carrots (*Daucus carota* L.) as influenced by development and a circadian rhythm. J Plant Physiol 127:111–121

Street HE, Withers LH (1974) The anatomy of embryogenesis in culture. In: Street HE (ed) Tissue culture and plant science. Academic Press, London, pp 71–100

Sujatha M (2008) Genetic improvement of Castor (*Ricinus communis* L.) through tissue culture and biotechnological tools. In: Kumar A, Spory S (eds) Recent advances in plant biotechnology. I.K. International, New Delhi, pp 185–197

Tiwari RK, Trivedi M (2010) *In vitro* multiplication in Guava (*Psidium guajava*). In: Kumar A, Sopory S (eds) Applications of plant biotechnology: *in vitro* propagation, plant transformation and secondary metabolite production. I.K. International, New Delhi, 106–114 pp

van Hengel AJ, Tadesse Z, Immerzeel P, Scols H, van Kammen A, de Vries SC (2001) N-Acetylglucosamin and glucosamin-containing arabinogalactan proteins control somatic embryogenesis. Plant Physiol 125:1880–1890

Vergauwe A, Van Geldre E, Inzé D, Van Montagu M, Van den Eeckhout E (1998) Factors influencing Agrobacterium tumefaciens-mediated transformation of Artemisia annua L. Plant Cell Rep 18:105–110

Viloria Z, Grosser J, Bracho B (2005) Immature embryo rescue and seedling development of acid citrus fruit derived from interploid hybridization. Plant Cell Tissue Organ Cult 82:159–167

von Arnold S, Sabala I, Bozhkov P, Dyachok J, Filonova L (2002) Developmental pathways of somatic embryogenesis. Plant Cell Tissue Organ Cult 69:233–249

Walker KA, Sluis CJ (1983) In vitro plant production-costs and methodologies. In: Proc Biotechnol Conf, Batelle Columbus Laboratories

Wang R, Guegler K, LaBrie ST, Crawford NM (2000) Genomic analysis of a nutrient response in Arabidopsis reveals divers expression patterns and novel metabolic and potential regulatory genes induced by nitrate. Plant Cell 12:1491–1510

Wang MB, Abbott DC, Upadhyaya NM, Jacobsen JV, Waterhouse PM (2001) *Agrobacterium tumefaciens*-mediated transformation of an elite Australian barley cultivar with virus resistance and reporter genes. Aust J Plant Physiol 28:149–156

Wang RF, Huang FL, Zhang J, Zhang QY, Sun LN, Song XS (2016a) Establishment of a high-frequency regeneration system in *Cerasus humilis*, an important economic shrub. J For Res. https://doi.org/10.1007/s10310-016-0535-4

Wang J, Nie J, Pattanaik S, Yuan L (2016b) Efficient Agrobacterium-mediated transformation of Artemisia annua L. using young inflorescence. In Vitro Cell Dev Biol-Plant 52:204–211. https://doi.org/10.1007/s11627-015-9744-3

Watt MP (2012) The status of temporary immersion system (TIS) technology for plant micropropagation. Afr J Biotechnol 11:4025–14035

Widholm JM (1992) Properties and uses of photoautotrophic plant cell cultures. Int Rev Cytol 132:109–175

Winton LL (1970) Shoot and tree production from aspen tissue cultures. Am J Bot 57:904–909

Yang JF, Piao XC, Sun D, Lian ML (2010) Production of protocorm- like bodies with bioreactor and regeneration in vitro of Oncidium 'Sugar Sweet'. Sci Hortic 125:712–717

Zhao Y, Sun W, Wang Y, Saxena PK, Liu CZ (2012) Improved mass multiplication of Rhodiola crenulata shoots using temporary immersion bioreactor with forced ventilation. Appl Biochem Biotechnol 166:1480–1490

Some Endogenous and Exogenous Factors in Cell Culture Systems

The performance of defined processes of differentiation forms the basis to use cell and tissue cultures for propagation and the production of valuable compounds on a commercial scale. To ensure reliability in both these domains, a thorough understanding of the procedure is a prerequisite. The core of this is an understanding of cellular growth and differentiation, and based on this, to develop ways and means to exert influences on productivity. In commercial production, the systems should work reliably and reproducibly every day. As long as more knowledge on differentiation is not available, our only option is empirical assessments based on trial and error. Indeed, from the newly emerged fields of genomics, proteomics, and metabolics, to date only very limited contributions have been made to achieve a better understanding of growth and differentiation. Still, these new approaches are in their infancy.

The many parameters exerting an influence on growth, development, and the biochemical performance of cells can be tentatively grouped in terms of "endogenous and exogenous factors." Genetic influences the developmental status of the "mother plant" used to obtain primary explants (or the state of the subculture used as origin) and also the developmental status or age of the plant organ from which primary explants obtained are all endogenous factors. Nutrition, the hormonal supplement, and physical environmental factors like light, temperature, or humidity of the ambient air are grouped as exogenous factors. Certainly, the list, especially of endogenous factors, is not complete yet. An all-encompassing discussion of these factors also will not be attempted here within the limited space available in this volume and in view of the tremendous wealth of literature available on the Internet. Still, some examples, mainly from our own research program, will be given to indicate tendencies in the significance of such factors.

© Springer Nature Switzerland AG 2020
K.-H. Neumann et al., *Plant Cell and Tissue Culture – A Tool in Biotechnology*,
https://doi.org/10.1007/978-3-030-49098-0_8

8.1 Endogenous Factors

8.1.1 Genetic Influences

In callus growth performance, it is difficult to distinguish between genetic influences and those stemming from the status of the organ serving as origin of the explants. Nevertheless, clear genetic influences can usually be observed by comparing the growth performance of explants from a given organ in a given developmental status in different varieties of a given species. One example of such strong influences is the ability to perform somatic embryogenesis in *Daucus* (Table 8.1) as has also recently been described for, e.g., *Medicago truncatula* following proteomic analysis of recalcitrant and readily embryogenic lines (Imin et al. 2005, see above).

In Table 8.1 some examples on the differentiation of cultured root explants from three carrot varieties cultivated under identical conditions are given. The explants of one variety produced only callus, those of the two others differentiated roots, and one variety could additionally be induced to somatic embryogenesis.

Differences can also be observed in pith explants of *Datura* plants from two different species, derived by androgenesis using anthers of a given flower of each. As a result of meiosis, these strains would differ in their genetics, and due to this, variations in growth and in the compactness of the developing callus material can indeed be observed (Table 8.2).

For a more detailed discussion of the topic, see Chap. 13.

8.1.2 Physiological Status of "Mother Tissue"

Often, clear relations of the physiological status of the original tissue and the mode of reactions of explants taken thereof can be observed. This could be shown for callus growth of explants obtained from different parts of the stele of tobacco (Table 8.3). Best growth was obtained in explants from the upper third of the tobacco plant, which would represent the physiologically youngest part. A position effect can also be observed for differentiation (Fig. 8.1). With increasing distance to the apex, the ability of the explants to produce flower buds is reduced (Van Tran Than 1973).

Table 8.1 Growth (number of cells $\times$ 10^3/explant) and development (rhizogenesis, somatic embryogenesis) in NL medium (see Table 3.3) of cultured root explants (cambium) of some carrot varieties

Variety	No horm.		IAA+inositol	Rhiz.	S. e.	IAA+inositol+kinetin
	t0					
Frühbund	73.0	270.1	320.6	–	–	1394.9
Zino	43.2	217.0	257.1	+	–	913.1
Rotin	41.6	124.8	109.2	+	+	1157.1

rhiz. adventitious roots, *s. e.* somatic embryogenesis

Table 8.2 Fresh weight (mg/explant) of stem sections of some strains of haploid plantlets of *Datura innoxia* and *Datura meteloides* (six strains each) cultured in NL+IAA+inositol+kinetin (3 weeks of culture; Kibler 1978)

Strain	Basis	Middle	Upper third	Growth characteristics of callus
Datura innoxia				
i1	115	162	97	Compact callus
i2	137	130	136	Friable callus
i3	203	196	168	Compact callus
i4	115	77	–	Friable callus
i5	120	92	98	Friable callus
i6	195	–	–	Sec. callus formation
Datura meteloides				
m1	67	72	54	Compact callus
m2	65	38	38	Sec. callus formation
m3	66	57	45	Compact callus
m4	–	68	54	Sec. callus formation
m5	34	37	49	Friable callus
m6	80	92	92	Compact callus

Table 8.3 Fresh weight (mg/explant) of some sections of the shoot of haploid plants (eight to ten leaves) of *Nicotiana tabacum* var. Xanthi

Growth regulator applied	Basis	Middle	Upper third
0	11.3	9.3	37.3
Inositol+IAA	23.2	16.8	76.2
Inositol+IAA+kinetin	33.6	39.1	123.6

2 weeks of culture in NL medium

Variation was also observed in explants of various tissues from the same organ of carrot plants (Tables 8.4 and 8.5). Using identical conditions of culture, somatic embryogenesis was observed 2 weeks after prior rhizogenesis in cambium explants, in explants of the secondary phloem after 4–6 weeks, and in explants of the xylem area after 10–12 weeks. Also explants of different organs of the same plant vary in growth, as shown in Table 8.6 for young poppy plants. These variations are certainly related to the number of meristematic cells and of parenchyma cells of an explant inducible to cell division, at least as far as callus growth is concerned.

Particularly the differentiation of the explants and also "plain" growth of cultured explants are strongly influenced by the phytohormone supplement to the nutrient medium (see callus cultures in Chaps. 3 and 11). Therefore, some relation of the endogenous hormonal status to the reaction of explants in culture would be expected. Evidence of this is rather scarce, and some examples will be discussed later.

The physiological status of cell suspensions is also important for the growth performance of subcultures derived thereof. In Fig. 8.2, the influence of duration of pre-culture (before setting up subcultures) on the growth of haploid and diploid

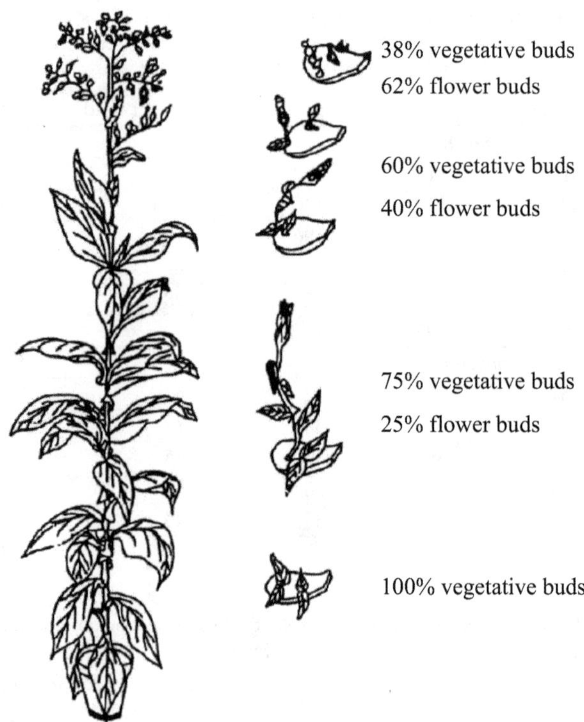

Fig. 8.1 Influences of the region of tissue used for explantation on the development of flower buds in culture (*Nicotiana tabacum*; Van Tran Than 1973)

callus cultures is presented. Particularly for the haploid cultures, two clear maxima can be observed.

8.1.3   Growth Conditions of the "Mother Plant"

The reaction of explants often correlates with the growth conditions of the mother plant used to obtain explants for culture. In our laboratory, rhizogenesis in an IAA-containing, cytokinin-free nutrient medium (as described in Chap. 3) could be induced only if the mother plants grew for several weeks under short-day conditions (Fig. 8.3). This unexpected result was repeated in 3 successive years. In temperate climatic zones, the sowing of carrots is usually done at the end of February or in March, and therefore during early development under natural conditions, the carrots obtained for investigation pass through several weeks of short-day conditions. This agrees with the formation of adventitious roots in the NL medium. Unfortunately, systematic investigations of influences of growth conditions of the "mother plant" on the reaction of explants in culture are hardly available.

This example shows how important the physiological status of cells of explants for reaction in culture can be. Neglecting this may cause problems in repeating experiments. Seemingly, the various tissues used for explantation, with their individual molecular and biochemical architecture, vary in their competence to receive

Table 8.4 Influence of kinetin on the fresh weight, number of cells per explant, root formation, and somatic embryogenesis of cultured explants of various root tissues of *Daucus carota*

	mg F. wt./explant		Cells $\times 10^3$/explant	
Tissue	Exp. I	Exp. II	Exp. I	Exp. II
Secondary phloem (4–6)[a]				
T0	2.0	2.0	8.9	9.1
NL	13.0	12.0	37.8	20.5
NL+I+IAA	26.4 R	62.0 R	80.0	156.6
NL+I+IAA+K	106.0	236.0	488.4	792.0
Cortex				
T0	2.0	2.0	9.4	8.8
NL	10.0	7.0	16.7	23.1
NL+I+IAA	24.0 R	35.0 R	66.2	113.1
NL+I+IAA+K	111.0	213.0	440.0	1183.2
Xylem (10–12)[a]				
T0	2.0	2.0	6.3	7.3
NL	10.0	37.0	36.2	76.7
NL+I+IAA	11.0	38.0	54.5	168.0
NL+I+IAA+K	142.0	201.0	399.6	2450.0
Cambium (2–3)[a]				
T0	2.0	2.0	11.1	9.8
NL	16.0	16.0	30.0	30.9
NL+I+IAA	23.0 R	22.0 R	59.5	42.6
NL+I+IAA+K	95.0	214.0	279.0	1484.0

I, II denote tissue of two carrot roots; NL medium, 21 days of culture
R adventitious roots; *I* 50.0 ppm m-inositol; *IAA* 2.0 ppm; *K* 0.1 ppm kinetin
[a]Duration of pre-culture for somatic embryogenesis, in weeks

Table 8.5 Influence of iron, manganese, and molybdenum on the fresh weight, number of cells per explant, and average cell weight of carrot callus cultures

	0	Fe	Mn	Fe +Mn	Mo	Fe +Mo	Mo +Mn	Fe+Mn +Mo
mg fresh weight	8	94	18	150	20	152	24	175
Number of cells $\times 10^3$ per explant	18.6	650.0	78.6	742.0	68.6	486.0	68.3	951.2
µg per cell	0.43	0.14	0.23	0.20	0.29	0.31	0.35	0.18

BM medium, see Table 3.3, with 10% coconut milk, 3 ppm Fe, 3.6 ppm Mn, 0.25 ppm Mo; 3 weeks of culture; Neumann and Steward (1968)

and respond to the stimuli associated with explantation and in vitro culture. To which extent such variation determines the response in culture has been discussed in Chap. 7, dealing with somatic embryogenesis in cultured petiole explants.

Here, only a short recapitulation shall be given. Under suitable conditions (NL2, B5), cultured petiole explants are able to differentiate adventitious roots and shoots,

Table 8.6 Influence of kinetin on the fresh weight and number of cells per explant of cultured explants of various tissues of *Papaver somniferum* L. (var. Scheibes Ölmohn) in NL3 medium

Tissue	Original tissue		No kinetin		+0.1 ppm kinetin	
	F. wt.[a]	No. cells	F. wt.	No. cells	F. wt.	No. cells
Root	0.3	2.9	1.0	22.2	9.0	275.2
Hypocotyl	1.0	2.1	7.0	132.2	24.0	368.9
Cotyledons	1.0	6.6	1.0	6.6	11.0	122.1
Leaves	0.4	3.1	1.0	16.0	17.0	168.9

See Table 3.3, 73 days of culture

[a]*F. wt.* mg fresh weight per explant, *no. cells* number of cells $\times 10^3$ per explant

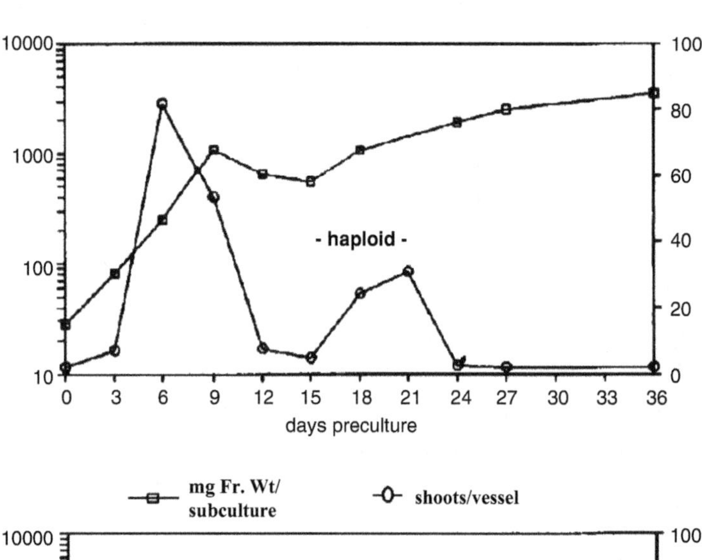

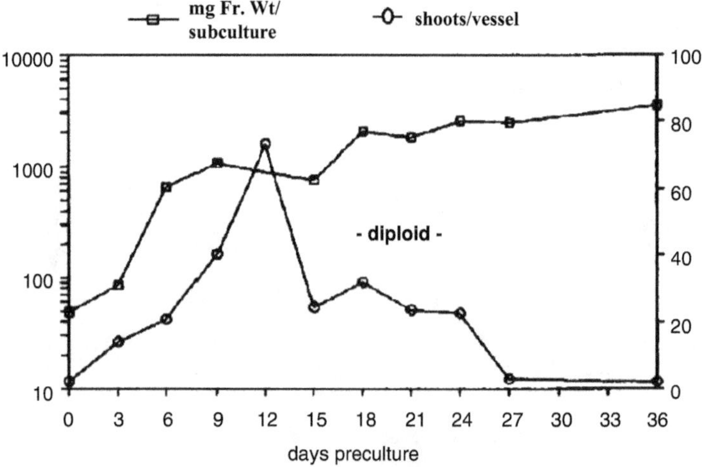

Fig. 8.2 Fresh weight and shoot differentiation of haploid and dihaploid callus cultures of *Datura innoxia* as a function of time of transfer from an MS medium with 2.4D to an MS medium with 10 ppm kinetin (induction medium, 54 days of culture, 22 °C, continuous illumination; Forche et al. 1981)

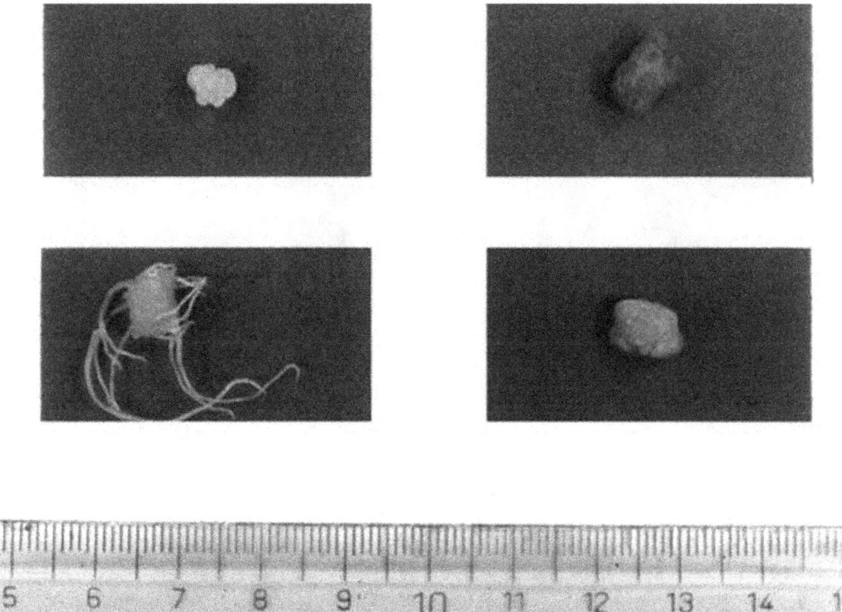

Fig. 8.3 Influence of the photoperiod during early development of carrot plants (until 6 weeks after seedling emergence) on the development of cultured explants of mature plants: NL2 with IAA and inositol; NL3 with IAA, inositol, and kinetin (*top row*) long-day conditions, (*bottom row*) short-day conditions

as well as somatic embryos. As summarized in Sect. 7.3, these different tissues serve as origin. An additional important factor is time. Indeed, 2–3 days after initiation of the experiment, adjacent to the conductive cells, and between the conductive elements and the glandular channel, after vigorous production of cytoplasm, cell divisions are initiated in some cells, which develop first into root primordials and eventually into adventitious roots. After 5–6 days of culture, cell division is initiated in the large parenchymatous cells after a prior growth of cytoplasm, and later the differentiation of adventitious shoots can often be seen. After about 2–3 weeks, the differentiation of somatic embryos from originally vacuolated subepidermal cells can be observed. Here again, the initial histological indication is a vigorous growth of cytoplasm (Fig. 8.4, Table 8.7).

A careful peeling of the epidermis connected to two or three subepidermal cell layers cultured under the same cultural conditions results also in the initiation of somatic embryogenesis. This indicates the capacity of these subepidermal cells to differentiate somatic embryos independently of other parts of the petiole, which would be related to the differential status of these cells at explantation. Important is the increase in cytoplasm in all three cases, as the first cytologically observable sign. The different morphogenic processes would subsequently be related to differences in the composition of the newly produced cytoplasm. It would be of interest to investigate the significance of the glandular channel for these processes.

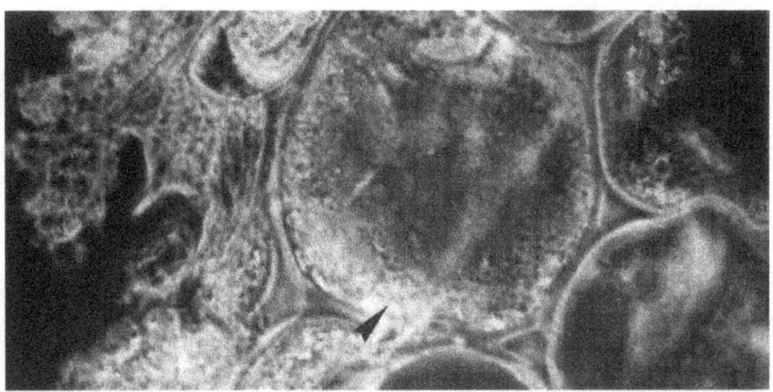

Fig. 8.4 Some histological images of cultured petiole explants (*Daucus carota*). Cytoplasm-rich subepidermal cell

Table 8.7 Flow sheet of somatic embryogenesis in cultured petiole explants of *Daucus carota* (Li and Neumann 1985)

Days after explantation to the next dev. stage	Developmental stage	Hormonal supplement at transition
t0	Somatic cells	High auxin conc., 2 ppm IAA
3–5	Meristematic cells near conductive elements	High auxin conc., 2 ppm IAA
ca. 10	Adventitious roots	Less auxin, 0.1 ppm 2.4D
ca. 15	Embryogenic cells (subepidermal region), densely filled with cytoplasm[a]	Low auxin conc., 0.01 ppm 2.4D
18–20	Four-cell stage of embryogenic cells, pre-globular stage	Low auxin conc., 0.01 ppm 2.4D
ca. 24	Globular stage	Low auxin conc., 0.01 ppm 2.4D
ca. 28	Heart-shaped stage	Low cytokinin conc. (0.02 ppm zeatin), low auxin conc., 0.01 ppm 2.4D
30–40	Torpedo-shaped stage[a]	No growth regulators
50–60	Mature embryo	No growth regulators
80–90	Young plant	

[a]Transfer to a new medium with the concentration of growth regulators indicated

8.2 Exogenous Factors

In this section, phytohormones and growth regulators, the mineral nutrition of cell cultures, and influences of light and temperature will be discussed. Most literature currently available deals with the significance of phytohormones and growth regulators, and also here only some examples will be given to indicate tendencies—for more information, the Internet is recommended. Again, some empirical ideas of more general significance will be considered, exemplified by research results mostly from our own laboratory. The same approach is taken for the significance of nutrition and physical factors.

8.2.1 Growth Regulators

Let us start with some remarks on terminology. In the literature, some confusion exists on the use of the terms phytohormones and growth regulators. In this book, mainly phytohormones are defined as natural occurring regulators of growth and development native to plants; the term growth regulators includes phytohormones and synthetic substances with influences similar to those of phytohormones.

Nutritional factors are generally rather unspecific with respect to growth and differentiation and predominately recognizable in quantitative terms. However, growth regulators exert rather specific influences usually at low concentrations in the medium. Some exceptions to this have been discussed in Chap. 7 and will be discussed in Chap. 11.

As a general principle, cell division and cellular differentiation are counteracting processes. Independently of the classification of a compound as, e.g., an auxin or a cytokinin, if its application promotes high cell division activity, then usually differentiation will be inhibited at the same concentration. The application of IAA, a native auxin, to cultured carrot root explants induces the differentiation of adventitious roots within about 2 weeks. Under these conditions, its activity to promote cell division is relatively low. After a simultaneous application of kinetin, a synthetic cytokinin, high cell division is induced, and root formation is either prevented or sometimes delayed for about 3 weeks. The same delay can be observed for an equimolar application of 2.4D, a synthetic auxin that strongly promotes cell division at suitable concentrations.

Another important factor is the concentration of the growth regulators applied. As an example, if kinetin is applied at 0.1 ppm to the nutrient medium of *Datura* explants, a strong stimulation of cell division activity can be observed; an application of 10 ppm inhibits cell division, and the differentiation of shoots is induced; brushing a solution of 30 ppm onto isolated leaves slows down senescence. Very important are interactions of the various growth regulators. As demonstrated in Fig. 8.5, a separate application of IAA, kinetin, or m-inositol induces only small growth responses, and even a combination of any two of these increases growth only slightly. A growth rate of callus cultures comparable to that recorded with a supplement of coconut milk is achieved only by a combination of all three components. There is evidence suggesting an enhanced multiple interaction of these growth regulators, rather than simply a summation of individual effects (see Chap. 11).

An important factor in such relations is certainly an endogenous hormonal system that evolves during culture of the explants, which will be dealt with in Chap. 11. Also genetic influences have to be considered, and it is open to which extent there

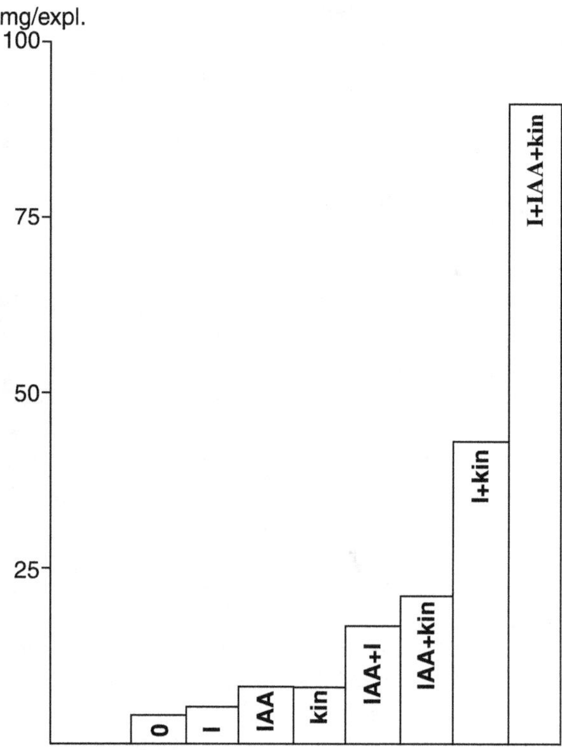

Fig. 8.5 Multiple interactions of several growth regulators influencing the fresh weight of cultured explants of the secondary phloem of the carrot root (NL, see Table 3.3, 22 °C, continuous light at ca. 4000 lux; Bender and Neumann 1978)

exist relations between these and the endogenous hormonal system of cultured tissue.

Such relations of cell division and differentiation, as described for the growth and development of cultured cells, can be also observed for biochemical differentiation and consequently for the production of components of secondary metabolism that could be of commercial interest (Chap. 10). Compared with highly active, proliferating cell populations, the development of the secondary metabolism usually requires a certain age of cells, i.e., a longer interphase in the cell cycle.

8.2.2   Nutritional Factors

As can be seen from the composition of nutrient media (Sect. 3.4), cell cultures require all the mineral nutrients as intact plants for optimal growth and development. Also in terms of growth performance, dose/response relations tend to be similar to those known for intact plants since a long time (cf. Figs. 8.6, 8.7 and 8.8). If a tangent is projected on the ascending curve, an angel of ascent can be observed that is characteristic for each nutrient in cell cultures as well as for intact plants, certainly due to the specific function of the nutritive element investigated. Here, differences

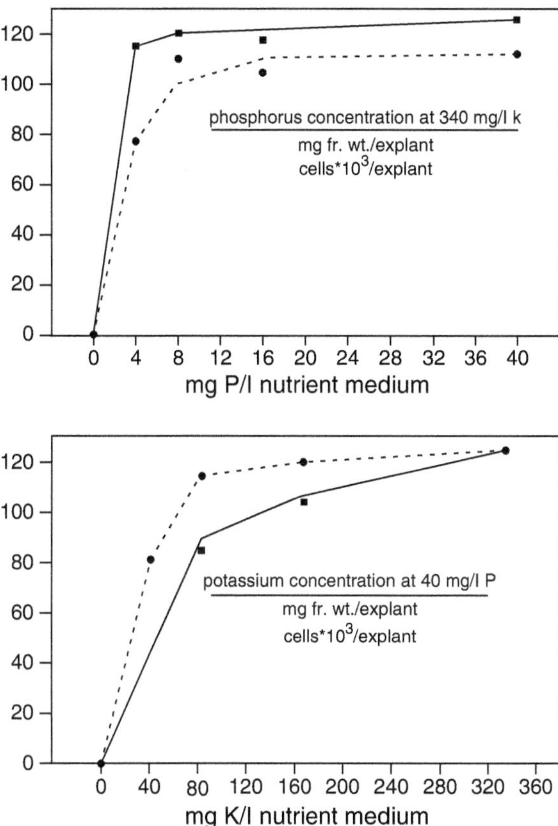

Fig. 8.6 Influence of potassium and of phosphorus in the nutrient medium on the fresh weight and number of cells per explant of cultured explants of the secondary phloem of the carrot root (NL, see Table 3.3, supplied with 50 ppm m-inositol, 2 ppm IAA, 0.1 ppm kinetin)

can be observed for callus growth and for the number of cells per explant. Evidently, cell division activity and cellular growth are influenced differently by the nutrient.

For phosphorus, the angel of ascent for callus growth is greater than for the number of cells per explant, the reverse being the case for potassium.

Clear influences on the equilibrium of cell division and cellular growth can also be observed for micronutrients. Especially iron promotes cell division, whereas Mn and Mo (at the concentrations applied) seem to preferentially promote cellular growth (Table 8.5). Some data on the consequences of nutrient deficiencies for metabolism will be discussed in Chap. 9, dealing with primary metabolism.

To characterize the function of an individual nutrient element in terms of yield production of cereals, a so-called C-value was introduced over 50 years ago by Mitscherlich (1954):

$$\mathrm{d}y/\mathrm{d}x = \mathrm{k}(A - y)$$

or, after integration,

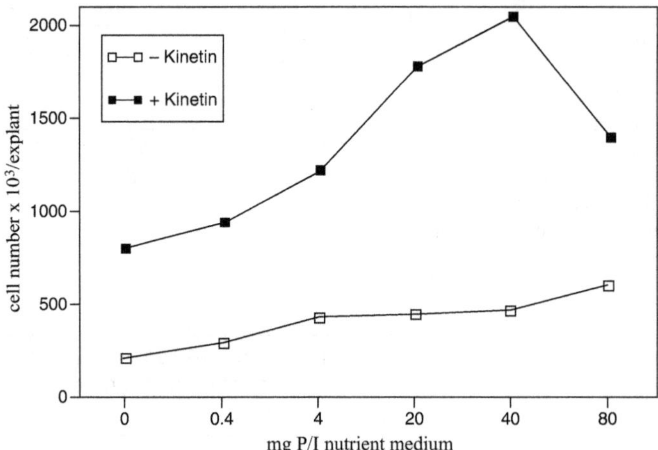

Fig. 8.7 Influence of various phosphorus concentrations and kinetin (0.1 ppm) on the cell number of explants of cultured carrot root explants after 3 weeks of culture (cell number at t0 = 15 × 10³/explant). The nutrient efficiency rate was derived by interpolation of the increment of cells per explant/ng of nutrient between the two lowest nutrient concentrations and amounts to 131 × 10³ for minus kinetin and 343 × 10³ for plus kinetin treatments (Stiebeling and Neumann 1987)

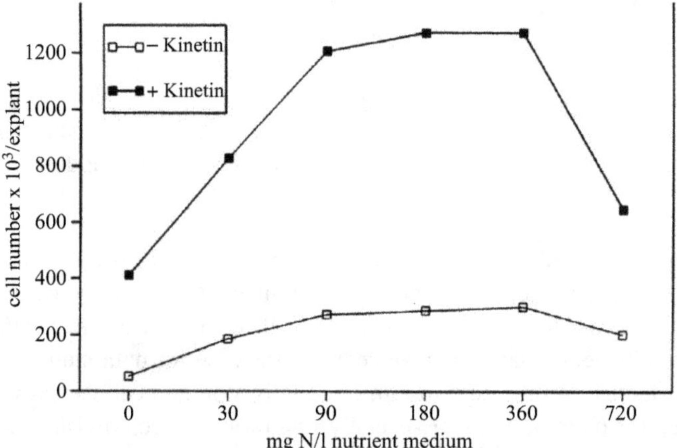

Fig. 8.8 Influence of various nitrogen concentrations and kinetin (0.1 ppm) on the cell number of explants of cultured carrot root explants after 3 weeks of culture (cell number at t0 = 15 × 10³/explant)

$$\ln (A - y) = c - kx$$

where y is the fresh weight, dry weight, or cell number per explant, x is the variable of the experimental system (e.g., the concentration of the nutrient), A is the maximum achievable growth, and k is a constant.

Table 8.8 Influence of casein hydrolysate (CH, 200 ppm) on the fresh weight and number of cells per explant of cultured carrot root tissue (secondary phloem) in NL medium

	Fresh weight (mg/explant)	No. of cells (cells $\times 10^3$/ explant)	Aver. cell weight (μg/cell)
Without CH	99.00	866.13	0.13
With CH	206.00	1683.00	0.10

See Table 3.3, supplemented with 50 ppm m-inositol, 2 ppm IAA, and 0.1 ppm kinetin (21 days of culture)

Here, c is represented by an integration constant of invariable components of a system, except for k. After transformation into Brigg logarithms, we have

$$\log\,(A - y) = \log\,A - cx$$

and c is proportional to k, which is based on the transformation to Brigg logarithms ($c = k \times 0.434$). Often, the Mitscherlich formula is written in a non-logarithmic form:

$$y = A(1 - 10^{-cx})$$

The value c describes the angle of ascent of the tangent in experiments on mineral nutrition. Such calculations can also be applied to cell and tissue cultures. Comparing C-values of mineral nutrients for intact plants in pot experiments with those calculated for cell cultures, the latter is considerably higher. This would be due to the meristematic character of cell and tissue cultures; in intact plants, meristematic areas are "diluted" by tissue with low or no proliferation and consequently low or no requirements for mineral nutrients (Stiebeling and Neumann 1987).

Besides influences of individual nutrients on growth, also interactions of these have to be considered for macro- as well as for micronutrients. In Table 8.5, examples are given for interactions of Fe with Mn and Mo. A supplement of the latter two, either alone or in combination, in addition to Fe clearly increases growth more than when summing their individual effects.

Nitrogen nutrition is satisfied by providing nitrate or ammonia, mostly as salts (Sect. 3.4). Starting with White's nutrient medium, glycine, as an organic nitrogen source, contains reduced nitrogen. Nowadays, organic nitrogen in reduced form is supplied to most nutrient media as casein hydrolysate. Carrot root explants grow quite well on only nitrate in the NL medium (Table 8.8), but callus weight is nearly twice as high following an application of casein hydrolysate. All amino acids of this mixture can be utilized by cell cultures, but often a selective preference in uptake can be observed—in carrot cultures, this is for leucine. Thus, this mixture of several amino acids, obtained by hydrolysis of the naturally occurring protein casein, can be replaced by one (usually glutamic acid) or a few amino acids.

Compared to media containing only ammonia as nitrogen source, growth of tobacco cell suspensions is higher in nutrient media containing only nitrate as nitrogen source (Table 8.9). For both, average cell weight is essentially identical,

Table 8.9 Influence of nitrogen form (360 mg N/l) on the growth, total nitrogen content, pH of the nutrient medium, and concentration of nicotine for tobacco cell cultures (var. Xanthi 8/11, NL medium, see Table 3.3), supplemented with 50 ppm m-inositol, 2 ppm IAA and 0.1 ppm kinetin (28 days of culture)

	g Dry wt./ 250 ml NL	Number of cells $\times 10^3$/ ml	Cell wt. (μg)	pH[a]	N uptake (mg/100 g)[b]	Nicotine (μg/g dry wt.) in cell material	Nicotine (μg/g dry wt.) in nutrient medium
Nitrate (NaNO$_3$)	2.628	348	0.030	5.8	461	25.5	8.75[c]
Ammonia (NH$_4$Cl)	1.235	184	0.027	4.4	486	11.7	6.90[c]

Elsner, unpublished results of our institute
[a]pH 5.6 at t0
[b]Kjeldahl-N
[c]Difference highly significant

and therefore differences would be due to a reduced cell division activity in the ammonia treatment, in which also the concentration of nicotine is at a considerably lower level. Ammonia is taken up by plant cells as a cation in exchange for protons, which accounts for the lowering of the pH of the medium. Nitrogen uptake was at the same level for both nitrogen forms, suggesting that the differences in growth performance are due to differences in metabolism of the two and possibly to the differences in pH (see later).

Only reduced nitrogen can be utilized by heterotrophic plant cells. The high energy requirement to reduce nitrate is fulfilled by photosynthesis in intact plants and for cell cultures usually by some carbohydrate in the medium, commonly sucrose. Besides influences on growth, the nitrogen form exerts influences on morphogenesis. Already in the mid-1960s, Halperin and Wetherell (1965) reported a requirement for ammonia in addition to nitrate, i.e., reduced nitrogen, in the nutrient medium to induce somatic embryogenesis. This was later confirmed using other species (e.g., Gleddie et al. 1982 for *Solanum melongena*). In this system, the nitrate to ammonia ratio of 2 is optimal up to a concentration of 60 mM of total nitrogen in the medium. With the exception of the NL medium, all other media used to induce somatic embryogenesis contain ammonia in addition to nitrate. In the NL medium, nitrate is the only source of inorganic nitrogen. Reduced nitrogen, however, is supplied as amino acids in casein hydrolysate. As will be reported elsewhere in detail, here a strict requirement of ammonia to induce somatic embryogenesis does not exist.

Employing the more recent methods of proteomics, some investigations using intact plants may shed more light on the differences in the function of the two nitrogen sources. A nitrate supply to nitrogen-starved tomato plants results in an upregulation of 115 genes, including nitrate transporters, nitrate and nitrite reductase, and also some of those involved in general metabolism, like transaldolase and transketolases, malate dehydrogenase, asparagine synthase, and histidine decarboxylase (Y.H. Wang et al. 2001). Similar results have been reported for *Arabidopsis* (R. Wang et al. 2000). Here, besides an upregulation, also repressions of some genes were observed, like for *AMT1;1*, encoding an ammonium transporter. Evidently, as of its first entry into metabolism, a molecule as small as nitrate is able to initiate a whole family of genes with possibly remote functions.

The function of mineral nutrients depends on the supplement of growth regulators to the medium. In Fig. 8.8, results of an experiment on the influence of kinetin on the growth of carrot root explants at various nitrogen concentrations are summarized. It is obvious that a kinetin supplement induces a higher efficiency of nitrogen for callus growth. Similar results can be obtained for phosphate with, however, some variation, possibly specific for this nutrient (Fig. 8.7). Nitrogen and phosphorus in casein hydrolysate were not considered in the two nutrient media given in the tables. These and similar results indicate an influence of growth regulators on the nutrient efficiency rate. It remains to be seen to which extent this influence, here of phytohormones, exists also for intact plants—some preliminary results dealing with this aspect are positive.

The nutrient efficiency rate was derived by interpolation of the increment of cells per explant/ng of nutrient between the two lowest nutrient concentrations. This amounts to 2.2×10^3 for minus kinetin and 13.6×10^3 for plus kinetin treatments (Stiebeling and Neumann 1987).

The influence of growth regulators on nutrient efficiency cannot be explained easily. Each amino acid requires its characteristic number of nitrogen atoms, and each nucleotide needs at least one phosphorus. Although no explanation is available for nitrogen, some first lines of evidence exist for phosphorus (see also Chap. 9). Although the explants of the experiments in Figs. 8.7 and 8.8 can be considered as mixotrophic, photosynthesis contributes considerably to fulfill the demands in energy and carbon. The export of assimilates from chloroplasts, however, is mediated mainly by a phosphate translocator that requires inorganic phosphate in the cytoplasm for operation. At phosphorus deficiency, the transport of assimilates through chloroplast membranes is less; assimilates will accumulate in the chloroplasts and be initially stored as starch. The starch storage capacity of the plastome is limited, and if this is exhausted, then the assimilates reduce the activity of the Calvin cycle enzymes to fix carbon dioxide by feedback. Neither NADPH nor ATP can be transported through the chloroplast membranes directly to the cytoplasm. To the Calvin cycle, a second route of assimilate export exists that is independent of inorganic phosphate—a dicarboxylate shuttle. A main function of assimilates in the cytoplasm is to provide substrates to produce reduction equivalents, mostly NADPH and NADH. This function can be at least partly substituted by the dicarboxylate shuttle. Some first results indicate a dependence of this shuttle on kinetin (Neumann and Bender 1987). Should these be confirmed, the influence of kinetin on phosphate efficiency could find some explanation. At low phosphate concentrations in the medium, kinetin could promote the operation of this shuttle, as supplement for a low activity of the phosphate translocator, thereby increasing the assimilate export of chloroplasts.

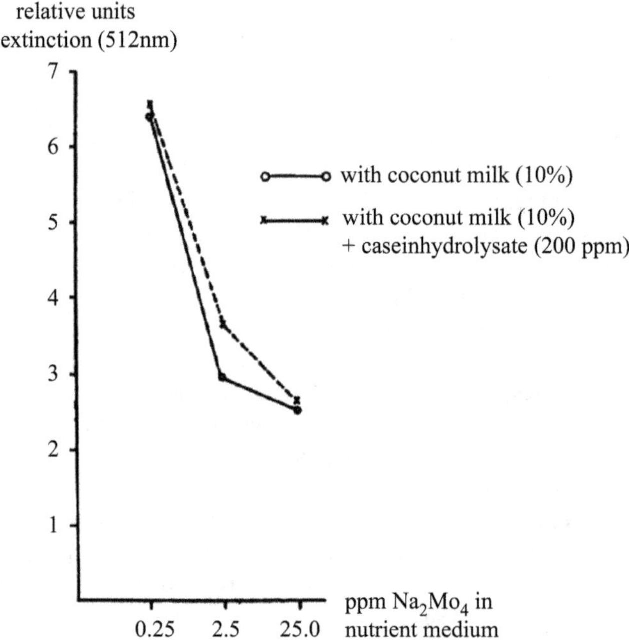

Fig. 8.9 Influence of various concentrations of molybdenum on the concentrations of anthocyanin in cultured carrot root explants (BM, see Table 3.3, with 10% coconut milk; after Neumann 1962)

It has to be checked to which extent such conditions could influence also secondary metabolism. As could be expected, generally many influences of the nutritional status of the culture can be observed on the concentration and the composition of the protein, concentrations of free amino acids, as well as of carbohydrates and other components of primary and secondary metabolism. For influences of nutrients on secondary metabolism, the anthocyanin concentration in cultured carrot root explants as influenced by Mo shall serve as an example (Neumann 1962). Cultured explants of some carrot roots are able to synthesize and accumulate anthocyanins. An increase results from higher iron concentrations in the medium. By contrast, a dramatic decrease is associated with high molybdenum levels (Fig. 8.9).

The synthesis of anthocyanin is closely related to an interaction of carbohydrate and nitrogen metabolism, and often its accumulation can be observed in situations favoring an accumulation of carbohydrates. Iron increases the uptake of sugars from the nutrient medium, and molybdenum, as a cofactor to nitrate reductase, should increase the synthesis of amino acids and other nitrogen-containing compounds (Neumann 1962; Neumann and Steward 1968). Consequently, due to requirements of carbohydrates for amino acid synthesis, the concentration of carbohydrates in cultures would be reduced by molybdenum. Also at phosphorus deficiency, usually anthocyanin will accumulate, which is used for diagnosis of phosphorus deficiency

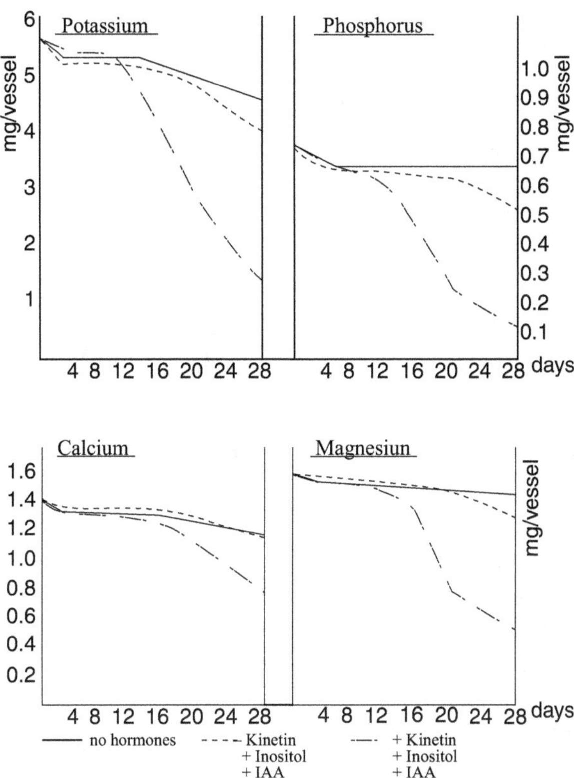

Fig. 8.10 Influence of kinetin (0.1 ppm), inositol (50 ppm), and IAA (2 ppm) on the concentrations of potassium, phosphorus, calcium, and magnesium in the nutrient medium of cultured carrot root explants during 28 days of culture

in intact plants. One explanation could be the requirements of phosphate for optimal operation of the phosphate translocator. A disturbance of this endogenous transport system would promote an accumulation of carbohydrates in the cells, and anthocyanin would accumulate. As expected, an accumulation of anthocyanin can also be induced by elevated sucrose levels in the nutrient medium.

The courses of uptake of the two twin nutrient pairs potassium/phosphorus and calcium/magnesium are quite similar, and also the influence of kinetin is comparable (Fig. 8.10).

The nutrients of the former pair are used up to a greater extent than those of the latter. This can be observed also for intact plants. The highest rate of uptake, at least for K, P, and Mg, takes place during the log phase from the 10th to the 20th day of culture (see also Chap. 3). During the stationary phase, uptake is slowed down again. The uptake follows growth intensity, and at least for P and Mg, a "luxury" consumption can be excluded. Although the concentrations of all four nutrients are lower if compared to those at t0, a deficiency of these should not be responsible for the transition of the cultures from the log to the stationary phase.

In the experiments described above, nutrient uptake was estimated by determination of the concentration of the nutrients in the nutrient medium at various stages of culture. If, however, the nutrient concentration is calculated on the basis of cell

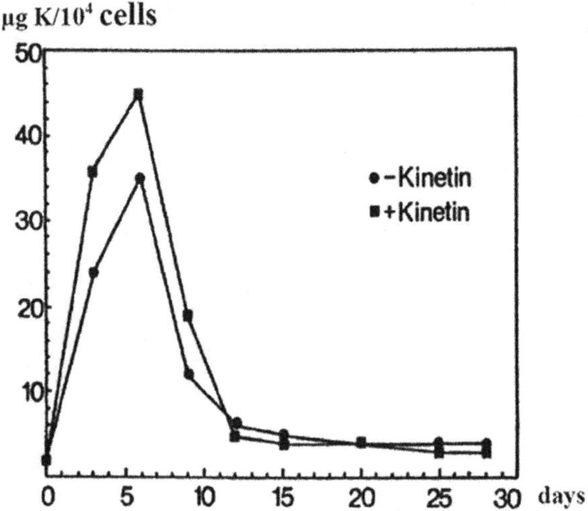

Fig. 8.11 Influence of kinetin (0.1 ppm) on the potassium concentration of carrot callus cultures during a 28-day culture period (Krömmelbein, unpublished results of our institute)

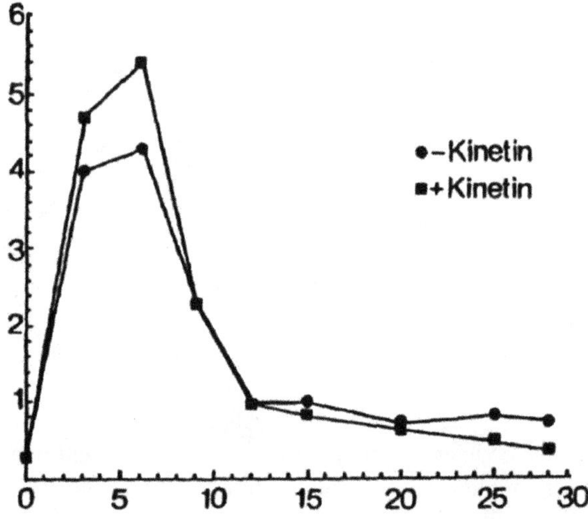

Fig. 8.12 Influence of kinetin (0.1 ppm) on the concentration of phosphorus (µg P/10 four cells, Y axis) in carrot callus cultures during a 28-day culture period (X axis; Krömmelbein, unpublished results of our institute)

number of the cultures, as in Fig. 8.11 for potassium and in Fig. 8.12 for phosphorus, then a high accumulation occurs during the lag phase of callus growth up to the 6th or 7th day of culture. A kinetin supplement increases the concentration of both nutrients on the 6th day by about 25%. At this day, cell number is approximately the same as in the kinetin-free treatment. Therefore, the nutrient uptake rate would not be related to the growth-promoting capacity of kinetin but rather to changes in metabolism initiated by kinetin during the lag phase.

This maximum of K and P is followed by a steep decrease in concentration, which certainly would at least be partly due to a "dilution" induced by the strong increase in cell number per explant during the log phase of callus growth. From the 12th day onward, the concentration of these two mineral nutrients remains more or less constant, and no influence of kinetin can be observed. In other experiments, sometimes the concentration is somewhat elevated in the kinetin treatment.

Although the experiments on mineral nutrients discussed above clearly show their significance for cell cultures, only a limited number of investigations are known dealing with this aspect. Except for the early investigation performed at the time to establish the mineral composition of nutrient media, systematic studies of the significance of mineral nutrition in cell cultures are not available. Also, it is only rarely that investigations on the influences of mineral nutrients on metabolism and the composition of cultured cells can be found in the literature. This should be also of commercial interest. The importance of such investigations shall be demonstrated by, e.g., the results of Fujita and Tabata (1987). A nitrogen supplement as ammonia reduces the production of shikonin by *Lithospermum* cultures; this is in agreement with results obtained for nicotine production by cultures of tobacco (see above). Another example is an increase of products of the secondary metabolism by a general reduction of nitrogen in the nutrient medium. A reduction of nitrogen results also in an increase in the production of capsaicin in pepper cultures, and the formation of serpentine and ajmalicine is increased in *Catharanthus* cultures by lowering the phosphate level. An increase of rosmarinic acid could be the result of a reduced growth rate by cell division in the cultures and the accumulation of older cells. In Chap. 10, more details will be discussed. Influences of mineral nutrients on morphogenesis have already been discussed in Chap. 7.

8.2.2.1 Improvement of Nutrient Uptake by Transgenic Carrot Cultures

Phosphorous (P) is an essential nutrient for plant growth, development, and production, as part of key molecules such as nucleic acids, phospholipids, ATP, and other biologically active compounds. The total amount of P in the soil may be high, but often it is unavailable for plant uptake.

To adapt to phosphate (Pi) deficiency, plant roots release citrate or malate, or both, which mobilizes Pi from sparingly soluble Pi sources (Penaloaza et al. 2005). Phosphoenolpyruvate carboxylase (PEPCase) is an important enzyme that regulates the generation of some organic acids, such as oxalacetic acid and malic acid, by carboxylation of phosphoenolpyruvate (PEP, see below). The transcriptional activation of PEPCase genes is also regulated by P deficiency (Toyota et al. 2003).

As described later, at phosphorous deficiency the activity of the phosphate translocator is reduced, and the energy export from the chloroplasts is substituted by a dicarboxylate shuttle. The initial step of this shuttle is a carboxylation of PEP by PEPCase; oxalacetic acid is produced, which is reduced to malate after uptake by the chloroplast (see also Chap. 9). Malate and also citrate accumulate in the cells and in the nutrient medium. To improve the utilization of phosphorus by increasing the production of malate and citrate in the nutrient medium, we have generated transgenic carrot cultures containing an additional PEPCase gene (ppcA, Accession Z48966) from *Flaveria pringlei* (C3 plant; Chu et al. 1997; Westhoff et al. 1997), under control of the MAS promoter with the methods described later (Sect. 13.6.3). In nutrient media, usually water-soluble Na-bisphosphate is supplied as phosphorus

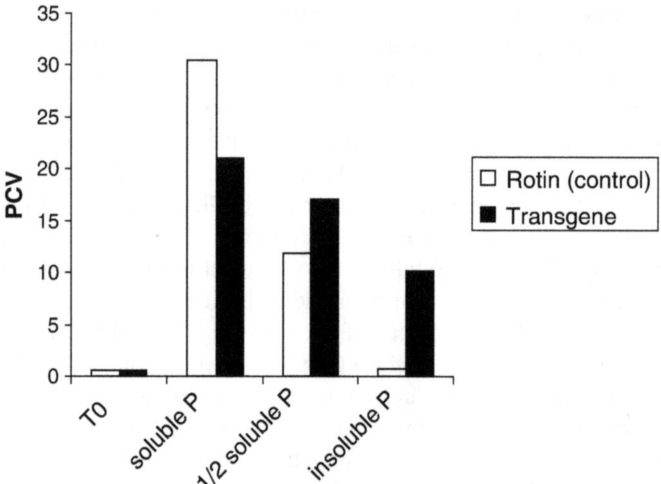

Fig. 8.13 Growth of carrot cell suspension during 5 weeks of culture in B5 medium with different P sources (NaH$_2$ PO$_4$.H$_2$O as soluble P and Thomas phosphate as insoluble P). Cell density is shown as pcv (packed cell volume, ml cells/100 ml suspension)

source. To check the efficiency of this foreign gene, bisphosphate was substituted by Thomas phosphate in which phosphate is hardly water-soluble. As shown in Fig. 8.13, only transformed cells carrying the second PEPCase gene are able to grow in the nutrient medium with Thomas phosphate as only P source (Natur et al., unpublished data of our laboratory). No data are available yet for transgenic carrot plants growing on P-deficient soil.

By transferring an embryogenic carrot cell suspension into a hormone-free B5 medium, the development of somatic embryos is decreased by P deficiency. Figure 8.14 shows that the transgenic cells as well as the control are growing normally in B5 with water-soluble hydrogen-P, whereas only the transgenic cells can grow in B5 with water-insoluble Thomas phosphate.

8.3   Physical Factors

Here, some examples on influences of temperature and illumination on cultured cells shall be discussed. Although marked influences of these factors can be expected for the performance of cell cultures, data of systematic studies on this aspect are rather rare.

Like all biological systems, also for cell culture, there exists a profound influence of temperature on growth and development. Up to 30–35 °C, an increase in growth performance of cell cultures of a number of species has been described. If possible, the optimal temperature for each cell culture system should be determined, and this could be expected to range between 20 and 30 °C. Usually, the temperature is kept

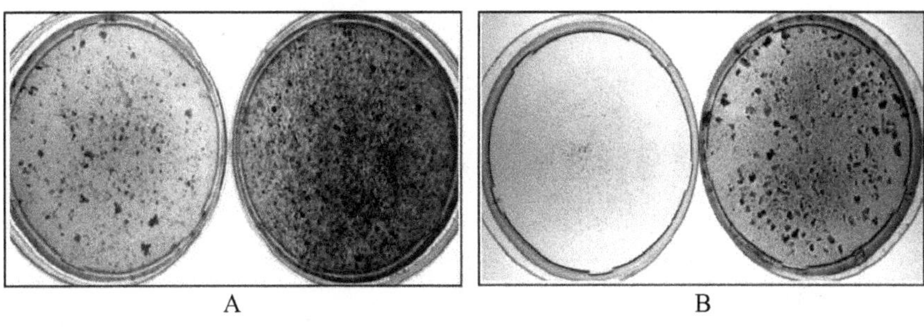

| A | B |
| Culture in B5 containing soluble Phosphate (Na-bis-phosphate) | Culture in B5 containing insoluble Phosphate (Thomas phosphate) |

Fig. 8.14 Development of somatic embryos during 28 days of culture in the hormone-free B5 medium with Na-bisphosphate or Thomas phosphate: (*A, left*) Rotin, (*right*) transgenic strain; (*B, left*) Rotin, (right) transgenic strain

Table 8.10 Influence of temperature on callus growth (mg fresh weight/explant) of *Nicotiana tabacum* (var. Xanthi 8/11 = n, 8/12 = 2xn) on MS medium, and 0.8% agar, 0.2 ppm 2.4D, and 0.1 ppm kinetin, 21 days of culture

	6 °C	22 °C	28 °C
8/11	10	170	270
8/12	10	270	360

Zeppernick, unpublished results of our institute

constant during an experiment. As an example, growth of tobacco shoot cultures at three temperature levels is given in Table 8.10. Also morphogenetic processes can be controlled by temperature, as shown for caulogenesis of cultured lily bulb explants (van Aartrijk and Blom-Barnhoorn 1983), which is strongly increased by elevating the temperature from 15 to 25 °C (Fig. 8.15). Even relatively small genetic differences, as between varieties of the same species, will be significant in determining the optimal temperature. The anthers of the tobacco variety "Wisconsin" produce abundant haploid plantlets at 22 °C, a temperature at which androgenesis could not be induced using anthers of "Xanthi." Here, temperatures of 27–28 °C are required for androgenesis. Also, the positive influence of a short storage at low temperatures to induce androgenesis described in Chap. 6 should be recalled. Eventually, the optimal temperature for rhizogenesis and caulogenesis could be different. As an example, influences of temperature on callus growth are given in Table 8.10.

For light, several factors have to be considered. Besides light intensity, which can vary between darkness and continuous illumination by 8000–10,000 lux, also light quality, and the variation in the daily duration of illumination are of significance.

Green cultures with photosynthetic activity will improve growth at higher light intensities. Also the response to growth regulators will be modified by illumination. In the experiment in Table 8.11, however, it is difficult to distinguish between light influences on photosynthetic activities and direct influences on the function of the growth regulators. Sometimes an increase in growth can also be observed in the

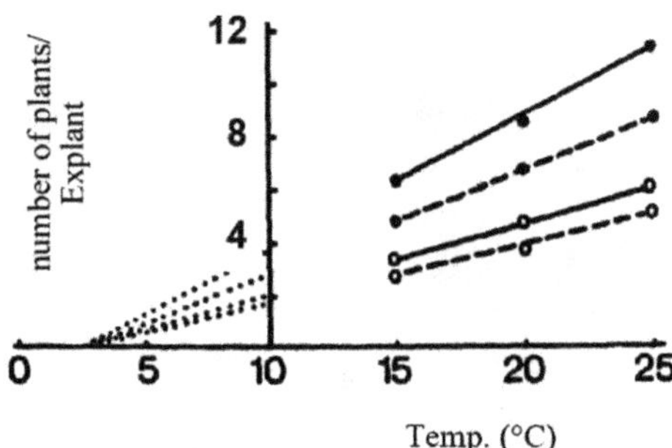

Fig. 8.15 Interactions of temperature, NAA, and wounding in the formation of adventitious shoots of explants of lily buds (van Aartrijk and Blom-Barnhoorn 1983): (*closed circles*) application of 0.5 µM NAA; (*open circles*) no NAA; (*dotted lines*) no wounding; (*continuous lines*) wounded explants

Table 8.11 Influence of light on the cell division activity (cell number $\times$ 10^3/explant) of cultured carrot root explants, cultured with m-inositol (50 ppm), IAA (2 ppm), and kinetin (0.1 ppm) in NL medium

	Dark	Light
Original tissue	8.0	18.0
No growth regulators	12.5	35.2
M-inositol+IAA	25.4	146.1
M-inositol+IAA+kinetin	444.0	752.4

See Table 3.3, 3 weeks of culture; Neumann and Raafat (1973)

dark, and rhizogenesis in some systems can be promoted in darkness or at low light intensities. The significance of illumination for protoplast cultures was discussed before (Chap. 5). With respect to light quality, Fluora lamps would be preferred to the usual fluorescent lamps, because of a light spectrum close to that of sunlight. In our laboratory, however, only Osram lamps are used (15 W/21, Lumilux White), with success.

References

Bender L, Neumann KH (1978) Investigations on indole-3-acetic acid metabolism of carrot tissue cultures. Z Pflanzenphysiol 88:209–217

Chu C, Qu N, Bassüner B, Bauwe H (1997) Genetic transformation of the C3–C4 intermediate plant, Flaveria pubescens (Asteraceae). Plant Cell Rep 16:715–718. https://doi.org/10.1007/s002990050308

Forche E, Kiebler R, Neumann K-H (1981) The influence of developmental stages of haploid and diploid callus cultures of Datura innoxia on shoot initiation. Z Pflanzenphysiol 101:257–262

Fujita Y, Tabata M (1987) Secondary metabolites from plant cells: pharmaceutical applications and progress in commercial production. In: Green CE, Somers DA, Hackett WP, Biesboer DD (eds) Proc. 6th Int. congress plant tissue and cell culture. Ar Liss, New York, pp 169–185

Gleddie S, Keller W, Geeterfield T (1982) In vitro morphogenesis from tissue, cells and protoplasts of *Solanum melongena* L. In: Fujiwara A (ed) Plant tissue culture 1982. Japanese Association for Plant Tissue Culture, Tokyo, pp 139–140

Halperin WS, Wetherell DF (1965) Ammonium requirement for embryogenesis *in vitro*. Nature 205:519–520

Imin N, Nizamidin M, Daniher D, Nolan KE, Rose RJ, Rolfe BG (2005) Proteomic analysis of somatic embryogenesis in *Medicago truncatula*. Explant cultures under 6-benzylaminopurin and 1-naphthaleneacetic acid treatments. Plant Physiol 137:1250–1260

Kibler R (1978) Untersuchungen zum Einfluß des Ploidieniveaus auf Wachstum, Differenzierung und Alkaloidsyntheseleistung von *Datura innoxia*-Zellsuspensionskukturen. Dissertation, Justus Liebig Universität, Giessen

Li T, Neumann KH (1985) Embryogenesis and endogenous hormone content of cell cultures of some carrot varieties (*Daucus carota* L.). Ber Dtsch Bot Ges 98:227–235

Mitscherlich EA (1954) Bodenkunde. Parey, Berlin

Neumann KH (1962) Untersuchungen über den Einfluß essentieller Schwermetalle auf das Wachstum und den Proteinstoffwechsel von Karottengewebekulturen. Dissertation, Justus Liebig Universität, Giessen

Neumann K-H, Bender (1987) Photosynthesis in plant cell and tissue culture system. In: Green CE, Somers DA, Hackett WP, Biesboer DD (eds) Proc. 6th Int. congress plant tissue and cell culture. Ar Liss, New York, pp 151–168

Neumann K-H, Raafat A (1973) Further studies on the photosynthesis of carrot tissue cultures. Plant Physiol 51:685–690

Neumann KH, Steward FC (1968) Investigation on the growth and metabolism of cultured explants of *Daucus carota* I. Effects of iron, manganese and molybdenum on growth. Planta 81:333–350

Penaloaza E, Monaz K, Salvo-Garrido H, Silva H, Corcurea L (2005) Phosphate deficiency regulates phosphoenolpyruvate carboxylase expression in proteoid root clusters of white lupin. J Exp Bot 56:145–153

Stiebeling B, Neumann KH (1987) Identification and concentration of endogenous cytokinins in carrots (*Daucus carota* L.) as influenced by development and a circadian rhythm. J Plant Physiol 127:111–121

Toyota K, Koizumi N, Sato F (2003) Transcriptional activation of phosphoenolpyruvate carboxylase by phosphorous deficiency in tobacco. J Exp Bot 54:961–969

van Aartrijk J, Blom-Barnhoorn GJ (1983) Adventitious bud formation from bulb-scale explants of *Lilium speciosum* Thunb. in vitro: effects of wounding, TIBA, and temperature. Z Pflanzenphysiol 110:355–363

Van Tran Than M (1973) Direct flower neoformation from superficial tissue of small explants of *Nicotiana tabacum* L. Planta 115:87–92

Wang R, Guegler K, LaBrie ST, Crawford NM (2000) Genomic analysis of a nutrient response in Arabidopsis reveals divers expression patterns and novel metabolic and potential regulatory genes induced by nitrate. Plant Cell 12:1491–1510

Wang MB, Abbott DC, Upadhyaya NM, Jacobsen JV, Waterhouse PM (2001) *Agrobacterium tumefaciens*-mediated transformation of an elite Australian barley cultivar with virus resistance and reporter genes. Aust J Plant Physiol 28:149–156

Westhoff P, Svensson P, Ernst K, Bläsing O, Burscheidt J, Stockhaus J (1997) Molecular evolution of C4 phosphoenolpyruvate carboxylase in the genus *Flaveria*. Aus J Plant Physiol 24:429–436

Primary Metabolism

9

Here, only a short sketch of some reactions considered as part of primary metabolism, concentrating on carbon assimilation from organic and inorganic sources, and some remarks on nitrogen metabolism will be given. Most culture systems are illuminated, and therefore interactions of carbon from sugar of the nutrient medium with products of light-dependent CO_2 fixation, i.e., photosynthesis, will be considered in more detail.

9.1 Carbon Metabolism

Cell cultures, like any other plant material, are able to fix carbon dioxide in the dark (Table 9.1). Compared to light fixation, however, this is very low, and using radioactive carbon as carbon dioxide shows labeling after a short exposure only in some organic acids and aspartic acid (coming from OAA), probably due to PEPCase activity.

Many, possibly most tissue culture studies are performed in the light. Consequently, generally, chloroplasts develop, and photosynthesis contributes to a varying degree to metabolism. Despite the fact that most plants depend upon photosynthesis to use light energy to obtain carbon and energy for growth, investigations on photosynthesis of plant tissue and cell cultures are relatively rare. One of us (Neumann 1962, 1966), while working at Cornell University, was probably one of the first to show the presence of chloroplasts in cultured explants and light-dependent carbon dioxide fixation of carrot callus cultures using a culture system of the Steward laboratory (cf. Steward et al. 1952). The carrot secondary phloem explants also provide basic material to study the primary metabolism in cultured cells grown in a medium supplemented with sugar as source of carbohydrates, and some understanding of the development of the photosynthetic apparatus was developed in our laboratory. Here, extensive research into various aspects of primary metabolic processes has been carried out over four decades (e.g., Neumann 1962, 1966, 1968, 1969, 1995; Neumann and Raafat 1973; Neumann et al. 1982; Kumar

© Springer Nature Switzerland AG 2020
K.-H. Neumann et al., *Plant Cell and Tissue Culture – A Tool in Biotechnology*,
https://doi.org/10.1007/978-3-030-49098-0_9

Table 9.1 Integration of carbon from carbon dioxide into various metabolites of photosynthetically active carrot callus cultures in the light and in darkness (21 days of culture in NL3 with m-inositol, IAA, and kinetin; see Table 3.3), after different lengths of incubation in NL3 as before supplied with $NaH^{14}CO_3$ (pH 7, μM C atoms $\times 10^{-4}$/ mg chlorophyll; Bender and Neumann 1978)

	Light, 15 s	Light, 4 min	Dark, 4 min
PGA	10.4	88	0
Gluc. P	0	214	0
Fruc. P	0	72	0
Glucose	0	108	0
Fructose	0	74	0
Sucrose	0	0	0
PEP	0	22	0
Glycerate	Traces	47	0
Glycine+serine	0	20	0
Alanine	0	19	0
Malate	7	188	17
Citrate	2	81	13
Aspartate	3	198	21
Glutamate	0	17	2

et al. 1983a, b, 1984, 1987, 1989, 1999; Bender et al. 1985; Kumar and Neumann 1999).

Most plant tissues grow in media containing 2–3% sucrose, through which energy and carbon needs are satisfied. The nitrogen source is provided in the form of nitrate, ammonia, and casein hydrolysate, separately or in combination. However, as shown in Fig. 9.1, during culture, there is gradual depletion of amino acids of casein hydrolysate from the medium. Often, a preferential uptake of amino acids can be observed, in many situations of leucine (Neumann et al. 1978). The heterotrophic phase correlates with the lag phase and leads to the log phase of cell division activity.

Detailed light, scanning, and electron microscope studies have indicated the development of a photosynthetic apparatus in these cultured secondary phloem explants of carrots, which could serve as model for the development of a photosynthetic apparatus in cultured plant tissues. Most experiments were performed using NL3, i.e., containing kinetin, and distinct deviations were observed. After about 10–12 days in culture, the explants pass from a heterotrophic to a mixotrophic phase, where the young chloroplasts (Neumann et al. 1982; Kumar et al. 1983a, b, 1984, 1999; Kumar and Neumann 1999) already play a role in carbon fixation and energy supply. This phase coincides with the log phase of cell division and lasts up to 20–25 days from the beginning of the experiment. This is followed by an autotrophic phase, in which the carbon supply and energy requirements are met through photosynthetic processes carried out by the cultured cells (Neumann et al. 1978; Nato et al. 1985). The plant cells are also able to utilize nitrate as nitrogen source in this phase (Fig. 9.1).

The development of chloroplasts and of chlorophyll has also been reported for other callus cultures (see review by Widholm 1992, 2000). The most extensively studied cultures include tobacco, *Chenopodium*, and carrots. However, there are also several reports on *Arachis*, *Gossypium*, *Glycine*, etc. (Bergmann 1967; Kumar 1974a, b). In contrast to this, cell cultures of several other plant species produce

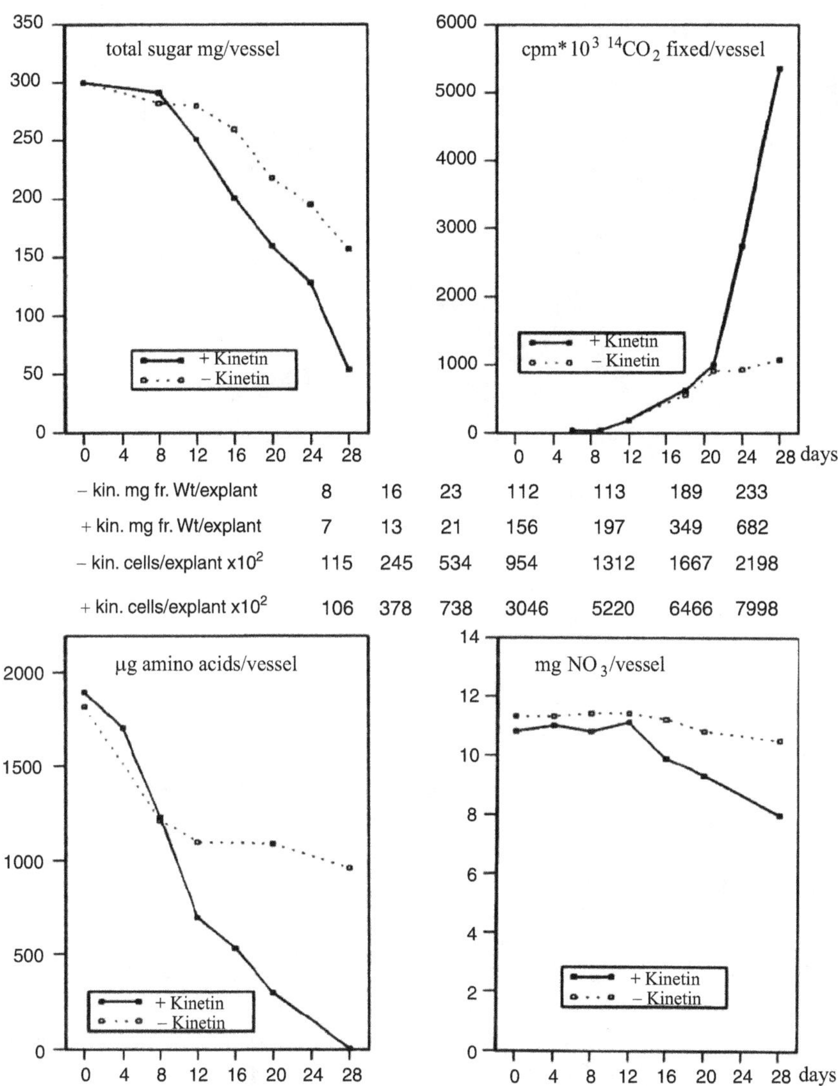

– kin. mg fr. Wt/explant	8	16	23	112	113	189	233
+ kin. mg fr. Wt/explant	7	13	21	156	197	349	682
– kin. cells/explant x10^2	115	245	534	954	1312	1667	2198
+ kin. cells/explant x10^2	106	378	738	3046	5220	6466	7998

Fig. 9.1 Concentrations of sugar, amino acids, and nitrate in the nutrient medium and CO_2 fixation of cultured carrot root explants during a 4-week culture period (Neumann et al. 1978)

only poorly developed chloroplasts (e.g., *Papaver*) or none at all. Growth is thus mainly heterotrophic or at the most mixotrophic in many culture systems.

The development of chloroplasts and a photosynthetic apparatus in carrot tissue cultures, as an example, has been reported in detail by Kumar et al. (1984). The original carrot secondary phloem explants show the presence of chromoplasts and some amyloplasts. The differentiation of these plastids present in the original

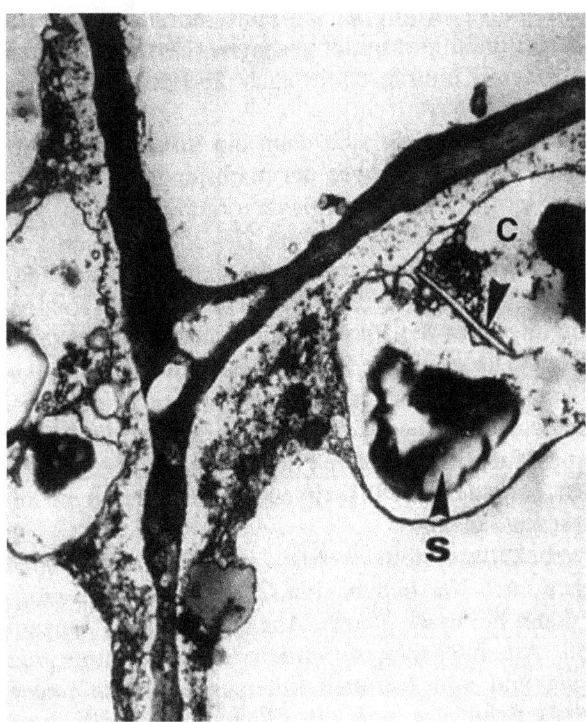

Fig. 9.2 "Amylochromoplast" in a cell of a cultured carrot root explants after 3 days of culture (NL, Table 3.3), containing starch (*S*) and carotene (*C*) crystals (Photograph A. Kumar)

explants is influenced by the availability of a carbohydrate source, some growth regulators, and light supply.

The chromoplasts develop into amyloplasts, and also intermediate structures named "amylochromoplasts" occur that contain starch deposits, in addition to carotene crystals (Fig. 9.2). However, after 6 days of culture, the "amylochromoplasts" are no longer detectable. This suggests that the original chromoplasts as well as the amyloplasts are now differentiating into chloroplasts. Various stages of chloroplast development have been documented for that period (Fig. 9.3; Kumar et al. 1984, 1999). The development of plastids in the cultured cells has been compared to the development of chloroplasts in the apical meristem (Kumar and Neumann 1999). In the dark, the *Daucus* leaf cells develop etioplasts, which could not be detected in dark/light cultures. This suggests that in the cultivated cells, proplastids play no role in the development of chloroplasts. Possibly, the main source of chloroplast multiplication in cultured cells is the division of chloroplasts or their precursors, though among ca. 1000 sections examined under EM, only about 30 showed the division of plastids. The propagation of plastids would either follow a strict circadian rhythm with a maximum that has to date not been identified or there would be some other propagation mechanism in operation, unknown at present.

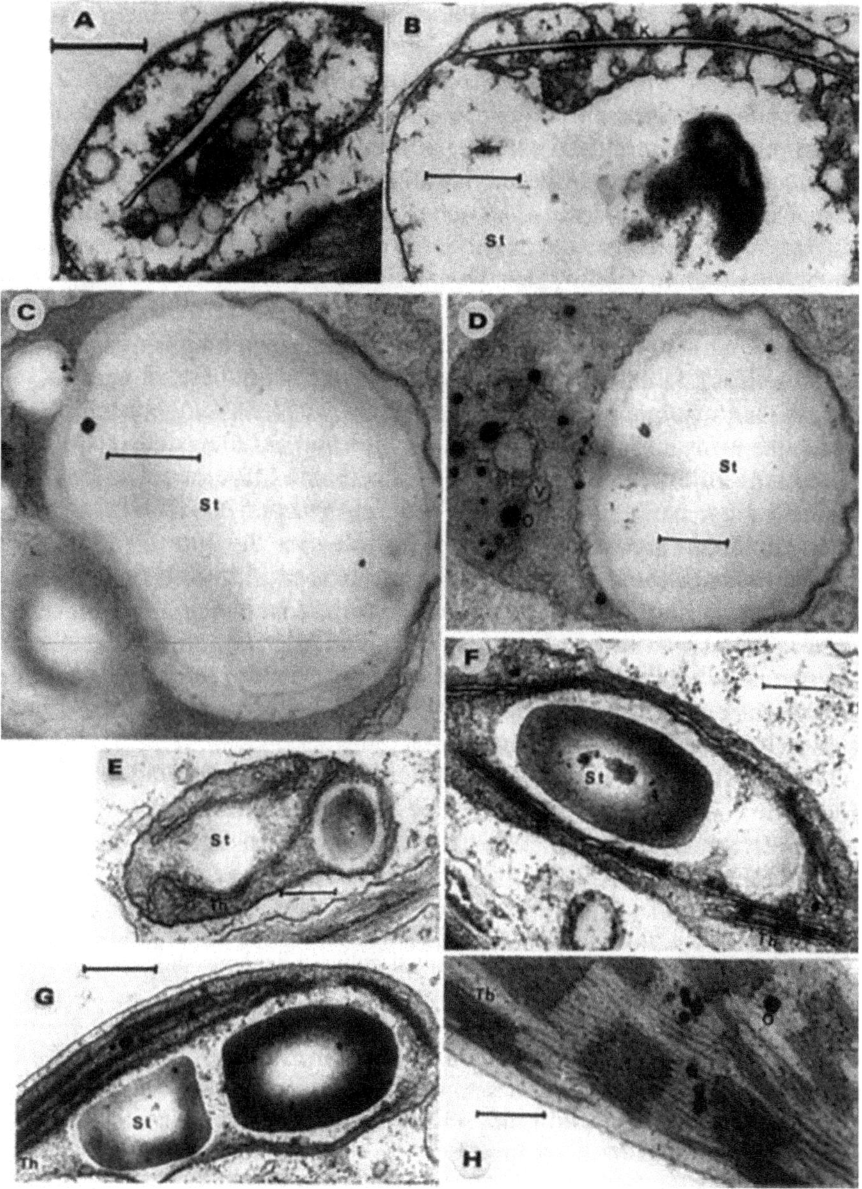

Fig. 9.3 Some stages of plastid development in cultured explants of the secondary phloem of carrot roots observed during a culture period of 4 weeks (developmental scheme, A–G callus cultures, H carrot leaf) (Photograph A. Kumar)

The developing chloroplasts in callus cultures attain the stage G developmental stage, as characterized in our investigations (see Fig. 9.3, Table 9.2), and only carrot leaves show stage H, which represents mature chloroplasts.

Table 9.2 Distribution of developmental stages of plastids (C–G) in cultured carrot root explants during a 4-week culture period (NL medium, supplied with m-inositol, IAA, and kinetin, see Table 3.3)

Developmental stage of plastids	Days of culture						
	3	6	10	12	14	18	28
C	x	x	(x)				
D	x	x	x	x			
E		x	x	x	x	x	
F				x	x	x	(x)
G							x

The development of the photosynthetic apparatus in cultured plant cells is influenced by several exogenous factors like physicochemical conditions, nutritional status, and growth regulators (Fig. 9.4). Bender et al. (1985) described the significance of some exogenous factors in chloroplast development in carrot callus cultures.

The development of chloroplasts is initiated by light, while in dark conditions, only chromoplasts are seen that develop into amyloplasts. However, light alone is not enough for the development of chloroplasts. In the presence of light only, chromoplasts and some amyloplasts could be seen.

The quality of light was found to play a significant role in cell cultures of *Chenopodium rubrum* (Hüsemann et al. 1989). Blue light has been shown to have positive effects on chloroplast development.

The number of plastids in carrot cultures (in relative terms) was positively influenced by the supplementation of sugars to the medium. Even in the absence of exogenous growth regulators, the number of plastids in the cells exposed to light alone was positively influenced by the supply of sugars in the medium. The sugar regulated the number of plastids per cell (Fig. 9.4). A combination of light and exogenous growth regulators, however, regulated the qualitative development of the chloroplasts. Although kinetin-supplemented cultured explants showed maximum greening, influences on the ultrastructural development of chloroplasts was not significant. The biochemical explanations of the regulation of subcellular dedifferentiation and redifferentiation (transdifferentiation) are lacking in such cultured plant cell systems. The carrot system developed here, however, could serve as a model system to study the regulation of such subcellular differentiation processes. The light compensation point of *Arachis* cultures was determined at 40 μmol O_2, which is less than that of leaves.

Although the deep temperature absorption spectra in the blue light region show some differences, in the red light region, the absorption spectra of photosynthetic callus cultures and leaves were at least qualitatively comparable (Fig. 9.5). This was also applicable to the electron flow through PS II and PS I, as indicated by fluorescence induction kinetics or Kautsky measurements. The influence of sugar supplementation, as well as of a supply of growth regulators on fluorescence induction kinetics, was correlated with other parameters of chloroplast development.

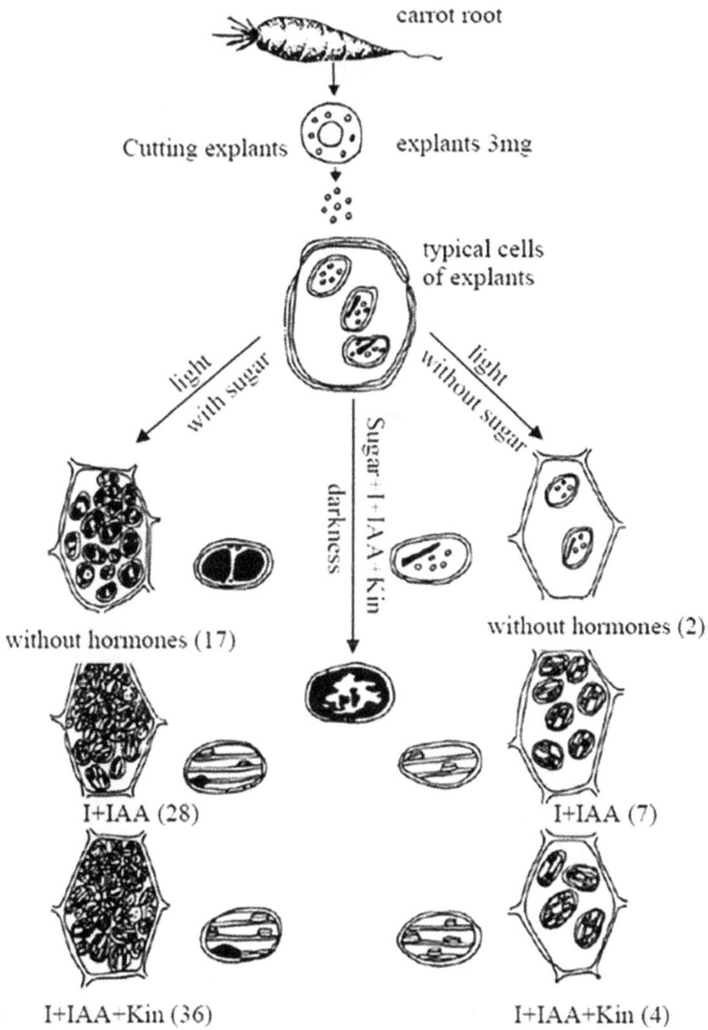

Fig. 9.4 Influence of various factors (light, sucrose, kinetin) on the development of plastids in cultured explants of the secondary phloem of carrot roots (drawing by A. Kumar)

A profile comparable to that of leaves could be seen only in the medium supplemented with kinetin. These results indicate that under optimal conditions, at least qualitatively, the light reaction system of chloroplasts from cultured plant cells and leaves is quite comparable.

Besides oxygen, the main products of the light reaction system are NADPH and ATP. Neither ATP nor NADPH pass through the chloroplast envelope directly, to be used for many reactions in the cytoplasm. To this end, organic compounds of the CO_2 fixation systems are employed as carriers. Based on contemporary knowledge,

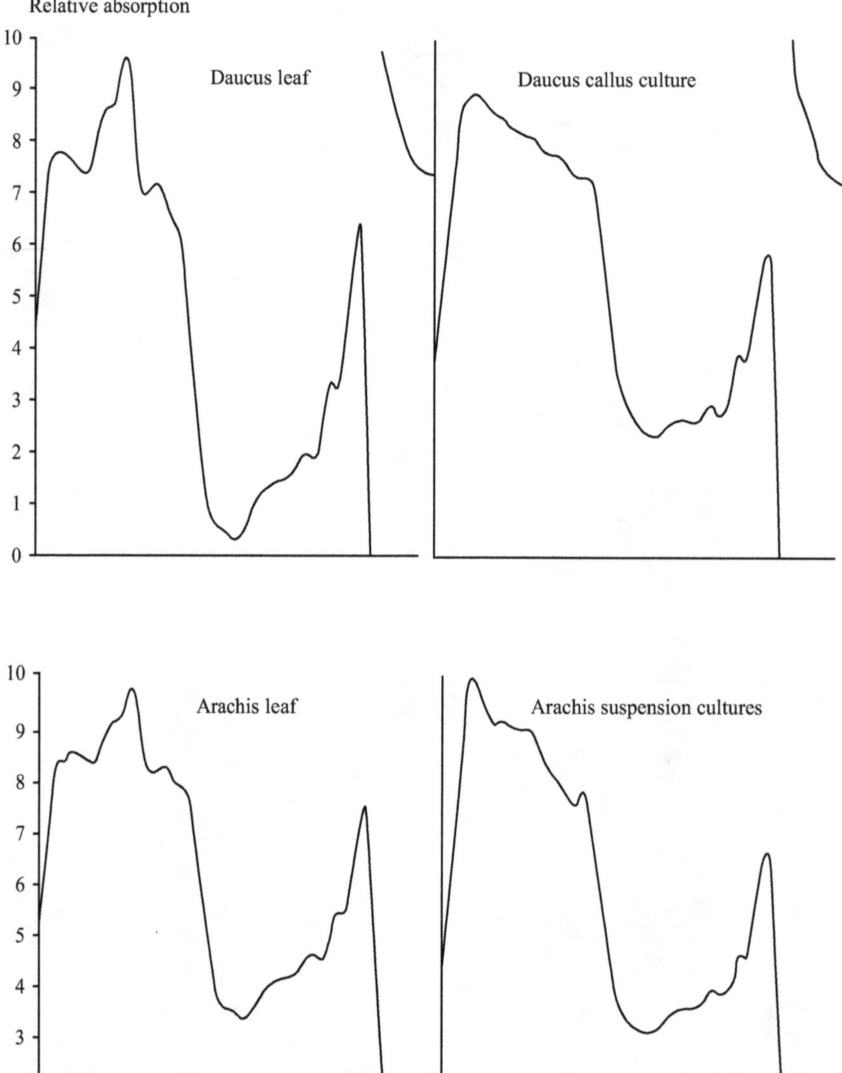

Fig. 9.5 Deep temperature spectra of leaves and cell cultures of *Daucus carota* and *Arachis hypogaea*

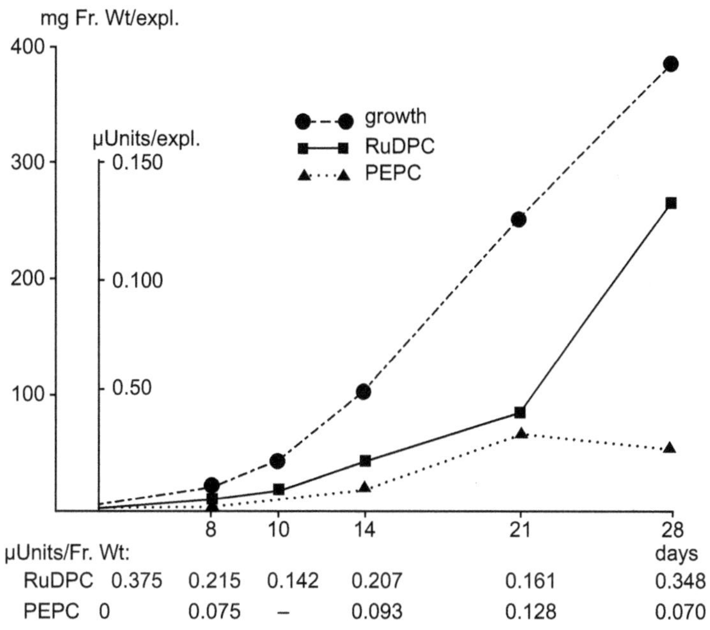

Fig. 9.6 Fresh weight and activity of ribulose-bisphosphate carboxylase/oxygenase (RuBisCO) and phosphoenolpyruvate carboxylase (PEPCase) of cultured explants of the secondary phloem of the carrot root during 4 weeks of culture (Kumar et al. 1983b)

the carbon assimilation system operates through two enzyme systems (Fig. 9.6). In the ribulose-bisphosphate carboxylase/oxygenase (RuBisCO) system, ribulose-bisphosphate is carboxylated, and the first stable products are two molecules of phosphoglyceric acid (PGA). In the second system, phosphoenolpyruvate carboxylase (PEPCase) uses phosphoenolpyruvate, an intermediate of glycolysis, as primary acceptor for carbon dioxide, and oxaloacetate is the first stable product.

The oxaloacetate pool of the cells is very small and subject to substantial turnover; therefore, its concentration is rather difficult to determine. OAA produced via the carboxylation process, however, is rapidly converted into malate in the plastids. Oxaloacetate is also a precursor for the synthesis of aspartic acid, and it should be regarded as a representative of OAA. As mentioned before, malate is produced in plastids from OAA (or aspartate after desamination) by an NADP (or NAD)-dependent malate dehydrogenase (Fig. 9.7).

A comparison of ^{14}C label of malate and aspartate in light and darkness indicates that in the light, malate clearly dominates (Fig. 9.8). Obviously, light, via the provision of NADPH, promotes the reduction of OAA to malate. Malate can be transferred to the cytoplasm where it can be utilized to fuel metabolism. Combined, these reactions represent a dicarboxylate shuttle (Heber and Heldt 1981).

In cultured carrot explants, both carboxylating systems are in operation (Fig. 9.6) at a varying extent at different stages of the culture. Up to about 20 days of culture,

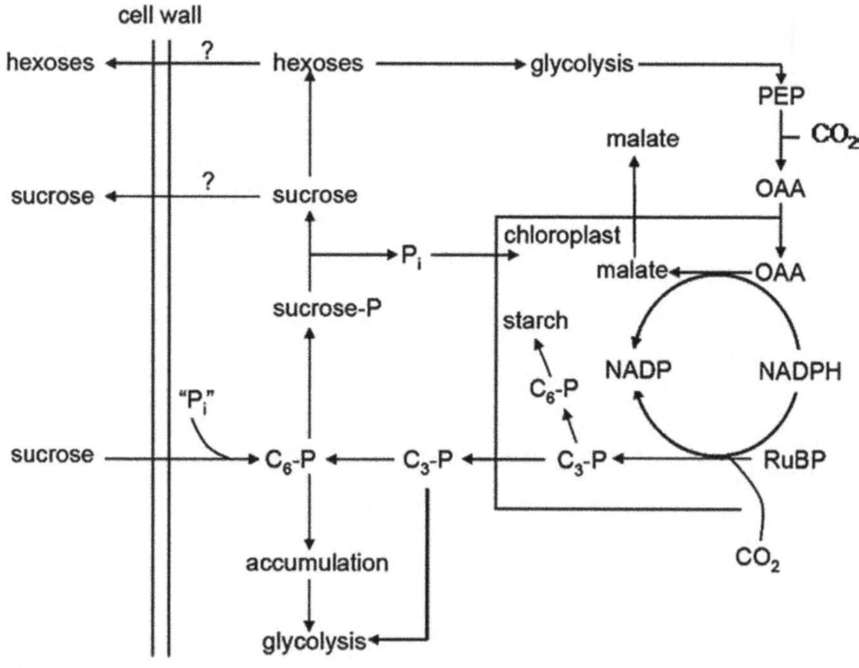

Fig. 9.7 Schematic summary of some metabolic reactions of carbon metabolism of cultured carrot root explants (Neumann et al. 1989)

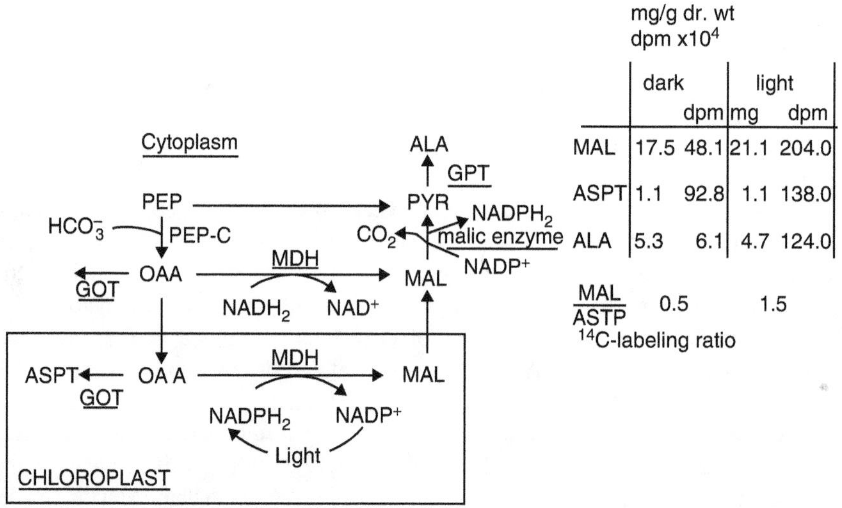

	dark		light	
mg/g dr. wt dpm x10⁴	mg	dpm	mg	dpm
MAL	17.5	48.1	21.1	204.0
ASPT	1.1	92.8	1.1	138.0
ALA	5.3	6.1	4.7	124.0
MAL/ASTP	0.5		1.5	

¹⁴C-labeling ratio

Fig. 9.8 Schematic summary of a partial C4 fixation and a dicarboxy shuttle in cultured carrot root tissue (Bender et al. 1985)

both RuBisCO and PEPCase exhibit more or less the same activities. This coincides with the heterotrophic and mixotrophic phases of the cultures. In the following stationary phase, RuBisCO by far dominates. This pattern has been confirmed by similar studies using *Chenopodium* cultures (Herzbeck and Hüsemann 1985). At this stage of the cultures, the sugar concentration in the medium is already quite low (Fig. 9.1) and may have less negative influences on the RuBisCO system, as described below. Another factor could be an increase in the number of functional chloroplasts at that stage.

The export of products of the RuBisCO system is governed by a triose phosphate/phosphate translocator with a requirement for inorganic phosphate taken up by the chloroplasts from the cytoplasm, in exchange for triose phosphates. This translocator belongs to a whole set of similar translocators that has been described in more detail for *Arabidopsis* (Knappe et al. 2003). PEP carboxylation is localized in the cytoplasm, and the products of the PEPCase system are transferred to the chloroplasts. Oxaloacetate (and aspartate, after deamination) can be reduced in the chloroplasts by NADPH to malate, which can be transferred to the cytoplasm. Here, following oxidation of malate by the malic enzyme, reducing equivalents originating from the Hill reaction for the many endergonic reactions of the cell are made available.

The phosphate translocator operates by stoichiometric exchange of plastid-produced organic phosphorylated compounds for inorganic phosphate from the cytoplasm. Consequently, the concentration of inorganic phosphate in the cytoplasm is an important factor in the regulation of the activity and efficiency of the export of photosynthetic products.

As an example of regulation at another level, the expression of the triose phosphate/phosphate translocator gene in wheat could be also inhibited by glucose (Sun et al. 2006), and its regulation is dependent on a hexokinase-dependent pathway. Later, influences of sugar on CO_2 fixation will be discussed again.

The requirement for phosphate suggests that phosphate deficiency promotes a dicarboxylate shuttle as an anaplerotic reaction to the Calvin cycle (Table 9.3), and additionally, a supplementary reaction to transfer reduction equivalents generated in excess in the Hill reaction into the cytoplasm (Bender et al. 1985; Neumann 1995). Such shuttles are actually more extensively characterized for mitochondria than for chloroplasts. Here, some reference is made to the section on influences of the transfer of an additional PEPCase gene by gene technology, as described elsewhere (Chap. 8).

The assimilation of inorganic carbon is also regulated through sucrose supplied to the nutrient medium (Table 9.4). As shown in Table 9.4, 2% sugar reduces $^{14}CO_2$ fixation by up to 50%. Before an explanation is sought to this observation, some of the results on the metabolism of carbohydrates taken up from the nutrient medium need to be discussed (Neumann 1995). Sucrose is the most commonly employed carbohydrate in cell and tissue culture. The invertase probably localized in cell wall breaks down the disaccharide within a few days of culture, into the monosaccharides glucose and fructose. The localization of the enzyme in cell walls is not totally clear yet, and protoplast cultures were used to further investigate this aspect. In protoplast culture, the sucrose molecule remains intact in the medium. In some of the

Table 9.3 Influence of restricted phosphate supply on the growth, P uptake, and some parameters of photosynthesis (μm carbon $\times$ 10^4/mg chlorophyll) for explants of the secondary phloem of carrot roots after 3 weeks of culture in the light (NL3 medium, see Table 3.3; Bender et al. 1985)

	mg P/l nutrient medium	
	11.6	46.5
Dry matter (mg/vessel)	56.6 ± 6.8	79.1 ± 6.5
P uptake (mg/vessel)	0.17 ± 0.02	0.62 ± 0.03
P conc. in tissue (mg/g dry wt.)	3.14 ± 0.14	7.27 ± 0.25
Chlorophyll (mg/g dry wt.)	0.75 ± 0.04	0.91 ± 0.03
CO_2 fix. light	1.428	1.338
CO_2 fix. dark	206	90
Carbon in soluble fract. (light)	1.301	1.278
Carbon in insoluble fract. (light)	127	61.00
CO_2 fix. RuBisCO (C3)	557.0	880.0
CO_2 fix. PEPCase (C4)	871.00	458.00

Table 9.4 Influence of sucrose on ^{14}C fixation in light (dpm $\times$ 10^4/g dry weight) and its distribution in C3 and C4 metabolites in cultured carrot explants after 3 weeks of culture (30-min fixation) in NL3 medium with m-inositol, IAA, and kinetin

	$-$Sucrose	$+$Sucrose (2%)	
C3 metabolites	335	191	-43%
C4 metabolites	505	288	-23%
Total	840	479	

See Table 3.3; Bender et al. (1985)

investigations, no invertase activity could be detected, although this was present in the callus cultures.

After the uptake by the cells, there is a preferential utilization of glucose compared to that of fructose, although both hexoses are phosphorylated and can be interconverted through hexose isomerase.

The Michaelis–Menten constant of the enzymes for fructose-6-phosphate was somewhat lower than for glucose-6-phosphate (unpublished results of our institute), so the rate of reaction for building fructose-6-phosphate is higher. At least in a 2% sucrose-supplemented medium, only less than 10% of the sugar taken up is released as CO_2 (Neumann et al. 1989). Interestingly, in a short-term labeling experiment, a predominant part of hexose phosphate is used in reconstituting sucrose (see Fig. 9.7). These molecules apparently act as some storage carbohydrates in carrot cultures. The high $^{14}CO_2$ labeling in the pools of free glucose or fructose indicates that endogenously synthesized sucrose is again split up into both hexoses (Neumann et al. 1989).

In these cultures, endogenous sucrose as well as the two hexoses, and interestingly also citrate and malate, can be exported into the nutrient medium. A more detailed summary on low molecular carbohydrates is given by Neumann (1995), and here only some examples will be discussed. The free hexoses represent a pool of carbohydrates. The physiological function of such a metabolic pathway with the splitting of sucrose, and its new synthesis in the cells with ultimate hydrolysis, is difficult to understand, especially seeing that ATP-requiring phosphorylation is essential. Possibly, there is a process in the cells of intact plants that is of great importance for the carbohydrate metabolism, with no significant or no function in cultured cells. Certainly, there could be more examples of this type of phenomenon in cultured plant tissues.

As summarized in Fig. 9.7, the phosphorylation of exogenous hexoses broadly requires phosphate. Consequently, an excess of sugar supply to the nutrient medium can lead to some metabolically induced "phosphate deficiency" with the reduction of activity of the phosphate translocators, and subsequently chloroplast metabolism, as presented above. Then, the dicarboxylate shuttle could, to a limited extent, provide the transport of energy of reducing equivalents from the chloroplasts to the cytoplasm. As mentioned above, glucose could regulate the expression of the gene of the triose phosphate/phosphate translocator (Sun et al. 2006). This is another area of influence by carbohydrates on this system.

Interestingly, the function of the dicarboxylate shuttle is dependent on kinetin supply and therefore is regulated through the growth regulator system (Bender et al. 1985). The physiological basis of kinetin function is not yet known.

The carrot is regarded as a C3 plant, and although PEPCase fixation of carbon dioxide has a prominent significance in cultured cells, no Kranz anatomy could be detected. More recently, some data became available indicating that both parts of carbon fixation—C4 fixation by PEPCase and C3 fixation by RuBisCO—can simultaneously occur in the same cell of C4 species. Investigations by Gowik et al. (2006) and others indicate a phylogenetic relation of C4 PEPCase with a C3 PEPCase from which it may have developed.

In photosynthetically active cultures, the starch deposits accumulated under continuous light and were dependent mainly on light fixation. However, starch could also be synthesized from the carbohydrates supplemented to the medium. It is interesting to note that embryogenic cultures have higher starch contents than is the case for non-embryogenic cultures cultivated under similar conditions. Tobacco as well as carrot cultures show starch formation in the cells differentiating shoots and roots. During organogenesis, this starch is broken down. Apparently, carbohydrates accumulate in such differentiating cells, and during the process of differentiation, the accumulated starch is utilized to obtain energy and substrate for the synthesis of various compounds associated with differentiation.

Little is known on the regulation of starch synthesis in cultured cells. As an example, in sweet potato cultures, it has been shown that gibberellic acid has a distinct and specific influence. While NAA, 6-BA, and ABA supplementation had no influence, GA3 application resulted in a marked reduction in starch content. Simultaneously, the activity of starch synthetase was also reduced, so that the regulation through gibberellic acid takes place probably at the translation or transcription level. In this investigation, glucose or glycerin was applied as carbohydrate source to the nutrient medium. The application of GA also reduces the possibility of

Table 9.5 Influence of sucrose (2% in nutrient medium) on the fresh weight production per day and dry weight content of *Arachis hypogea* and *Daucus carota* bioreactor cultures (grown for 6 weeks at continuous illumination of ca. 7000 lux)

	No sucrose		With 2% sucrose	
	mg F. wt.	% Dry wt.	mg F. wt.	% Dry wt.
Arachis hypogea	250	3.88	1409	7.9
Daucus carota	932	6.04	3264	6.28

At the beginning of the experiment, the fermenter was filled with 3 l NL3 supplied with IAA, m-inositol, and kinetin and with 7–8 g of callus material inoculated

starch formation from excess of carbohydrates given to the medium (Sasaki and Kainuma 1982). There is also an increase in the nutrient medium of exported carbohydrates (galactose, arabinose, galactouric acid, etc.), possibly due to a failure to store excess carbohydrates as starch.

Although several tissue cultures grow in light under mixotrophic nutritional conditions, attempts have also been made during the last 2–3 decades to obtain photoautotrophic cultures. There are several reports of autotrophic cultures for about 15 plant systems, where tobacco, carrots, *Arachis*, and *Chenopodium* have been studied in some detail (see review by Widholm 1992; Neumann 1995). The establishment of such cultures could be through the selection of cell lines characterized by high chlorophyll concentrations. Long-term autotrophic cultures over several years have been obtained by increasing the CO_2 concentrations of the atmosphere (Husemann and Barz 1977). For around 20 years, however, at our Institut für Pflanzenernährung der Justus Liebig Universität, Giessen, Germany, *Arachis* cultures could be grown at ambient CO_2 concentration without carbohydrate supply to the medium. Nevertheless, as shown in Table 9.5 for fermenter cultures, the growth of those cultures without sucrose supplementation is much lower than for those supplemented with sucrose to the medium. Autotrophic growth at ambient CO_2 concentration has also been shown in cotton cultures. In this case, an increase in CO_2 concentration resulted in a six- to sevenfold increase in growth. Here, the relative role of PEPCase in terms of total CO_2 fixation is also enhanced (Widholm 1989).

In connection with the regeneration of cell walls in protoplasts, several experiments on cell wall structure have been reported, but little is known about the intact cell. The cultures grow with high cell division activity, and it can be concluded that the primary cell wall will dominate in most systems. Apparently, for the primary cell wall, hydroxyl-rich glycoproteins are derived from the Golgi. At least in tobacco cultures, it could be shown that arabinose is bound to hydroxyproline to form a glycoprotein. This glycoprotein is described as arabinogalactan-rhamnogalacturonan protein.

In contrast to the secondary walls, the primary cell wall of many plants is similarly structured. Thus, similar molecules can be expected from cell cultures of different plants. Already around 50 years ago, the Steward group reported high hydroxyproline concentrations in fast-A cell cultures with domination of the primary wall, contrasting with slow- or nongrowing cultures (Steward et al. 1958).

In the absence of molybdenum in the medium, the concentration of hydroxyproline declines in proteins. However, a deficiency of other micronutrients like iron or manganese does not reduce hydroxyproline concentration in the cultures (Neumann 1962). This indicates specific influences of molybdenum in the medium on OH-proline-containing proteins, largely of the primary wall. Apparently, changes in the cell wall also occur during the cell cycle (Amino and Komamine 1982). During cell division, the first middle lamella built consists mainly of pectin; during cytokinesis, an increase in this cell wall fraction occurs. During the G1 phase, the concentration of all cell wall fractions is increased, and in the 5% KOH-soluble cell wall fractions, the proportion of galactose (percent concentration) is preferentially increased. The 24% KOH-soluble fraction remains stable during the cell division cycle. During the G2 phase, an increase in the galactose concentration takes place, together with a reduction in arabinose concentration; the G1 phase shows the reverse. Furthermore, during the entire cell cycle, more ^{14}C from glucose is built into pectin and hemicellulose than into the cellulose fraction.

In cotton cultures, it could be shown that cellulose synthesis is influenced by the substrate. When supplying UDP, glucose beta-1,3-glucane is preferentially synthesized, and ca. 10% of ^{14}C is found in cellulose. When supplying ^{14}C glucose, beta-1,4-glucane dominates, and ^{14}C cellulose levels are between 20 and 50% (Widholm 1992).

The biochemistry of respiration consists in the transfer of electrons from NADH$_2$ equivalents generated in the citric acid cycle, to oxygen in the final stage. The energy thereby liberated is utilized in the building of ATP from ADP and phosphorus. The energy stored in ATP thus remains available for the many phosphate-coupled endergonic reactions of the cell. The extent of oxygen use indicates the intensity of respiration. This electron transport system is localized in the inner mitochondrial membrane, and from flavoproteins and different cytochromes, a respiratory chain is established. In plant cells, an additional, alternative electron transport chain exists, as yet not well understood. Here, the electrons bypass the cytochrome system and are transferred directly from the flavoproteins (FMN and FAD) to oxygen. This system yields less free energy in the form of ATP (one ATP, compared to three in the respiratory chain). Maybe it represents an "outlet" in case of excess of electrons. The difference in the energy level between NADH and oxygen is released as heat. In many members of Araceae, this transport way is utilized to produce heat.

While the cytochrome-dependent electron transport chain could be blocked by HCN or acid, this is not the case with the alternative electron transport chain. This enables simple measurements of the relative contribution of the two pathways to respiration. The alternative transport route can be disturbed by rhodanid. Both the electron transport pathways are localized in the inner mitochondrial membrane.

The relationships between the two electron transport systems in tissue cultures have been studied in potato callus cultures (van der Plas and Wagner 1982). Here, the cytochrome-dependent electron transport capacity of freshly isolated potato explants during the culture period was compared to that of the original explant and was found to be 3–4 times higher during the first week of culture. The alternative transport system was lacking in the fresh explants; its development was detectable only during the first week of culture, when its capacity was independent of the supply of sucrose to the medium. With some supply of sucrose to the medium the following week, the ratio of the activity of the two transport pathways remained constant. In the absence of sugar, however, the capacity of both was reduced. At ample sugar supply to the medium in the following week, the cytochrome-dependent pathway remained constant, and the alternative pathway reached a maximum, followed by a decline. By providing an increasing level of sugar (5% sucrose), the maximum was higher than the corresponding value at lower concentrations. Here, the alternative pathway apparently provides an outlet for an excess of electrons resulting from an excess supply of sugar. A recent paper by Costa et al. (2008), with extensive literature citations, reports results on the occurrence of the alternative oxidase of *Daucus* as related to stress and functional reprogramming.

Lipids are important, firstly, for the building of cell membranes, and, secondly, in their function as reserve material. Most young cell cultures with high cell division activity have a high demand for lipids as membrane-building material, and the latter function is of lesser importance. Triglycerides can differ strongly between cultures and are influenced by the culture conditions and the developmental status of the cultures. During studies with carrots, it was found that triglycerides represent about two thirds of the total lipid fraction but only one third in rapeseed and poppy cultures. In general, the composition of the fatty acids in cultured cells is comparable to that of leaves. The use of fats as energy source has not been investigated in these cultured cells.

Detailed investigations on rapeseed cultures were made by Kleinig et al. (1982), in which ^{14}C-labeled acetate was used for 24 h. Acetate binds with coenzyme A to form acetyl CoA, which is the starting point for fatty acid synthesis. In the cytoplasm, acetyl CoA also comes from mevalonate for synthesis of isopentenyl pyrophosphate, the key substance for the synthesis of steroids, carotenoids, gibberellins, abscisic acid, and other essential components of the cell (see also Chap. 10). In the cytoplasm, isopentenyl pyrophosphate is involved in the synthesis of lipids, which are then transferred to different cell compartments. Fatty acid synthesis also takes place in the chloroplasts. The cultures used for the investigations were in their mixotrophic phase of growth. Also PEP could be exported by plastids isolated from embryos of *Brassica napus* L. and used for fatty acid synthesis, accounting for about 30% of all fatty acids synthesized in vivo (Kubis et al. 2004).

^{14}C labeling of phospholipids in rapeseed cultures indicated a maximum of fatty acids with considerable specific activity. Differences were determined for different compartments. In plastids that contain specifically phosphatidylglycerin, this phospholipid has the same specific activity as the universally synthesized phosphatidylcholine, phosphatidyl-inositol, and phosphatidylethanolamine. The plastid characteristic galactolipids are, however, less labeled, as is the typical mitochondrial cardiolipin. Interestingly, the phytol chain of chlorophyll has around 80% less activity than the ring structure.

A high labeling was found in the cytoplasmic steroids and steroid glycosides. The high level of labeling was influenced by the type of nutrition. In heterotrophic cultures, an application of labeled acetate to *Daucus* and *Papaver* resulted in around 50% of ^{14}C in cytoplasmic steroids and their derivatives, whereas this was only 10–15% in total lipids.

In another experiment performed by Yamada et al. (1982), the transfer of fatty acids into other lipids was followed; only results on oleic acid (18:1) will be discussed here. This fatty acid was applied together with diethylene glycol monoethylether. After a pulse of the ^{14}C-labeled compounds for 1 h, there followed an unlabeled chase of 30 h. At the end of the pulse period, more than 50% of the label was found in phosphatidylcholine, and 10–15% in neutral lipids and phosphatidylethanolamine. During the chase period, a continual decrease of labeling was observed in phosphatidylcholine, as well as a proportional increase in neutral lipid. Evidently, there was a transformation of oleic acid into the neutral fraction via the formation of phosphatidylcholine. The formation of the second double bond occurs as phosphatidylcholine in the endoplasmatic reticulum. After this, it is taken up by chloroplasts, and the transformation to monogalactosyldiacylglycerin follows. Here, the formation of the third double bond takes place.

9.1.1 Improving Photosynthesis

Photosynthesis is the basis of plant growth, and improving photosynthesis can contribute toward greater food security in the coming decades as world population increases. Multiple targets have been identified that could be manipulated to increase crop photosynthesis (Evans 2013). According to Weber and Brautigam (2013), photosynthetic efficiency can be improved by (1) increasing the efficacy of photosynthetic carbon assimilation and through (2) optimizing sink strength, thereby relieving feedback inhibition of photosynthesis (Huetsch et al. 2016).

Major efforts are currently underway to introduce carbon concentrating systems into C3 crops plants, such as wheat and rice, with special emphasis on converting C3 plants to C4 plants (Caemmerer von et al. 2012). C4 plants have greater transpiration efficiency, gaining more carbon per unit of water transpired than C3 plants. The combination of these three attributes means that C4 plants fix more carbon per unit of light, per unit of nitrogen, and per unit of water than C3 plants in many situations (Ghannoum et al. 2010). Phosphoenolpyruvate carboxylase (PEPC) is an important enzyme that catalyzes the carboxylation reaction of phosphoenolpyruvate into oxaloacetate, which is then used by the citric cycle (Peng et al. 2012). Why, then, have C4 plants not taken over the world? As the C4 pathway is virtually absent from woody plants, they are unable to displace forest biomes apart from resorting to fire (Osborne 2011). With a few exceptions, C4 plants are also less competitive in colder climates.

Engineering the transport capacity of cellular membranes is an important and frequently underappreciated aspect in efforts to engineer plant primary metabolism. As an example, converting C3 to C4 photosynthesis requires a massively increased metabolic flux for some metabolites across the chloroplast envelope by at least one order of magnitude (Braeutigam and Weber 2011). Proteomic and transcriptomic approaches contributed to identifying candidate proteins carrying this flux (Furumoto et al. 2011; Braeutigam et al. 2011) which now enables the implementation of a complete C4 carbon concentration mechanism. Plant tissue culture can provide a tool to study this in detail (Bender and Kumar 2001).

9.2   Nitrogen Metabolism

Nitrogen is usually supplied to cultured plant cells in three forms—the inorganic ions ammonium, and nitrate and amino nitrogen in the form of amino acids. These amino acids can be applied to the nutrient medium singly or in combination as a mixture of amino acids in casein hydrolysate. Also amides like urea or glutamine have been applied as sources of nitrogen to the medium.

In Fig. 9.9 as a general survey, the concentrations of RNA and of protein (including the activity of some enzymes) during a 4-week culture period of primary carrot callus cultures are summarized. Already in the early 1960s, the Steward group at Cornell University reported a steep increase in protein concentration immediately after initiation of culture of carrot explants grown on BM with coconut milk. This could be confirmed later for cultures in the NL3 medium. This strong initial increase would be related to the formation of threads of cytoplasm traversing cells induced to cell division (cf. Chap. 3). The protein concentration level calculated on a cellular basis, as well as the time of occurrence of the maximum level, is influenced by kinetin. This maximum is followed by a decline, with a minimum about 2 weeks after culture initiation, and it is probably related to the high cell division activity during the log phase. After this, an increase in protein concentration can be seen again until the end of the experiment after 4 weeks of culture. The values on enzyme activity indicate also qualitative differences in the protein synthesized at different stages of culture (Neumann 1995).

A given protein concentration level represents the mass balance between protein synthesis and protein breakdown. Pulse–chase experiments using ^{14}C-labeled metabolites indicated a faster turnover of protein in cultures in a nutrient medium without kinetin. Also this may account for the higher protein concentration of cultures grown with kinetin in the medium (Neumann 1968, 1972, 1995). At all stages, Fe seems to play a central role.

Ammonium or nitrate could be used as sole sources of nitrogen. As shown in the experiment with tobacco callus cultures, growth was better in the medium supplemented with nitrate, compared to that with ammonium. The number of cells per explant, and calculations of average cell size, indicate that the slow growth with NH_4 is due mainly to poor cell division.

Cellular growth is nearly similar with both inorganic sources of nitrogen. Ammonium is taken up in exchange of H^+ ions from the cultures, resulting in a decline in pH in the medium during culture. This leads to a decline of growth, due to changes in the availability of other ions in the medium. Exact studies to this end have not been

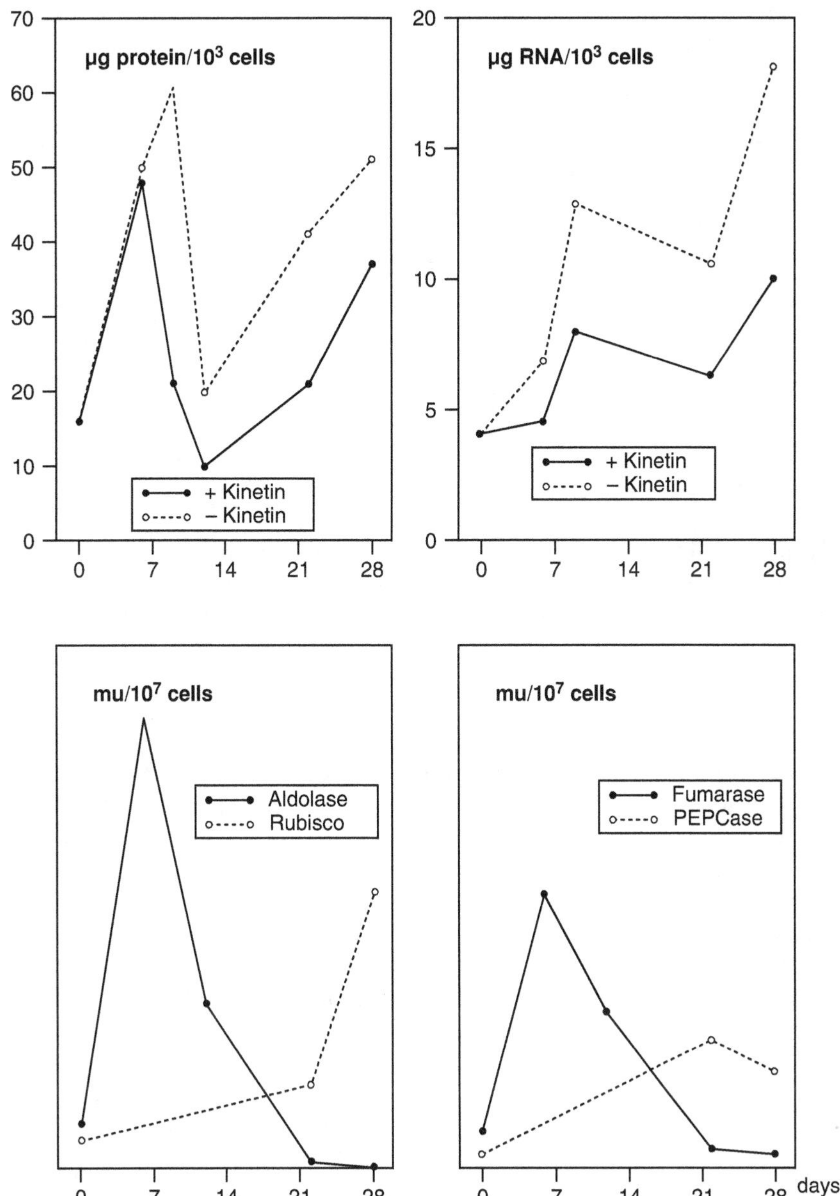

Fig. 9.9 Contents of RNA and protein and the activity of some enzymes during a 4-week carrot callus culture period in NL3

reported. In most media, e.g., the MS medium, both inorganic sources of nitrogen, are supplied. Here, ammonium is utilized first and later nitrate. The excretion of protons following a supplement of ammonia may be the reason for a liberation of

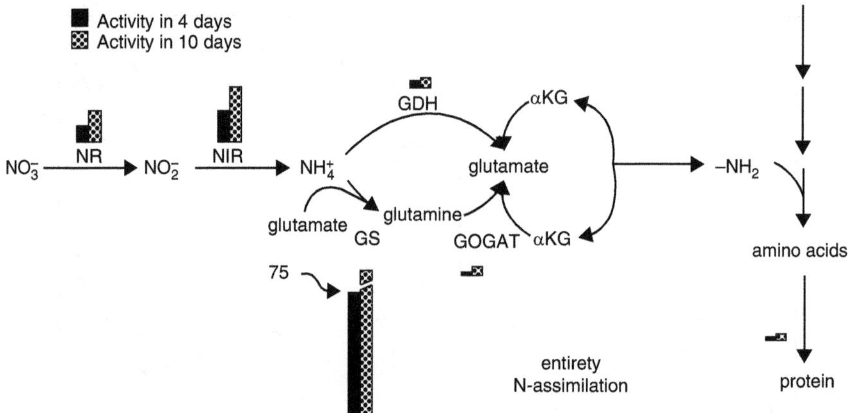

Fig. 9.10 Enzyme activities and nitrogen assimilation of cell cultures (Paul's Scarlet rose; after Fletcher 1982)

cells from callus cultures, to be isolated as a cell suspension for further investigations. If a requirement for such isolated cells exists to produce cell suspensions from callus material for further investigations, then a supplement of ammonia to the medium can often be helpful.

Nitrogen in ammonia is already a reduced form that can be used directly for amino acid synthesis, but nitrate must be reduced in the cells to synthesize amino acids. The reduction takes place via two enzymatic reactions. By means of the nitrate reductase located in the cytoplasm, nitrate is reduced to nitrite. Further reduction from nitrite to NH_4 takes place by the activity of plastid-localized nitrite reductase. This NH_4 is further metabolized either through glutamate dehydrogenase with the formation of glutamic acid by an amino group transfer to oxoglutarate or via glutamine synthetase and the glutamate synthetase reactions (GS–GO–GAT), and glutamic acid is synthesized again. In experiments with rose suspension cultures supplemented with only nitrate, during the first 2 days of culture, measurable amounts of ammonium (0.4 μmol/g f. wt.) and nitrite (1.2 μmol/g f. wt.) were detected in the cells. On the fourth day, cellular nitrate concentration reached 2.3 μmol/g f. wt., and on the fifth day, the amide concentration was at 5.9 μmol/g f. wt., its maximum (Fig. 9.10; Fletcher 1982). More data on the function of the two inorganic nitrogen sources and of casein hydrolysate in differentiation was discussed earlier.

References

Amino SI, Komamine A (1982) Dynamic aspects of cell walls during the cell cycle in a synchronous culture of Vinca rosea L. In: Fujiwara A (ed) Plant tissue culture 1982. The Japanese Association for Plant Tissue Culture, Tokyo, pp 65–66

Bender L, Kumar A (2001) From soil to cell: a broad approach to plant life. Giessen Electron. Library GEB, pp 1–5. http://geb.uni-giessen.de/geb/volltexte/2006/3039/pdf/estschriftNeumann-2001.pdf

Bender L, Neumann KH (1978) Investigations on indole-3-acetic acid metabolism of carrot tissue cultures. Z Pflanzenphysiol 88:209–217

Bender L, Kumar A, Neumann KH (1985) On the photosynthetic system and assimilate metabolism of *Daucus* and *Arachis* cell cultures. In: Neumann KH, Barz W, Reinhard E (eds) Primary and secondary metabolism of plant cell cultures. Springer, Berlin, pp 24–42

Bergmann L (1967) Wachstum grüner Suspensionskulturen von *Nicotiana tabacum* var. Samsun mit CO_2 als Kohlenstoffquelle. Planta 74:243–249

Braeutigam A, Weber APM (2011) Do metabolite transport processes limit photosynthesis? Plant Physiol 155:43–48

Braeutigam A, Kajala K, Wullenweber J, Sommer M, Gagneul D, Weber KL, Carr KM, Gowik U, Maß J, Lercher MJ et al (2011) An mRNA blueprint for C4 photosynthesis derived from comparative transcriptomics of closely related C3 and C4 species. Plant Physiol 155:142–156

Caemmerer von S, Quick WP, Furbank RT (2012) The development of C4 rice: current progress and future challenges. Science 336:1671–1672

Costa JH, Cardoso HC, Campos MD, Zavattieri A, Frederico AM, Fernandes de Molo D, Arnholdt-Schmitt B (2008) *Daucus carota* L.—an old model for cell reprogramming gains new importance through a novel expansion pattern of alternative oxidase (AOX) genes. Plant Physiol Biochem. (in press)

Evans JR (2013) Improving photosynthesis. Plant Physiol 162(4):1780–1793. https://doi.org/10.1104/pp.113.219006

Fletcher JS (1982) Control of nitrogen assimilation in suspension cultures of Paul's scarlet rose. In: Fujiwara A (ed) Plant tissue culture 1982. Japanese Association for Plant Tissue Culture, Tokyo, pp 225–230

Furumoto T, Yamaguchi T, Ohshima-Ichie Y, Nakamura M, Tsuchida-Iwata Y, Shimamura M, Ohnishi J, Hata S, Gowik U, Westhoff P et al (2011) A plastidial sodium-dependent pyruvate transporter. Nature 476:472–475

Ghannoum O, Evans JR, von Caemmerer S (2010) Chapter 8. Nitrogen and water use efficiency of C_4 plants. In: Raghavendra A, Sage R (eds) C4 photosynthesis and related CO2 concentrating mechanisms. Advances in photosynthesis and respiration, vol 32. Springer, Dordrecht

Gowik U, Engelmann S, Blasing OE, Raghvendra AS, Westhoff P (2006) Evolution of C4 for Phosphoenolpyruvate carboxylase in the genus Alternanthera: gene families and the enzymatic characterstics of the C4 isoenzyme and its orthologues in C3/C4 Alternantheras. Planta 223:359–268

Heber U, Heldt HW (1981) The chloroplast envelope: structure, function and role in leaf metabolism. Annu Rev Plant Physiol 32:139–168

Herzbeck H, Hüsemann W (1985) Photosynthesis and carbon metabolism of photoautotrophic cell suspension cultures of *Chenopodium rubrum*. In: Neumann KH, Barz W, Reinhard E (eds) Primary and secondary metabolism of plant cell cultures. Springer, Berlin, pp 15–23

Hüsemann W, Barz W (1977) Photoautotrophic growth and photosynthesis in cell suspension cultures of *Chenopodium rubrum*. Physiol Plant 40:77–81

Hüsemann W, Fischer K, Mittelsbach I, Hübner S, Richter G, Barz W (1989) Photoautotrophic plant cell cultures for studies on primary and secondary metabolism. In: Kurz WGW (ed) Primary and secondary metabolism of plant cell cultures, vol II. Springer, Berlin, pp 35–46

Hütsch BW, Osthushenrich T, Faust F, Kumar A, Schubert S (2016) Reduced sink activity in growing shoot tissues of maize (Zea mays) under salt stress of the first phase can be compensated by increased PEP-carboxylase activity. J Agron Crop Sci 202:384–393

Kleinig H, Hara S, Schuchmann R (1982) Lipid metabolism in plant tissue culture cells. Acetate incorporation, triacylglycerol accumulation. In: Fujiwara A (ed) Plant tissue culture 1982. Japanese Association for Plant Tissue Culture, Tokyo, pp 257–258

Knappe S, Flugge UI, Fischer K (2003) Analysis of the plastidic phosphate translocator gene family in Arabidopsis and identification of new phosphate translocator homologous transporters classified by their putative substrate-binding site. Plant Physiol 131:1178–1190

Kubis SE, Pike MJ, Everett CJ, Hill MN, Rawsthorne S (2004) The import of phosphoenolpyruvate by plastids from developing embryos oilseed rape (*Brassica napus* L.) and the potential as a substrate for fatty acid synthesis. J Exp Bot 55:1455–1462

Kumar A (1974a) Effect of iron and magnesium on the growth and chlorophyll development of the tissues grown in culture. Indian J Exp Biol 12:595–596

Kumar A (1974b) In vitro growth and chlorophyll formation in mesophyll callus tissues on sugar free medium. Phytomorphology 24:96–101

Kumar A, Neumann KH (1999) Comparative investigations on plastid development of meristematic regions of seedlings and tissue cultures of *Daucus carota* L. J Appl Bot Angew Bot 73:206–210

Kumar A, Bender L, Neumann KH (1983a) Photosynthesis in cultured plant cells and their autotrophic growth. J Plant Physiol Biochem 10:130–140

Kumar A, Bender L, Pauler B, Neumann KH, Senger H, Jeske C (1983b) Ultrastructural and biochemical development of the photosynthetic apparatus during callus induction in carrot root explants. Plant Cell Tissue Organ Cult 2:161–177

Kumar A, Bender L, Neumann KH (1984) Growth regulation, plastid differentiation and the development of photosynthetic system in cultured carrot root explants as influenced by exogenous sucrose and various phytohormones. Plant Cell Tissue Organ Cult 3:11–28

Kumar A, Bender L, Neumann KH (1987) Some results of the photosynthetic system of mixotrophic carrot cells (*Daucus carota*). In: Biggins J (ed) Progress in photosynthesis research, vol III(4). Martin Nijhoff, Dordrecht, pp 363–366

Kumar A, Roy S, Neumann KH (1989) Activities of carbon dioxide fixing enzymes in maize tissue cultures in comparison to young seedlings. Physiol Plant 76:185

Kumar A, Bender L, Neumann K-H (1999) Characterization of the photosynthetic system (ultra-structure of plastid, fluorescence induction profiles, low temperature spectra) of *Arachis hypogaea* L. callus cultures as influenced by sucrose and various hormonal treatments. J Appl Bot Angew Bot 73:211–216

Nato A, Hoarau J, Brangeon J, Hirel B, Suzuki A (1985) Regulation of carbon and nitrogen assimilation pathways in tobacco cell suspension cultures in relation with ultrastructural and biochemical development of the photosynthetic apparatus. In: Neumann KH, Barz W, Reinhard E (eds) Primary and secondary metabolism of plant cell cultures. Springer, Berlin, pp 43–57

Neumann KH (1962) Untersuchungen über den Einfluß essentieller Schwermetalle auf das Wachstum und den Proteinstoffwechsel von Karottengewebekulturen. Dissertation, Justus Liebig Universität, Giessen

Neumann KH (1966) Wurzelbildung und Nukleinsäuregehalt bei Phloem-Gewebekulturen der Karottenwurzel auf synthetischem Nährmedium. Congr Coll Univ Liege 38:96–102

Neumann KH (1968) Über Beziehungen zwischen hormonalgesteuerter Zellteilungsgeschwindigkeit und Differenzierung, ein Beitrag zur Physiologie der pflanzlichen Ertragsbildung. Habilitation, Justus Liebig Universität, Giessen

Neumann KH (1969) Der Eintritt der Elemente Kohlenstoff, Wasserstoff und Sauerstoff in den Stoffwechsel. In: Linser H (ed) Handbuch der Pflanzenernährung und Düngung. Springer, Wien, pp 301–374

Neumann KH (1972) Untersuchungen über den Einfluß des Kinetins und des Eisens auf den Nukleinsäure- und Proteinstoffwechsel von Karottengewebekulturen. Zeitschr Pflanzenernäh Bodenkunde 131:211–220

Neumann KH (1995) Pflanzliche Zell- und Gewebekulturen. Eugen Ulmer, Stuttgart

Neumann KH, Raafat A (1973) Further studies on the photosynthesis of carrot tissue cultures. Plant Physiol 51:685–690

Neumann KH, Pertzsch C, Moos M, Krömmelbein C (1978) Untersuchungen zur Charakterisierung des Ernährungssystems von Karottengewebekulturen, *Daucus carota* L. (Photosynthese,

Kohlenstoff- und Stickstoffernährung, Mineralstoffverbrauch) unter dem Einfluß des Kinetins. Zeitschr Pflanzenernähr Bodenk 141:298–311

Neumann KH, Bender L, Kumar A, Szegoe M (1982) Photosynthesis and pathways of carbon in tissue cultures of *Daucus* and *Arachis*. In: Proceedings of the international congress of plant tissue and cell culture, Tokyo, pp 251–252

Neumann KH, Gross U, Bender L (1989) Regulation of photosynthesis in *Daucus carota* and *Arachis hypogea* cell cultures by exogenous sucrose. In: Kurz WGW (ed) Primary and secondary metabolism of plant cell cultures, vol II. Springer, Berlin, pp 281–294

Osborne CP (2011) The geologic history of C4 plants. In: Raghavendra AS, Sage RF (eds) C4 photosynthesis and related CO2 concentrating mechanisms, vol 32. Springer, Dordrecht, pp 339–357

Peng Y, Cai J, Wang W, Su B (2012) Multiple inter-kingdom horizontal gene transfers in the evolution of the phosphoenolpyruvate carboxylase gene family. PLoS One 7(12):e51159. https://doi.org/10.1371/journal.pone.0051159

Sasaki T, Kainuma K (1982) Regulation of starch synthesis and external polysaccharide synthesis by gibberellic acid in cultured sweet potato cells. In: Fujiwara A (ed) Plant tissue culture 1982. Japanese Association for Plant Tissue Culture, Tokyo, pp 255–256

Steward PC, Mapes MO, Mears K (1952) Investigation on growth and metabolism of plant cells. I. New techniques for the investigation of metabolism, nutrition and growth in undifferentiated cells. Am J Bot 16:57–77

Steward FC, Mapes MO, Mears K (1958) Growth and organized development of cultured plant cells. Am J Bot 45:705–708

Sun JY, Chan YM, Wang OM, Chen J, Wang XC (2006) Glucose inhibits the expression triose phosphate/phosphate translocator gene in wheat via hexokinase-dependent mechanism. Int J Biochem Cell Biol 38:1102–1113

van der Plas LHW, Wagner MJ (1982) Respiratory physiology of potato tuber callus cultures. In: Fujiwara A (ed) Plant tissue culture 1982. Japanese Association for Plant Tissue Culture, Tokyo, pp 259–260

Weber APM, Brautigam A (2013) The role of membrane transport in metabolic engineering of plant primary metabolism. Curr Opin Biotechnol 24:256–262. https://doi.org/10.1016/j.copbio.2012.09.010

Widholm JM (1989) Initiation and characterization of photoautotrophic suspension cultures. In: Kurz WGW (ed) Primary and secondary metabolism of plant cell cultures, vol II. Springer, Berlin, pp 3–13

Widholm JM (1992) Properties and uses of photoautotrophic plant cell cultures. Int Rev Cytol 132:109–175

Widholm JM (2000) Plant cell cultures: photosynthetic. In: Encyclopedia of cell technology. Wiley, New York, pp 1004–1009

Yamada Y, Sato F, Watanabe K (1982) Photosynthetic carbon metabolism in cultured photoautotrophic cells. In: Fujiwara A (ed) Plant tissue culture 1982. Japanese Association for Plant Tissue Culture, Tokyo, pp 259–260

10

10.1 Introduction

The phenomenon of secondary metabolism was already recognized in the early phases of modern experimental botany. In his textbook published in 1873, Julius Sachs, one of the great pioneers of plant physiology, gave the following definition:

"Als Nebenprodukte des Stoffwechsels kann man solche Stoffe bezeichnen, welche während des Stoffwechsels entstehen, aber keine weitere Verwendung für den Aufbau neuer Zellen finden. Irgend eine Bedeutung dieser Stoffe für die innere Ökonomie der Pflanze ist bis jetzt nicht bekannt" (Sachs 1873, p. 641). Translation: "We can designate as by-products of metabolism such compounds that are formed by metabolism, but that are no longer used for the formation of new cells. Any importance of these compounds for the inner economy of the plant is as yet unknown." This clear statement is still valid. Sachs did not refer to any functions of the by-products, today known as secondary products (see review by Hartmann 1996).

Plants form an important part of our everyday diet, and their constituents have been intensively studied for decades. In addition to essential primary metabolites (e.g., carbohydrates, lipids, and amino acids), higher plants are able to synthesize a wide variety of low-molecular-weight compounds—the secondary metabolites (Fig. 10.1). The production of these compounds is often low (less than 1% of dry weight) and depends strongly on the physiological and developmental stage of the plant.

Although plant secondary metabolites seem to have no recognized role in the maintenance of fundamental life processes of the plants that synthesize these, they do have an important role in the interaction of the plant with its environment (Isah 2019).

To study secondary metabolism per se is an exciting area of plant physiology or actually of botany in general. Moreover, many of its constituents are important substances of medical interest and other areas of human life, and therefore in vitro studies on this topic were soon of commercial interest. Investigations focused on

K.-H. Neumann et al., *Plant Cell and Tissue Culture – A Tool in Biotechnology*,
https://doi.org/10.1007/978-3-030-49098-0_10

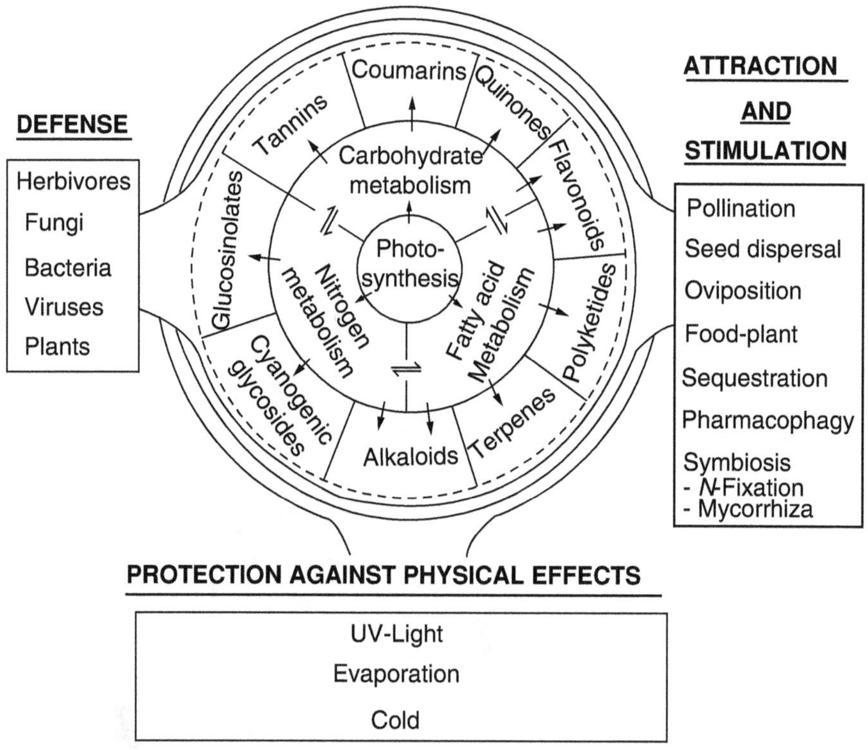

Fig. 10.1 Secondary metabolites originate from common precursors as products of primary metabolism. Three basic metabolic processes governed by photosynthesis, i.e., nitrogen metabolism, fatty acid metabolism, and carbohydrate metabolism, are responsible for the synthesis of secondary metabolites like alkaloids, terpenes, polyketides, flavonoids, quinones, coumarins, tannins, glucosinolates, and cyanogenic glycosides. Their functions involve all aspects of a plant's chemical interactions with the environment (from Hartmann 1996)

metabolites to be produced by cultured cells of some plant species producing commercially highly valuable chemicals (Zárate and Yeoman 2001). These investigations included, e.g., the characterization of several hundred enzymes also as a contribution to basic interests. Still, due to the economic importance of this topic, in the following, commercial aspects will dominate.

At least one fourth of all prescribed pharmaceuticals in industrialized countries contain compounds that are directly or indirectly, via semi-synthesis, derived from plants. Many of these pharmaceuticals are still in use today, and often no useful synthetic substitutes have been found that possess the same efficacy and pharmacological specificity. Furthermore, 11% of the 252 basic and essential drugs considered by WHO are exclusively derived from flowering plants (Rates 2001). Misawa (1991) reviewed the production of secondary metabolites in plant tissue culture in an FAO bulletin. Indeed, prescription drugs containing phytochemicals were valued at more than US$30 billion in 2002 in the USA (Raskin et al. 2002).

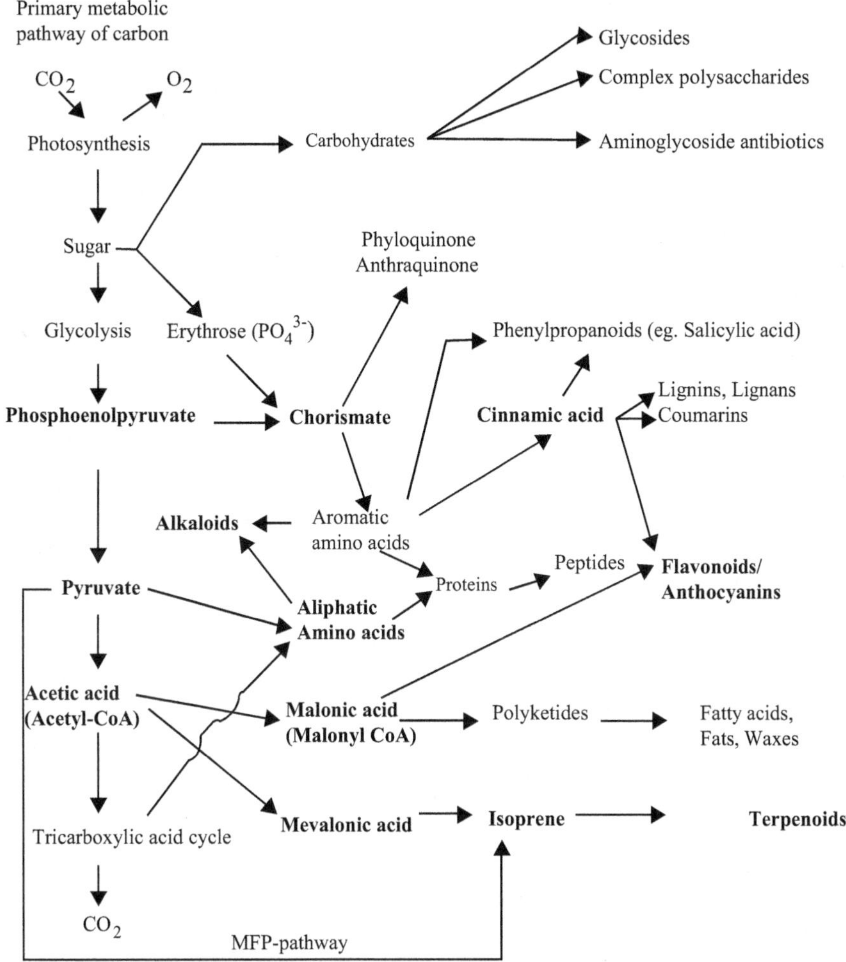

Fig. 10.2 Major pathways of biosynthesis of secondary metabolites, viz., polyketides, isoprenoids (e.g., terpenoids), alkaloids, phenylpropanoids, and flavonoids (after Verpoorte et al. 2000)

Based on their biosynthetic origins, plant secondary metabolites can be structurally subdivided into five major groups (Fig. 10.2): polyketides, isoprenoids (e.g., terpenoids), alkaloids, phenylpropanoids, and flavonoids (Kitaoka et al. 2015). Some carbohydrates can also be classified as secondary metabolites (McCranie and Bachmann 2014) and saccharides that are unique to secondary metabolism often decorate the core structure of many secondary metabolites:

1. The polyketides are produced via the acetate–mevalonate pathway.
2. The isoprenoids (terpenoids and steroids) are derived from the five-carbon precursor isopentenyl diphosphate (IPP), produced via the classical mevalonate pathway or the novel MEP (non-mevalonate or Rohmer) pathway.

3. The alkaloids are synthesized from various amino acids.
4. Phenylpropanoids having a C6–C3 unit are derived from aromatic amino acids, phenylalanine, or tyrosine.
5. Flavonoids are synthesized by the combination of phenylpropanoids and polyketides (Verpoorte et al. 2000).

1. *Polyketides* are a large family of natural products found in bacteria, fungi, and plants and include many clinically important drugs such as tetracycline, daunorubicin, erythromycin, rapamycin, and lovastatin. Polyketides are synthesized from acetate units and are also subdivided into several classes (McDaniel et al. 2005). In plants, polyketide biosynthesis intersects with aromatic amino acid biosynthesis to generate phenylpropanoid structures, such as flavonoids, flavonones, stilbenes, and anthocyanins (Vogt 2010). The polyketides are produced via the acetate–mevalonate pathway e.g., *erythromycin A* and *erythromycin C*.
2. *Isoprenoids* in plants occur in the essential oils, which are found in the gummy exudates (oleoresins and latices) of many trees and shrubs. The isoprenoids (terpenoids and steroids) are derived from the five-carbon precursor isopentenyl diphosphate (IPP) known as isoprene produced via the classical mevalonate pathway or the novel MEP (non-mevalonate or Rohmer or methylerythritol phosphate) pathway (Kitaoka et al. 2015). They include terpenes: triterpene *Avenacin A* 1. Diterpene *Pacliaxel (Taxol)*. Thus in plants, terpene biosynthesis generally occurs in either the cytosol via the mevalonic acid (MVA) pathway to make sesquiterpenes and triterpenes or in the plastid via the (MEP) pathway to make monoterpenes and diterpenes. Plant isoprenoids affect growth (e.g., the hormone gibberellic acid) and contribute to red, yellow, and orange pigments carotenoids.
3. Alkaloids encompass a broad range of metabolites with the sole commonality of a basic nitrogen atom in a heterocyclic ring, occurring chiefly in flowering plants (Ashihara et al. 2008; Ziegler and Facchini 2008; Duge et al. 2014; Xu et al. 2014). *Alkaloids* are organic compounds basic in nature. The alkaloids are synthesized from various amino acids, e.g., alkaloids—*morphine* and *vinblastine*. Plant-derived alkaloids currently in clinical use include the anticancer agents camptothecin (CPT), Taxol, vincristine, and vinblastine, the analgesics codeine and morphine, the gout suppressant colchicine, the muscle relaxant (+)-tubocurarine, the antiarrhythmic ajmalicine, the antimalarial quinine, the antiamoebic emetine, the antibiotic sanguinarine, the sedative scopolamine, and the topical analgesic capsaicin to mention some of the most representative examples (Raskin et al. 2002). Many alkaloids, such as nicotine, quinine, and cocaine are known for their poisonous or medicinal attributes. Isoquinoline alkaloids include, among others, the important medicines morphine and codeine (Mora-Palea et al. 2013).
4. *Phenylpropanoids* are synthesized by shikimate pathway. Phenylpropanoids having a C6–C3 unit are derived from aromatic amino acids, phenylalanine, or

tyrosine, e.g., flavonoid, anthocyanin, and stilbene. Phenylpropanoid-based polymers, like lignin, suberin, or condensed tannins, contribute substantially to the stability and robustness of gymnosperms and angiosperms toward mechanical or environmental damage, like drought or wounding. Its intracellular, plastidial location and complex regulation have been investigated for more than a decade (Schmid and Amrhein 1995).

5. *Flavonoids* are synthesized by the combination of phenylpropanoids and polyketides (Verpoorte et al. 2000). Flavonoids are polyphenolic molecules containing 15 carbon atoms and are soluble in water. They consist of two benzene rings connected by a short three carbon chain. The flavonoids can be divided into six major sub types, which include chalcones, flavones, isoflavonoids, flavanones, anthoxanthins, and anthocyanins.

Flavonoids are essential pigments for producing colors in plants that will attract pollinating insects. They are therefore found in a variety of fruit and vegetables. Catechins are important flavonoids abundant in the leaves of the tea plant. Green tea is rich in flavonoids that possess antioxidant properties, which can protect against heart disease and cancer.

Isoflavones have an affinity to estrogen β receptors in humans and are reported to exhibit numerous health-promoting effects, including the alleviation of menopausal symptoms, the prevention of osteoporosis and cardiovascular diseases, and lowering the risk of breast cancer (Patisaul and Jefferson 2010).

10.2 Mechanism of Production of Secondary Metabolites

There are some basic metabolic pathways for the synthesis of secondary metabolites, as shown in Fig. 10.3. These metabolites form five major groups, as mentioned above.

One possible way to classify the 12,000 known alkaloids is to further subdivide these into the following 15 subclasses: proto-, piperidine, pyrrolidine, pyridine, quinolizidine, tropane, pyrrolizidine, imidazole, purine, quinoline, isoquinoline, quinazoline, indole, terpenoid, and steroidal alkaloids.

Secondary metabolism is an integral part of the developmental program of plants, and the accumulation of secondary metabolites can demarcate the onset of developmental stages. However, only a few pathways (e.g., flavonoids and terpenoid indole and isoquinoline alkaloids) in plants are well understood today, after many years of classical biochemical research (e.g., Street 1977; Staba 1980; Dixon and Steele 1999; Hashimoto and Yamada 2003; Vanisree and Tsay 2004; Vancanneyt et al. 2004; Vanisree et al. 2004; Ramirez-Estrada et al. 2016).

Detailed biosynthetic pathways of these metabolites are beyond the scope of this book. Thus, a brief outline of various key compounds within plants and of their biosynthetic pathways will be given.

Secondary metabolites belonging to a given subclass are not always synthesized from the same primary metabolites, but their chemical structures share the same basic skeleton. Cinnamic acid and its simple derivatives are the common precursors

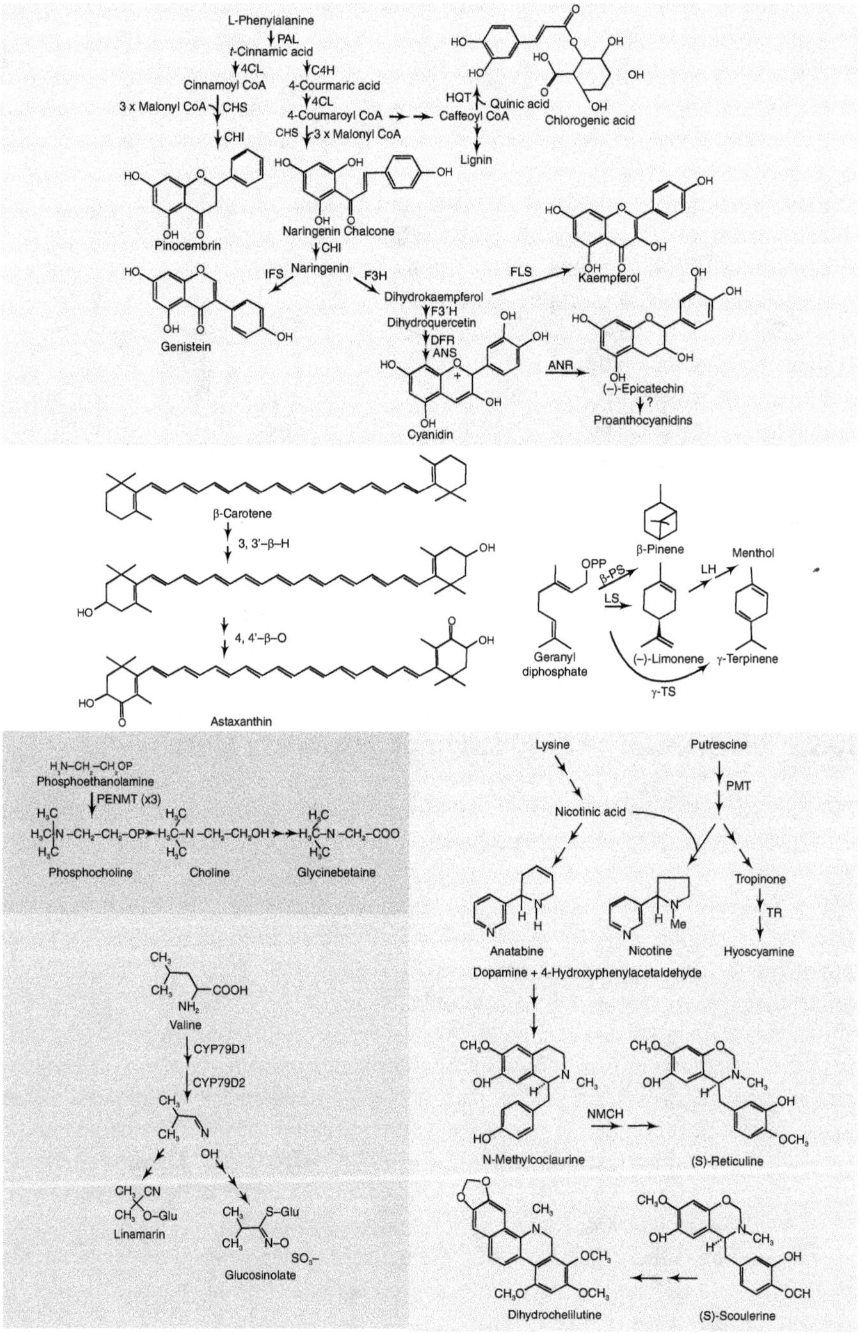

Fig. 10.3 Schematics of pathways for the production of natural products

of key intermediates of the various phenylpropanoid classes illustrated. In turn, the class-specific key intermediates are structurally diversified to yield 1000s of individual compounds (Hartmann 1996).

Because of the activity of enzymes with different substrate- and stereospecificity, the chemical diversity and biological activity of the molecules belonging to a given subclass can be enormous (Tulp and Bohlin 2002). For example, various types of cyclic monoterpenes are synthesized from the common precursor geranyl diphosphate by action of specific monoterpene cyclases. Some subclasses are found only in a few plant families (e.g., medicinal tropane alkaloids are found only in the Solanaceae and Erythroxylaceae), whereas flavonoids, for example, are widely distributed throughout the plant kingdom. The concept of combinatorial biochemistry is based on the fact that different plants, either closely or more distantly related, synthesize structurally similar but nevertheless diverse molecules. As such, it can be expected that an enzyme with a certain substrate specificity isolated from one plant might encounter new but related substrates when introduced into another plant. This has been experimentally proved, as given below (Li et al. 2017). Thus, by introducing genes involved in the biosynthesis of a given compound isolated from one plant into another plant synthesizing related molecules, new chemical structures not previously found in nature may be obtained.

Successful attempts of insertion of more than one gene of a known pathway into a host organism have also been reported. For instance, following particle bombardment of tobacco leaves and plant regeneration, the expression of two consecutive genes involved in the terpenoid indole alkaloid pathway of *Catharanthus roseus* has been reported; *C. roseus* is a well-known species able to accumulate the two potent anticancer drugs vincristine and vinblastine encoding tryptophan decarboxylase (TDC) and strictosidine synthase (STR1) in tobacco plants (Leech et al. 1998). TDC and STR1 are two adjacent pathway enzymes that together form strictosidine, which is an important intermediate of over 3000 indole alkaloids (Fig. 10.4), many of which possess important pharmaceutical properties. Both TDC and STR1 genes are absent in tobacco plants. Analysis of transgenic plants at the RNA and DNA levels demonstrated a range of integration events and steady-state transcript levels for both transgenes, besides a 100% co-integration of both transgenes (Zárate and Yeoman 2001). Similarly, a gene involved in the terpenoid indole alkaloid pathway of *C. roseus*, SGD (cf. strictosidine β-D-glucosidase; Fig. 10.4), has been introduced via *Agrobacterium tumefaciens* and expressed in suspended tobacco cells (Zárate 1999).

There is also an account where a whole heterologous secondary metabolic pathway was expressed in a host plant (Ye et al. 2000) following *A. tumefaciens*-mediated transformation. Ye et al. (2000) introduced the entire β-carotene biosynthetic pathway, vitamin A precursor, into rice endosperm in a single transformation effort with three vectors harboring four transgenes: psy, plant phytoene synthase; crt-1, bacterial phytoene desaturase; lcy, lycopene β-cyclase; and tp, transient peptide. In most cases, the transformed endosperms were yellow, indicating carotenoid formation, and in some lines β-carotene was the only carotenoid detected. This elegant report illustrates how the nutritional value of a major staple food may be augmented by recombinant DNA technology.

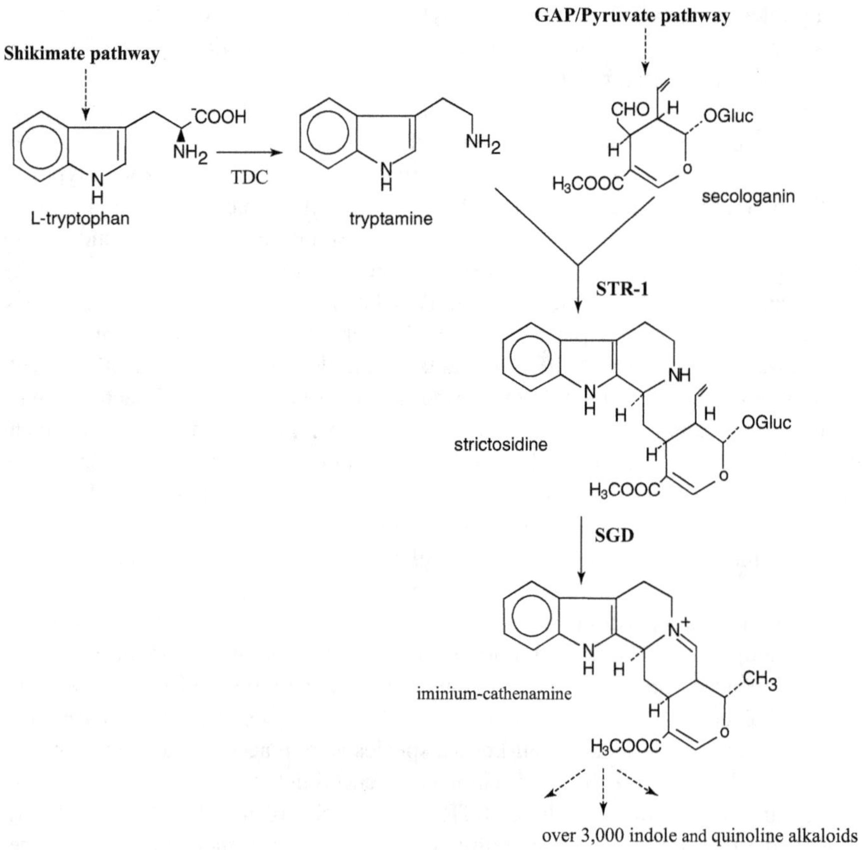

Fig. 10.4 Partial illustration of the biosynthetic pathway of terpenoid indole alkaloids in *Catharanthus roseus* leading to the formation of the intermediate strictosidine, central precursor of over 3000 indole and quinoline alkaloids. *TDC* tryptophan decarboxylase, *STR-1* strictosidine synthase, *SGD* strictosidine glucosidase, *GAP* glyceraldehyde-3-phosphate (Zárate and Yeoman 2001)

Genetic transformation of the Indian medicinal plant, *Bacopa monnieri*, using a gene encoding cryptogein, a proteinaceous elicitor, via Ri and Ti plasmids, was established and induced bioproduction of bacopa saponins in crypt-transgenic plants was obtained (Majumdar et al. 2012).

For further details on genetic transformation, the reader is referred to Sect. 13.6.3.

10.3   Historical Background

In contrast to primary metabolism of cell cultures where only limited investigations have been carried out, the literature is full of investigations on secondary metabolism. This difference in the variability of information is due to the fact that the intermediate products and end products of primary metabolism can be obtained from agriculture in huge amounts at low costs, in contrast to secondary plant products of high value that fetch high prices for even small amounts to be used in cosmetic or pharmaceutical industries (Charlwood et al. 1990; Misawa 1991; Komamine et al. 1991; Neumann 1995; Bender and Kumar 2001; Alfermann et al. 2003; Vanisree and Tsay 2004; Vanisree et al. 2004; see also Kumar and Roy 2006, 2011; Kumar and Sopory 2008; Kumar and Shekhawat 2009). Bourgaud et al. (2001) reviewed the historical perspective of plant secondary metabolite production (see also Kumar and Sopory 2010).

To date, there has been continuous increase of patents filed for products based on tissue culture by commercial companies. These include additives to food and pigments. These substances were often obtained from raw materials imported from tropical and subtropical regions. To ensure continuous production, storage of significant amounts of these raw materials is required, associated with considerable costs and risks. In addition, they can vary strongly in quality, depending on the year of production and the regions of export, and also in price, depending on economic considerations like changes in world market prices. All these factors have stimulated the production of secondary products under controlled conditions in plant tissue culture laboratories near the commercial unit, to produce the final product for the market.

By the beginning of the 1970s, plant cell culture had attained a developmental status employing methods of microbial fermentation techniques—e.g., antibiotic production to be used for large-scale cultures from plants, in order to avoid the above mentioned problems of imports of raw materials. Today, up to 30% of medical prescriptions are based on plants or contain plant components. Traditional medicinal systems utilize plant-based medicines and are experiencing a revival worldwide. This has resulted in enormous pressures on biodiversity and the destruction of valuable biotopes particularly in developing countries involved in meeting the demands of global markets. Tissue culture could provide alternatives.

Among the plant-derived compounds are two drugs derived from the Madagascar periwinkle (*Catharanthus roseus*): vinblastine and vincristine. Other examples of important drugs derived directly or indirectly from plants include the anticancer drugs paclitaxel (Taxol), podophyllotoxin, and camptothecin, the analgesic drug morphine, and semi-synthetic drugs such as the vast group of steroidal hormones derived from diosgenin. There is revival of interest in plant secondary metabolites, as there has been only limited success of combinatorial chemistry or computational drug design to deliver novel pharmaceutically active compounds (Müller-Kuhrt 2003).

The products of highest market interest are based on glycosides and alkaloids. Besides these, steroids, enzymes, and pigments are of considerable interest. Table 10.1 provides some of the important plants and their products that have a potential for use in tissue culture.

Only few plant materials were used at the beginning, i.e., systems for the production of heart alkaloids from *Digitalis*, and atropine and scopolamine from *Datura* cultures. *Lithospermum* produces antimicrobial agents, shikonin being of

Table 10.1 Compounds of industrial interest produced in plant tissue culture

S. no.	Effects	Plants
1	Antimicrobial effects (virus) (protozoan) (bacteria) (bacteria)	*Agrostemma/Phytolacca* *Catharanthus* *Lithospermum* *Ruta*
2	Antitumor effects	*Camptotheca, Antharanthus, Maytenus, Podophyllum,* *Taxus, Tripterygium*
3	Painkillers	*Chamomilla, Valeriana, Papaver*
4	Enzymes for proteolysis	*Papaya, Scopolia, Ananas*
5	Enzymes for biotransformation	*Cannabis, Digitalis, Lupinus, Mentha, Papaver*
6	Appetizers or taste enhancers	*Asparagus, Apium graveolens, Allium, Capsicum, Sinapis*
7	Hydrocarbon-yielding	*Asclepias, Euphorbia*
8	Sweeteners	*Glycyrrhiza, Hydrangea, Stevia*
9	Tonics	*Bluperrum, Cinchona, Coptis, Phellodendron, Panax*
10	Insecticides	*Derris, Pyrethrum*

particular importance (Yamamura et al. 2003). The synthesis of methyldigoxin by hydroxylation of methyldigitoxin was another goal (Alfermann et al. 1985). *Coptis* is used for making tonics of berberines.

Due to the increased appeal of natural products for medicinal purposes, metabolic engineering can have a significant impact on the production of pharmaceuticals and help in the design of new therapies. The candidate plant cell cultures are generally chosen by screening from medicinal and aromatic plants already used in drug production. At present, research and development are focused on plants producing substances with immunomodulating, antiviral, antimicrobial, antiparasite, antitumor, anti-inflammatory, hypoglycemic, tranquilizer, and antifeedant activity (Yamada 1991).

The last 15 years have produced a large quantity of results on the biosynthetic pathways leading to secondary metabolites. Concomitantly, at the beginning of the 1990s, a new discipline called metabolic engineering appeared. According to Bailey (1991), metabolic engineering is "the improvement of cellular activities by manipulation of enzymatic, transport, and regulatory functions of the cell with the use of recombinant DNA technology." In many cases, this approach relies on the identification of limiting enzyme activities after successful pathway elucidation and metabolite mapping (metabolomics). Such limiting steps are improved with an appropriate use of genetic transformation. Most of the strategies developed so far are based on the introduction of genes isolated from more efficient organisms, promoters that enhance the expression of a target gene, or antisense and co-suppression techniques for the obtainment of plants with the desired traits. In addition to their synthesis as such, the transport of metabolites within the plant system and its localization play a key role in optimizing the yield. Recently, attempts have been made to understand

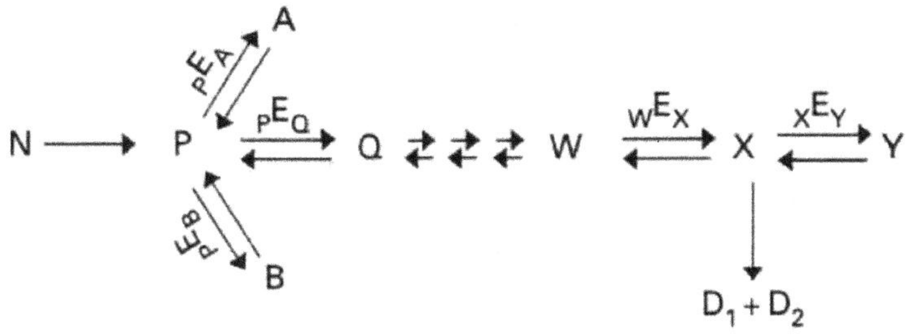

Fig. 10.5 Model to regulate secondary metabolism (Yeoman and Yeoman 1996)

the regulation of transport (Yazaki 2005). An emerging approach is to synthesize secondary metabolites in compartments where the metabolic process does not normally occur (Wu et al. 2012). For example, localizing terpene biosynthetic pathways to the plastids of plants has been shown to result in high levels of product (Heinig et al. 2013).

Quite some time ago, Yeoman et al. (1980) suggested an interesting model to influence the synthesis of secondary products (Fig. 10.5). In this model, W is the immediate precursor of the substance X to be produced, and P an unspecific precursor from which X can be derived following the production of Q, the first specific intermediate in the pathway eventually producing X.

Based on this model, there are several possibilities to promote the synthesis of X as the desired product. For a start, optimizing the metabolic intensity of the cultures will establish the basic production of X. Moreover, P can be diverted to alternative pathways symbolized as A and B. The entrance of P into the metabolic pathway specific for the synthesis of X can be limited by a low activity of the enzymes pEQ, or the following enzymes. Finally, also X could simply be an intermediate of the synthesis of Y. Consequently, its concentration would be determined by the activity of the two enzymes WEX and XEY as an equilibrium of the synthesis of X and Y, and at a given time, a given concentration of X would be determined. The concentration of X will also be influenced by direct breakdown (D1 + D2) or by fixation as a conjugate (K with other molecules). Especially the formation of conjugates has been investigated quite extensively these recent years.

Based on this (certainly too) simple model, some conclusions can already be drawn to initiate more detailed investigations. One possibility to promote the reaction chain P–Q–W–X is the application of Q to the nutrient medium, this being the first pathway-specific intermediate. In terms of simple enzyme kinetics, it can be assumed that the reaction P to Q will be inhibited by an excess of P in the cells. Another possibility to promote the pathway to produce X is a supplement of A, B, or Y. Making use of various possibilities to influence the production of the target substance requires knowledge of the metabolism of this compound, as well as of the pool size of the various molecules and the equilibrium conditions of the enzymes involved. As described, such information is available for some cell culture systems (see also below). The Yeoman group used this model as a basis to optimize capsaicin

Table 10.2 Influences of some precursors of the synthesis of tropane alkaloids on alkaloid concentration (μg/g dry wt.) of haploid cell suspensions of *Datura innoxia* Mill[a]

	Tropine	Atropine
Control	15	0
+Leucine and glycine	10	0
+Ornithine and phenylalanine	90	Traces
+Tropine and tropic acid	175	75

[a]Application of precursors for 1 week after 3 weeks pre-culture

production. Into this scheme, it would be of interest to include changes in enzyme availability following gene technological manipulations of the cells.

Basically, the assumptions of the model have been confirmed also in our own studies to produce atropine and scopolamine in *Datura* cultures. These are the two main alkaloids of this species, synthesized from the two amino acids phenylalanine and ornithine and symbolized as P in the model. The latter is transformed via tropine and tropic acid into atropine and finally into scopolamine. Tropine and tropic acids are symbolized as Q/W in the model. At a supplement of ornithine or phenylalanine, or both, to the *Datura* cultures (symbolized as P), only the concentration of tropine is increased, i.e., of Q/W. An application of tropic acid and tropine, however, results in an increase in the atropine concentration (Table 10.2) in other experiments also of scopolamine (Forche, unpublished results of our institutes; see Neumann 1995).

Several other laboratories have reported secondary metabolite production from plant tissue cultures (Carew and Staba 1965; Khanna and Staba 1968; Khanna 1977; Barz et al. 1977; Kibler and Neumann 1980; Neumann et al. 1985; Alfermann and Reinhard 1986; Furuya 1988; P.R. Holden et al. 1988; Holden 1990; Vasil 1991; Abe et al. 1993; Neumann 1995; Datta and Srivastava 1997; Jain et al. 1998; Jacob and Malpathak 2006; Narula et al. 2006; Hiroaka and Bhatt 2008; Kukreja and Garg 2008; Sonderquist and Lee 2008; Jacob et al. 2008; Sharada et al. 2008; Srivastava et al. 2008). Vanisree et al. (2004) and Dixon (2005) reviewed the production of secondary metabolites in tissue culture and engineering of natural product pathways, respectively.

More than 50 years ago, Routien and Nickel (1956) suggested the potential for the production of secondary metabolites in culture and received the first patent. Later, the National Aeronautics and Space Administration (NASA) started to support research on plant cell cultures for regenerative life-support systems (Krikorian and Levine 1991; Krikorian 2001). Indeed, since the early 1960s, experiments with plants and plant tissue cultures have been performed under various conditions of microgravity in space (one-way spaceships, biosatellites, space shuttles and parabolic flights, the orbital stations Salyut and Mir), accompanied by ground studies using rotating clinostat vessels (http://www.estec.esa.nl/spaceflights).

10.4   Plant Cell Cultures and Pharmaceuticals and Other Biologically Active Compounds

Plant cells have been successfully used as "factories" to produce high-value secondary metabolites under economically viable conditions, in some notable cases. Since Tabata et al. (1974) first described the production of shikonin pigments by callus cultures of *Lithospermum erythrorhizon*, intensive efforts have been made to identify the regulatory factors controlling shikonin biosynthesis. As a result, shikonin represents the first example of industrial production of a plant-derived pharmaceutical (Tabata and Fujita 1985). Shikonin is a red naphthoquinone pigment that is used in traditional dyes, another major application being for lipsticks. Shikonin acyl esters exhibit various pharmacological properties including anti-inflammatory and antitumor activity (Chen et al. 2002). Other examples are berberine production by cell cultures of *Coptis japonica*, rosmarinic acid production by cell cultures of *Coleus blumei*, and sanguinarine production by cell cultures of *Papaver somniferum* (Eilert et al. 1985; Ulbrich et al. 1985). An example of a high-value drug produced partially from plant cell cultures is paclitaxel, an anticancer drug originally extracted from the bark of 50–60 year old Pacific yew trees (*Taxus brevifolia*; http://www.phyton-inc.com; Zenk et al. 1988; Ketchum et al. 1999; Tabata 2004). Recent advances in the molecular biology, enzymology, and fermentation technology of plant cell cultures suggest that these systems will become a viable source of important secondary metabolites (Vanisree et al. 2004).

A brief description of some important secondary metabolites, their structure, and production in plant tissue culture is given below.

Alkaloids are a group of nitrogen-containing bases. They are physiologically active in humans (e.g., cocaine, nicotine, morphine, strychnine) and chemotherapeutics (vincristine, vinblastine, camptothecin derivatives, and paclitaxel). Some of the important alkaloids are nicotine of *Nicotiana*, the tropane alkaloids of *Hyoscyamus*, *Datura*, and *Atropa*, the isoquinoline alkaloids of *Coptis* and *Eschscholzia californica*, and the terpenoid indole alkaloids of *Catharanthus roseus* and *Rauvolfia serpentina* (Rates 2001; Hughes and Shanks 2002).

Papaver somniferum L. (opium poppy) is a traditional commercial source of codeine and morphine. Two tyrosine rings condense to form the basic structure of morphine. During this process, the first important intermediate is dopamine, which is also the starting substance of the biosyntheses of berberine, papaverine, and morphine. Production of morphine and codeine in morphologically undifferentiated cultures has been reported by Li and Doran (1991).

Berberine is an isoquinoline alkaloid that occurs in roots of *Coptis japonica* and the cortex of *Phellodendron amurense*. Berberine chloride is used for intestinal disorders in the Orient. However, it takes 5–6 years to produce *Coptis* roots as the raw material. Berberine has been reported from a number of cell cultures—e.g., *C. japonica*, *Thalictrum* spp., and *Berberis* spp. Sato and Yamada (1984) improved the productivity of berberine in cell cultures by optimizing the nutrients in the growth medium and the levels of phytohormones.

L-DOPA, L-3,4-dihydroxyphenylalanine, is the precursor of the alkaloids betalain, melanin, and others. It is also a precursor of catecholamines in animals and is being used as a potent drug for Parkinson's disease, a progressive disabling disorder associated with a deficiency of dopamine in the brain. The widespread application of this therapy has created a demand for large quantities of L-DOPA at an economical price level, and this has led to the introduction of cell cultures as an alternative means for enriched production. Brain (1976) found that the callus tissue of *Mucuna pruriens* accumulated 25 mg DOPA/l medium containing relatively high concentrations of 2.4D. The DOPA synthesized by plant tissues is secreted mostly into the medium.

Scopolamine and hyoscyamine are tropane alkaloids that are used in anesthetic and antispasmodic drugs. Ornithine is one of the starting materials for their synthesis, and methylornithine is the first intermediate. These alkaloids occur in leaves of solanaceous plants including *Datura* sp., *Atropa*, *Hyoscyamus*, and *Scopolia* sp.

Capsicum frutescens produces the alkaloid capsaicin in nature, used as a pungent food additive largely in the Eastern world. The sharp taste of the *Capsicum* fruit is caused by this substance. Suspension cultures of *C. frutescens* produce low levels of capsaicin. Yeoman and his group (Yeoman 1987) developed culture conditions for immobilizing the cells in reticulated polyurethane foam that could yield the same amounts of capsaicin as those obtained under natural conditions (see Sect. 3.3). M.A. Holden et al. (1988) reported elicitation of capsaicin in cell cultures of *C. frutescens* by spores of *Gliocladium deliquescens*. Biotransformation of externally fed protocatechuic aldehyde and caffeic acid to capsaicin in freely suspended cells and immobilized cell cultures of *C. frutescens* has also been reported (Rao and Ravishankar 2000). Jones and Veliky (1981) studied the effect of medium constituents on the viability of immobilized plant cells.

Withania somnifera Dunal (Solanaceae) is used as Indian ginseng in traditional Indian medicine. The active pharmacological components of *W. somnifera* are steroidal lactones of the withanolide type. Withanolides are known to have important pharmacological properties (antitumor, immunosuppressive), but they are also antimicrobial agents, insect deterrents, and ecdysteroid receptor antagonists. The principal withanolides in Indian *W. somnifera* are withaferin A and withanolide D. Both leaves and roots of the plant are used for the drug, and steroidal lactones occur in both parts. Ray and Jha (1999) reported production of withanolide D in roots transformed with *A. rhizogenes* but withaferin A was not detected in the transformed root cultures, although both compounds are present in the leaves and roots of field-grown plants.

Steroids form a group of compounds comprising the sterols, bile acids, heart poisons, saponins, and sex hormones. Saponins constitute a group of structurally diverse molecules consisting of glycosylated steroids, steroidal alkaloids, and triterpenoids. However, one common feature shared by all saponins is the presence of a sugar chain attached to the aglycone at the C-3 hydroxyl position. The sugar chains differ substantially between saponins but are often branched and may consist of up to five sugar molecules (usually glucose, arabinose, glucuronic acid, xylose, and rhamnose). Sapogenins constitute the aglycone part of saponins, with well-known detergent properties. They are oxygenated C27 steroids with a hydroxyl group in C-3. Diosgenin is an example of these compounds.

Diosgenin is a saponin aglycone obtained from the roots of *Dioscorea* species. It is very similar to cholesterol, progesterone, and dehydroepiandrosterone (DHEA)—the precursor to testosterone. Diosgenin provides about 50% of the raw material for the manufacture of cortisone, progesterone, and many other steroid hormones and is a multibillion dollar industry. The steroid synthesis pathway is cholesterol→pregnenolone→DHEA→testosterone→estrogen. However, the supply of diosgenin cannot currently satisfy the demands of the ever-growing steroid industry, and therefore new plant species and new production methods, including biotechnological approaches, are being researched (Verpoorte et al. 2000).

Several other groups have successfully obtained cell cultures for diosgenin production (Heble et al. 1967; Heble and Staba 1980; Jain et al. 1984; Huang et al. 1993). Kaul et al. (1969) studied the influence of various factors on diosgenin production by *Dioscorea deltoidea* callus and suspension cultures.

Dioscorea spp. (Dioscoreaceae) are frequently used as a tonic in traditional Chinese medicine, e.g., *Dioscorea doryophora*. Yeh et al. (1994) have established a cell suspension culture of *D. doryophora* Hance. Cell suspension cultures were obtained from microtuber- and stem node-derived callus in liquid culture medium supplemented with 0.1 mg 2.4D/l, 3% sucrose, and incubated in a rotary shaker at 120 rpm. Although 6% sucrose was found to be optimal for the growth of cell suspension culture, cells cultured in a 3% sucrose medium produced more diosgenin. Analysis by HPLC revealed that both stem node- and microtuber-derived suspension cells contained diosgenin. The microtuber-derived cell suspension culture contained 3.2% diosgenin per gram dry weight, the stem node-derived cultures only 0.3%. This is another example of influences of the origin of explants on the performance of cultured cells. As the amount of diosgenin obtained from a tuber-derived cell suspension is high, and similar to that found in the intact tuber (Chen 1985), a cell suspension culture can conveniently be used to produce diosgenin.

Cardenolides are naturally occurring glycosides that are widely distributed in plants. They are also called cardiac glycosides, because they exhibit the ability to strengthen the contraction of heart muscles. The best known cardiac glycosides come from *Digitalis*, including the drug digoxin. The aglycone is the non-sugar component of a glycoside molecule that results from hydrolysis of the molecule. Today, the sole source of the extensively used *Digitalis* drugs is the commercial harvesting of flowering *Digitalis* (foxglove) plants. The active compounds obtained from *Digitalis* include cardiac glycosides, digoxin, digitoxin, strophanthin, and ouabain. The structure of digitoxigenin is shown in Fig. 10.6, as a typical example of cardenolides.

A digoxin product, Lanoxin, is the brand name of a Burroughs Wellcome product and has the largest market of the company's cardiovascular drugs. The major markets of Lanoxin are in the USA and Italy, and the total sales are approximately 6000 kg/year at US$50 million. Other companies, such as Boehringer Mannheim, Merck Darmstadt, and Beiersdorf AG in Germany, also sell cardiac glycosides.

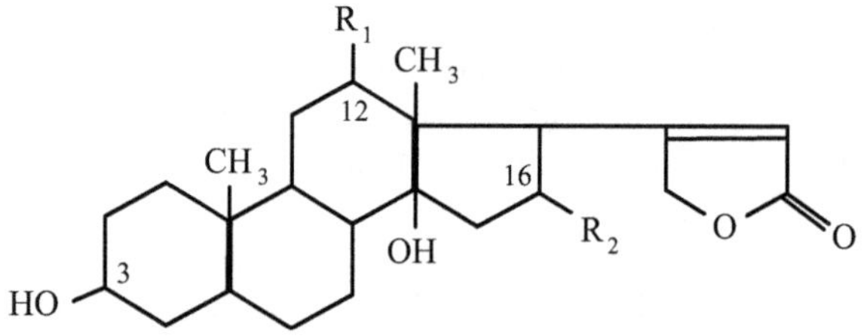

Fig. 10.6 Chemical structures of principal cardioactive glycosides of *Digitalis* species. Changes in R1 and R2 result in several new compounds

Digitalis lanata and *Digitalis purpurea* are commonly used for the production of cardiac glycosides (Fig. 10.6). Muir et al. (1954) were among the first to work in this field. Staba (1962) investigated the nutritional requirements of tissue cultures of *D. lanata* and *D. purpurea*.

Plumbagin, a naphthoquinone compound (5-hydroxy-2-methyl-1,4-naphthoquinone, PL) occurring mainly in *Plumbago* species (family Plumbaginaceae), was well-known for its use in traditional medicines. It is mainly produced in the roots of *Plumbago zeylanica* (Plumbaginaceae), which is distributed in Southeast Asia, India, and China. However, among all *Plumbago* species, *Plumbago indica* (syn. *rosea*) is the best source for harvesting plumbagin (Gangopadhyay et al. 2008). According to several current reports (Chetia and Handique 2000), *Plumbago indica* is becoming rare in several parts of India. *P. indica* hairy root culture provides an attractive system for the production of plumbagin: a secondary metabolite with high economical and medicinal relevance (Gangopadhyay et al. 2008).

The medicinal properties of Valerian (*Valeriana officinalis*) root preparations are attributed to the anxiolytic sesquiterpenoid valerenic acid and its biosynthetic precursors valerenal and valerenadiene, as well as the anti-inflammatory sesquiterpenoid β-caryophyllene. In order to study and engineer the biosynthesis of these pharmacologically active metabolites, a binary vector cotransformation system was developed for *V. officinalis* hairy roots (Ricigliano et al. 2016).

The presence of pyridine alkaloids, such as nicotine, nornicotine, anabasine, and anatabine, is characteristic for *Nicotiana* species. The first committed step in nicotine biosynthesis is the N-methylation of putrescine to N-methylputrescine (Hashimoto and Yamada 1994) catalyzed by putrescine methyltransferase (PMT). *Putrescine* is directly derived from *ornithine* or indirectly from arginine.

Azadirachtin ($C_{35}H_{44}O_{16}$) obtained from *Azadirachta indica* (neem) is a high-value secondary metabolite commercially used as a broad-spectrum biopesticide. Allan et al. (2002) established hairy root cultures from stem and leaf explants of *Azadirachta indica* A. Juss (neem) following infection with *Agrobacterium rhizogenes*. Srivastava and Srivastava (2013) reported batch cultivation of *Azadirachta indica* hairy roots in different liquid-phase bioreactor configurations (stirred tank, bubble column, bubble column with polypropylene basket, and polyurethane foam disc as root supports) to investigate possible scale-up of the *A. indica* hairy root culture for in vitro production of the biopesticide, azadirachtin. The hairy roots failed to grow in the conventional bioreactor designs (stirred tank and bubble column). They reported batch cultivation of *A. indica* hairy roots in modified bubble column reactor (with polypropylene mesh support). The incorporation of a PUF disc as a support for the hairy roots inoculated inside the bubble column reactor facilitated increased biomass production and azadirachtin accumulation in hairy roots (Srivastava and Srivastava 2013).

Phenylpropanoids from plants, which are composed of stilbenes, flavonoids, flavonols, flavonones, and anthocyanins, have beneficial health properties (Putignani et al. 2013; Trantas et al. 2015). Some of the best known phytoestrogens are genistein, genistin, daidzein, and puerarin (Patisaul and Jefferson 2010).

Biotechnological production of isoflavones, mainly from family Fabaceae is based on suspension cultures of *Pueraria* sp. (Goyal and Ramawat 2008a, b; Sharma et al. 2009), *Psoralea* sp. (Shinde et al. 2009), and *Glycine max* (Terrier et al. 2007). *Calycosin, formononetin,* and pseudobaptigenin are also present in the more widespread legume *Trifolium pretense* (Kokotkiewicz et al. 2013).

Benzylisoquinoline alkaloids (BIAs) are derived from aromatic amino acid tyrosine. BIAs alkaloids include narcotic analgesic *morphine*, the cough suppressant *codeine*, the muscle relaxants papaverine and (+)-*tubocurarine*, the antimicrobial compound *sanguinarine*, and the cholesterol-lowering drug berberine (Kong et al. 2004). Recently, efforts have been made to assemble BIA biosynthetic pathways in microorganisms through the heterologous expression of multiple alkaloid biosynthetic genes (Hawkins and Smolke 2008; Nakagawa et al. 2011).

Farrow et al. (2012) suggested that species-specific metabolite accumulation is influenced by the presence or absence of key enzymes and perhaps by the substrate range of these enzymes. They provided a valuable functional genomics platform to test these hypotheses through the continued discovery of BIA biosynthetic enzymes. Besides this, the cough suppressant and promising anticancer agent *noscapine* (Dumontet and Jordan 2010), BIA biosynthetic enzymes from a number of related plant species have been characterized using *EST*. The integration of transcript and metabolite profiles predicts the occurrence of both functionally redundant and novel enzymes.

10.4.1 Antitumor Compounds

Several antitumor compounds have been isolated from higher plants, but the concentrations of these active compounds in plants are generally low (Table 10.3). Some of the higher plant products, such as vinblastine, vincristine, podophyllotoxin derivatives including etoposide, and camptothecin and its derivatives, are marketed as very important anticancer drugs. Taxol, from *Taxus brevifolia* and related plants, is one of the most exiting compounds and was marketed in 1992. Besides being controlled by the slow growth rate of these plants, the accumulation pattern of these compounds is dependent on geographical and environmental conditions. Large-scale harvesting of antitumor drug-yielding native plants is becoming a serious problem in terms of possible extinction, and steps are needed for environmental preservation.

Table 10.3 Antitumor compounds isolated from higher plants

Antitumor compounds	Plant (dry wt. %)
Baccharin	2.0×10^{-2}
Bruceantin	1.0×10^{-2}
Camptothecin	5.0×10^{-3}
Ellipticine	3.2×10^{-5}
Homoharringtonine	1.8×10^{-5}
Maytansine	2.0×10^{-5}
Podophyllotoxin	6.4×10^{-1}
Taxol	5.0×10^{-1}
Tripdiolide	1.0×10^{-3}
Vinblastine, vincristine	5.0×10^{-3}

Sesquiterpenoids have been demonstrated to possess important pharmacological activity including anticancer, anti-inflammatory, antimicrobial, and antimalarial properties (Zhang et al. 2005). Plants of the family Valerianaceae—e.g., *Nardostachys jatamansi*, *Valeriana wallichii*, and *Valeriana officinalis* L. var. *angustifolia*—have been used as folk medicines in India, Bhutan, and Nepal. *Nardostachys chinensis* has been employed in China for hundreds of years. These plants contain a group of compounds characterized by, e.g., sedative, tranquilization, cytotoxicity, and antitumor activities, and they are collectively called "valepotriates."

Valeriana officinalis root preparations contain *Valerian* having anxiolytic sesquiterpenoid valerenic acid and its biosynthetic precursors valerenal and valerenadiene, as well as the anti-inflammatory sesquiterpenoid β-caryophyllene. The immediate precursor to all plant sesquiterpenoids is farnesyl diphosphate (FPP, 4); a 15-carbon branched chain hydrocarbon tethered to a high-energy ionizable diphosphate group FPP is itself derived from the successive condensation of the universal 5-carbon building blocks isopentenyl diphosphate (IPP) and dimethylallyl pyrophosphate (DMAPP) (Merfort 2011).

Becker and Chavadej (1988) induced callus tissues of nine different species of Valerianaceae on MS media and found that *Fedia cornucopiae* and *Valerianella locusta* cells produced higher levels of the compounds than did the intact plants.

The dimeric terpenoid indole alkaloids, the anticancer drugs vincristine and vinblastine, are obtained from cultivated *Catharanthus roseus* (Apocynaceae) plants. However, the process is not efficient, because of very low concentrations of the alkaloids in the plant. It was reported that the concentration of both vinblastine and vincristine was only 0.0005% on a dry weight basis.

The vinblastine molecule is derived from two monomeric alkaloids, catharanthine and vindoline (Fig. 10.7a). The concentration of vindoline in the intact *C. roseus* plant is approximately 0.2% on a dry weight basis, which is much higher than that of catharanthine. The cost of vindoline is less than that of catharanthine and vinblastine. The lack of vinblastine and vincristine in *C. roseus* hairy roots has been ascribed to an absence of vindoline (Bhadra et al. 1998). This may be due to the undetectable expression of the D4H and DAT genes in the transgenic hairy roots.

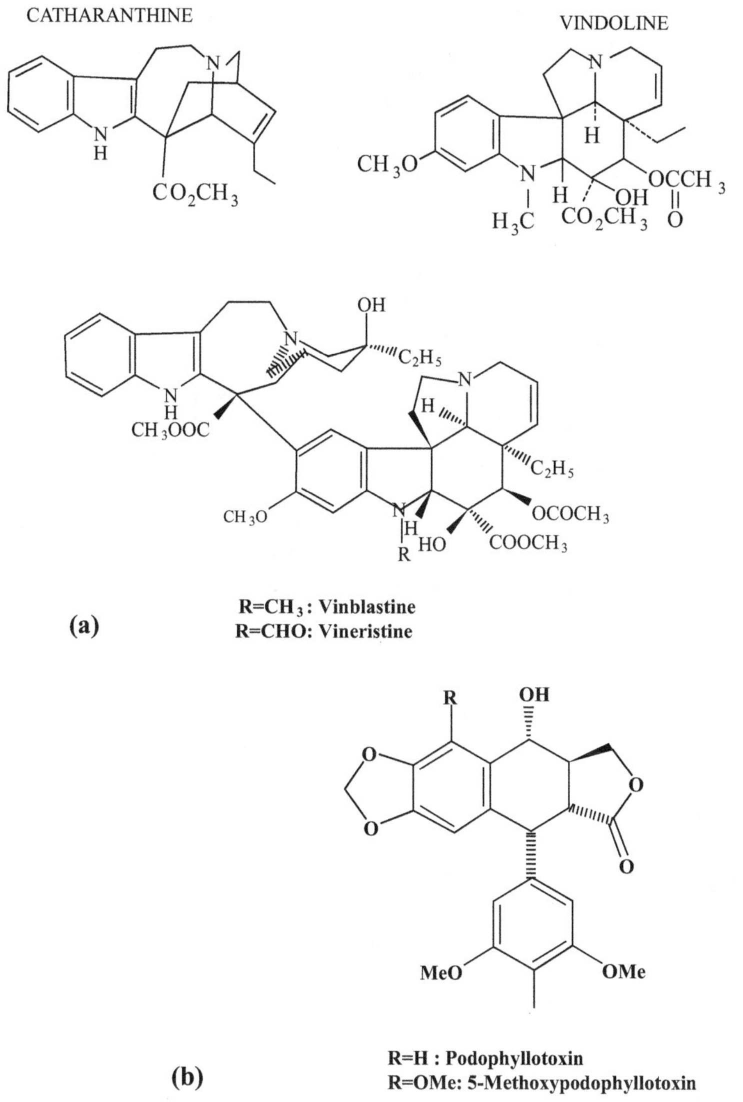

CATHARANTHINE

VINDOLINE

R=CH$_3$: Vinblastine
R=CHO: Vineristine

(a)

R=H : Podophyllotoxin
R=OMe: 5-Methoxypodophyllotoxin

(b)

Fig. 10.7 (a) Chemical structures of catharanthine, vindoline, vinblastine, and vincristine. (b) Chemical structure of podophyllotoxin and 5-methoxypodophyllotoxin

Deacetylvindoline-4-O-acetyltransferase (DAT) gene, which is responsible for the terminal step of vindoline biosynthesis in *C. roseus*, has been overexpressed in *C. roseus* hairy roots. Interestingly, overexpression of DAT did not increase vindoline production but improved the accumulation of another monoterpenoid indole alkaloid, horhammericine (Magnotta et al. 2007).

Camptotheca acuminata, a native of northern China, was found to produce a potent antitumor alkaloid, camptothecin (Wall et al. 1966). Camptothecin (CPT), a monoterpene indole alkaloid, has been found in several plant species including *Camptotheca acuminata*, *Nothapodytes foetida*, and *Ophiorrhiza pumila* (Lorence and Nessler 2004). Since it possesses topoisomerase I poisoning properties, its semi-synthetic derivatives, topotecan and irinotecan, have been developed to be clinically used as anticancer drugs. Li et al. (2017) have established a hairy root culture of *Ophiorrhiza pumila* which has already been shown to be a desirable experimental system to study the biosynthesis of camptothecin, since the culture produces a high level of CPT and excretes it into the culture medium.

In 2000, Wall obtained a US patent for a method of treating pancreatic cancer in humans with water-insoluble S-camptothecin of the closed lactone ring form and derivatives thereof (Wall 2000). Sakato and Misawa (1974) induced *C. acuminata* callus on MS medium containing 0.2 mg 2.4D and 1 mg kinetin per liter and developed liquid cultures in the presence of gibberellin, L-tryptophan, and a conditioned medium, which yielded camptothecin at about 0.0025% on a dry weight basis. In the cultures grown on MS medium containing 4 mg NAA/l, accumulation of camptothecin reached 0.998 mg/l (Van Hengel et al. 1992).

Podophyllotoxin is an antitumor aryltetralin lignan found in *Podophyllum peltatum* and *Podophyllum hexandrum*. It serves as a starting material for the preparation of its semi-synthetic derivatives, etoposide and teniposide, widely used in antitumor therapy of small-cell lung cancer, testicular cancer, acute lymphatic leukemia, and children's brain tumors (Issell et al. 1984). These slow-growing plants are collected from the wild and are thus becoming increasingly rare. This limits the supply of podophyllotoxin and necessitates a search for alternative production methods. Cell cultures of *P. peltatum* (Kadkade 1982) and *P. hexandrum* (Chattopadhyay et al. 2002) have been reported for production of podophyllotoxin (see also Arroo et al. 2002).

To increase the yield of podophyllotoxin, Woerdenbag et al. (1990) used a complex composed of a precursor, coniferyl alcohol, and β-cyclodextrin in *P. hexandrum* cell suspension cultures. Kadkade (1982) reported production of podophyllotoxin by *P. peltatum* cell cultures for the first time, and he found that a combination of 2.4D and kinetin in the medium yielded the highest production. Red light also stimulated the production. Since 5-ethoxypodophyllotoxin, an analog of podophyllotoxin (Fig. 10.7b), has strong cytostatic activity, many researchers have tried to improve its yield through tissue cultures (e.g., Oostdam et al. 1993; see review by Ionkova 2007).

10.4.1.1 Taxol

Taxol, a diterpene amide obtained from *Taxus brevifolia*, is considered as the prototype of a new class of cancer chemotherapeutic agents. Some other plants, such as *Taxus canadensis* and *Taxus cuspidata* also contain Taxol. Pure Taxol was first isolated in 1969, and its chemical structure was disclosed in 1971 (Fig. 10.8; Wani et al. 1971). Taxol has a unique mode of action, because it stabilizes microtubules and inhibits depolymerization; consequently, cell division is inhibited at the M-phase of the cell cycle. Taxol is used for curing breast and lung cancer and has shown positive results in curing ovarian cancer also. The FDA in the USA has

Fig. 10.8 Chemical structure of Taxol

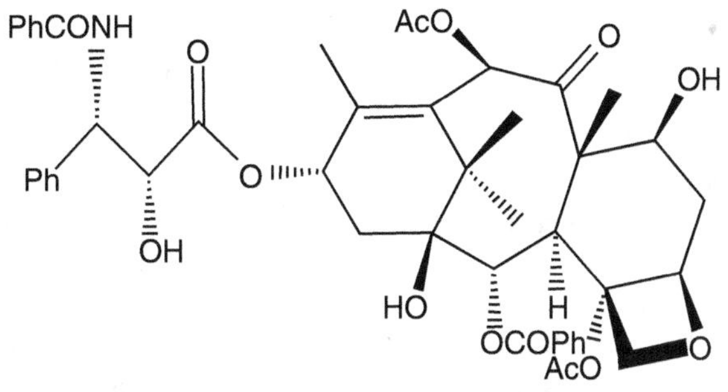

approved Taxol (generically known as paclitaxel) at the end of 1992 for clinical treatment of ovarian and breast cancer.

The thin bark of the yew tree contains 0.001% Taxol on a dry weight basis. A century-old tree yields an average of 3 kg of bark, corresponding to 300 mg Taxol, which is approximately a single dose in the course of a cancer treatment. The bark of 2000–3000 *T. brevifolia* trees is estimated to be required for commercial production of 1 kg of Taxol; this is equivalent to harvesting a 100-year-old tree to extract one dose of the drug (Horwitz 1994; Jennewein and Croteau 2001; Tabata 2006). Due to low concentrations of Taxol in *Taxus*, the commercial production of Taxol poses a serious threat to these trees.

The plant cell culture of *Taxus* sp. is also considered as one of the approaches available to provide a stable supply of Taxol and related taxane derivatives (Slichenmyer and Von Horf 1991). In 1989, Christen et al. reported for the first time the production of Taxol (paclitaxel) by *Taxus* cell cultures. They filed a US patent describing that the tissue of *T. brevifolia* had been successfully cultured to produce Taxol-related alkaloids and alkaloid precursors (Christen et al. 1991). Fett-Neto et al. (1995) have studied the effects of nutrients and other factors on paclitaxel production by *T. cuspidata* cell cultures (0.02% yield on dry weight basis). Srinivasan et al. (1995) have examined the kinetics of biomass accumulation and paclitaxel production by *T. baccata* cell suspension cultures. Paclitaxel was found to accumulate at high yields (1.5 mg/l) exclusively in the second phase of growth. Kim et al. (1995) established a similar level of paclitaxel from *T. brevifolia* cell suspension cultures, following 10 days in culture with optimized medium containing 6% fructose. Addition of carbohydrate during the growth cycle increased the production rate of paclitaxel, which accumulated in the culture medium (14.78 mg/l; Ketchum et al. 1999). Biotic and abiotic elicitors also improved the production and accumulation of Taxol through tissue cultures.

Factors influencing the stability and recovery of paclitaxel from suspension cultures and the media have been studied in detail by Nguyen et al. (2001). The effects of rare earth elements and gas concentrations on Taxol production have also been reported.

Shuler (1994) at Cornell University showed that a cell line of *T. brevifolia*, provided by the USDA Agriculture Research Station and Python Catalytic, produced Taxol in the medium after 26 days in suspension culture. It is of interest that all the Taxol produced was secreted into the medium, which is very unusual for plant cell cultures.

Some other species of *Taxus* have been assessed for the production of Taxol. Vanisree et al. (2004) reported Taxol production from *Taxus mairei* calli induced from needle and stem explants on Gamborg's B5 medium supplemented with 2 mg 2.4D or NAA per liter. Different cell lines were established using stem- and needle-derived callus. One of the cell lines, after precursor feeding and 6 weeks of incubation, produced 200 mg Taxol per liter of cell suspension culture.

Two commercial Taxol production system using plant cell cultures have been reported (Malik et al. 2011). However, large-scale production of secondary metabolites (e.g., paclitaxel) through plant cell suspension cultures is still challenging because the levels and patterns of secondary metabolite production in cell cultures are often unstable and unpredictable (Hirasuna et al. 1996; Kim et al. 2004). Li et al. (2009) found that after 5 years of subculturing, paclitaxel could no longer be detected in *Taxus chinensis* cells. Based on these it was proposed that accumulation of paclitaxel in Taxus cell cultures might be regulated by DNA methylation (Fu et al. 2012).

Ginseng

Ginseng (*Panax ginseng* C.A. Meyer), a classical herb widely used in East Asia, provides resistance to stress, disease, and exhaustion. Beveridge et al. (2002) have analyzed the phytosterol content in American ginseng seed oil. The root contains various saponins and sapogenins.

Among these, ginsenoside-Rb acts as a sedative, while Rg is stimulatory.

Recent progress in large-scale gene analysis (Jung et al. 2003), and proteome analysis (Kim et al. 2003; Nam et al. 2003) revealed that *P. ginseng* is one of the suitable sources for the study of dammarane-type triterpene saponin biosynthesis. This was the first result of molecular breeding to show the hyperaccumulation of triterpene saponins.

In recent years, ginseng cell culture has been explored as a potentially more efficient method of producing ginsenosides (Fig. 10.9). Medium components like carbon, nitrogen, phosphate, potassium ion, and plant growth hormones influence the production of ginsenosides (Wu and Ho 1999; Zhang and Zhong 2004). Other types of tissue cultures, such as embryogenic tissues (Asaka et al. 1993) and hairy roots transformed by *Agrobacterium*, have been examined. Yu et al. (2002) reported ginsenoside production using elicitor treatments. These developments indicate that ginseng cell culture is still an attractive area for commercial development around the world, and it possesses great potential for mass industrialization.

Recent advances with large-scale production have successfully produced ginseng roots in a 10,000 l bioreactor establishing the feasibility of the root system to accommodate industrial processes (Sivakumar et al. 2006). Numerous studies have

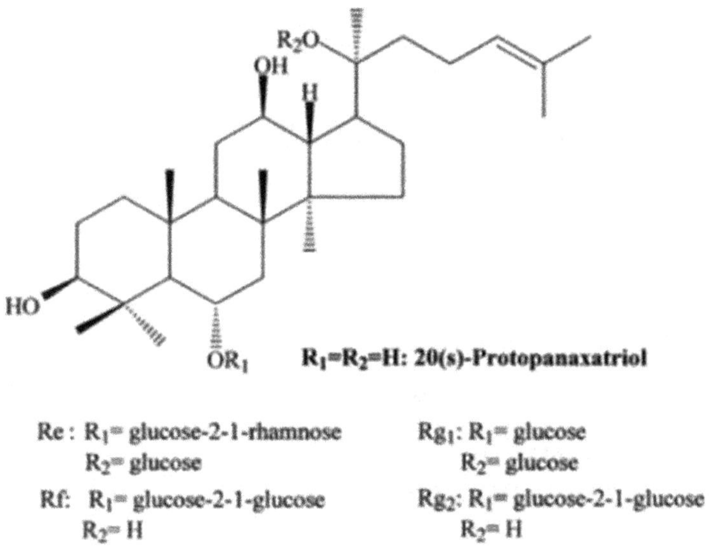

R$_2$O

OH

H

HO

OR$_1$ R$_1$=R$_2$=H: 20(s)-Protopanaxatriol

Re : R$_1$= glucose-2-1-rhamnose
R$_2$= glucose

Rf: R$_1$= glucose-2-1-glucose
R$_2$= H

Rg$_1$: R$_1$= glucose
R$_2$= glucose

Rg$_2$: R$_1$= glucose-2-1-glucose
R$_2$= H

Fig. 10.9 Structure of ginsenosides, among which RG1 is the most important

applied bioreactors in plant cell (Huang and McDonald 2012) and organ culture to obtain specific metabolites (Srivastava and Srivastava 2012).

Modern bioreactor culture systems provide a more advanced technology to produce higher secondary metabolites from plant cell, tissue, or organ using artificial nutrients with MeJA. Yu et al. (2002) found that the *ginsenoside* content was obviously enhanced by the addition of 100 μM MeJA during adventitious root culture of *Panax ginseng*; Donnez et al. (2011) examined that 0.2 mM MeJA was optimal for the efficient production and high accumulation of *resveratrol* in grape cell.

Han et al. (2014) established the dammarenediol-II production via a cell suspension culture of transgenic tobacco overexpressing PgDDS. Transgenic tobacco plants overexpressing PgDDS (AB122080) under the control of the CaMV35 promoter were constructed. Dammarenediol-II is biologically active tetracyclic triterpenoid, which is a basic compound of ginsenoside saponin and is a useful candidate with potentially biologically active triterpenes.

Triterpenoids are a large class of natural isoprenoids present in higher plants that exhibit a wide range of biological activities. Changes in triterpenoid content during the growth cycle of cultured plant cells have been demonstrated (Kamisako et al. 1984).

The isoprenoid biosynthetic pathway plays an important role in plant metabolism. Sterols and triterpenes are widely distributed isoprenoids. Plant sterols, so-called phytosterols, have important pharmacological activities, including cholesterol-lowering and antitumor effects (Lee et al. 2004).

Ginkgo

Ginkgo produces important terpenoids. The root bark and leaves of *Ginkgo biloba* L. contain diterpenoids (ginkgolides) and a sesquiterpenoid (bilobalide) that have interesting pharmacological properties. Some studies have been made on undifferentiated cell cultures of *G. biloba* with the aim of producing ginkgolides in vitro. Enieux and Van Beekt (1997) studied ginkgolide in transformed and gametophyte-derived cell cultures of *G. biloba*.

10.4.2 Anthocyanin Production

Besides pharmaceutical compounds and food additives, perfumes and dyes have been produced in cultures of plant cells. Here, anthocyanin production serves as a model system to explain the basic mechanism of biosynthesis of secondary metabolites, their transport, and storage in plant tissue.

Anthocyanins are the large group of water-soluble pigments responsible for many of the bright colors seen in flowers and fruit. They are also used in acidic solutions in order to impart a red color to soft drinks, sugar confectionary, jams, and bakery toppings. The major source of anthocyanins for commercial purposes is grape pomaces and wastes from juice and wine industries. Crude preparations of anthocyanins, which are relatively inexpensive, are used extensively in the food industry. The pure anthocyanins, however, are priced at US$1250–2000 per kg.

Cell suspension cultures of *Vitis vinifera* produce anthocyanins after cessation of cell division (Kakegawa et al. 1995), and anthocyanin biosynthesis is regulated by the endogenous level of phenylalanine that is accumulated within the cells (Sakuta et al. 1994). Focusing on the fundamental understanding of the complex metabolic pathway and regulation of secondary metabolism in plant cell cultures, Zhang et al. (2004) reviewed advanced knowledge of biosynthesis as well as post-biosynthesis pathways of anthocyanins from the genetic to the metabolite level. To illustrate this approach, they presented some data on the functional analysis of metabolic pathways for the biosynthesis of anthocyanins, from the profiling of gene expression and protein expression to metabolic profiling in *Vitis vinifera* cell culture as a model system. Emphasis was placed on a global correlation at three molecular levels—gene transcript, enzyme, and metabolite, as well as on the interactions between the biosynthetic pathway and post-biosynthetic events that have been largely overlooked in earlier work (Zhang et al. 2004; Fig. 10.10).

End products of the flavonoid biosynthesis pathway include the anthocyanin pigments. Pigment extracts from plant sources generally contain mixtures of different anthocyanin molecules, which vary by their levels of hydroxylation, methylation, and acylation. The major anthocyanins (see Fig. 10.11) that accumulate in *V. vinifera* cell culture are cyanidin 3-glucoside (Cy3G), peonidin 3-glucoside (Pn3G), and malvidin 3-glucoside (Mv3G) and the acylated versions of these,

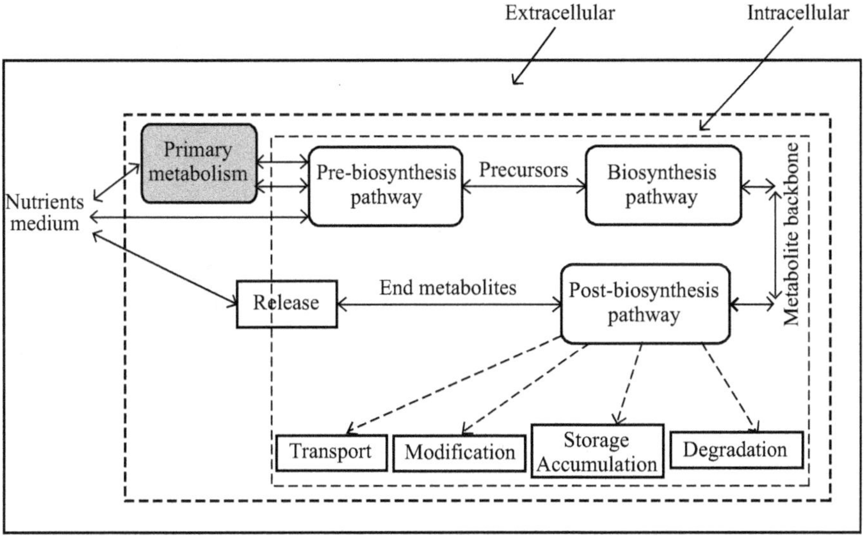

Fig. 10.10 Pathway events involved in the biosynthesis of a metabolite in plant cells: primary metabolism and secondary metabolism (pre-biosynthetic, biosynthetic, and post-biosynthetic pathways; after Zhang et al. 2004)

cyanidin 3-p-coumaroylglucoside (Cy3CG), peonidin 3-p-coumaroylglucoside (Pn3CG), and malvidin 3-coumaroylglucoside (Mv3CG; Conn et al. 2003).

Anthocyanins are synthesized in the cytoplasm and transported into the vacuole, where they bind with a protein matrix and form anthocyanic vacuolar inclusions (AVIs; Fig. 10.12; Conn et al. 2003). AVIs were proposed to be the storage sites of anthocyanins. Anthocyanins assume their distinct color after transport to the vacuole, concomitantly diminishing feedback inhibition of cytosolic biosynthetic enzymes. Spherical pigmented inclusions are present in the vacuoles of specific cells in over 70 anthocyanin-producing species and bind anthocyanins in a noncovalent manner (Markham et al. 2000). These insoluble protein matrices have been called anthocyanic vacuolar inclusions (Markham et al. 2000). It is thought that the anthocyanins are sequestered by AVIs primarily to increase their stability but also to reduce inhibition of certain vacuolar enzymes (Conn et al. 2003).

The production of anthocyanins using cultured cells has been assessed in various plant species, and most studies use an anthocyanin-producing cell line as model system for secondary product production, because of the color that enables production to be easily visualized. Yamamoto et al. (1982), of Nippon Paint Co. in Japan, have studied the production of anthocyanins intensively. High osmotic potential in *Vitis vinifera* L. (grape) cell suspension cultures enhanced anthocyanin production. The addition of sucrose or mannitol in the medium increased the osmotic pressure,

Fig. 10.11 Anthocyanin species present in *V. vinifera* suspension cultured cells: (**a**) 3-glucoside, (**b**) 3-p-coumaroylglucoside anthocyanin, and a summary table of modifications giving rise to the grape variants (Conn et al. 2003)

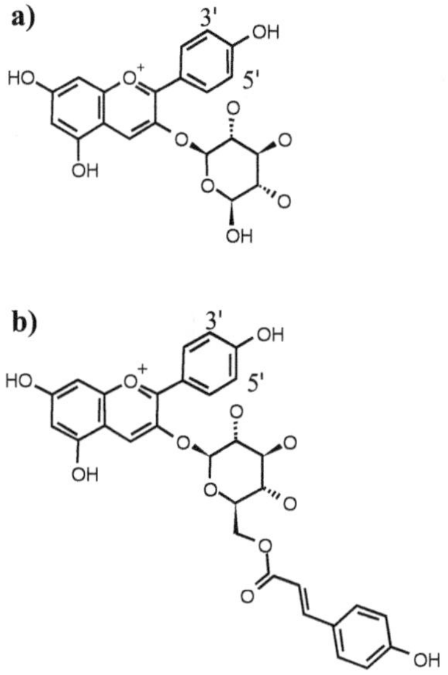

Compound	3′ Substitution	5′ Substitution
Cyanidin	-OH	-H
Delphinidin	-OH	-OH
Peonidin	-OCH3	-H
Petunidin	-OCH3	-OH
Malvidin	-OCH3	-OCH3

and the level of anthocyanins accumulated was increased. Similar observations were recorded on carrot cultures from our laboratory. The carrot secondary phloem explants grown on 4% sucrose produced excessive anthocyanins, compared to 2%, and were colored red (see Chap. 9; Kumar and Neumann, unpublished data; see also influences of micronutrients on anthocyanin production above and Neumann 1962; Ozeki and Komamine 1986).

Vitis vinifera produces *resveratrol* (3,4′,5-trihydroxystilbene). In plant cell suspension cultures, the resveratrol is secreted more in to the medium than within the cell (Cai et al. 2012). This secretion of resveratrol in growth medium could be related to active transport mechanisms involving ABC transporters or H+-gradient-dependent mechanisms (Yazaki 2005; Hashimoto and Yamada 2003; Donnez et al. 2011).

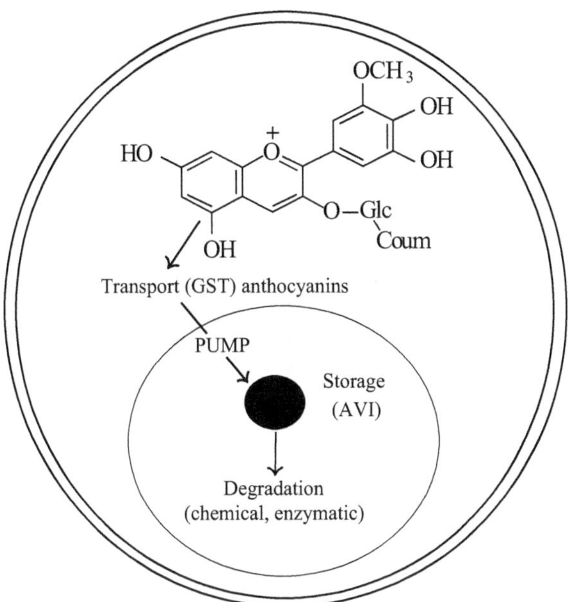

Fig. 10.12 Schematic summary of anthocyanin post-biosynthetic events. Anthocyanins are synthesized in the cytoplasm and transported into the vacuole

Saffron

Stamens of *Crocus sativus* give saffron, which is prized for use as a flavoring additive and as colorant. The stigma of the plant contains crocin (a yellow pigment), safranal (a fragrance), and picocrocin (a bitter substance). The plant is grown mainly in Spain and India, and about 30,000–35,000 handpicked blooms are required for the production of 1 lb of dry saffron. Crocin, being a glycoside, is water-soluble and is not soluble in oils and fats. It is used in baked goods, soups, meat and curry products, cheese, confectionary, and as a condiment for rice in Indian foods. It also has medicinal value for stomach ailments.

Ajinomoto of Japan has attempted propagation of stigma-like saffron structures in vitro (Sano and Himeno 1987). They showed that crocin and picrocrocin were present, and after heat treatment (as done with field-grown stigmas), safranal was produced. The composition of these phytochemicals corresponded with that of similarly treated, young intact stigmas (Sano and Himeno 1987).

Safflower Yellow

Florets of the safflower plant (*Carthamus tinctorius* L.) give Mexican saffron or American saffron a yellow pigment that has no relation to genuine saffron. The major pigment is carthamin, which exists at levels of up to 30% in the flowers, and there is also a red pigment in concentrations of about 0.5% (Wakayama et al. 1994).

Madder Colorants

Rubia tinctorum (Rubiaceae) is a perennial plant, and its roots have been used as red dyes in Western Europe. The major components of the pigment are alizarin, purpurine, and its glycoside, ruberythric acid. Pure alizarin is an orange crystal soluble at 1 part to 300 in boiling water and other solvents. Due to its high resistance to heat and light, it is suitable for the food industry. The callus of *R. tinctorum* induced from the root at San-Ei Chemical Industries of Japan was used to produce the pigment. After 21 days of cultivation in a 100 l jar fermenter, it produced approximately 1.5 g of the pigment (Odake et al. 1991).

Ashwagandha (*Withania somnifera* Dunal, Ashwagandha, Solanaceae) has ginseng-like properties and also known as Indian ginseng. *Withania somnifera* contains a variety of glycosylated steroids called withanosides that possess neuroregenerative, adaptogenic, anticonvulsant, immunomodulatory, and antioxidant activities (Saema et al. 2016). The withanolides have been elucidated to possess significant and specific pharmacological actions like immunomodulation, anticancer and neuroprotection, reversal of Alzheimer's and Parkinson's disease, etc. (Mondal et al. 2010; Sehgal et al. 2012). Present knowledge about the withanolide biosynthesis is very limited (Gupta et al. 2013). *Withaferin A, withanolide A, withanolide D,* and *withanone* are the most predominant withanolides of the plant (Chatterjee et al. 2010). Withanolides are C28 steroidal lactones which are biosynthesized by five carbon precursor isopentenyl diphosphate (IPP) and its isomer dimethylallyl diphosphate (DMPP) via cytosolic mevalonate (MVA) pathway and plastid localized methyl-D-erythritol-4-phosphate (MEP) pathway leading to biosynthesis of 24-methylene cholesterol (Gupta et al. 2013; Gloster 2014; Singh et al. 2016).

10.5 Strategies for Improvement of Metabolite Production

As discussed at length above, cell cultures have been established from many plants, but often do not produce sufficient amounts of the required secondary metabolites (Rao and Ravishankar 2000). Nevertheless, in many cases the production of secondary metabolites can be enhanced by treating the undifferentiated cells with elicitors such as methyl jasmonate, salicylic acid, chitosan, and heavy metals (DiCosmo and Misawa 1985; Barz et al. 1988; Gundlach et al. 1992; Ebel and Cosio 1994; Poulev et al. 2003). In some cases, secondary metabolites are produced only in organ cultures such as hairy root or shooty teratoma (tumor-like, see below) cultures; e.g., hairy roots produce high levels of alkaloids (Sevo'n and Oksman-Caldentey 2002), whereas shooty teratomas produce monoterpenes (Spencer et al. 1993).

In terms of cell growth kinetics, which usually incorporates an exponential curve phase, most secondary metabolites are produced during the stationary or plateau phase. This lack of production during the early stages can be explained by carbon allocation being mainly to primary metabolism during the active phase of growth. When growth stops, carbon is no longer needed in large quantities for primary metabolism, and secondary compounds are more actively synthesized. This is one explanation—others are possible (e.g., Chap. 12). It has been frequently observed that many new enzymatic activities, absent during the lag or log phases, appear during the plateau phase. This has led many authors to propose a possible biochemical differentiation of the cells when growth stops (e.g., Payne et al. 1987; Charlwood et al. 1990). However, some secondary plant products are known to be growth associated with undifferentiated cells, such as betalains and carotenoids.

The biosynthesis and accumulation of a number of secondary metabolites take place in specialized cells during specific developmental stage(s), i.e., differentiation and pigmentation in the organs and/or whole plants (Roja and Heble 1996).

10.5.1 Addition of Precursors and Biotransformations

Biotransformation involves region selective and stereospecific chemical transformations mediated by plant enzymes, which are variable in nature (e.g., aromatic, steroid, alkaloid, coumarin, terpenoid, and lignin). For example, podophyllotoxin, a precursor of a semi-synthetic anticancer drug is generally extracted from its source plant, *Podophyllum* species. Ardalani et al. (2017) demonstrated that a cell line of *Podophyllum peltatum*, active in the biosynthesis of podophyllotoxin, was able to maintain repeated biotransformation of butanolide to the podophyllotoxin analog. Ramachandra Rao and Ravishankar (2000) used freely suspended and immobilized cells of *Capsicum frutescens* for conversion of protocatechuic aldehyde and caffeic acids to vanillin and capsaicin. Moreover, Li et al. (2005) used ginseng cultured cells and roots for the bioconversion of paeonol into its glycosides, which have radical scavenging effects.

Baque et al. (2013) reported improve root growth and production of bioactive compounds such as anthraquinones (AQ), phenolics, and flavonoids by *adventitious root cultures* of *Morinda citrifolia*. They studied effects of aeration rate, inoculum density, and Murashige and Skoog (MS) medium salt strengths using a balloon-type bubble bioreactor.

Cell culture of *Weigela* "Styriaca" is capable of secologanin biosynthesis, and overexpression of the TDC and STR genes produced small amounts of ajmalicine and serpentine. This shows that indole alkaloid biosynthesis in otherwise non-alkaloid-producing plants are feasible via expression of one or a few heterologous pathway genes (Facchini et al. 2000). Another pharmaceutically important group of plant secondary metabolites comprises the isoquinoline alkaloids, which include, among others, the important medicines morphine and codeine (Mora-Palea et al. 2013).

An exogenous supply of biosynthetic precursor to the culture medium, as discussed above, may increase the yield of the final product when productivity is limited by lack of precursor.

Table 10.4 Biotransformation of β-methyldigitoxin into β-methyldigoxin by *Digitalis lanata* cells in a 20 l reactor

Budget parameters	Weight	Proportion
B-methyldigitoxin added	17.24 g	(100%)
Unconverted β-methyldigitoxin	2.04 g	(11.8%)
B-methyldigoxin formed	14.36 g	(81.7%)
By-product	0.28 g	(1.4%)
Yield	94.90%	

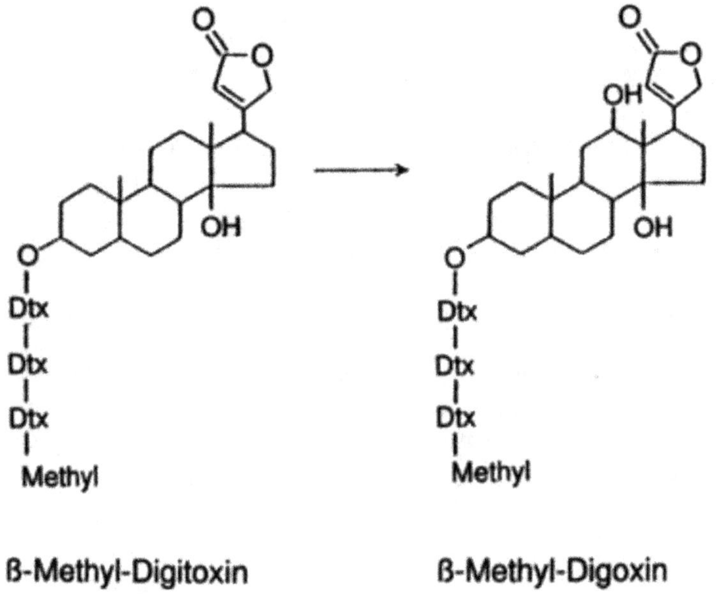

ß-Methyl-Digitoxin   **ß-Methyl-Digoxin**

Fig. 10.13 Transformation of β-methyldigitoxin into β-methyldigoxin (Alfermann et al. 1985). *Dtx* digitoxose

The production of tropane alkaloids has been markedly increased by the addition of tropic acid, a direct precursor (see above; Tabata et al. 1971).

In contrast to the de novo synthesis, the biotransformation process with *Digitalis* plant cells seems to be more promising from a commercial point of view (see also Table 10.4, Figs. 10.13 and 10.14). Graves and Smith (1967) reported that *D. lanata* and *D. purpurea* callus cultures rapidly transformed progesterone into pregnane.

Some cultures are able to transform cheap precursors into costlier chemicals. In the leaves of *Digitalis* are digoxin and digitoxin and their derivatives, where digitoxin is around one fourth of the concentration of digoxins. From a medical point of view, digoxin receives priority over digitoxin; the transformation of digitoxin into digoxin has been carried out successfully in undifferentiated cultures. The reaction takes place by attachment of a hydroxyl group in C-12 of the digoxins, and

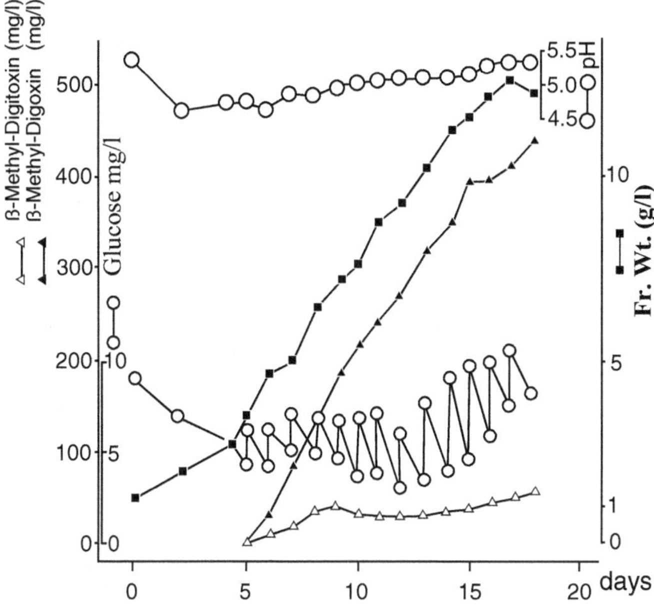

Fig. 10.14 Biotransformation of β-methyldigitoxin into β-methyldigoxin by cell cultures of *Digitalis lanata* in a 200 l airlift reactor (Alfermann et al. 1985)

these "wastes" of digitoxin production can be utilized by the pharmaceutical industry (Alfermann et al. 1983, 1985; Figs. 10.13 and 10.14).

Digitoxin is diverted into different products. One of this is methyldigitoxin. Certain cell clones have been isolated that transform methyldigitoxin into methyldigoxin (Figs. 10.13 and 10.14). For industrial usage, generally digoxin obtained from the leaves is methylated through a chemical process.

At the beginning of the experiment, the substrate is fed into the fermenter, and after 13 days about 70% of methyldigitoxin is transformed into methyldigoxin and is excreted to the nutrient medium; 20% could be located in cells in the form of digitoxin and digoxin, and some remains of the substrates were still in the nutrient medium. Using this technique, within 2 weeks of experimentation, 430 mg methyldigoxin was produced in a 200 l fermenter culture. This approach was developed further into a semi-continuous process in which a relatively long "scale-up period" is reduced to 17 days, and thus the cost of production is reduced. Here, after 14 days only 85% of the contents of the fermenter is taken for the extraction, and the rest remains as inoculum for the next culture. The fermenter is provided with fresh nutrient medium, and by providing the substrate, the manufacture of digoxin could be achieved afresh. This subculture was employed six times without loss of transformation capacity. During the 3 months of investigation, 500 g of methyldigoxin was produced, and based on a report from Alfermann et al. (1985), this would suffice to treat 1000 patients for more than 7 years.

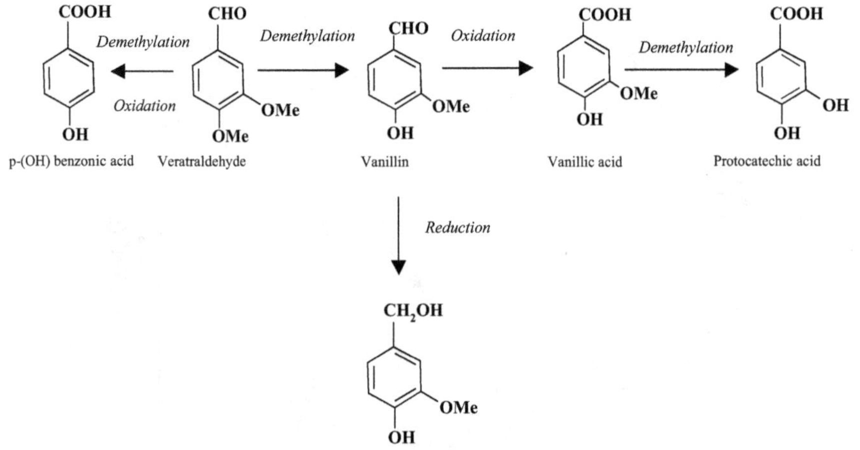

COOH   CHO   CHO   *Oxidation*   COOH   COOH

Demethylation   *Demethylation*   *Demethylation*

Oxidation   OMe   OMe   OMe   OH

OH   OMe   OH   OH   OH

p-(OH) benzonic acid   Veratraldehyde   Vanillin   Vanillic acid   Protocatechic acid

Reduction

CH$_2$OH

OMe

OH

Fig. 10.15 Probable biosynthetic pathway of the vanilla-flavored metabolite veratraldehyde via biotransformation (after Suresh and Ravishankar 2005)

Such "biotransformations" are also possible in other systems where the chemically synthesized compounds are obtained from cell cultures of either the same or another plant. One expects new, unknown molecules produced to replace economically important chemical substances.

Normal root cultures of *Capsicum frutescens* biotransform externally fed precursors like caffeic acid and veratraldehyde into vanillin and other related metabolites. The bioconversion of caffeic acid into further metabolites—viz., vanillin, vanillylamine, vanillic acid—was shown to be elicited by treating the cultures with 10 µM methyl jasmonate (Suresh and Ravishankar 2005). Root cultures treated with MeJA accumulated 1.93 times more vanillin (20.2 µM on day 3) than did untreated ones. Among all the precursors studied for the biotransformations, *Capsicum* root cultures could biotransform veratraldehyde most efficiently, leading to highest production of vanillin (78 µM on day 6 after veratraldehyde addition) than for any other phenylpropanoid precursor. The probable biosynthetic pathway of veratraldehyde biotransformation is indicated in Fig. 10.15, which shows that the precursor veratraldehyde would be enzymatically demethylated to vanillin and parahydroxybenzoic acid. Upon oxidation, vanillin would be converted into vanillic acid, which in turn would be demethylated to yield protocatechuic acid. The formation of vanillyl alcohol would be by reduction of vanillin.

10.5.2 Immobilization of Cells (See Also Sect. 3.3)

In immobilization (Yeoman 1987; Holden and Yeoman 1987), plant cells or microaggregates are encapsulated in polymers (alginate, carrageenans, etc.), and this usually enhances the production of secondary metabolites (Gontier et al. 1994).

The immobilization of plant cells with suitable matrices has been utilized to overcome problems of low shear resistance and cell aggregation (Dornenburg and Knorr 1997). The matrix effect of the polymers around the cells perhaps mimics the tissue organization. This reportedly gives rise to the so-called biochemical differentiation that favors the synthesis of secondary products (Yeoman 1987; Gontier et al. 1994).

Biotransformation of codeinone into codeine with immobilized cells of *Papaver somniferum* has been reported by Furuya et al. (1972). The conversion yield was 70.4%, and about 88% of the codeine converted was excreted into the medium. Ishida (1988) established *Dioscorea* immobilized cell cultures in which reticulated polyurethane foam was shown to stimulate diosgenin production, increasing the cellular concentration by 40% and the total yield by 25%.

The advantages of immobilization include the following: (1) extended viability of cells in the stationary stage, thus, enabling the maintenance of biomass over a prolonged time period; (2) simplified downstream processing (if products are secreted); (3) high cell density within relatively small bioreactors showing reduced cost and risk of contamination; (4) reduced shear stress; (5) increased product accumulation; (6) flow-through reactors that can be used to enable greater flow rates; and (7) minimization of fluid viscosity, which causes mixing and aeration problems in cell suspensions (DiCosmo and Misawa 1985).

10.5.3 Differentiation and Secondary Metabolite Production

Differentiated cell cultures are reported to have a higher biochemical potential (Yeoman and Yeoman 1996). There is a tight link between morphological differentiation and differentiation in metabolite biosynthesis in plant cells. Characterization of such metabolic differentiation at the molecular level is an important step in the development of effective methods to induce high levels of secondary metabolite production in cultured plant cells (Krisa et al. 1999). As an example, established hairy root cultures following infection with *Agrobacterium rhizogenes* displayed an enhanced production of those secondary metabolites that occur naturally in untransformed roots, resulting in amounts of secondary compounds comparable to or even higher than those present in intact roots (Sharp and Doran 1990; Zárate 1999).

The degree of cellular differentiation and organization of the tissue, which is implied in tissue and organ culture, favors the accumulation of the secondary compounds (Flores 1992). An increased secondary metabolite production is correlated with a slow cell division rate in cell suspension cultures (Lindsey and Yeoman 1983).

Several chemical and physical factors have been recognized, which could influence biomass accumulation and synthesis of secondary metabolites in plant cell and organ cultures. Some of the key constituents include the type of culture medium, suitable salt strength of the medium, types and levels of carbohydrates, nitrate levels, phosphate levels, and growth regulator levels (Dornenburg and Knorr 1997; Kumar and Sopory 2010; see review Murthy et al. 2008, 2014).

In some cases, complete differentiation may not be required, and tissue differentiation could increase secondary metabolite production. As an example, the development of xylem differentiation in calli of *Duboisia myoporoides* R. Br. led to the expression of stable tropane alkaloid biosynthesis without the need to regenerate differentiated organs. This finding may enable commercial tropane alkaloid production from calli with differentiated xylem, but not requiring organ development (Khanam et al. 2000).

Although cell cultures have been attempted from several *Digitalis* spp., those obtained from *Digitalis lanata* are most extensively utilized. As with the other systems, the undifferentiated callus culture is not able to produce the secondary metabolism-based metabolite glycosides. This has been demonstrated using highly sensitive RIA tests. Others, like the steroid testosterone, were detected using sensitive analytical methods. The glycosides were detectable as soon as the process of differentiation started (Luckner and Diettrich 1985, 1987). Such differentiation could be either the formation of compact green globuli in cell suspensions, shoot differentiation, or differentiation of embryos. These different types of differentiation patterns are associated with variations in the auxin/cytokinin ratio in the nutrient medium. In the green globuli, the cardenolid concentration was very low, i.e., 0.01 mg/g dry matter. In unspecialized cell cultures, or in the dark, the concentrations were even lower (less than 0.001 mg/g f. wt.). Light can increase the cardenolidic concentrations, though light is not essential for the biosynthesis of cardenolides. This is only of quantitative importance here (Luckner and Diettrich 1985, 1987). Induction of embryogenesis in these cultures significantly enhanced the cardenolide concentrations (0.7 mg/g dry matter). The concentration in the shoot apex was somewhat lower, at 0.4 mg/g dry matter. Thus, morphogenesis induced higher concentrations of secondary metabolites in general.

The carbohydrate and nitrogen composition of the medium also influences the production of cardenolides. Maltose was found to be most suitable. It is composed of two molecules of glucose as a disaccharide. Sucrose and also both the monosaccharide glucose and fructose lead to lower levels of embryogenic differentiation and also lower levels of cardenolide concentration. Also the source of nitrogen supply influences embryogenesis and the cardenolide concentration; the optimum for both processes is achieved through a mixture of ammonium and nitrate in ratios of 1:5–1:10. By contrast, glutamine and ammonium salts of different organic acids have negative effects.

In different experiments to isolate high-yielding strains from suspension cultures based on the selection of cardenolide levels, a broad variability in cardenolide concentration was documented in the colonies. In such cultures, variations in cardenolide concentrations are not due to genetic factors but rather to the developmental stages of the cultures (Luckner and Diettrich 1985).

The synthesis of the tropane alkaloids hyoscyamine and scopolamine in *Atropa belladonna* and *Catharanthus roseus* is developmentally regulated, the highest levels occurring in younger, faster dividing regions of the plant (De Luca and St-Pierre 2000). Physical factors that influence differentiation also affect secondary metabolite production; e.g., light has been reported to be necessary for the synthesis of vindoline, an important precursor of vincristine and vinblastine in *C. roseus* (De Luca and St-Pierre 2000). The pathways of interest often involve multiple organellar compartments, resulting in transport limitations and sequestered pools of metabolites. The lack of differentiation in cell cultures has often also been a barrier to successful alkaloid production. As an example, the synthesis of strictosidine, the precursor to the indole alkaloids, requires three organellar compartments. Tryptophan and the terpenoid precursor geraniol are synthesized in the plastids, tryptophan is then decarboxylated in the cytosol, and the two moieties are condensed in the vacuole (De Luca and St-Pierre 2000).

Duboisia myoporoides R. Br., an Australian member of the Solanaceae family, contains different groups of alkaloids and is cultivated in Australia for its high scopolamine content. In the complete *D. myoporoides* plant, alkaloid biosynthesis takes place in the root cells (Hashimoto and Yamada 1994). While different classes of alkaloids have been detected in cultured roots of this species (Yukimune et al. 1996; Khanam et al. 2000), tropane alkaloids are found in the cultured shoot only after root initiation (Kukreja et al. 1986; Lin and Tsay 2004).

Alkaloid production has also been studied in *Corydalis ambigua* (Papaveraceae). Corydaline and cavidine were accumulated in the leaf, tuber, and somatic embryos, whereas corybulbine was detectable only in tubers. Callus cultures and immature seeds, which lack embryos, contain only trace amounts of these alkaloids, suggesting the necessity of organ differentiation for alkaloid production in *C. ambigua*. Somatic embryos of *C. ambigua* that were cultured in liquid Linsmaier and Skoog medium, supplemented with 0.1 M IAA and 3% sucrose, produced two tetrahydroprotoberberine alkaloids, corydaline (0.03% of dry cell weight), and cavidine (1.09%; Hiraoka et al. 2004). These investigations show that the differentiation status can increase the cardenolide concentration. Still, the yield lies below the level of profitable commercial utilization.

10.5.4 Elicitation

Secondary metabolite accumulation in plant cells is in response to various biotic (e.g., insects or pathogens) and abiotic (e.g., salinity, temperature, radiation, water, heavy metal, and mineral) stresses (Ramakrishna and Ravishankar 2011). Elicitation is usually one of the most successful strategies to increase secondary metabolite production. Elicitors of bacterial, fungal, and yeast origins (e.g., inactivated enzymes, glycoproteins, xanthan, and chitosan) and salts of heavy metals have been widely used for the overproduction of secondary metabolites in plant cell and organ cultures (Dornenburg and Knorr 1997; Ramakrishna and Ravishankar 2011). Elicitors are defined as molecules that stimulate defense or stress-induced responses in plants (Van Etten et al. 1994). By applying elicitors one can trigger the production of secondary metabolites normally not produced or only at low concentrations. Elicitors like jasmonade and its derivatives are known to stimulate the production of secondary metabolites in plants (Sanz et al. 2000).

Signaling molecules such as jasmonic acid (JA), methyl jasmonate (MJA or MeJA), and salicylic acid (SA) have also been widely used for the increased accumulation of secondary metabolites in cell and organ cultures, e.g., hypericin production in *Hypericum perforatum* L. (St. John's Wort) (Bais et al. 2002; Thanh et al. 2005; Liu et al. 2014).

Besides enhancing the production of some of the desired secondary metabolites, elicitor treatment also activates the genes involved in the biosynthesis of such compounds. This has been recently demonstrated in studies targeting nicotine biosynthesis in tobacco cells (Breyne and Zabeau 2001; Goossens et al. 2003a).

Furanocoumarins, the phytoalexins having medicinal and industrial value, are restricted to the four families Leguminosae, Apiaceae, Umbelliferae, and Rutaceae. As their name suggests, these compounds are based on a skeleton formed by a furan ring fused to a coumarin unit. At present, furanocoumarins are produced from essential oils of bergapten (*Citrus bergamia*), but this production is insufficient to meet the increasing demands for these molecules. Elicited response was demonstrated very clearly for cell cultures of *Ruta graveolens* exposed to autoclaved culture homogenate of *Rhodotorula rubra*. The increased accumulation of these compounds was preceded by induction of specific secondary-product enzymes like L-phenylalanine ammonia lyase (PAL), cinnamate-4-hydroxylase (C4H), and 4-coumarate, CoA ligase (4CL) and S-adenosyl-L-methionine, bergaptol-O-methyltransferase (BMT), and S-adenosyl-L-methionine and xanthtoxol-O-methyltransferase (XMT). The enzymes showed sequential changes in activity with time for PAL and 4CL (Diwan and Malpathak 2007).

10.5.4.1 Jasmonic Acid

Jasmonic acid (JA) and its methyl ester (MJ) have also been studied as elicitors of secondary metabolites in plants with hairy roots (Rao and Ravishankar 2000). Exogenous MJ has been shown to mimic the effects of wounding through the induction of proteinase inhibitors (Xu et al. 1993), vegetative storage proteins (Berger et al. 1995), and secondary metabolites such as nicotine (Baldwin et al. 1994). MJ fed along with fungal elicitors was reported to activate the enzyme phenylalanine ammonia lyase (PAL), resulting in higher production of scopoletin and scopoline in tobacco cell cultures (Sharan et al. 1998), Taxol accumulation in cell suspension cultures of *Taxus chinensis* (Wu and Lin 2003), and ginsenoside production by cell suspension cultures of *Panax ginseng* in 5 l balloon-type bubble bioreactors (Thanh et al. 2005). It was also shown that MJ induces the expression of polyphenol oxidase (ppo) genes and markedly increases the level of the enzyme. Hayashi et al. (2003) studied upregulation of soyasaponin biosynthesis by methyl jasmonate in cultured cells of *Glycyrrhiza glabra*.

Increased levels of enzymes induced by MJ promoted the formation of secondary metabolites for a broad range of plant species (Blechert et al. 1995). In *Catharanthus roseus* (Madagascar periwinkle), methyl jasmonate induces terpenoid indole alkaloid (TIA) production. ORCA (octadecanoid-responsive catharanthus AP2/ERF domain) transcription factors have been shown to regulate the JA-responsive activation of several TIA biosynthesis genes (Endt et al. 2002; Memelink and Gantet

2007). A promoter element involved in jasmonate- and elicitor-responsive gene expression (JERE) was identified in the TIA biosynthetic gene strictosidine synthase (STR; Menke et al. 1999).

Interaction of MJ, wounding, and fungal elicitation influenced the production of sesquiterpenes in *Agrobacterium*-transformed root cultures of *Hyoscyamus muticus* (Choi et al. 2001, 2005). These results indicate that signaling, in addition to MJ, is required for the induction of these phytoalexins. Jasmonic acid also altered the accumulation of major anthocyanins in *Vitis vinifera* cell cultures (see above). Peonidin 3-glucoside content at day 3 was increased from 0.3 to 1.7 mg/g dry cell wt., while other major anthocyanins were increased less. Light further enhanced anthocyanin accumulation induced by jasmonic acid elicitation (Curtin et al. 2003).

MJ treatment increases the levels of ginsenoside. Choi et al. (2005) analyzed the ESTs (expressed sequence tags) derived from MJ-treated ginseng hairy roots and attempted to identify the genes involved in the MJ-induced biosynthesis of various secondary metabolites, including ginsenosides.

Hypericum perforatum L. (St. John's wort) produces hypericin, a photosensitive napthodianthrone considered to be responsible for the reversal of depression symptoms. Production levels and localization of hypericin in cell suspension cultures are entirely different from those of an intact plant (Bais et al. 2002).

JA elicitation of *Hypericum* cells increased the accumulation of phenylpropanoids and naphtodianthrones (Gadzovska et al. 2007). Earlier, Walker et al. (2002) reported that an administration of 250 mM JA induced an increased accumulation of hypericin in cultured cells of *H. perforatum* L. grown under dark conditions (0.318–0.02 mg/g dry wt.), compared to JA-elicited cultures under light conditions (0.089–0.006 mg/g dry wt.), and their respective controls (Bais et al. 2002). It is likely that secondary metabolites, in particular phenolic compounds, can constitute a photoblock resulting in hindered photoconversion under continuous light conditions (Hahlbrock and Scheel 1989).

10.5.4.2 Salicylic Acid

Salicylic acid (SA) a ubiquitous phytohormone is an elicitor widely used to promote the synthesis of secondary metabolites. According to Vicente and Plasencia (2011), SA is involved not only in the regulating plant growth and development but also in mediating the plant defense response against pathogen invasion (see also Liu et al. 2016; Li et al. 2016). Salicylic acid (SA) is an endogenous growth regulator that has phenolic properties and plays an important role in regulating plant growth, development, and interaction with other organisms, such as tobacco (Khan et al. 2012). SA is also an important signaling molecule for plants to cope with biotic or abiotic stress (Khan et al. 2015) and involved in endogenous signal-mediated local and systemic plant defense responses against pathogens. In plants, cytosolic acidification not only presages the defense response against pathogen invasion but also regulates the synthesis of specific secondary metabolites (Benouaret et al. 2014).

10.5.4.3 Effect of UV on Production of Secondary Metabolites in Cultured Tissues

Resveratrol and piceatannol have various beneficial health effects—e.g., moderate intake of resveratrol or resveratrol-containing food, such as red wine, may diminish the risk of cardiovascular diseases (Maxwell et al. 1994). Many studies have linked the antitumor activities of resveratrol and piceatannol to their abilities to inhibit cell proliferation and arrest cells in the S-phase (Jang et al. 1997; Joe et al. 2002).

Several research groups have demonstrated the ability to produce stilbenoid compounds from cultured plant tissues under normal or induced conditions. For example, resveratrol has been isolated from suspended cell cultures of grape, *Vitis vinifera*, and peanut, *Arachis hypogaea* (Schöppner and Kindl 1984; Ku et al. 2005). In the callus of *A. hypogaea*, isopentenyl resveratrol was induced via UV irradiation (Fritzemeier et al. 1983).

10.6　Organ Cultures

Plant organs represent an interesting alternative to cell cultures for the production of plant secondary products (Fig. 10.16). The production of secondary metabolites by cell suspension culture is not always guaranteed. Thus, organ culture methods (e.g., root, embryo, and shoot culture methods) have been developed for various plant species as alternatives for use in the production of secondary metabolites (Verpoorte and Memelink 2002; Murthy et al. 2008, 2014; Espinosa-Leal et al. 2018). Two

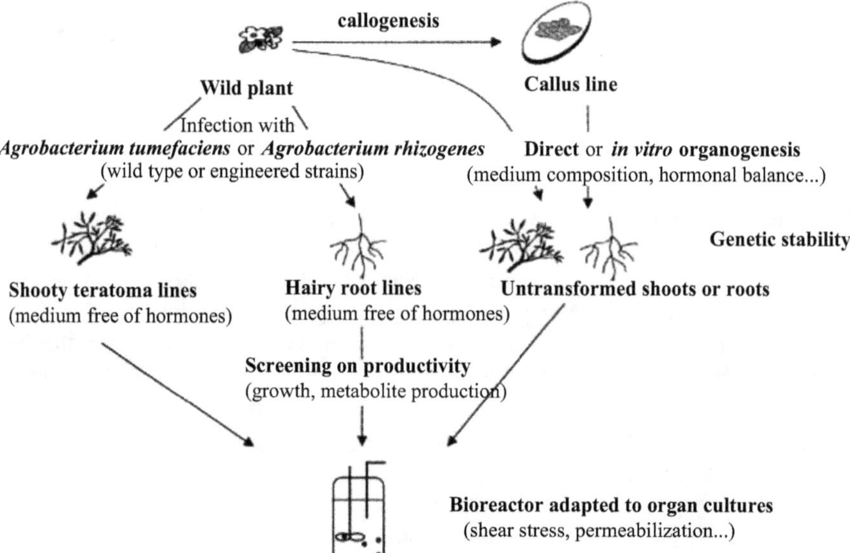

Fig. 10.16 Guidelines for the production of secondary metabolites from plant organ cultures (after Bourgaud et al. 2001)

types of organs are generally considered for this purpose: hairy roots and shoot cultures (Fig. 10.16).

Hairy root lines producing valuable phytochemicals have been developed from various plant species (Dehghan et al. 2012; Kim et al. 2012; Syklowska-Baranek et al. 2012; Cai et al. 2012). Hairy root cultures along with or without use of jasmonic acid have also been reported to considerably enhance secondary metabolite production.

10.6.1 Shoot Cultures

Shoot cultures are often superior generators of secondary metabolism products than are callus or suspension cultures.

Although no trace of the dimeric alkaloids anhydrovinblastine and leurosine were detected in suspension cultures of *C. roseus*, these have been detected in shoot cultures. Similarly, much higher levels of quinine and related alkaloids are present in shoot cultures than in suspension cultures of *Cinchona ledgeriana*. More cardiac glycosides are accumulated in shoot cultures of *Digitalis* than in undifferentiated cultures of this species. To date, however, techniques for shoot cultures are underdeveloped for mass cultivation. Nevertheless, there is a promising possibility to produce products associated specifically with shoots or leaves. Shoot cultures can easily be established simply by removing the top from sterile seedlings and placing it in B5 or MS media.

10.6.2 Root Cultures

Until recently, only a very limited number of plant species had been recorded for yield in root cultures, which could be grown indefinitely with an acceptable growth rate.

However culture of *adventitious roots* in bioreactors offers several advantages such as faster growth rates, tremendous quantities of metabolite accumulation, and stable production year-round. This can also reduce production costs and time, and final product quality can be more easily controlled (Lee et al. 2011). Baque et al. (2013), using large-scale bioreactors, raised adventitious root culture as an efficient and attractive alternative to cell, hairy root, or whole-plant cultivation for biomass and metabolite production.

The possibility of genetic transformation for root culture has been mentioned earlier. *Agrobacterium rhizogenes* are soilborne bacterial pathogens of plants, and these bacteria can enter into any wounded part of plant cells (Sect. 13.6.3). The Ri plasmid of *A. rhizogenes* induces rhizogenesis in inoculated cells of numerous roots. Such roots can be cultured and, unlike undifferentiated culture cells, stably maintain the biosynthetic characteristics of the original plant. This system has therefore been used as a means of culturing cells that will synthesize and accumulate secondary metabolite characteristics of the roots of the intact plant.

Based on this technique, hairy root cultures have already made an impact on the production of certain secondary metabolites. Considerable increases in biomass and alkaloid accumulation in root cultures can be achieved by RI-TDNA. Jung and Tepfer (1987) reported that transformation improved root growth in agitated flasks, and these roots produced tropane alkaloids at levels similar to those recorded for roots of the corresponding intact plants.

Hairy roots, the result of genetic transformation by *Agrobacterium rhizogenes*, have attractive properties for secondary metabolite production (Fig. 10.16; Park and Facchini 2000; Kim et al. 2002; Pavlov and Bley 2005). In some cases, secondary metabolites are produced only in organ cultures, such as hairy root or shooty teratoma (tumor-like) cultures. For example, hairy roots produce high levels of alkaloids (Sevo'n and Oksman-Caldentey 2002).

The greatest advantages of hairy roots is that they often exhibit about the same or greater biosynthetic capacity for secondary metabolite production as do their mother plants and are able to grow on growth regulator-free media. Many valuable secondary metabolites are synthesized in roots in vivo, and often synthesis is linked to root differentiation (Flores et al. 2000). Even in cases where secondary metabolites accumulate only in the aerial part of an intact plant, hairy root cultures have been shown to accumulate the metabolites as well. For example, lawsone normally accumulates only in the aerial part of the plant, but hairy roots of *Lawsonia inermis* grown in half- or full-strength MS medium (Table 3.3) can produce lawsone under dark conditions (Bakkali et al. 1997). Likewise, artimicin accumulates only in the aerial part of *Artemisia annua* plants, but several laboratories have shown that hairy roots can produce artemisinin (Liu et al. 1997). Genetic stability is another characteristic of hairy roots.

Hairy root cultures of the endangered species *Atropa baetica* display high accumulation of the major tropane alkaloids, atropine (±hyoscyamine) and scopolamine, with atropine levels similar to those of intact non-transformed roots. Surprisingly, scopolamine levels were fourfold higher than for intact roots, suggesting a much higher H6H activity (hyoscyamine 6-β-hydroxylase; Hashimoto and Yamada 1994; Zárate et al. 2006), this being the enzyme responsible for the conversion of hyoscyamine into scopolamine.

Catharanthus roseus hairy roots, in a fast-growing, differentiated tissue culture generated by *Agrobacterium rhizogenes* infection, accumulated higher levels of alkaloids than was the case for undifferentiated cell and callus cultures (Moreno-Valenzuela et al. 1998). The biosynthesis of vindoline was reported to be significant only in shooty teratomas (O'Keefe et al. 1997) or shoots regenerated from calli (Miura et al. 1988). Shimomura et al. (1991) established a hairy root culture of *Lithospermum erythrorhizon* with *A. rhizogenes*. The hairy root culture did not produce shikonin on solid MS medium but did produce the pigment in the root culture medium and also secreted it into the medium.

Solasodine present in Solanaceae plants has gained significant importance globally. Certain fast-growing hairy root clones of *Solanum khasianum* are reported to be high producers of solasodine (Aird et al. 1988). The effect of nitrogen on growth and solasodine production, when nitrate and ammonia are used as nitrogen source, has been demonstrated by Jacob and Malpathak (2005).

Hypericin is a traditional medicinal plant for the treatment of depression and wound healing, and hypericin is one of the main effective active substances. Wu et al. (2014) reported use of balloon-type airlift bioreactors for producing hypericin in adventitious root. They investigated the effect of air volume, inoculation density, indole-3-butyric acid (IBA) concentration, and methyl jasmonate (MeJA) concentration on hypericin content and productivity during adventitious root culture. MeJA efficiently elicited the hypericin synthesis of *H. perforatum* adventitious roots (Wu et al. 2014).

Hairy root cultures provide novel opportunities for production of valuable phytochemicals that are synthesized in roots. Hairy roots are developed by infecting plant leaf or stem tissue with *Agrobacterium rhizogenes* that transfers genes that encode hormone biosynthesis enzymes into the plants (Bhansali and Kumar 2014). Hairy root culture methodologies are common practice in many plant species, yet cotransformation with additional recombinant T-DNAs is comparatively underexplored as a means of metabolic engineering (Chandra 2012). For example, shoot cultures were established in *Bacopa monnieri* for the production of bacoside A, and regenerated shoots resulted in a threefold increase in bacoside A when compared to field-grown plants (Praveen et al. 2009). Due to an increased appeal of natural products for medicinal purposes, metabolic engineering can have a significant impact on the production of pharmaceuticals and help in the design of new therapies (Neumann et al. 2009; Kumar and Sopory 2010; Kumar et al. 2014).

Hairy root lines producing valuable phytochemicals have been developed from various plant species (Dehghan et al. 2012; Kim et al. 2012; Syklowska-Baranek et al. 2012; Cai et al. 2012). Recently, successful efforts have been made using hairy root cultures to improve secondary metabolism compounds in *Hyoscyamus niger* (Zhang et al. 2004, 2007), express foreign proteins or vaccine in tobacco (Shadwick and Doran 2007), and produce "unnatural" products in *C. roseus* (Runguphan and O'Connor 2009). Several TIAs' biosynthesis genes have also been overexpressed in *C. roseus* hairy root cultures (Wang et al. 2010). Hairy root cultures of *Nicotiana tabacum* are a better alternative for tropane alkaloid production than cell suspension cultures, mainly because they are stable, both genetically and in alkaloid production during long subculture periods (Maldonado-Mendoza et al. 1993).

In addition, hairy root cultures are efficient producers of secondary metabolites; for example, the production of the terpenoid compound withanolide A is more than adequate in the hairy root cultures of *W. somnifera* (Murthy et al. 2008). The hairy roots showed high production capacity, and withanolide A was present in amounts that were 2.7-fold higher than those observed in non-transformed.

10.6.3   Adventitious Root Culture

Culture of *adventitious roots* in bioreactors offers several advantages such as faster growth rates, tremendous quantities of metabolite accumulation, and stable production year-round. This can also reduce production costs and time and final product quality can be more easily controlled (Lee et al. 2011). Baque et al. (2013), using large-scale bioreactors, raised adventitious root culture as an efficient and attractive alternative to cell, hairy root, or whole-plant cultivation for biomass and metabolite production.

Baque et al. (2013) reported improve root growth and production of bioactive compounds such as anthraquinones (AQ), phenolics, and flavonoids by *adventitious root cultures* of *Morinda citrifolia*. They studied effects of aeration rate, inoculum density, and Murashige and Skoog (MS) medium salt strengths using a balloon-type bubble bioreactor.

Isoflavones are among the most extensively studied groups of natural compounds of polyphenolic character, which is mostly due to their affinity to estrogen β receptors in humans. Some of the best known phytoestrogens are genistein, genistin, daidzein, and puerarin (Patisaul and Jefferson 2010). The growing demand for isoflavonoid derivatives resulted in numerous research projects focused on the in vitro cultures of selected plant species of Fabaceae which is rich in isoflavones.

10.7 Genetic Engineering of Secondary Metabolites (See Also Sect. 13.6.3)

Production of secondary metabolites is under strict regulation in plant cells, due to coordinate control of the biosynthetic genes by transcription factors. The aim of metabolic engineering is usually to increase the accumulation of a desirable target product because it is naturally produced in small quantities, if at all (Farre et al. 2014). The potential of metabolically engineered plant-derived secondary metabolites is high and has been well documented by modifying anthocyanin and flavonoid pathways, leading to changes in flower color, or increased levels of antioxidative flavonol production in tomato (Muir et al. 2001). This may involve direct intervention to boost the accumulation of the target product by increasing flux through the pathway and/or by reducing flux through competing or catabolic pathways that remove intermediates or the target product itself (Capell and Christou 2004). The introduction of multiple genes into plants was initially achieved using iterative processes, such as successive rounds of crosses between transgenic lines (Datta et al. 2002) or sequential retransformation (Lapierre et al. 1999; Qi et al. 2004).

Metabolic engineering may also be used to produce metabolites that are not usually made naturally in the host cell because the corresponding metabolic steps are missing. This requires the import of a partial or complete heterologous pathway to increase the metabolic capabilities of the host (Zhou et al. 2011; Caspi et al. 2013; Farre et al. 2014). Each of these approaches is compatible with *Agrobacterium*-mediated transformation and direct DNA transfer (Twyman et al. 2002), and in the latter case, it is also possible to introduce genes into the plastid genome (Bock 2013).

Based on this success, genetic transformation of medicinal plants has been attempted, primarily to enhance the production of various pharmaceuticals but also flavors and pigments. Metabolic engineering of the biosynthetic pathways of interest has been successfully applied to produce high-value natural products, such as scopolamine (Bopana and Saxena 2010; Wang et al. 2012) and artemisinin (Lu et al. 2019). Lu et al. (2019) suggested that artemisinin might be used to treat neurodegenerative disorders by decreasing oxidation, inflammation, and amyloid beta protein (Aβ). They highlight the possible mechanisms of their neuroprotective activities, suggesting that artemisinins might have therapeutic potential in neurodegenerative disorders. *A. annua* remains as the main commercial source of artemisinin. Current advances in genetic and metabolic engineering drives to more diverse approaches and developments on improving in plant production of artemisinin, both in *A. annua* and in other plants (see review Ikram and Simonsen 2017).

Metabolic engineering approaches could optimize the biosynthesis of specific metabolites (Farre et al. 2014). Generally, the cytosolic pH will be maintained at a relatively constant level in plant cells for keeping normal physiological function (Monshausen et al. 2009). Cytoplasmic pH variation is one of the important signals of plant responses to external stimuli (Monshausen et al. 2009). However, the change of intracellular pH can affect many metabolic processes, such as cytosolic acidification (Kumar et al. 2015a b; Hütsch et al. 2016), protein phosphorylation (Isfort et al. 1993), and biosynthesis of secondary metabolites. Engineering of secondary metabolites, which aims to enhance yield, increase chemical diversity, and/or generate species with enhanced phenotypes, poses both substantial challenges and opportunities (O'Connor 2015).

Transgenic cultures and plants have been reported some time ago for about 70 species (Bajaj and Ishimaru 1999). Hashimoto et al. (1993) reported increased production of tropane alkaloids in genetically engineered root cultures. There are several strategies that can be used to enhance the production of desired pharmaceuticals by genetic engineering (Verpoorte et al. 2000; Sumner et al. 2003). Oksman-Caldentey and Inzé (2004) have reviewed the work on the production of designer metabolites in the post-genomic domain.

Functional genomics approaches (transcriptomics, proteomics, and metabolomics) are powerful tools for accelerating comprehensive investigations of biological systems (Fig. 10.17). Because no genomic tools are available for most plants producing interesting secondary metabolites (e.g., terpenoid indole alkaloids, and paclitaxel), it is not surprising that virtually no such comprehensive studies have been reported yet.

Recent years have witnessed a true revolution in the profiling of primary, but also secondary metabolites (Fiehn et al. 2001). Several genes in the biosynthetic pathways for scopolamine, nicotine, and berberine have been cloned, making the metabolic engineering of these alkaloids possible (Fig. 10.18). Expression of two branching-point enzymes was for engineered putrescine N-methyltransferase (PMT) in transgenic plants of *Atropa belladonna* and *Nicotiana sylvestris*, and (S)-scoulerine 9-O-methyltransferase (SMT) in cultured cells of *Coptis japonica* and *Eschscholzia californica*. Overexpression of PMT increased the nicotine content in *N. sylvestris*, whereas suppression of endogenous PMT activity severely decreased

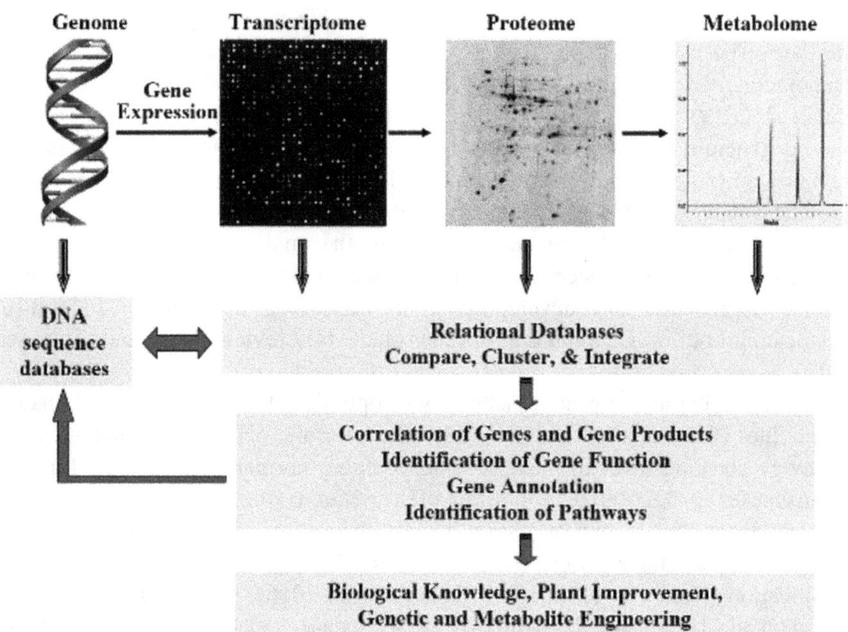

Fig. 10.17 Functional genomics approaches (transcriptomics, proteomics, and metabolomics) are powerful tools for accelerating comprehensive investigations of biological systems (Yeoman et al. 1980)

the nicotine content and induced abnormal morphologies. Ectopic expression of SMT caused the accumulation of benzylisoquinoline alkaloids in *E. californica*. However, they explore solutions to such challenges; metabolic engineers must nevertheless take care in recognizing the limitations inherent in designing plant systems (Hughes and Shanks 2002).

Based on the positive correlation between PMT activity and nicotine synthesis, Sato et al. (2001) expressed tobacco PMT using the CaMV 35S promoter in *N. sylvestris* plants. In the overexpressing lines, a 40% increase in nicotine content was noted over controls.

Another example is the observation that antisense-mediated downregulation of putrescine N-methyltransferase in transgenic tobacco plants resulted in a concomitant reduction in nicotine content but surprisingly also in elevated levels of the secondary metabolite anatabine (Chintapakorn and Hamill 2003). These examples, as well as others comprehensively reviewed elsewhere (Sato et al. 2001; Verpoorte and Memelink 2002; Hashimoto and Yamada 2003; Magnotta et al. 2007), show that engineering a single functional gene has considerable value for metabolic engineering but also some limitations. The ability to switch on entire pathways by ectopic expression of transcription factors suggests new possibilities for engineering secondary metabolite pathways.

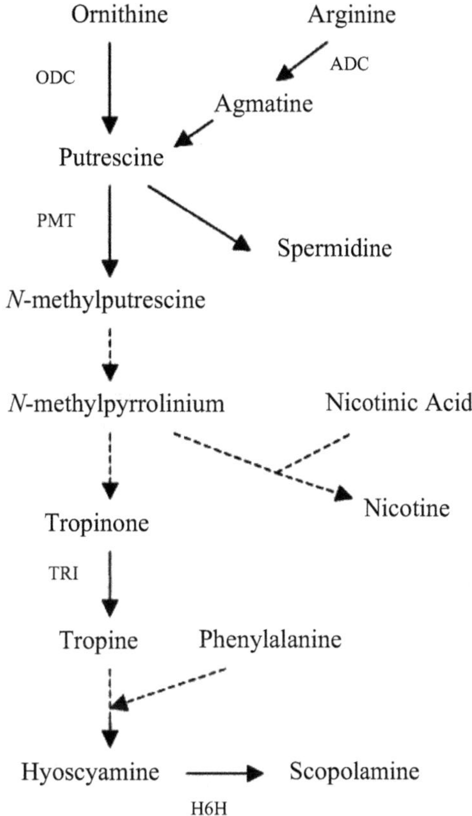

Fig. 10.18 Cloned genes in the nicotine and tropane alkaloid pathways. *ODC* ornithine decarboxylase, *ADC* arginine decarboxylase, *PMT* putrescine N-methyltransferase, *TRI* tropinone reductase I, *H6H* hyoscyamine 6-β-hydroxylase (after Hughes and Shanks 2002)

The first step in approaching the engineering of the pathways has been an attempt to quantify the relative importance of the terpenoid and indole pathways with precursor feedings. Although the terpenoid pathway has generally been found to be limiting in hairy root and cell cultures (Morgan and Shanks 2000), a few cell lines responded to indole feeding. Results for hairy roots also demonstrate that growth stages play a key role in determining the relative importance of the two pathways (Morgan and Shanks 2000). Thus, tryptophan decarboxylase (TDC) activity coincides with alkaloid accumulation, while strictosidine synthase (STR) activity is relatively stable (Meijer et al. 1993a, b).

In manipulating the alkaloid contents of *Cinchona officinalis*, TDC overexpression was ineffective. TDC has, however, proved useful in other systems, like canola, as a means of diverting flux to a metabolic sink to reduce undesired products (Chavadej et al. 1994). It has also been used to manipulate the alkaloid contents of *C. officinalis* (Geerlings et al. 1999).

Although metabolic engineering of alkaloid production is still in its infancy, the field offers great promise. Other papers have extensively reviewed the strategies of engineering pathways, methods for cloning genes, means to quantify flux, tools to characterize pathways at the enzymatic level, and techniques for pathway elucidation at the metabolite level (Morgan et al. 1999; Ratcliffe and Shachar-Hill 2001).

Triterpene saponins are important bioactive compounds in many other medicinal plants—for example, glycyrrhizin in *Glycyrrhiza* sp. (Dixon and Summer 2003) and saikosaponins in *Bupleurum falcatum* (Aoyagi et al. 2001). Squalene synthase (SS; EC 2.5.1.21) catalyzes the first enzymatic step from the central isoprenoid pathway toward sterol and triterpenoid biosynthesis (Abe et al. 1993).

From EST data analysis, Devarenne et al. (2002) identified several genes involved in the biosynthesis of 2,3-oxidosqualene, such as SS and SE that are upregulated in MJ-treated hairy roots. SS is involved in the biosynthesis of squalene from farnesyl diphosphate. This reaction is the first step in the transfer of carbon from the isoprenoid pathway toward triterpene biosynthesis and may be a potential point of triterpene biosynthesis regulation (Devarenne et al. 2002). SE converts squalene into 2,3-oxidosqualene, which serves as the substrate for the synthesis of protopanaxadiol. Devarenne et al. (2002) also identified three transcripts encoding SE, which suggests that SE forms a small multigene family in the ginseng genome. Moreover, they identified genes encoding two 3-hydroxy-3-methylglutaryl CoA reductases (HMGR) and a farnesyl diphosphate synthase (FPS). These enzymes are considered to constitute potential regulatory points in the isoprenoid pathway (Devarenne et al. 2002). Most of the ginseng genes described above, which were identified on the basis of their function in other plants, had never previously been identified. The first committed step in the biosynthesis of the triterpenoid saponin, ginsenoside, involves the cyclization of oxidosqualene by oxidosqualene cyclase (OSC), which produces one of several different triterpenoids, including protopanaxadiol (Figs. 10.19 and 10.20).

Genetic Transformation of Ornamental Plants

Genetic engineering may provide a solution through gene transfer to overcome the barriers of hybridization. Secondly, genetic engineering can introduce traits that could not be generated by conventional breeding, because the genes of interest do not exist in the natural gene pool. Third, genes that improve pest and disease resistance in food crops can also be used in ornamentals (Suprasanna et al. 2017).

10.8   Membrane Transport and Accumulation of Secondary Metabolites

Plants produce a large number of secondary metabolites, and their subcellular localization is highly regulated according to their biosynthetic routes and structural features. To achieve their function, such as protection against UV light or pathogens, they are generally accumulated in specific tissues or cell types. Some examples of secondary metabolite production and storage are given in Table 10.5. The majority of secondary metabolites are hydrophilic and are stored in the vacuole (Marinova et al. 2007; Martinoia 2018). However, hydrophobic secondary metabolites typically accumulate in membranes, vesicles, dead cells, or extracellular sites such as the cell

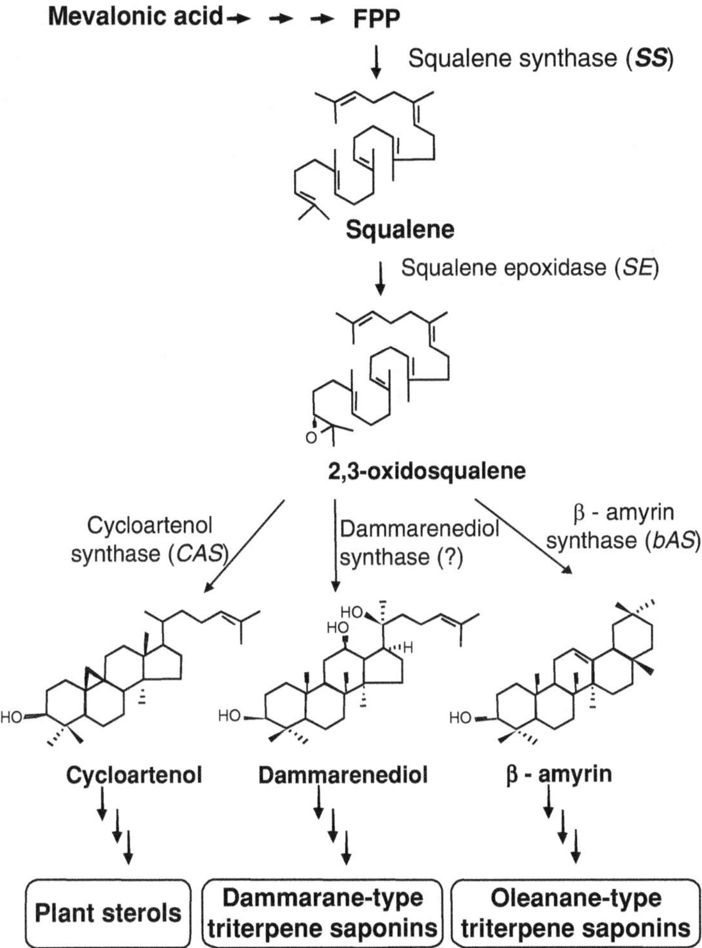

Fig. 10.19 Metabolism of squalene

wall (Cai et al. 2012). In addition to the linear design of metabolic pathways, successful engineering must consider this third dimension (Glenn et al. 2013). Many eukaryotic biosynthetic pathways, particularly in plants, are highly compartmentalized at the inter- and intracellular levels (Guirimand et al. 2011; Heinig et al. 2013). An emerging approach is to synthesize secondary metabolites in compartments where the metabolic process does not normally occur (Wu et al. 2012; Glenn et al. 2013). For example, localizing terpene biosynthetic pathways to the plastids of plants has been shown to result in high levels of product (Wu et al. 2012; Heinig et al. 2013).

Secondary metabolites are often transported from source cells to neighboring cells or even further to other tissues or remote organs (Kunze et al. 2002). Yazaki (2005) studied transporter proteins for these natural products in plants. Storage

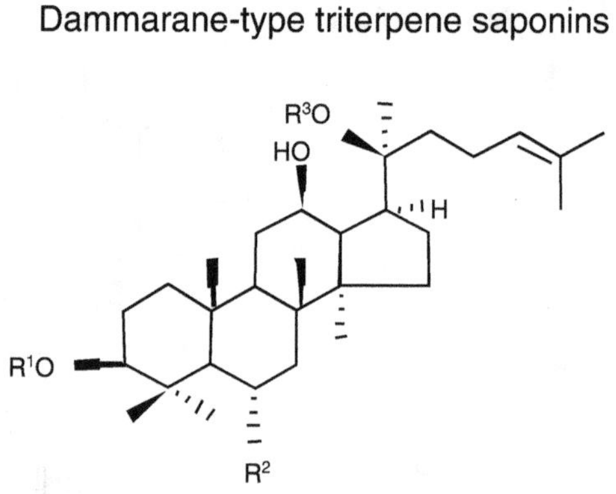

Fig. 10.20 Biosynthetic pathway of phytosterols and triterpenes in *P. ginseng*. Triterpenes undergo oxidation and glycosylation and are converted into triterpene saponins. Dammarenediol synthase activity was detected in the microsome fraction of *P. ginseng* (Kushiro et al. 1997, 1998)

Dammarane-type triterpene saponins

	R^2	R^1	R^3
Ginsenoside Rb$_1$	-H	- Glc^2-Glc	-Glc^6-Glc
Ginsenoside Rb$_2$	-H	- Glc^2-Glc	-Glc^6-Ara(pyr)
Ginsenoside Rc	-H	- Glc^2-Glc	-Glc^6-Ara(fur)
Ginsenoside Rd	-H	- Glc^2-Glc	-Glc
Ginsenoside Re	-O-Glc^2-Rha	-H	-Glc
Ginsenoside Rf	-O-Glc^2-Glc	-H	-H
Ginsenoside Rg$_1$	-O-Glc	-H	-Glc

vacuoles, which often occupy 40–90% of the inner volume of plant cells, play a pivotal role in the accumulation of secondary metabolites in plants (Fig. 10.21). The accumulation of secondary metabolites in vacuoles has at least two positive roles: the sequestration of biologically active endogenous metabolites inside the cells and the protection of such metabolites from catabolism (Gunawardena et al. 2004). Two major mechanisms are proposed for the vacuolar transport of secondary metabolites, these being H+ gradient-dependent secondary transport via the H+-antiport, and directly energized primary transport by ATP-binding cassette (ABC) transporters (Martinoia et al. 2002).

Membrane transport is fairly specific and highly regulated for each secondary metabolite, and recent progress in genome and expressed sequence tag (EST) databases has revealed that many transporters and channels exist in the plant genome. Studies of the genetic sequences that encode these proteins and of phenotypes caused by the mutation of these sequences have been used to characterize the membrane transport of plant secondary metabolites (Yazaki 2005). Such studies have clarified that not only genes that are involved in the biosynthesis of secondary metabolites, but also genes that are involved in their transport would be important for systematic metabolic engineering aimed at increasing the productivity of valuable secondary metabolites in plants (Yazaki 2005).

The mechanism for the long-distance transport of alkaloids is well elucidated in Solanaceae. Nicotine biosynthetic enzymes are expressed specifically in the root tissues, which is advantageous for the xylem transport of nicotine (Hashimoto and

Table 10.5 Production of metabolites, and their storage in plant systems[a]

Local compartment	Plant species	Secondary metabolite	Synthesis and storage site
Cellular	Asteraceae	Benzofurane Benzopyrane	Specific oil cells
	Catharanthus roseus	Alkaloids	Synthetic capacity depends on the number of storage cells
	Nicotiana rustica	Nicotine	Synthesis in the roots and storage in the cytosol of leaf cells
	Digitalis lanata	Digitoxin	Mesophyll cells of leaves
	Euphorbia lathyris	DOPA	Synthesis in leaves, storage in the latex
Subcellular/cell structures	*Coptis japonicum*	Anthraquinones	Rough endoplasmic reticulum (ER)
	Juniperus communis	Tannins	Vacuoles
	Lithospermum erythrorhizon	Naphthoquinones	Naphthoquinone vesicles: rough endoplasmic reticulum
	Papaver somniferum	Alkaloids	Alkaloid vesicles
	Pinus elliottii	Tannins	ER and Golgi vesicles
Subcellular/ membrane formation/ multi-enzyme complexes	*Haplopappus gracilis*	Naringenin Eriodictyol	Endoplasmic reticulum
	Sorghum bicolor	Coumaric acid p-hydroxy-mandelicnitrile	Endoplasmic reticulum, microsomes
Subcellular/various precursor pools		Malic acid	Mitochondria, vacuoles

[a]Example of compartmentalization in the formation of plant secondary metabolites

Yamada 2003). The transporter that is involved in the translocation of nicotine has not yet been identified, but a multidrug resistance protein (MDR)-like transport activity was recorded in the Malpighian tubules of tobacco hornworm, *Manduca sexta*.

Plant alkaloids are often effluxed by ABC transporters in microorganisms and herbivorous insects, but only a few of these transporters are currently known to be responsible for alkaloid transport in plants. Recent studies show that the uptake of an isoquinoline alkaloid, berberine, by *Coptis japonica* cells, is mediated by an ABC transporter (Sakai et al. 2002). Functional analyses of CjMDR1 using *Xenopus* oocytes showed that this protein recognized berberine as its substrate and transported it in an inward direction (Shitan et al. 2003), although most eukaryotic ABC transporters are known to function as efflux carriers. Because berberine is biosynthesized in root tissues, this alkaloid is translocated to the rhizome where it is trapped.

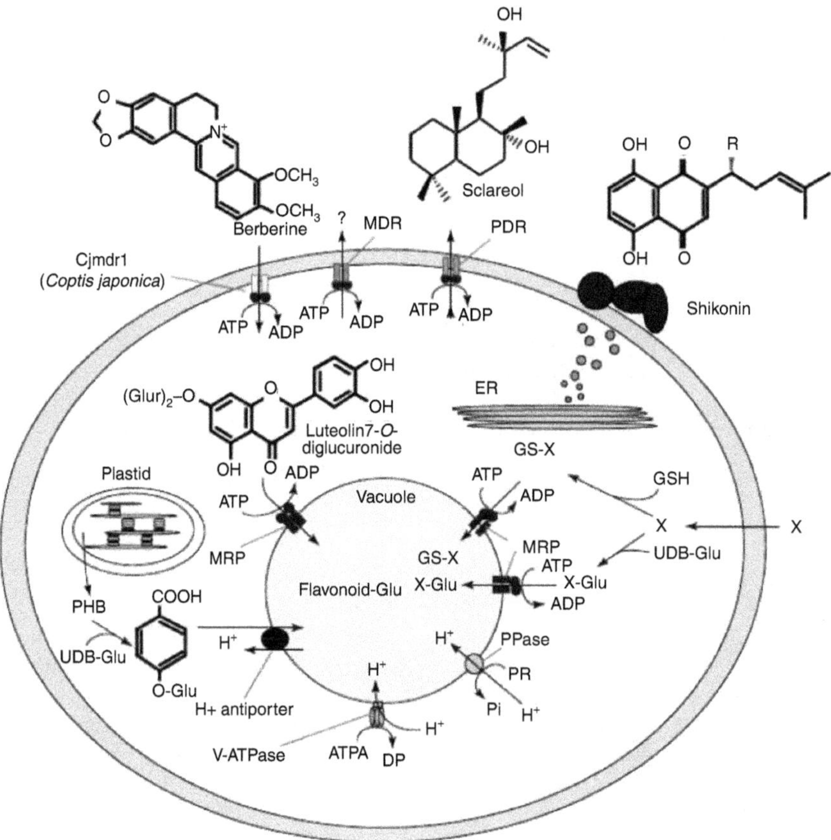

Fig. 10.21 Model of transport processes for secondary metabolites in a plant cell (Yazaki 2005)

Transport of alkaloids across the tonoplast and other cell membranes has also been shown to involve active mechanisms. Secretion of berberine from cell cultures of *Thalictrum minus* (Yamamoto et al. 1987) and benzophenanthridine alkaloids from *Eschscholzia californica* were both found to be vanadate-sensitive. A more recent study isolated an ABC transporter involved in the secretion of an antifungal terpenoid alkaloid from a cell culture of *Nicotiana plumbaginifolia* (Jasinski et al. 2001).

Tropane alkaloids are localized in the vascular region where large cells in the secondary xylem are reported to be present. In some alkaloid-producing plant species, alkaloid transport is mediated by carrier proteins. Wink (1985) reported significant alkaloid production when gene expression for alkaloid biosynthesis and transport (which is related to accumulation) took place at the same time. Guern et al. (1987) reported that gene expression for both alkaloid biosynthesis and transport takes place in the presence of a suitable storage site where accumulation without degradation occurs. Since large xylem cells are dead, biosynthesis of enzymes is not possible in these cells. The gene expression for alkaloid biosynthesis and the carrier proteins thus takes place after formation of the large cells in the secondary xylem. Formation of alkaloids in the non-rooted shoots may also be related to other cells. Further investigations related to other cell differentiation are therefore necessary.

When the concentration of a highly toxic secondary metabolite is increased by genetic engineering, does its intrinsic toxicity become a limiting factor? Recent experiments suggest that this might be the case. Nicotine and also other alkaloids are highly toxic to plant cells, but overexpression of the yeast ABC transporter PDR5 in transgenic tobacco cells was recently demonstrated to decrease the cellular toxicity (Goossens et al. 2003b).

In plants, terpene biosynthesis generally occurs in either the cytosol via the mevalonic acid (MVA) pathway to make sesquiterpenes and triterpenes or in the plastid via the methylerythritol phosphate (MEP) pathway to make monoterpenes and diterpenes. In a pioneering study, a sesquiterpene synthase, along with an upstream enzyme that makes the terpene precursor, was expressed in tobacco plastids, resulting in up to 1000-fold higher accumulation of sesquiterpene levels when compared to levels achieved when the enzymes were expressed in the cytosol (Wu et al. 2006; Kumar et al. 2012; Lu et al. 2019).

10.9 Bioreactors (See Also Sect. 3.2)

Generally, the plant products of commercial interest are secondary metabolites belonging to three main categories, i.e., essential oils, glycosides, and alkaloids. This categorization differs from the one given at the beginning of this chapter. Whereas the earlier one is oriented more on chemical definitions, here the definition is more from a point of view of application. The essential oils consist of a mixture of terpenoids, which are used as flavoring agents, perfumes, and solvents. The glycosides include flavonoids, saponins, phenolics, tannins, cyanogenic glycosides, and mustard oils, which are utilized as dyes, food colors, and medicinals (e.g., steroid hormones, antibiotics). The alkaloids are a diverse group of compounds with over 4000 structures known. Almost all naturally occurring alkaloids are of plant origin. Alkaloids are physiologically active in humans (e.g., cocaine, nicotine, morphine, strychnine) and therefore of great interest for the pharmaceutical industry (Shuler 1981). However, various problems associated with low cell productivity, slow growth, genetic instability of high-producing cell lines, poor control of cellular differentiation, and inability to maintain photoautotrophic growth have limited the application of plant cell cultures (Sajc et al. 2000).

As described above, in 1983 for the first time, a dye, shikonin, with anti-inflammatory and antibacterial properties, was produced by plant cell cultures on an industrial scale by Mitsui Petrochemical Industries Ltd (Fujita et al. 1982). Although this was thought to be a major breakthrough, shikonin was still up to the 1990s the only plant compound to be produced on a commercial scale by cell cultures.

There are several means of increasing the production of secondary metabolites by plant cell cultures or suspensions, as has been discussed before:

- Use of biotic or abiotic elicitors
- Addition of a precursor of the desired compound
- Secondary metabolite production or inducing changes in the flux of carbon to
- Favor the expression of pathways leading to the target compound
- Production of new genotypes by means of protoplast fusion or genetic
- Engineering
- Use of mutagens to increase the variability already existing in living cells
- Use of root cultures

At the European level, some years ago, all these issues were discussed comprehensively at the symposium on "Primary and Secondary Metabolism of Plant Cell Cultures" (Neumann et al. 1985) and at a seminar on "Bioproduction of Metabolites by Plant Cell Cultures" held in Paris in September 1988, organized by the International Association of Plant Tissue Culture and the French Association pour la Promotion Industrie-Agriculture (APRIA, Association for the Promotion of Industry-Agriculture). However, constrains still remain that need high investments and finding solutions to the problems of raising plant tissues in bioreactors.

Economic considerations govern the importance attached to the production of natural substances and biochemicals by cell cultures. Some additional information to those given earlier can be obtained from the data in Table 10.6. The estimated annual market value of pharmaceutical products of plant origin in industrialized countries was over US$20 billion in the mid-1980s. The annual market value of codeine and of the antitumor alkaloids vinblastine and vincristine has been estimated at about US$100 million per product (Pétiard and Bariaud-Fontanel 1987). The worldwide

Table 10.6 Economic data for some substances of plant origin

Substance and use	Annual demand	Industrial cost (US$ per kg)	Estimated annual market value (in US$ million)
Pharmacy			
Ajmalicine	3–5 t	1500	4.5–7.5
Codeine	80–150 t	650–900	52–135
Digoxin	6 t	3000	18
Diosgenin	200 t	20–40	4–8
Vinblastine, vincristine	5–10 kg	5 million	25–50
Food additives and fragrances			
Jasmine oil	100 kg	5000	0.5
Mint oil	3000 t	30	90
Natural vanillin	30 t	2500	75
Cosmetics			
Shikonin	150 kg	4000	0.6

t tons

market value of aromas and fragrances was expected to rise to US$6 billion in 1990 (Rajnchapelmessai 1988).

In 1988, the estimated annual market value of shikonin (for details, see below) was about US$600,000, which is far from the US$20–50 million investment of the original research and development work. However, the final cost of the product fell to US$4000 per kg, which is similar to US$4500 per kg for the substance extracted from the roots of *Lithospermum erythrorhizon* (Sasson 1991).

It should be noted that Kanebo, the Japanese cosmetics corporation that developed lipsticks containing shikonin, realized a turnover of about US$65 million over 2 years in Japan through the sale of 5 million lipsticks, each selling for US$13. In the Republic of Korea and in China, Mitsui Petrochemicals Ltd today intends to sell at over US$4 billion (as estimated).

In the Federal Republic of Germany, Alfermann et al. (1985), in collaboration with Boehringer Mannheim AG, were able to grow cells of *Digitalis lanata* in 200 l bioreactors and obtain 500 g of beta-methyldigoxin in 3 months (see also above); the bioconversion rate of beta-methyldigitoxin was very high, up to 93.5%, if the non-used substrate was recycled. Ulbrich et al. (1985) cultured *Coleus blumei* cells in a 42 l bioreactor fitted with the module spiral stirrer, using this system with aeration. They reported high yields of rosmarinic acid (5.5 g/l), representing 21% dry weight of cells. Heble and Chadha (1985a, b) reported the successful cultivation of *Catharanthus roseus* cells in 7 to 20 l capacity bioreactors, modified to provide airlift and agitation, in single and multiple stages.

The cells produced high levels of total alkaloids, comprising ajmalicine and serpentine as the major components. It was shown that plant cells could withstand shear to some extent and that judicious use of airlift and low agitation were advantageous. Researchers at Ciba-Geigy AG, Basel, Switzerland, have produced the alkaloid scopolamine from cell cultures of *Hyoscyamus aegypticus* grown in airlift bioreactors.

10.9.1 Technical Aspects of Bioreactor Systems

Bioreactor studies represent the final step that leads to a possible commercial production of secondary metabolites from plant cell cultures. This is an important phase, as numerous problems arise when scaling up the work realized in Erlenmeyer flasks. For example, growth is considerably modified when cells are cultivated in large tanks, and the production of cell biomass remains a critical point for bioreactor productivity. Here, some of the guidelines for upscaling will be given (Figs. 10.22, 10.23 and 10.24).

Despite potential advantages in the production of secondary metabolites in plant cell cultures, other than shikonin, only ginsenosides and berberine are today produced on a large scale, and all three processing plants are located in Japan (Hara 1996). The anticancer drug Taxol (registered trademark of Bristol-Myers Squibb) is under consideration for large-scale production (Seki et al. 1995, 1997; Roberts and Shuler 1997).

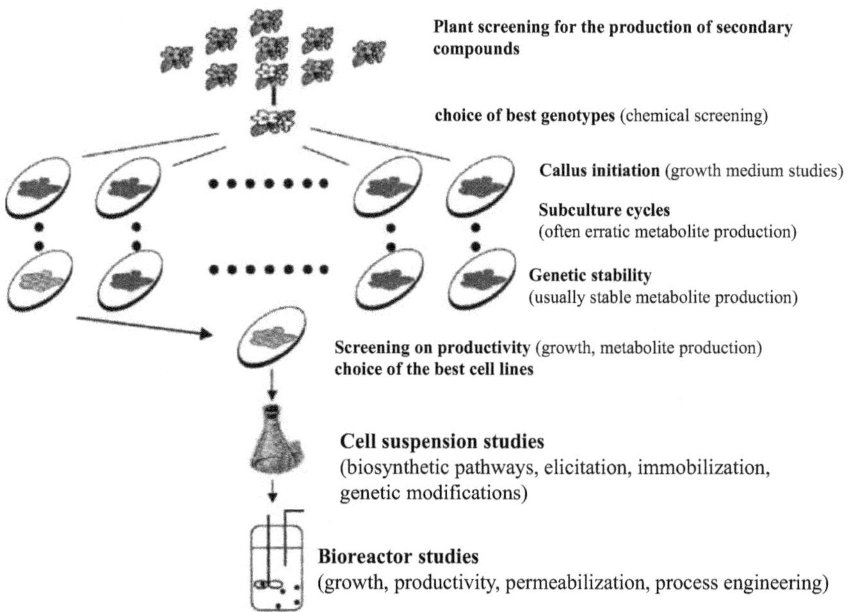

Fig. 10.22 Guidelines for the production of secondary metabolites from plant cells (from Bourgaud et al. 2001)

Problems are due mainly to mass transfer limitations of oxygen (Jones and Veliky 1981; Hulst et al. 1985), as well as inhomogeneous culture systems that cause cell sedimentation and death. Recent studies have confirmed the low percentage of viable cells (approx. 50%) generally present in such liquid systems, except for the first days of culture (Steward et al. 1999).

Often, another strong limitation of growth is due to plant cell sensitiveness to shear stress, which is responsible for extensive cell death. This lysis is a consequence of the agitation of the culture medium.

Fundamental studies of bioreactors with plant cells involve three important scientific and practical issues related to bioreactor design and operation: (1) assessment of cell growth and product formation; (2) analysis and modeling of the culture dynamics, including the integration of biosynthesis and product separation; and (3) studies of flow, mixing, and mass transfer between the phases, in order to define criteria for bioreactor design and scale-up. For a given application, the culture conditions can be optimized with respect to cell support, medium composition and renewal rate, mass transfer of chemical substances, and bioreactor fluid dynamics, in order to define the conditions that are permissive for or even designed to promote selected cell functions.

Large-scale suspension culture of ginseng cells was first reported by Yasuda et al. (1972). Later, industrial-scale culturing was initiated by Nitto Denko Corporation (Ibaraki, Osaka, Japan) in the 1980s, using 2000 and 20,000 1 stirred tank fermenters to achieve productivities of 500–700 mg/l/day (Furuya 1988; Ushiyama 1991). This

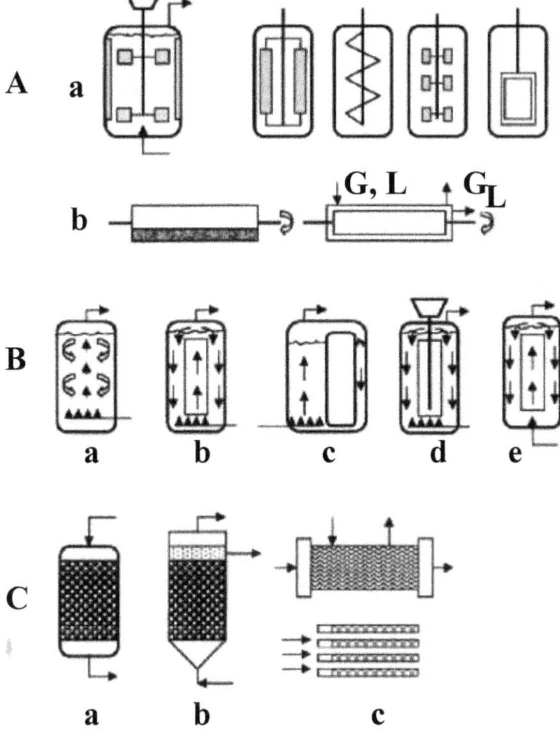

Fig. 10.23 Bioreactor types for plant cell, tissue, and organ cultures. *A* Mechanically agitated bioreactors: (**A**) stirred tank reactor equipped with various propellers (spin, helix, bladed, paddle), (**B**) rotary drum tank reactor. (**C**) Air-driven bioreactors: (*a*) bubble column, (*b*) concentric tube airlift reactor (IL ALR), (*c*) external loop airlift

process is considered an important landmark in the commercialization of plant tissue and cell culture on a large scale.

Prenosil and Pedersen (1983), Payne et al. (1987), Panda et al. (1989), and Scragg (1991) reviewed different reactor configurations for plant cell suspensions, plant tissue, and organ cultures (Fig. 10.23). The relative advantages and selection criteria for various reactor configurations were discussed for specific process applications. In particular, bioreactors that integrate biosynthesis with product release and separation were most extensively studied in Japan (Uozumi et al. 1991; Honda et al. 1993).

Numerous modifications of the conventional stirring tank reactor (STR) with bubble aeration have been developed by employing a variety of impeller designs. The controllability and flexibility of the STR, in terms of independent adjustment of mixing and aeration, makes it the most frequently chosen configuration, despite several limitations such as high power consumption, high shear, and problems with sealing and stability of shafts in tall bioreactors. Although membrane reactors and packed bed reactors (Fig. 10.23) are advantageous, in that a large amount of cells can be immobilized per unit volume (see above), diffusional limitations of mass transfer to the immobilized cells, as well as the difficulties in supplying and removing gaseous components, can limit the use of both configurations to biotransformations. Airlift bioreactors (ALR) using low-density beads with immobilized cells or enzymes are currently under research for a variety of applications in bioprocess

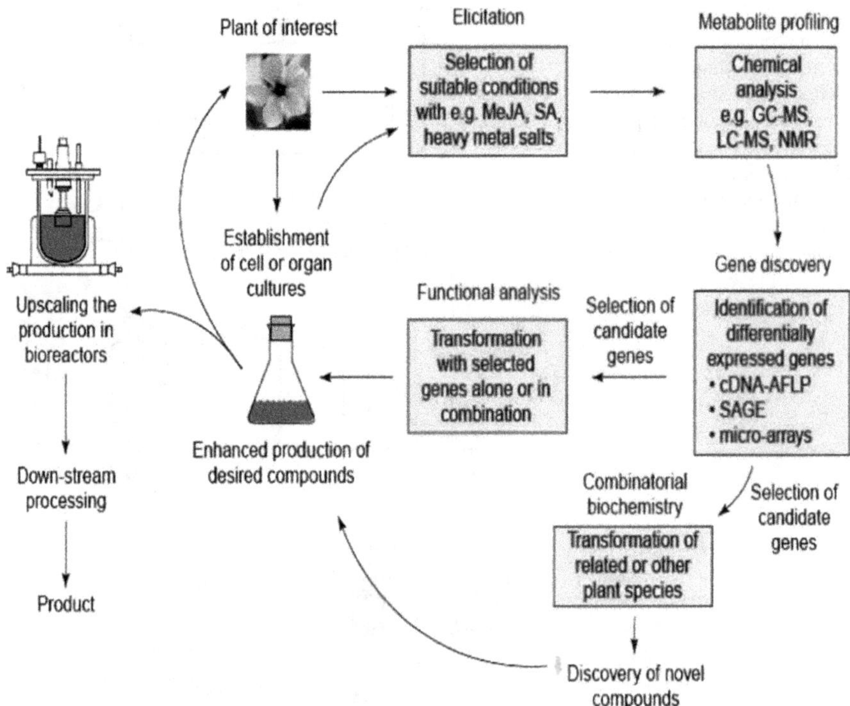

Fig. 10.24 Outline of how functional genomics could contribute to enhancing the production of known and novel secondary metabolites in plant cells. *AFLP* amplified fragment length polymorphism, *GC–MS* gas chromatography–mass spectrometry, *LC–MS* liquid chromatography–mass spectrometry, *MeJA* methyl jasmonate, *NMR* nuclear magnetic resonance, *SA* salicylic acid, *SAGE* serial analysis of gene expression (after Oksman-Caldentey and Inze 2004)

engineering, and they have several advantages over alternate bioreactor designs. Airlift bioreactors combine high loading of solid particles and good mass transfer, which are inherent for three-phase fluidized beds (Fig. 10.23). Efficient mixing in the liquid phase is generated by air bubbles, using internal (IL ALR) or external (EL ALR) recirculation loops (Fig. 10.23).

A typical recovery process involves four separation steps. The starting feed stream contains particulate material that must be removed by, e.g., centrifugation and/or filtration. Dilute solution of the product is then concentrated using, e.g., nonselective separation techniques (ultrafiltration, precipitation, liquid–liquid extraction, and adsorption). The subsequent steps involve a series of purifications to capture the product and remove trace contaminants. Chromatography, in its various forms (molecular size, charge, hydrophobicity, and molecular recognition), has proved to be the only general separation technique that can simultaneously achieve high purity, retain biological activity, and be scaled up to an appropriate production capacity.

A great deal of attention has been given to methods of capturing products directly from dilute and particulate-laden feed streams, thereby eliminating the need for all concentration steps. Affinity chromatography is such a technique, and in cases where high specificity and sensitivity are required, antibodies are an ideal choice for the separation of biomolecules, because of their high binding strength and selectivity (Birnbaum and Mosbach 1991). Although extremely versatile, packed bed separations (adsorption, absorption, ion exchange, and gel and affinity chromatography) are limited to batch operation and are not capable of handling cells or particulate material. Continuous separations can be performed using a magnetically stabilized fluidized bed (MSFB) of ferromagnetic particles or a mixture of nonmagnetic and magnetically susceptible support. MSFB has the flow and mass transfer properties of a packed bed, in conjunction with the solid phase fluidity of a fluidized bed. Continuous separations can thus be performed by countercurrent contact of solid and liquid phases without significant longitudinal mixing (i.e., in plug flow) at a low operating pressure drop. Recent applications of MSFB in biotechnology include continuous protein recovery, plant cell filtration, and plant cell cultivation.

The synthesis and excretion of secondary products are often coupled and associated with membrane transport of the product (see above). An artificial accumulation site can therefore reproduce similar transport phenomena in vitro. Luckner (1980) indicated that the productivity of plant cells can be improved by integration of biosynthesis and product recovery in the extractive phase. Additional advantages of integrated production and separation include enhanced rates of mass transfer, decreased level of product inhibition, facilitated product recovery, and reduced reaction volume for a given amount of product.

Selecting a proper bioreactor type and cultivation mode is crucial for obtaining productive in vitro system (Georgiev et al. 2009, 2013).

Kokotkiewicz et al. (2013) used a stirred tank bioreactor and batch cultivation mode in the scale-up experiments. In order to prevent cell damage, the growth vessel was equipped with plate type impeller generating low shear stress (Georgiev et al. 2013).

Numerous studies have applied bioreactors in plant cell (Huang and McDonald 2012) and organ culture to obtain specific metabolites (Srivastava and Srivastava 2012). Recent advances with large-scale production have successfully produced ginseng roots in a 10,000 l bioreactor establishing the feasibility of the root system to accommodate industrial processes (Gao et al. 2013).

Modern bioreactor culture systems provide a more advanced technology to produce higher secondary metabolites from plant cell, tissue, or organ using artificial nutrients with MeJA. Yu et al. (2002) found that the *ginsenoside* content was obviously enhanced by the addition of 100 µM MeJA during adventitious root culture of *Panax ginseng*; Donnez et al. (2011) examined that 0.2 mM MeJA was optimal for the efficient production and high accumulation of *resveratrol* in grape cell.

It is clear from these studies that temporary immersion bioreactor culture systems are appropriate for shoot multiplication and regeneration (Watt 2012) and the continuous immersion system is suitable for the proliferation of propagules (without leaves) such as bublets (Kim et al. 2004), PLBs (Park and Facchini 2000; Yang et al. 2010) and rhizomes (Jin et al. 2007). Gao et al. (2013) reported micropropagation of *Cymbidium* sp. using continuous and temporary air lift bioreactor system. In addition, a considerable number of researchers have cultured plant propagules in bioreactors to produce high-quality seedling (Zhao et al. 2012).

Besides this a considerable number of researchers have cultured plant propagules in bioreactors to produce high-quality seedling (Zhao et al. 2012). It is clear from these studies that temporary immersion bioreactor culture systems are appropriate for shoot multiplication and regeneration and the continuous immersion system is suitable for the proliferation of propagules (without leaves) such as bublets (Kim et al. 2004), PLBs (Yang et al. 2010), and rhizomes (see Gao et al. 2013). Kong et al. (2014) were constructed and tested airlift bioreactors (ALBs) for their potential to enhance chestnut embryogenic tissue proliferation for genetic transformation and mass propagation.

10.10 Prospects

With the onset of the 1990s, only Japan and, to a lesser extent, the Federal Republic of Germany were really engaged in the industrial production of secondary metabolites by plant cell cultures.

The only marketed product (as of 1990) remains shikonin. In Japan, seven private corporations have created a common subsidiary in research and development on plant cell cultures. Plant Cell Culture Technology (PCC Technology) has been set up with the support of the Japan Key Technology Centre (JKTC) by Kyowa Hakko Kogyo Co., Mitsui Petrochemical Industries Ltd, Mitsui Toatsu Chemical Inc., Hitachi Ltd, Suntory Ltd, Toa enryo Kogyo Co., and Kirin Breweries Co. Ltd. By contrast, most North American and European companies have not been enthusiastic about the prospects of profitable industrial production (Rajnchapelmessai 1988).

References

Abe I, Rohmer M, Prestwich GD (1993) Enzymatic cyclization of squalene and oxidosqualene to sterols and triterpenes. Chem Rev 93:2189–2206

Aird ELF, Hamill JD, Rhodes MJC (1988) Cytogenetic analysis of hairy root cultures from a number of plant species transformed with agrobacterium rhizogenes. Plant Cell Tissue Org Cult 15:47–57

Alfermann AW, Spieler H, Reinhard E (1985) Biotransformation of cardiac glycosides by Digitalis cell cultures in air lift reactors. In: Neumann KH, Barz W, Reinhard E (eds) Primary and secondary metabolism of plant cell cultures. Springer, Berlin, pp 316–322

Alfermann AW, Reinhard E (1986) Biotransformation of cardenolides by special cell strain of Digitalis lanata. In: VI IAPTC, Minnesota, p 350

Alfermann AW, Bergmann W, Figur C, Helmbold U, Schwantag D, Schuller I, Reinhard E (1983) Biotransformation of O-methyldigitoxin to p-methyldigoxin by cell cultures of Digitalis lanata. In: Mantell SH, Smith H (eds) Plant biotechnology. Cambridge University Press, Cambridge, pp 67–74

Alfermann AW, Petersen M, Fuss F (2003) Production of natural products by plant cell biotechnology: results, problems and perspectives. In: Laimer M, Rücker W (eds) Plant tissue culture. 100 Years since Gottlieb Haberlandt. Springer, Wien, pp 153–166

Allan E et al (2002) Induction of hairy root cultures of Azadirachta indica A. Juss. and their production of azadirachtin and other important insect bioactive metabolites. Plant Cell Rep 21 (4):374–379

Aoyagi H, Kobayashi Y, Yamada K, Yokoyama M, Kusakari K, Tanaka H (2001) Efficient production of saikosaponins in Bupleurum falcatum root fragments combined with signal transducers. Appl Microbiol Biotechnol 57:482–488

Ardalani H, Avan A, Ghayour-Mobarhan M (2017) Podophyllotoxin: a novel potential natural anticancer agent. Avicenna J Phytomed 7(4):285–294

Ashihara H, Sano H, Crozier A (2008) Caffeine and related purine alkaloids: biosynthesis, catabolism, function and genetic engineering. Phytochemistry 69:841–856

Arroo RRJ, Alfermann AW, Medarde M, Petersen M, Pras N, Woolley JG (2002) Plant cell factories as a source for anti-cancer lignans. Phytochem Rev 1:27–35

Asaka I, Ii I, Hirotani M, Asada Y, Furuya T (1993) Production of ginsenosides saponins by culturing ginseng (Panax ginseng) embryogenic tissues in bioreactors. Biotechnol Lett 15:1259–1264

Bailey JE (1991) Toward science of metabolic engineering. Science 252:1668–1675

Bais HP, Walker TS, McGrew JJ, Vivanco JM (2002) Factors affecting the growth of cell suspension cultures of *Hypericum perforatum* L. (St. John's Wort) and production of hypericin. In Vitro Growth Dev Plant 38:1–9

Bajaj YPS, Ishimaru K (1999) Genetic transformation of medicinal plants. In: Bajaj YPS (ed) Transgenic medicinal plants. Biotechnology in agriculture and forestry, vol 45. Springer, New Delhi, pp 1–29

Bakkali AT, Jaziril M, Foriers A, Van der Heyden Y, Vanhaelen M, Homès J (1997) Lawsone accumulation in normal and transformed cultures of henna, Lawsonia inermis. Plant Cell Tissue Org Cult 51:83–87

Baldwin IT, Schmelz EA, Ohnmeiss TE (1994) Wound-induced changes in root and shoot jasmonic acid pools correlate with induced nicotine synthesis in Nicotiana sylvestris spegazzini and comes. J Chem Ecol 20:2139–2157

Baque MA et al (2013) Production of biomass and bioactive compounds by adventitious root suspension cultures of *Morinda citrifolia* L. in a liquid-phase airlift balloon-type bioreactor. In Vitro Cell Dev Biol Plant 49(6):737–749

Bender L, Kumar A (2001) From soil to cell—a broad approach to plant life. Giessen Electronic Library GEB. http://geb.uni-giessen.de/geb/ebooks_ebene2.php

Benouaret R, Goujon E, Goupil P (2014) Grape marc extract causes early perception events, defence reactions and hypersensitive response in cultured tobacco cells. Plant Physiol Biochem 77(2):84–89. https://doi.org/10.1016/j.plaphy.2014.01.021

Bhadra R, Morgan JA, Shanks JV (1998) Transient studies of light-adapted cultures of hairy roots of *Catharanthus roseus*: growth and indole alkaloid accumulation. Biotechnol Bioeng 60:670–678

Bhansali S, Kumar A (2014) Hairy root culture of Eclipta Alba (L.) Hassk. J Acad (NY) 4(1):3–9

Bock R (2013) Strategies for metabolic pathway engineering with multiple transgenes. Plant Mol Biol 83:21–31

Bopana N, Saxena S (2010) Biotechnological aspects of secondary metabolite production. In: Kumar A, Sopory S (eds) Applications of plant biotechnology: *in vitro* propagation, plant transformation and secondary metabolite production. I.K. International, New Delhi, pp 451–473

Bourgaud F, Gravot A, Milesi S, Gontier E (2001) Production of plant secondary metabolites: a historical perspective. Plant Sci 161:839–851

Barz W, Reinhard E, Zenk MH (eds) (1977) Plant tissue culture and is biotechnological applications. Springer, Berlin

Barz W, Daniel S, Hinderer W, Jaques U, Kessmann H, Koster J, Tiemann K (1988) Elicitation and metabolism of phytoalexins in plant cell cultures. In: Pais M, Mavituna F, Novais J (eds) Plant cell biotechnology. NATO ASI Series. Springer, Berlin, pp 211–230

Becker H, Chavadej S (1988) Valepotriates: production by plant cell cultures. In: Bajaj YPS (ed) Biotechnology in agriculture and forestry, vol 4. Springer, Berlin, pp 294–309

Berger S, Bell E, Sadka A, Mullet JE (1995) Arabidopsis thaliana Atvsp is homologous to soybean VspA and VspB, genes encoding vegetative storage protein acid phosphatases, and is regulated similarly by methyl jasmonate, wounding, sugars, light and phosphate. Plant Mol Biol 27:933–942

Beveridge THJ, Li TSC, Drover CG (2002) Phytosterol content in American ginseng seed oil. J Agric Food Chem 50:744–750

Birnbaum S, Mosbach K (1991) Perspectives of immobilized proteins. Curr Opin Biotechnol 2:44–51

Blechert S, Brodschelm W, Holder S, Kammerer L, Kutchan TM, Chappell J (1995) The biochemistry and molecular biology of isoprenoid metabolism. Plant Physiol 107:1–6

Brain KR (1976) Accumulation of L-DOPA in cultures from Mucuna pruriens. Plant Sci Lett 7:157–161

Breyne P, Zabeau M (2001) Genome-wide expression analysis of plant cell cycle modulated genes. Curr Opin Plant Biol 4:136–142

Cai ZZ, Kastell A, Knorr D, Smetanska I (2012) Exudation: an expanding technique for continuous production and release of secondary metabolites from plant cell suspension and hairy root cultures. Plant Cell Rep 31:461–477

Capell T, Christou P (2004) Progress in plant metabolic engineering. Curr Opin Biotechnol 15:148–154

Carew DP, Staba EJ (1965) Plant tissue culture, its fundamentals, applications and relationship to medicinal plant species. Lloydia 28:1–26

Caspi R, Dreher K, Karp PD (2013) The challenge of constructing, classifying and representing metabolic pathways. FEMS Microbiol Lett 345:85–93

Chandra S (2012) Natural plant genetic engineer *Agrobacterium rhizogenes*: role of T-DNA in plant secondary metabolism. Biotechnol Lett 34:407–415. https://doi.org/10.1007/s10529-011-0785-3

Charlwood BV, Charlwood KA, Molina-Torres J (1990) Accumulation of secondary compounds by organized plant cultures. In: Charlwood BV, Rhodes MJC (eds) Secondary products from plant tissue culture. Clarendon Press, Oxford, pp 167–300

Chatterjee S, Srivastava S, Khalid A, Singh N, Sangwan RS, Sidhu OP, Roy R, Khetrapal CL, Tuli R (2010) Comprehensive metabolic fingerprinting of Withania somnifera leaf and root extract. Phytochemistry 71:1085–1094

Chattopadhyay S, Srivastava AK, Bhojwani SS, Bisaria VS (2002) Production of podophyllotoxin by plant cell cultures of Podophyllum hexandrum in bioreactor. J Biosci Bioeng 93:215–220

Chavadej S, Brisson N, McNeil JN, De Luca V (1994) Redirection of tryptophan leads to production of low indole glucosinolate canola. Proc Natl Acad Sci U S A 91:2166–2170

Chen AH (1985) Study on application of diosgenin. I. Analysis of diosgenin constituent of plants from Taiwan. Sci Monthly 43:79–85

Chen X, Yang L, Oppenheim JJ, Howard OMZ (2002) Cellular pharmacology studies of shikonin derivatives. Phytother Res 16:199–209

Chetia S, Handique PJ (2000) High frequency in vitro shoot multiplication of Plumbago indica, a rare medicinal plant. Curr Sci 78(10):1187–1188

Chintapakorn Y, Hamill JD (2003) Antisense-mediated downregulation of putrescine N-methyltransferase activity in transgenic Nicotiana tabacum L. can lead to elevated levels of anatabine at the expense of nicotine. Plant Mol Biol 53:87–10534

Choi YE, Yang DC, Kusano T, Sano H (2001) Rapid and efficient Agrobacterium-mediated genetic transformation by plasmolyzing pre-treatment of cotyledons in Panax ginseng. Plant Cell Rep 20:616–621

Choi DW, Jung JD, Ha YI, Park HW, Su In D, Chung HJ, Liu JR (2005) Analysis of transcripts in methyl jasmonate-treated ginseng hairy roots to identify genes involved in the biosynthesis of ginsenosides and other secondary metabolites. Plant Cell Rep 23:557–566

Christen AA, Gibson DM, Bland T (1991) US Patent no 5019504

Connor SEO (2015) Engineering of secondary metabolism. Annu Rev Genet 49:71–94

Conn S, Zhang W, Franko C (2003) Anthocyanin vacuolar inclusions (AVIs) selectively bind acylated anthocyanins in Vitis vinifera L. (Grapewine) suspension cultures. Biotechnol Lett 25:835–839

Curtin C, Zhang W, Franco C (2003) Manipulating anthocyanin composition in Vitis vinifera suspension cultures by elicitation with jasmonic acid and light irradiation. Biotechnol Lett 25:1131–1135

Datta A, Srivastava PS (1997) Variation in vinblastine production by Catharanthus roseus during in vivo and in vitro differentiation. Phytochemistry 96:135–137

Datta K, Baisakh N, Thet KM, Tu J, Datta S (2002) Pyramiding transgenes for multiple resistance in rice against bacterial blight, yellow stem borer and sheath blight. Theor Appl Genet 106:1–8

De Luca V, St-Pierre B (2000) The cell and developmental biology of alkaloid biosynthesis. Trends Plant Sci 5:168–173

Dehghan E, Hakkinen ST, Oksman-Caldentey KM, Ahmadi FS (2012) Production of tropane alkaloids in diploid and tetraploid plants and in vitro hairy root cultures of Egyptian henbane (Hyoscyamus muticus L.). Plant Cell Tissue Organ Cult 110:35–44

Devarenne TP, Ghosh A, Happell J (2002) Regulation of squalene synthase, a key enzyme of sterol biosynthesis, in tobacco. Plant Physiol 129:1095–1106

DiCosmo F, Misawa M (1985) Eliciting secondary metabolism in plant cell cultures. Trends Biotechnol 3:318–322

Diwan R, Malpathak N (2007) A novel source of furanocoumarins: Ruta graveolens L. In: Kumar A, Sopory S (eds) Recent advances in plant biotechnology and its applications. I.K. International, New Delhi. (in press)

Dixon RA (2005) Engineering of plant natural product pathways. Curr Opin Plant Biol 8:329–336

Dixon RA and Summer LW (2003) Legume natural products: understanding and manipulating complex pathways for human and animal health. Plant Physiol. 131: 878–885

Dixon RA, Steele CL (1999) Flavonoids and isoflavonoids – a goldmine for metabolic engineering. Trends Plant Sci 4:394–400

Donnez D, Kim K-H, Antoine S, Conreux A, De Luca V, Jeandet P, Clement C, Courot E (2011) Bioproduction of resveratrol and viniferins by an elicited grapevine cell culture in a 2 L stirred bioreactor. Process Biochem 46:1056–1062

Dornenburg H, Knorr D (1997) Challenges and opportunities for metabolite production from plant cell and tissue cultures. Food Technol 51:47–54

Duge de Bernonville T, Clastre M, Besseau S, Oudin A, Burlat V et al (2014) Phytochemical genomics of the Madagascar periwinkle: unravelling the last twists of the alkaloid engine. Phytochemistry 113:9–23

Dumontet C, Jordan MJ (2010) Microtubule-binding agents: a dynamic field of cancer therapeutics. Nat Rev 9:790–803

Ebel J, Cosio EG (1994) Elicitors of plant defense responses. Int Rev Cytol 148:1–36

Eilert U, Kurz WGW, Constabel F (1985) Stimulation of sanguinarine accumulation in Papaver somniferum cell cultures by fungal elicitors. J Plant Physiol 119:65–76

Endt DV, Kijne JW, Memelink J (2002) Transcription factors controlling plant secondary metabolism: what regulates the regulators? Phytochemistry 61:107–114

Enieux G-C, Van Beekt T (1997) Production of ginkgolide and bilobalide in transformed and gametophyte derived cell cultures of Ginkgo biloba. Phytochemistry 46:127–130

Espinosa-Leal CA, Puente-Garza CA, García-Lara S (2018) In vitro plant tissue culture: means for production of biological active compounds. Planta 248(1):1–18. https://doi.org/10.1007/s00425-018-2910-1

Facchini PJ, Huber-Allanach KL, Tari LW (2000) Plant aromatic L-amino acid decarboxylases: evolution, biochemistry, cell biology, and metabolic engineering applications. Phytochemistry 54:121–138

Farre G, Blancquaert D, Capell T, Van Der Straeten D, Christou P, Zhu C (2014) Engineering complex metabolic pathways in plants. Plants Annu Rev Plant Biol 65:187–223

Farrow SC, Hagel JM, Facchini PJ (2012) Transcript and metabolite profiling in cell cultures of 18 plant species that produce benzylisoquinoline alkaloids. Phytochemistry 77:79–88

Fett-Neto AG, Pennington JJ, DiCosmo F (1995) Effect of white light on taxol and baccatin III accumulation in cell cultures of Taxus cuspidata Sieb and Zucc. J Plant Physiol 146:584–590

Fiehn O, Linke T, Willmitzer ARF (2001) Metabolic profiling allows comprehensive phenotyping of genetically or environmentally modified plant systems. Plant Cell 13:11–29

Flores HE (1992) Plant roots as chemical factories. Chem Indus 10:374–377

Flores BE, Sarrafi A, Fabre F, Alibert G (2000) Genotypic variation and chromosomal location of QTLs for somatic embryogenesis revealed by epidermal layers culture of recombinant inbred lines in sunflower (Helianthus annuus L.). Theor Appl Genet 101:1307–1312

Fritzemeier KH, Rolfs CH, Pfau J, Kindl H (1983) Action of ultraviolet C on stilbene formation in callus of Arachis hypogaea. Planta 159:25–29

Fu C et al (2012) Assessment of genetic and epigenetic variation during long-term Taxus cell culture. Plant Cell Rep 31(7):1321–1331

Fujita Y, Tabata M, Nishi A, Yamada Y (1982) New medium and production of secondary compounds with the two staged culture method. In: Fujiwara A (ed) Plant tissue culture 1982. Japanese Association for Plant Tissue Culture, Tokyo, pp 399–400

Furuya T (1988) Saponins (ginseng saponins). In: Vasil IK (ed) Cell cultures and somatic cell genetics of plants, vol 5. Academic Press, San Diego, CA, pp 213–234

Furuya T, Ikuta A, Syono K (1972) Alkaloids from callus cultures of Papaver somniferum. Phytochemistry 11:3041–3044

Gadzovska S, Maury S, Delaunay A, Spasenoski M, Joseph D, Hagege D (2007) Jasmonic acid elicitation of Hypericum perforatum L. cell suspensions and effects on the production of phenylpropanoids and naphtodianthrones. Plant Cell Tissue Org Cult 89:1–13

Gangopadhyay M et al (2008) Hairy root culture of Plumbago indica as a potential source for plumbagin. Biol Plant 52(3):533–537

Gao R, Wu S-Q, Piao X-C, Park S-Y, Lian M-L (2013) Micropropagation of Cymbidium sinense using continuous and temporary airlift bioreactor systems. Acta Physiol Plant 36(1):117–124

Geerlings A, Hallard D, Caballero AM, Cardoso L, van der Heijden R, Verpoorte R (1999) Alkaloid production by a Cinchona officinalis 'Ledgeriana' hairy root culture containing constitutive expression constructs of tryptophan decarboxylase and strictosidine synthase cDNAs from Catharanthus roseus. Plant Cell Rep 19:191–196

Georgiev MI, Weber J, Maciuk A (2009) Bioprocessing plant cell cultures for mass production of targeted compounds. Appl Microbiol Biotechnol 83:809–823

Georgiev MI, Eibl R, Zhong J (2013) Hosting the plant cells in vitro: recent trends in bioreactors. Appl Microbiol Biotechnol 97:3787–3800. https://doi.org/10.1007/s00253-013-4817-x

Glenn WS, Runguphan W, O'Connor SE (2013) Recent progress in the metabolic engineering of alkaloids in plant systems. Curr Opin Biotechnol 24:354–365

Gloster TM (2014) Advances in understanding glycosyltransferases from a structural perspective. Curr Opin Struct Biol 28:131–141

Gontier E, Sangwan SB, Barbotin JN (1994) Effects of calcium, alginate, and calcium-alginate immobilization on growth and tropane alkaloid levels of a stable suspension cell line of Datura inoxia Mill. Plant Cell Rep 13:533–536

Goossens A, Häkkinen ST, Laakso I, Seppänen-Laakso T, Biondi S, De Sutter V, Lammertyn F, Nuutila AM, Söderlund H, Zabeau M, Inzé D, Oksman-Caldentey KM (2003a) A functional genomics approach toward the understanding of secondary metabolism in plant cells. Proc Natl Acad Sci U S A 100:8595–8600

Goossens A, Häkkinen ST, Kirsi-Marja IL, Oksman-Caldentey KM, Inzé D (2003b) Secretion of secondary metabolites by ATP-binding cassette transporters in plant cell suspension cultures. Plant Physiol 131:1161–1164

Goyal S, Ramawat KG (2008a) Ethrel treatment enhanced isoflavonoids accumulation in cell suspension cultures of Pueraria tuberosa, a woody legume. Acta Physiol Plant 30:849–853

Goyal S, Ramawat KG (2008b) Synergistic effect of morphactin on cytokinin-induced production of isoflavonoids in cell cultures of Pueraria tuberosa (Roxb. Ex. Willd.) DC. Plant Growth Regul 55:175–181

Graves JMH, Smith WK (1967) Transformation of pregnenolone and progesterone by cultured plant cells. Nature 214:1248–1249

Guern J, Renaudin JP, Brown SC (1987) The compartmentation of secondary metabolites in plant cell cultures. In: Constabel F, Vasil IK (eds) Cell culture and somatic cell genetics of plants, vol 4. Academic Press, San Diego, CA, pp 43–76

Guirimand G, Guihur A, Poutrain P, Hericourt F, Mahroug S et al (2011) Spatial organization of the vindoline biosynthetic pathway in Catharanthus roseus. J Plant Physiol 168:549–557

Gunawardena AH, Greenwood JS, Dengler NG (2004) Programmed cell death remodels lace plant leaf. Plant Cell 16:60–73

Gundlach H, Muller MJ, Kutchan TM, Zenk MH (1992) Jasmonic acid is a signal transducer in elicitor-induced plant cell cultures. Proc Natl Acad Sci U S A 89:2389–2393

Gupta P et al (2013) De novo assembly, functional annotation and comparative analysis of Withania somnifera leaf and root transcriptomes to identify putative genes involved in the withanolides biosynthesis. PLoS One 8:e62714

Hahlbrock K, Scheel D (1989) Physiology and molecular biology of phenylpropanoid metabolism. Annu Rev Plant Physiol Mol Biol 40:347–369

Han J-Y, Wang H-Y, Choi Y-E (2014) Production of dammarenediol-II triterpene in a cell suspension culture of transgenic tobacco. Plant Cell Rep 33(2):225–233

Hara Y (1996) Research on the production of useful compounds by plant cell cultures in Japan. In: Di Cosmo F, Misawa M (eds) Plant cell culture secondary metabolism—toward industrial application. CRC Press, Boca Raton, FL, pp 187–202

Hartmann T (1996) Diversity and variability of plant secondary metabolism: a mechanistic view. Entomol Exp Appl 80:177–188

Hashimoto T, Yamada Y (1994) Alkaloid biogenesis: molecular aspects. Annu Rev Plant Physiol Plant Mol Biol 45:257–285

Hashimoto T, Yamada Y (2003) New genes in alkaloid metabolism and transport. Curr Opin Biotechnol 14:163

Hashimoto T, Yun DJ, Yamada Y (1993) Production of tropane alkaloids in genetically engineered root cultures. Phytochemistry 32:713–718

Hawkins KM, Smolke CD (2008) Production of benzylisoquinoline alkaloids in Saccharomyces cerevisiae. Nat Chem Biol 4:564–573

Hayashi H, Huang P, Inoue K (2003) Up-regulation of soyasaponin biosynthesis by methyl jasmonate in cultured cells of Glycyrrhiza glabra. Plant Cell Physiol 44:404–411

Heble MR, Staba E (1980) Steroid metabolism in stationary phase cell suspensions of Dioscorea deltoidea. Planta Med 40(Suppl 1):124–128

Heble, MR and Chadha M S (1985a) Recent developments in the biotechnological application of plant tissue and organ cultures. In: Proc. 5th ISHS Symp. Mad. Aromatic and spice plants. Darjeeling, pp 67–74

Heble MR, Chadha MS (1985b) Recent developments in the biotechnology. Biotechnology in Healthcare, New Delhi, pp 55–64

Heble MR, Narayanaswamy S, Chadha MS (1967) Diosgenin production and beta-sitosterol isolation from Solanum xanthocarpum tissue cultures. Science 161:1145

Heinig U, Gutensohn M, Dudareva N, Aharoni A (2013) The challenges of cellular compartmentalization in plant metabolic engineering. Curr Opin Biotechnol 24:239–246

Hiraoka N. Bhatt ID, Change JI (2004) Alkaloid production by somatic embryo cultures of *corydalis ambigua*. Plant Biotechnol 21(5): 361–366

Hirasuna TJ, Pestchanker LJ, Srinivasan V, Shuler ML (1996) Taxol production in suspension cultures of Taxus baccata. Plant Cell Tiss Org 44:95–102

Hiroaka N, Bhatt ID (2008) Micropropagation and alkaloid production through somatic embryo cultures of Corydalis. In: Kumar A, Sopory S (eds) Recent advances in plant biotechnology and its applications. I.K. International, New Delhi, pp 622–630

Holden M (1990) Prospect for the genetic manipulation of metabolic pathways leading to secondary products. In: Sangwan RS, Sangwan-Norreel BS (eds) The impact of biotechnology in agriculture. Kluwer, Dordrecht, pp 403–417

Holden AM, Yeoman MM (1987) Optimization of product yield in immobilized plant cell cultures. In: Moody GW, Holden AM (eds) Bioreactors and biotransformation. Elsevier, London, pp 1–11

Holden MA, Holden PR, Yeoman MM (1988) Elicitation of secondary product formation in Capsicum frutescens cultures. In: Robins RJ, Rhodes MJC (eds) Manipulating secondary metabolism in culture. Cambridge University Press, Cambridge, pp 57–65

Honda H, Itoh T, Shiragini N, Unno H (1993) Phytohormone-control for plant cell culture using a bioreactor equipped with in-line feeding column. J Chem Eng Japan 26:291–296

Horwitz SB (1994) How to make taxol from scratch. Nature 367:593–594

Huang TK, McDonald KA (2012) Bioreactor systems for in vitro production of foreign proteins using plant cell cultures. Biotechnol Adv 30:398–409

Huang WW, Cheng CC, Yeh FT, Tsay HS (1993) Tissue culture of Dioscorea doryophora Hance. 1. Callus organs and the measurement of diosgenin content. Chin Med Coll J 2(2):151–160

Hughes EH, Shanks JV (2002) Metabolic engineering of plants for alkaloid production. Metab Eng 4:41–48

Hulst AC, Tramper J, Brodelius P, Eijkenboom LJC, Luyben KCAM (1985) Immobilised plant cells: respiration and oxygen transfer. J Chem Technol Biotechnol 35B:198–204

Hütsch BW, Osthushenrich T, Faust F, Kumar A, Schubert S (2016) Reduced sink activity in growing shoot tissues of maize under salt stress of the first phase may be compensated by increased PEPCarboxylase activity. J Agron Crop Sci 202(5):384–393. https://doi.org/10.1111/jac.12162

Ikram NKBK, Simonsen HT (2017) A review of biotechnological artemisinin production in plants. Front Plant Sci 8:1966. https://doi.org/10.3389/fpls.2017.01966

Ionkova I (2007) Biotechnological approaches for the production of lignans. Pharmacognosy Rev 1:57–68

Isah T (2019) Stress and defense responses in plant secondary metabolites production. Biol Res 52:39. https://doi.org/10.1186/s40659-019-0246-3

Isfort RJ et al (1993) Induction of protein phosphorylation, protein synthesis, immediate-early-gene expression and cellular proliferation by intracellular pH modulation Implications for the role of hydrogen ions in signal transduction. Eur J Biochem 357:349–357

Ishida BK (1988) Improved diosgenin production in Dioscorea deltoidea cell cultures by immobilization in polyurethane foam. Plant Cell Rep 7:270–273

Issell BF, Muggia FM, Carter SK (1984) Etoposide (VP-16) current status and new developments. Academic Press, Orlando, FL

Jacob A, Malpathak N (2005) Manipulation of MS and B5 components for enhancement of growth and solasodine production in hairy root cultures of Solanum khasianum Clarke. Plant Cell Tissue Org Cult 80:247–257

Jacob A, Malpathak N (2006) Steroid factories: prospects and practicalities in effective farming of solanum plants. In: Kumar A, Roy S (eds) Plant biotechnology and its applications in tissue culture. I.K. International, New Delhi, pp 205–241

Jacob A, Malpathak NP, Sudha G, Ravishankar GA (2008) A comparative study of growth kinetics in hairy root cultures of Solanum khasianum Clark grown in shake flasks and bioreactors. In:

Kumar A, Sopory S (eds) Recent advances in plant biotechnology and its applications. I.K. International, New Delhi, pp 610–621

Jain M, Rathore AK, Khanna P (1984) Influence of kinetin and auxins on the growth and production of diosgenin by Costus speciosus (Koen) Sm. Callus derived from rhizome. Agric Biol Chem 48:529

Jain SC, Sharma RA, Jain R, Mittal C (1998) Antimicrobial screening of Cassia occidentalis in vivo and in vitro. Phytother Res 12:200–204

Jang M, Cai L, Udeani GO, Slowing KV, Tomas CF, Beecher CW, Fong HH, Farnsworth NR, Kinghorn AD, Mehta RG, Moon RC, Pezzuto JM (1997) Cancer chemopreventive activity of resveratrol, a natural product derived from grapes. Science 275:218–220

Jasinski M, Stukkens Y, Degand H, Purnelle B, Marchand-Brynaert J, Boutry M (2001) A plant plasma membrane ATP binding cassettetype transporter is involved in antifungal terpenoid secretion. Plant Cell 13:1095–1107

Jennewein S, Croteau R (2001) Taxol: biosynthesis, molecular genetics, and biotechnological applications. Appl Microbiol Biotechnol 57:13–19

Jin H, Piao XC, Sun D, Xiu JR, Lian ML (2007) Mass production of rhizome and shoot of Cymbidium niveo-marginatum using simple bioreactor. J Northeast Forest Univ 35:44–48

Joe AK, Liu H, Suzui M, Vural ME (2002) Resveratrol induces growth inhibition, S-phase arrest, apoptosis, and changes in biomarker expression in several human cancer cell lines. Clin Cancer Res 8:893–903

Jones A, Veliky IA (1981) Effect of medium constituents on the viability of immobilized plant cells. Can J Bot 59:2095–2101

Jung G, Tepfer D (1987) Use of genetic transformation by the Ri T–DNA of Agrobacterium rhizogenes to stimulate biomass and tropane alkaloid production in Atropa belladonna and Calystegia sepium roots grown in vitro. Plant Sci 50:145–151

Jung JD, Park HW, Hahn Y, Hur CG, In DS, Chung HJ, Liu JR, Choi DW (2003) Discovery of genes for ginsenoside biosynthesis by analysis of ginseng expressed sequence tag. Plant Cell Rep 22:224–230

Kadkade PG (1982) Growth and podophyllotoxin production in callus tissues of Podophyllum peltatum. Plant Sci Lett 25:107–115

Kakegawa K, Suda J, Sugiyama M, Komamine A (1995) Regulation of anthocyanin biosynthesis in cell suspension cultures of Vitis in relation to cell division. Physiol Plant 94:661–666

Kamisako W, Morimoto K, Makino I, Isoi K (1984) Changes in triterpenoid content during the growth cycle of cultured plant cells. Plant Cell Physiol 25:1571–1574

Kaul B, Stohs SJ, Staba EJ (1969) Dioscorea tissue cultures. 3. Influence of various factors on diosgenin production by Dioscorea deltoidea callus and suspension cultures. Lloydia 32:347–359

Ketchum REB, Gibson DM, Croteau RB, Shuler ML (1999) The kinetics of taxoid accumulation in cell suspension cultures of Taxus following elicitation with methyl jasmonate. Biotechnol Bioeng 62:97–105

Khan MIR, Syeed S, Nazar R, Anjum NA (2012) An insight into the role of salicylic acid and jasmonic acid in salt stress tolerance. Springer, Berlin

Khan MI, Fatma M, Per TS, Anjum NA, Khan NA (2015) Salicylic acid-induced abiotic stress tolerance and underlying mechanisms in plants. Front Plant Sci 6:462

Khanam N, Khoo C, Close R, Khan AG (2000) Organogenesis, differentiation and histolocalization of alkaloids in cultured tissues and organs of Duboisia myoporoides R. Br. Ann Bot 86:745–752

Khanna P (1977) Tissue culture and useful drugs: a review of twenty plant species grown in vitro: cultivation and utilization of medicinal and aromatic plants. In: Atal CK, Kapoor BM (eds). RRL, Jammu Tawi, pp 495–500

Khanna P, Staba EJ (1968) Antimicrobials from plant tissue cultures. Lloydia 31:180–189

Kibler R, Neumann KH (1980) On cytogenetic stability of cultured tissues and cell suspensions of haploid and diploid origin. In: Sala F, Parisi B, Cella R, Ciferri O (eds) Plant cell cultures: results and perspectives. Elsevier/North Holland, Amsterdam, pp 59–65

Kim JH, Yun JH, Hwang YS, Byun SY, Kim DI (1995) Production of taxol and related taxanes in Taxus brevifolia cell cultures: effect of sugar. Biotechnol Lett 17:101–106

Kim Y, Wyslouzil BE, Weathers PJ (2002) Secondary metabolism of hairy root cultures in bioreactors. In vitro Cell Dev Bio Plant 38:1–10

Kim SI, Kim JY, Kim EA, Kwon KH, Kim KW, Cho K, Lee JH, Nam MH, Yang DC, Yoo JS, Park YM (2003) Proteome analysis of hairy root from Panax ginseng C.A. Meyer using peptide fingerprinting, internal sequencing and expressed sequence tag data. Proteomics 3:2379–2397

Kim BJ, Gibson DM, Shuler ML (2004) Effect of subculture and elicitation on instability of taxol production in Taxus sp. suspension cultures. Biotechnol Prog 20:1666–1673

Kim OT, Yoo NH, Kim GS, Kim YC, Bang KH, Hyun DY, Kim SH, Kim MY (2012) Stimulation of Rg3 ginsenoside biosynthesis in ginseng hairy roots elicited by methyl jasmonate. Plant Cell Tissue Organ Cult. https://doi.org/10.1007/s11240-012-0218-6

Kitaoka N, Lu X, Yang B, Peters RJ (2015) The application of synthetic biology to elucidation of plant mono-, sesqui-, and diterpenoid metabolism. Mol Plant 8:6–16

Kokotkiewicz A et al (2013) Isoflavone production in *Cyclopia subternata* Vogel (honeybush) suspension cultures grown in shake flasks and stirred-tank bioreactor. Appl Microbiol Biotechnol 97(19):8467–8477

Komamine A, Misawa M, DiCosmo FE (1991) Plant cell culture in Japan. Progress in production of useful plant metabolites by Japanese enterprises using plant cell culture technology. CMC Co., Tokyo

Kong W, Wei J, Abidi P, Lin M, Inaba S, Li C, Wang Y, Wang Z, Si S, Pan H, Wang S, Wu J, Wang Y, Li Z, Liu J, Jiang JD (2004) Berberine is a novel cholesterol-lowering drug working through a unique mechanism distinct from statins. Nat Med 10:1344–1351

Kong L, Holtz CT, Nairn CJ, Houke H, Powell WA, Baier K, Merkle SA (2014) Application of airlift bioreactors to accelerate genetic transformation in American chestnut. Plant Cell Tissue Organ Cult 117(1):39–50

Krikorian AD (2001) Novel application of plant tissue culture and conventional breeding techniques to space biology research. In: Bender L, Kumar A (eds) From soil to cell—a broad approach to plant life. Giessen Electronic Library, GEB. http://geb.uni-giessen.de/geb/ebooks_ebene2.php

Krikorian AD, Levin HG (1991) Development and growth in space. In: Bidwell RGS, Steward FC (eds) Plant physiology. A treatise, vol X. Academic Press, New York, pp 491–555

Krisa S, Vitrac X, Decendit A, Larronde F, Deffieux G, Mérillon JM (1999) Obtaining Vitis vinifera cell cultures producing higher amounts of malvidin-3-O-glucoside. Biotechnol Lett 21:497–500

Ku KL, Chang PS, Cheng YC, Lien CY (2005) Production of stilbenoids from the callus of Arachis hypogaea: a novel source of the anticancer compound piceatannol. J Agric Food Chem 53:3877–3881

Kukreja AK, Garg S (2008) Forskolin biosynthesis and invertase activity in Agrobacterium rhizogenes mediated transformed roots of Coleus forskohlii. In: Kumar A, Sopory S (eds) Recent advances in plant biotechnology and its applications. I.K. International, New Delhi, pp 517–533

Kukreja AK, Mathur AK, Ahuja PS (1986) Morphogenetic potential of foliar explants in Duboisia myoporoides R. Br. (Solanaceae). Plant Cell Rep 5:27–30

Kumar A (2015a) Metabolic engineering in plants. In: Bahadur B, Rajam MV, Sahijram L, Krishnamurthy KV (eds) Plant biology and biotechnology. II Plant genomics and biotechnology. Springer, New Delhi, pp 517–526

Kumar A (2015b) Improving secondary metabolite production in tissue cultures. In: Bahadur B, Rajam MV, Sahijram L, Krishnamurthy KV (eds) Plant biology and biotechnology. II Plant genomics and biotechnology. Springer, New Delhi, pp 397–406

Kumar A, Roy S (2006) Plant biotechnology and its applications in tissue culture. I.K. International, New Delhi, 307 pp

Kumar A, Roy S (2011) Plant tissue culture and applied plant biotechnology. Avishkar Publishers, Jaipur, 346 pp

Kumar A, Shekhawat NS (2009) Plant tissue culture and molecular markers: their role in improving crop productivity. I.K. International, New Delhi, 688 pp

Kumar A, Sopory S (2008) Recent advances in plant biotechnology and its applications. I.K. International, New Delhi, 718 pp

Kumar A, Sopory S (2010) Applications of plant biotechnology: *in vitro* propagation, plant transformation and secondary metabolite production. I.K. International, New Delhi, 606 pp

Kumar S, Hahn FM, Baidoo E, Kahlon TS, Wood DF et al (2012) Remodeling the isoprenoid pathway in tobacco by expressing the cytoplasmic mevalonate pathway in chloroplasts. Metab Eng 14:19–28

Kumar A, Sharma M, Basu SK, Asif M, Li XP, Chen X (2014) Plant molecular breeding: perspectives from plant biotechnology and marked assisted selection. In: Benkeblia N (ed) Omics technologies and crops improvement. CRC Press, Boca Raton, FL

Kunze R, Frommer WB, Flugge UI (2002) Metabolic engineering of plants: the role of membrane transport. Metab Eng 4:57–66

Kushiro T, Ohno Y, Shibuya Y, Ebizuka Y (1997) In vitro conversion of 2,3-oxidosqualene into dammarenediol by Panax ginseng microsomes. Biol Pharmcol Bull 20:292–294

Kushiro T, Shibuya M, Ebizuka Y (1998) Amyrin synthase: cloning of oxidosqualene cyclase that catalyzes the formation of the most popular triterpene among higher plants. Eur J Biochem 256:238–244

Lapierre C, Pollet B, Petit-Conil M, Toval G, Romero J et al (1999) Structural alterations of lignins in transgenic poplars with depressed cinnamyl alcohol dehydrogenase or caffeic acid methyltransferase activity have an opposite impact on the efficiency of industrial kraft pulping. Plant Physiol 119:153–163

Lee M-H, Jeong J-H, Seo J-W, Shin C-G, Kim Y-S, In J-G, Yang D-C, Yi J-S, Choi Y-E (2004) Enhanced triterpene and phytosterol biosynthesis in Panax ginseng—overexpressing squalene synthase. Gene Plant Cell Physiol 45:976–984

Lee OR, Yang DC, Chung HJ, Min BH (2011) Efficient in vitro plant regeneration from hybrid rhizomes of *Cymbidium sinense* seeds. Hortic Environ Biotechnol 52:303–308

Leech MJ, May K, Hallard D, Verpoorte R, De Luca V, Christou P (1998) Expression of two consecutive genes of a secondary metabolic pathway in transgenic tobacco: molecular diversity influences levels of expression and product accumulation. Plant Mol Biol 38:765–774

Li SC, Doran PM (1991) Enhanced codeine and morphine production in suspended Papaver somniferum cultures after removal of exogenous hormones. Plant Cell Rep 10:349–353

Li W, Koike K, Asada Y, Yoshikawa T, Nikaido T (2005) Biotransformation of paeonol by *Panax ginseng* root and cell cultures. J Mol Catal B Enzym 35:117–121

Li LQ, Fu CH, Zhao CF, Xia J, Yu LJ (2009) Efficient extraction of RNA and analysis of gene expression in long-term taxus cell culture using real-time RT-PCR. Z Naturforsch C 64:125–130

Li X, Guo H, Qi Y, Liu Y et al (2016) Salicylic acid-induced cytosolic acidification increases the accumulation of phenolic acids in *Salvia miltiorrhiza* cells. Plant Cell Tissue Organ Cult 126 (2):333–341. https://doi.org/10.1007/s11240-016-1001-x

Li M, Xu J, Algarra Alarcon A et al (2017) In planta recapitulation of isoprene synthase evolution from ocimene synthases. Mol Biol Evol 34(10):2583–2599. https://doi.org/10.1093/molbev/msx178

Lin Y, Tsay HS (2004) Studies on the production of some important secondary metabolites from medicinal plants by plant tissue cultures. Bot Bull Acad Sin 45:1–22

Lindsey K, Yeoman MM (1983) The relationship between growth rate, differentiation and alkaloid accumulation in cell cultures. J Exp Bot 34:1055–1065

Liu CZ, Wang YC, Ouyang F, Ye HC, Li GF (1997) Production of artemisinin by hairy root cultures of Artemisia annua L. Biotechnol Lett 19:927–930

Liu Y-R, Jiang Y-L, Huang R-Q, Yang J-Y, Xiao B-K, Dong J-X (2014) Hypericum perforatum L. preparations for menopause: a meta-analysis of efficacy and safety. Climacteric 17 (4):325–335. https://doi.org/10.3109/13697137.2013.861814

Liu L, Sonbol F, Huot B et al (2016) Salicylic acid receptors activate jasmonic acid signalling through a non-canonical pathway to promote effector-triggered immunity. Nat Commun 7:13099

Lorence A, Nessler CL (2004) Camptothecin, over four decades of surprising findings. Phytochemistry 65(20):2735–2749

Lu BW, Baum L, So KF, Chiu K, Xie LK (2019) More than anti-malarial agents: therapeutic potential of artemisinins in neurodegeneration. Neural Regen Res 14(9):1494–1498. https://doi.org/10.4103/1673-5374.255960

Luckner M (1980) Alkaloid biosynthesis in Penicillium cyclopium—does it reflect general features of secondary metabolism? J Nat Prod 43:21–40

Luckner M, Diettrich B (1985) Formation of cardenolides in cell and organ cultures of Digitalis lanata. In: Neumann KH, Barz W, Reinhard W (eds) Primary and secondary metabolism of plant cell cultures. Springer, Berlin, pp 154–163

Luckner M, Diettrich B (1987) Biosynthesis of caedenolides in cell cultures of Digitalis lanata—the results of a new strategy. In: Green CE, Somers DA, Hackett WP, Biesboer DD (eds) Plant tissue and cell culture. A.R. Liss, New York, pp 187–197

Magnotta M, Murata J, Chen J, De Luca V (2007) Expression of deacetylvindoline-4-O-acetyltransferase in *Catharanthus roseus* hairy roots. Phytochemistry 68:1922–1931

Majumdar S, Garai S, Jha S (2012) Use of the cryptogein gene to stimulate the accumulation of bacopa saponins in transgenic *Bacopa monnieri* plants. Plant Cell Rep 31:1899–1909. https://doi.org/10.1007/s00299-012-1303-3

Maldonado-Mendoza IE, Ayora-Talavera T, Loyola-Vargas VM (1993) Establishment of hairy root cultures of *Datura stramonium*: characterization and stability of tropane alkaloid production during long periods of subculturing. Plant Cell Tissue Organ Cult 33:321–329

Malik S, Cusido RM, Mirjalili MH, Moyano E, Palazon J, Bonfill M (2011) Production of the anticancer drug taxol in Taxus baccata suspension cultures: a review. Process Biochem 46:23–34

Markham KR, Gould KS, Winefield CS, Mitchell KA, Bloor SJ, Boase MR (2000) Anthocyanic vacuolar inclusions—their nature and significance in flower colouration. Phytochemistry 55:327–336

Martinoia E (2018) Vacuolar transporters – companions on a longtime journey. Plant Physiol 176 (2):1384–1407. https://doi.org/10.1104/pp.17.01481

Martinoia E, Klein M, Geisler M, Bovet L, Forestier C, Kolukisaoglu U, Muller-Rober B, Schulz B (2002) Multifunctionality of plant ABC transporters—more than just detoxifiers. Planta 214:345–355

Marinova K, Pourcel L, Weder B, Schwarz M, Barron D, Routaboul JM, Debeaujon I, Klein M (2007) The Arabidopsis MATE transporter TT12 acts as a vacuolar flavonoid/H+-antiporter active in proanthocyanidin-accumulating cells of the seed coat. Plant Cell 19:2023–2038

Maxwell S, Cruickshank A, Thorpe G (1994) Red wine and antioxidant activity in serum. Lancet 344:193–194

McCranie EK, Bachmann BO (2014) Bioactive oligosaccharide natural products. Nat Prod Rep 31 (8):1026–1042. https://doi.org/10.1039/c3np70128j

McDaniel R, Welch M, Hutchinson CR (2005) Genetic approaches to polyketide antibiotics. Chem Rev 105:543–558

Meijer AH, Verpoorte R, Hoge JHC (1993a) Regulation of enzymes and genes involved in terpenoid indole alkaloid biosynthesis in Catharanthus roseus. J Plant Res 3:145–164

Meijer AH, Cardoso MIL, Voskuilen JT, de Waal A, Verpoorte R, Hoge JHC (1993b) Isolation and characterization of a cDNA clone from Catharanthus roseus encoding NADPH: cytochrome P-450 monooxygenases in plants. Plant J 4:47–60

Memelink J, Gantet P (2007) Transcription factors involved in terpenoid indole alkaloid biosynthesis in Catharanthus roseus. Phytochem Rev 6:353–362

Menke FLH, Champion A, Kijne JW, Memelink J (1999) A novel jasmonate- and elicitor-responsive element in the periwinkle secondary metabolite biosynthetic gene Str interacts with a jasmonate- and elicitorinducible AP2-domain transcription factor, ORCA2. EMBO J 18:4455–4463

Merfort II (2011) Perspectives on sesquiterpene lactones in inflammation and cancer. Curr Drug Targets 12:1560–1573

Misawa M (1991) Plant tissue culture: an alternative for production of useful metabolites. FAO Agricult Serv Bull 108:1010–1365

Miura Y, Hirata K, Kurano N, Miyamoto K, Uccida K (1988) Formation of vinblastine in multiple shoot culture of Catharanthus roseus. Planta Med 54:18–20

Mondal S, Mandal C, Sangwan RS, Chandra S, Mandal C (2010) Withanolide D induces apoptosis in leukemia by targeting the activation of neutral sphingomyelinase-ceramide cascade mediated by synergistic activation of c-Jun N-terminal kinase and p38 mitogen-activated protein kinase. Mol Cancer 9:239–255

Monshausen GB, Bibikova TN, Weisenseel MH, Gilroy S (2009) Ca2+ regulates reactive oxygen species production and pH during mechanosensing in Arabidopsis roots. Plant Cell 21 (8):2341–2356. https://doi.org/10.1105/tpc.109.068395

Mora-Palea M, Sanchez-Rodrigueza SP, Linhardta RJ, Dordicka JS, Koffas MAG (2013) Metabolic engineering and in vitro biosynthesis of phytochemicals and non-natural analogues. Plant Sci 210(2013):10–24

Morgan JA, Shanks YV (2000) Determination of metabolic rate-limitations by precursor feeding in Catharanthus roseus hairy root cultures. J Biotechnol 79:137–145

Morgan J, Rijhwani SK, Shanks JV (1999) Metabolic engineering for the production of plant secondary metabolites. In: Lee SY, Papoutsakis ET (eds) Metabolic engineering. Dekker, New York, pp 325–352

Muir WH, Hildebrandt AC, Riker AJ (1954) Plant tissue cultures produced from single isolated plant cells. Science 119:877–878

Muir SR, Collins GJ, Robinson S, Hughes S, Bovy A, De Vos CHR, Van Tunen AJ, Verhoeyen ME (2001) Overexpression of petunia chalcone isomerase in tomato results in fruits containing increased levels of flavonols. Nat Biotechnol 19:470–474

Müller-Kuhrt L (2003) Putting nature back into drug discovery. Nature Biotechnol 21:602

Murthy HN, Dijkstra C, Anthony P, White DA, Davey MR, Power JB, Hahn EJ, Paek KY (2008) Establishment of Withania somnifera hairy root cultures for the production of withanolide A. J Integr Plant Biol 50:975–981

Murthy HN, Lee E-J, Paek K-Y (2014) Production of secondary metabolites from cell and organ cultures: strategies and approaches for biomass improvement and metabolite accumulation. Plant Cell Tissue Organ Cult 118(1):1–16

Nakagawa A, Minami H, Kim J-S, Koyanagi T, Katayama T, Sato F, Kumagai H (2011) A bacterial platform for fermentative production of plant alkaloids. Nat Commun 2:326. https://doi.org/10. 1038/ncomms1327

Nam MH, Heo EJ, Kim JY, Kim SI, Kwon KH, Seo JB, Kwon O, Yoo JS, Park YM (2003) Proteome analysis of the responses of Panax ginseng C.A. Meyer leaves to high light: use of electrospray ionization quadrupole-time of flight mass spectrometry and expressed sequence tag data. Proteomics 3:2351–2367

Narula A, Kumar S, Abdin MZ, Srivastava PS (2006) Biotechnology of medicinal plants. In: Kumar A, Roy SS (eds) Plant biotechnology and its applications in tissue culture. I.K. International, New Delhi

Neumann KH (1995) Pflanzliche Zell- und Gewebekulturen. Eugen Ulmer, Stuttgart

Neumann KH (1962) Untersuchungen über den Einfluß essentieller Schwermetalle auf das Wachstum und den Proteinstoffwechsel von Karottengewebekulturen. Dissertation. Justus Liebig Universität, Giessen

Neumann KH, Barz W, Reinhard E (eds) (1985) Primary and secondary metabolism of plant cell cultures. Springer, Berlin

Neumann K, Kumar A, Imani J (2009) Plant cell and tissue culture – a tool in biotechnology basics and application. Springer, Berlin, 333 pp

Nguyen T, Eshraghi J, Gonyea G, Ream R, Smith R (2001) Studies on factors influencing stability and recovery of paclitaxel from suspension media and cultures of Taxus cuspidata cv Densiformis by highperformance liquid chromatography. J Chromatogr A 911:55–61

O'Keefe BR, Mahady JJ, Gills CWW (1997) Beecher stable vindoline production in transformed cell cultures of Catharanthus roseus. J Nat Prod 60:261–264

Odake K, Ichi T, Kusahara K (1991) Production of madder colorants. In: Komamine A, Misawa M, DiCosmo FE (eds) Plant cell culture in Japan. CMC Co. Ltd, Tokyo, pp 301–325

Oksman-Caldentey K-M, Inzé D (2004) Plant cell factories in the post-genomic era: new ways to produce designer secondary metabolites. Trends Plant Sci 9:433–440

Oostdam A, Mol JNM, Van der Plas LHW (1993) Establishment of hairy root cultures of Linum flavum producing the lignan 5-methoxy podophyllotoxin. Plant Cell Rep 12:474–477

Ozeki Y, Komamine A (1986) Effects of growth regulators on the induction of anthocyanin synthesis in carrot suspension cultures. Plant Cell Physiol 27:1361–1368

Panda AK, Saroj M, Bisaria VS, Bhojwani SS (1989) Plant cell reactors: a perspective. Enzym Microb Technol 11:386–397

Park SU, Facchini PJ (2000) Agrobacterium rhizogenes-mediated transformation of opium poppy, Papaver somniferum L., and California poppy, Eschscholzia californica Cham., root cultures. J Exp Bot 347:1005–1016

Patisaul HB, Jefferson W (2010) The pros and cons of phytoestrogens. Front Neuroendocrinol 31:400–419

Pavlov A, Bley T (2005) Betalains biosynthesis by Beta vulgaris L. hairy root culture in different bioreactor systems. Scientific Works, University of Food Technologies 52:299–304

Payne GF, Shuler ML, Brodelius P (1987) Large scale plant cell culture. In: Lydersen BK (ed) Large scale cell culture technology. Hansen, New York, pp 193–229

Pétiard V, Bariaud-Fontanel A (1987) La culture des cellules végétales. Recherche (Paris) 18 (188):602–610

Poulev A, Neal JM, Logendra S, Pouleva RB, Garvey AS, Gleba D, Timeva V, Jenkins IS, Halpern B, Kneer R, Cragg GM, Raskin I (2003) Elicitation, a new window into plant chemodiversity and phytochemical drug discovery. J Med Chem 46:2542–2547

Praveen N, Naik PM, Manohar SH, Nayeem A, Murthy HN (2009) *In vitro* regeneration of brahmi shoots using semisolid and liquid cultures and quantitative analysis of bacoside A. Acta Physiol Plant 31:723–728. https://doi.org/10.1007/s11738-009-0284-5

Prenosil JE, Pedersen H (1983) Immobilized plant cell reactors. Enzym Microb Technol 5:323–331

Putignani L, Massa O, Alisi A (2013) Engineered Escherichia coli as new source of flavonoids and terpenoids. Food Res Int 54:1084–1095

Qi B, Fraser T, Mugford S, Dobson G, Sayanova O et al (2004) Production of very long chain polyunsaturated omega-3 and omega-6 fatty acids in plants. Nat Biotechnol 22:739–745

Rajnchapelmessai J (1988) Cellules végétales en quête de métabolites. Biofutur (Paris) 70:23–34

Ramakrishna A, Ravishankar GA (2011) Influence of abiotic stress signals on secondary metabolites in plants. Plant Signal Behav 6:1720–1731. https://doi.org/10.4161/psb.6.11.17613

Ramirez-Estrada K, Vidal-Limon H, Hidalgo D et al (2016) Elicitation, an effective strategy for the biotechnological production of bioactive high-added value compounds in plant cell factories. Molecules 21(2):182–189

Rao SR, Ravishankar GA (2000) Biotransformation of protocatechuic aldehyde and caffeic acid to vanillin and capsaicin in freely suspended and immobilized cell cultures of Capsicum frutescens. J Biotechnol 76:137–146

Raskin I, Ribnicky DM, Komarnytsky N, Ilic A (2002) Plants and human health in the twenty-first century. Trends Biotechnol 20:522–531

Ratcliffe RG, Shachar-Hill Y (2001) Probing plant metabolism with NMR. Annu Rev Plant Physiol Plant Mol Biol 52:499–526

Rates SMK (2001) Plants as sources of drugs. Toxicon 39:603–613

Ray S, Jha S (1999) Withanolide synthesis in cultures of Withania somnifera transformed with Agrobacterium tumefaciens. Plant Sci 146:1–7

Ricigliano V, Kumar S, Kinison S, Brooks C, Nybo SE, Chappell J, Howarth DG (2016) Regulation of sesquiterpenoid metabolism in recombinant and elicited Valeriana officinalis hairy roots. Phytochemistry 125:43–53

Roberts SC, Shuler ML (1997) Large scale plant cell culture. Curr Opin Biotechnol 8:154–159

Roja G, Heble MR (1996) Indole alkaloids in clonal propagules of Rauwolfia serpentiana. Plant Cell Tissue Org Cult 44:111–115

Routien JR, Nickel LG (1956) Cultivation of plant tissue. US Patent no 2,747,334

Runguphan W, O'Connor SE (2009) Metabolic reprogramming of periwinkle plant culture. Nat Chem Biol 5:151–153

Sachs J (1873) Lehrbuch er Botanik. W. Engelman, Leipzig

Saema S, Rahman LU, Singh R, Niranjan A, Ahmad IZ, Misra P (2016) Ectopic overexpression of WsSGTL1, a sterol glucosyltransferase gene in Withania somnifera, promotes growth, enhances glycowithanolide and provides tolerance to abiotic and biotic stresses. Plant Cell Rep 35 (1):195–211

Sajc L, Grubisic D, Vunjak-Novakovic G (2000) Bioreactors for plant engineering: an outlook for further research. Biochem Eng J 4:89–99

Sakai K, Shitan N, Sato F, Ueda K, Yazaki K (2002) Characterization of berberine transport into Coptis japonica cells and the involvement of ABC protein. J Exp Bot 53:1879–1886

Sakuta M, Hirano H, Kakegawa K, Suda J, Hirose M, Joy RW, Sugiyama M, Komamine A (1994) Regulatory mechanisms of biosynthesis of betacyanin and anthocyanin in relation to cell division activity in suspension cultures. Plant Cell Tissue Org Cult 38:167–169

Sano K, Himeno H (1987) In vitro proliferation of saffron (Crocus sativus L.) stigma. Plant Cell Tissue Org Cult 11:159–166

Sanz MK, Hernandez XE, Tonn CE, Guerreio E (2000) Enhancement of tessaric acid production in Tessaria absinthioides cell suspension cultures. Plant Cell Rep 19:821–824

Sakato K, Misawa M (1974) Effects of chemical and physical conditions on growth of *Camptotheca acuminata* cell cultures. Agric Biol Chem 38:491–497

Sasson A (1991) Options Méditerranéennes—Série Séminaires 14:59–74

Sato F, Yamada Y (1984) High berberine producing cultures of Coptis japonica cells. Phytochemistry 23:281–285

Sato F, Hashimoto T, Hachiya A, Tamura KT, Choi KB, Morishige T, Fujimoto HI, Yamada Y (2001) Metabolic engineering of plant alkaloid biosynthesis. Proc Natl Acad Sci U S A 98:367–372

Schmid J, Amrhein N (1995) The molecular organisation of the shikimate pathway in plants. Phytochemistry 39:739

Schöppner A, Kindl H (1984) Purification and properties of a stilbene synthase from induced cell suspension cultures of peanut. J Biol Chem 259:6806–6811

Scragg A (1991) Plant cell bioreactors. In: Stafford A, Warren G (eds) Plant cell and tissue culture. Open University Press, London, pp 220–234

Sehgal N, Gupta A, Valli RK, Joshi SD, Mills JT, Hamel E, Khanna P, Jain SC, Thakur SS, Rabindranath V (2012) Withania somnifera reverses Alzheimer's disease pathology by enhancing low-density lipoprotein receptor-related protein in liver. Proc Natl Acad Sci 109:3510–3515

Seki M, Takeda M, Furusaki S (1995) Continuous production of taxol by cell culture of *texas cuspidate*. J Chem Eng Japan 28(4): 488–490

Seki M, Ohzora C, Takeda M, Furusaki S (1997) Taxol (paclitaxel) production using free and immobilized cells of Taxus cuspidata. Biotechnol Bioeng 53:214–219

Sevo'n N, Oksman-Caldentey KM (2002) Agrobacterium rhizogenes-mediated transformation: root cultures as a source of alkaloids. Planta Med 68:859–868

Shadwick FS, Doran PM (2007) Propagation of plant viruses in hairy root cultures: a potential method for *in vitro* production of epitope vaccines and foreign proteins. Biotechnol Bioeng 96:570–583

Sharada M, Ahuja A, Vij SP (2008) Applications of biotechnology in Indian Ginseng (Ashwagandha): progress and prospects. In: Kumar A, Sopory S (eds) Recent advances in plant biotechnology and its applications. I.K. International, New Delhi, pp 645–667

Sharma V, Goyal S, Ramawat KG (2009) Scale up production of isoflavonoids in cell suspension cultures of Pueraria tuberosa grown in shake flasks and bioreactor. Eng Life Sci 9:267–271

Sharan M, Taguchi G, Gonda K, Jouke T, Shimosaka M, Hayashida N, Okazaki M (1998) Effects of methyl jasmonate and elicitor on the activation of phenylalanine ammonia-lyase and the accumulation of scopoletin and scopolin in tobacco cell cultures. Plant Sci 132:13–19

Sharp JM, Doran PM (1990) Characteristics of growth and tropane alkaloid synthesis in Atropa belladonna roots transformed by Agrobacterium rhizogenes. J Biotechnol 16:171–186

Shimomura K, Sudo H, Saga H, Kamada H (1991) Shikonin production and secretion by hairy root cultures of Lithospermum erythrorhizon. Plant Cell Rep 10:282–285

Shinde AN, Malpathak N, Fulzele DP (2009) Studied enhancement strategies for phytoestrogens production in shake flasks by suspension culture of Psoralea corylifolia. Bioresour Technol 100:1833–1839

Shitan N, Bazin I, Dan K, Obata K, Kigawa K, Ueda K, Sato F, Forestier C, Yazaki K (2003) Involvement of CjMDR1, a plant MDR type ABC protein, in alkaloid transport in Coptis japonica. Proc Natl Acad Sci U S A 100:751–756

Shuler ML (1981) Production of secondary metabolites from plant tissue culture—problems and prospects. Ann N Y Acad Sci 369:65–79

Shuler ML (1994) Bioreactor engineering as an enabling technology to tap biodiversity: the case of taxol. Ann N Y Acad Sci 745:455–461

Singh G, Tiwari M, Singh SP, Singh S, Trivedi PK, Misra P (2016) Silencing of sterol glycosyltransferases modulates the withanolide biosynthesis and leads to compromised basal immunity of Withania somnifera. Sci Rep 6:25562

Sivakumar G, Yu K, Lee J, Kang J, Lee L, Kim W, Paek K (2006) Tissue cultured mountain ginseng adventitious roots. Eng Life Sci 6:372–383

Slichenmyer WJ, Von Horf DD (1991) Taxol: a new and effective anticancer drug. Anti-Cancer Drugs 2:519–530

Sonderquist RG, Lee JM (2008) Enhanced production of recombinant proteins from plant cells. In: Kumar A, Sopory S (eds) Recent advances in plant biotechnology and its applications. I.K. International, New Delhi, pp 330–345

Spencer A, Hamill JD, Rhodes MJC (1993) In vitro biosynthesis of monoterpenes by agrobacterium transformed shoot cultures of two Mentha species. Phytochemistry 32:911–919

Srinivasan V, Pestchanker L, Moser S, Hirasuna TJ, Taticek RA, Shuler ML (1995) Taxol production in bioreactors: kinetics of biomass accumulation, nutrient uptake, and Taxol production by cell suspensions of Taxus baccata. Biotechnol Bioeng 47:666–676

Srivastava S, Srivastava AK (2012) In vitro azadirachtin production by hairy root cultivation of Azadirachta indica in nutrient mist bioreactor. Appl Biochem Biotechnol 166:365–378

Srivastava S, Srivastava AK (2013) Production of the biopesticide azadirachtin by hairy root cultivation of Azadirachta indica in liquid-phase bioreactors. Appl Biochem Biotechnol 171 (6):1351–1361. http://www.ncbi.nlm.nih.gov/pubmed/23955295. Accessed 17 Jan 2014

Srivastava PS, Sopory SK, Rajam MV, Narula A, Srivastava T, Das S (2008) Transgenics of some medicinal plants. In: Kumar A, Sopory S (eds) Recent advances in plant biotechnology and its applications. I.K. International, New Delhi, pp 594–603

Staba EJ (1962) Production of cardiac glycosides by plant tissue culture I. J Pharm Sci 51:249–254

Staba EJ (1980) Secondary metabolites and biotransformation. In: Staba EJ (ed) Plant tissue cultures as a source of biochemicals. CRC Press, Boca Raton, FL, pp 59–97

Steward N, Martin R, Engasser JM, Goergen JL (1999) Determination of growth and lysis kinetics in plant cell suspension cultures from the measurement of esterase release. Biotechnol Bioeng 66:114–121

Street HE (1977) Plant tissue and cell culture. University of California Press, Berkley, CA

Sumner LW, Mendes P, Dixon RA (2003) Plant metabolomics: large-scale phytochemistry in the functional genomics era. Phytochemistry 62:817–836

Suprasanna P, Ganapathi TR, Ghag SB, Jain SM (2017) Genetic modifications of horticultural plants by induced mutations and transgenic approach. Acta Hortic 1187:219–232. https://doi.org/10.17660/ActaHortic.2017.1187.22

Suresh B, Ravishankar GA (2005) Methyl jasmonate modulated biotransformation of phenylpropanoids to vanillin related metabolites using Capsicum frutescens root cultures. Plant Physiol Biochem 43:125–131

Syklowska-Baranek K, Pietrosiuk A, Gawron A, Kawiak A, Lojkowska E, Jeziorek M, Chinou I (2012) Enhanced production of antitumour naphthoquinones in transgenic hairy root lines of *Lithospermum canescens*. Plant Cell Tissue Organ Cult 108:213–219

Tabata H (2004) Paclitaxel production by plant-cell-culture technology. Adv Biochem Eng Biotechnol 87:1–23

Tabata H (2006) Production of paclitaxel and the related taxanes by cell suspension cultures of Taxus species. Curr Drug Targets 7:453–461

Tabata M, Fujita Y (1985) Production of shikonin by plant cell cultures. In: Zaitlin M, Day P, Hollaender A, Wilson CM (eds) Biotechnology in plant science: relevance to agriculture in the eighties. Academic Press, Orlando, FL, pp 207–218

Tabata MH, Yamamoto KA, Hirao M (1971) Alkaloid production in the tissue cultures of some Solanaceous plants. In: Cultures de tissue de plantes. CNRS, Paris, pp 389–402

Tabata M, Mizukami H, Hiraoka N, Konoshima M (1974) Pigment formation in callus cultures of Lithospermum erythrorhizon. Phytochemistry 13:927–932

Terrier B, Courtois D, Hénault N, Cuvier A, Bastin M, Aknin A, Dubreuil J, Pétiard V (2007) Two new disposable bioreactors for plant cell culture: the wave and undertow bioreactor and the slug bubble bioreactor. Biotechnol Bioeng 96:914–923

Thanh NT, Murthy HN, Yu KW, Hahn EJ, Paek LY (2005) Methyl jasmonate elicitation enhanced synthesis of ginsenoside by cell suspension cultures of *Panax ginseng* in 5l balloon type bubble bioreactors. Appl Microbiol Biotechnol 67:197–201

Trantas EA, Koffas MA, Xu P, Ververidis F (2015) When plants produce not enough or at all: metabolic engineering of flavonoids in microbial hosts. Front Plant Sci 6:7

Tulp M, Bohlin L (2002) Functional versus chemical diversity: is biodiversity important for drug discovery? Trends Pharmacol Sci 23:225–231

Twyman RM, Kohli A, Stoger E, Christou P (2002) Foreign DNA: integration and expression in transgenic plants. In: Setlow JK (ed) Genetic engineering: principles and methods, vol 24. Kluwer/Plenum, New York, pp 107–136

Ulbrich B, Wiesner W, Arens H (1985) Large scale production of rosmarinic acid from plant cell cultures of Coleus blumei Benth. In: Neumann KH, Barz W, Reinhard E (eds) Primary and secondary metabolism of plant cell cultures. Springer, Berlin, pp 293–303

Uozumi N, Kohketsu K, Kondo O, Honda H, Kobayashi T (1991) Fed-batch culture of hairy root cells using fructose as a carbon source. J Ferment Bioeng 72:457–460

Ushiyama K (1991) Large scale culture of ginseng. In: Komamine A, Misawa M, DiCosmo FE (eds) Plant cell culture in Japan. CMC Co. Ltd, Tokyo, p 97

Valenzuela Moreno OA, Galaz-Avalos RM, Minero-García Y, Loyola-Vargas VM (1998) Effect of differentiation on the regulation of indole alkaloid production in Catharanthus roseus hairy roots. Plant Cell Rep 18:99–104

van Etten HD, Mansfield JW, Bailey JA, Farmer EE (1994) Two classes of plant antibiotics: phytoalexins versus "phytoanticipins". Plant Cell 6:1191–1192

van Hengel AJ, Harkes MP, Wichers HJ, Hesselink PGM, Buitelaar RM (1992) Characterization of callus formation and camptothecin production by cell lines of Camptotheca acuminata. Plant Cell Tissue Org Cult 28:11–18

Vancanneyt G, Schmidt R, O'Connor-Sanchez A, Vanisree M, Tsay SH (2004) Plant cell cultures—an alternative and efficient source for the production of biologically important secondary metabolites. Int J Appl Sci Eng 2:1–29

Vanisree M, Tsay H-S (2004) Plant cell cultures—an alternative and efficient source for the production of biologically important secondary metabolites. Int J Appl Sci Eng 2:29–48

Vanisree M, Lee CY, Shu-Fung L, Nalawade SML, Yih C, Tsay HS (2004) Studies on the production of some important secondary metabolites from medicinal plants by plant tissue cultures. Bot Bull Acad Sin 45:1–22

Vasil IK (1991) Plant tissue culture and molecular biology as tools for understanding plant development and plant improvement. Curr Opin Biotechnol 2:158–163

Verpoorte R, Memelink J (2002) Engineering secondary metabolite production in plants. Curr Opin Biotechnol 13:181–187

Verpoorte R, van der Heijden R, Memelink J (2000) Engineering the plant cell factory for secondary metabolite production. Transgenic Res 9:323–343

Vicente MR-S, Plasencia J (2011) Salicylic acid beyond defence: its role in plant growth and development. J Exp Bot 62:3321–3338

Vogt T (2010) Phenylpropanoid biosynthesis. Mol Plant 3:2–20

Wall M (2000) Method for treating pancreatic cancer in humans with water-insoluble S-camptothecin of the closed lactone ring form and derivatives thereof. US Patent no 6080751

Wakayama S, Kusaka K, Kanehira T (1994) Kinobeon A, a novel red pigment produced in safflower tissue culture systems. Zeitschr Naturforsch C 49:1–5

Walker TS, Bais HP, Vivanco JM (2002) Jasmonic acid-induced hypericin production in cell suspension cultures of Hypericum perforatum L. (St. John's wort). Phytochemistry 60:289–293

Wall M, Wani MC, Cooke CE, Palmer KH, McPhail AT, Slim GA (1966) Plant antitumor agents. I. The isolation and structure of camptothecin, a novel alkaloidal leukemia and tumor inhibitor from Camptotheca acuminata. J Am Chem Soc 88:3888–3890

Wang C-T et al (2010) Overexpression of G10H and ORCA3 in the hairy roots of Catharanthus roseus improves catharanthine production. Plant Cell Rep 29(8):887–894

Wang Y et al (2012) Regeneration and Agrobacterium-mediated transformation of multiple lily cultivars. Plant Cell Tissue Organ Cult 111(1):113–122

Wani MC, Taylor HL, Wall ME, Coggon P, McPhail AT (1971) Plant antitumor agents VI. The isolation and structure of taxol, a novel antileukemic and antitumor agent from Taxus brevifolia. J Am Chem Soc 93:2325–2327

Wink M (1985) Metabolism of quinolizidine alkaloids in plants and cell suspension cultures: induction and degradation. In: Neumann K-H, Barz W, Reinhard E (eds) Primary and secondary metabolism of plant cell cultures. Springer, Berlin, pp 107–116

Watt MP (2012) The status of temporary immersion system (TIS) technology for plant micropropagation. Afr J Biotechnol 11:4025–14035

Woerdenbag HJ, Van Uden W, Frijlink HW, Lerk CF, Pras N, Malingre THM (1990) Increased podophyllotoxin production in Podophyllum hexandrum cell suspension cultures after feeding coniferyl alcohol as a b-cyclodextrin complex. Plant Cell Rep 9:97–100

Wu J, Ho KP (1999) Assessment of various carbon sources and nutrient feeding strategies for Panax ginseng cell culture. Appl Biochem Biotechnol 82:17–26

Wu J, Lin L (2003) Enhancement of taxol production and release in Taxus chinensis cell cultures by ultrasound, methyl jasmonate and in situ solvent extraction. Appl Microbiol Biotechnol 62:151–155

Wu S, Schalk M, Clark A, Miles RB, Coates R, Chappell J (2006) Redirection of cytosolic or plastidic isoprenoid precursors elevates terpene production in plants. Nat Biotechnol 24:1441–1447

Wu S, Jiang Z, Kempinski C, Eric Nybo S, Husodo S et al (2012) Engineering triterpene metabolism in tobacco. Planta 236:867–877

Wu S, Yu X, Lian M et al (2014) Several factors affecting hypericin production of *Hypericum perforatum* during adventitious root culture in airlift bioreactors. Acta Physiol Plant 36:975–981. https://doi.org/10.1007/s11738-013-1476-6

Xu D, McElroy D, Thornburg RW, Wu R (1993) Systemic induction of a potato pin2 promoter by wounding, methyl jasmonate, and abscisic acid in transgenic rice plants. Plant Mol Biol 22:573–588

Xu W, Gavia DJ, Tang Y (2014) Biosynthesis of fungal indole alkaloids. Nat Prod Rep 31:1474–1487

Yamada H (1991) Natural products of commercial potential as medicines. Curr Opin Biotechnol 2:203–210

Yamamoto Y, Mizuguchi R, Yamada Y (1982) Selection of a high and stable pigment-producing strain in cultured Euphorbia millii cells. Theor Appl Genet 61:113–116

Yamamoto H, Suzuki M, Suga Y, Fukui H, Tabata M (1987) Participation of an active transport system in berberine-secreting cultured cells of Thalictrum minus. Plant Cell Rep 6:356–359

Yamamura Y, Pinar Sahin F, Nagatsu A, Mizukami H (2003) Molecular cloning and characterization of a cDNA encoding a novel apoplastic protein preferentially expressed in a shikonin-producing callus strain of Lithospermum erythrorhizon. Plant Cell Physiol 44:437–446

Yang JF, Piao XC, Sun D, Lian ML (2010) Production of protocorm-like bodies with bioreactor and regeneration in vitro of *Oncidium* 'Sugar Sweet'. Sci Hortic 125:712–717

Yasuda S, Satoh K, Ishii T, Furuya T (1972) Studies on the cultural conditions of plant cell suspension culture. In: Terui G (ed) Fermentation technology today. Society for Fermentation Technology, Osaka, pp 697–703

Yazaki K (2005) Transporters of secondary metabolites. Curr Opin Plant Biol 8:301–307

Ye X, Al-Babili S, Klöti A, Zhang J, Lucca P, Beyer P, Potrykus I (2000) Engineering the provitamin A (βcarotene) biosynthetic pathway into (carotenoid free) rice endosperm. Science 287:303–305

Yeh FT, Huang WW, Cheng CC, Na C, Tsay HS (1994) Tissue culture of Dioscorea doryophora Hance. II. Establishment of suspension culture and the measurement of diosgenin content. Chin Agron J 4:257–268

Yeoman MM (1987) Techniques, characteristics, properties, and commercial potential of immobilized plant cells. In: Constabel F, Vasil IK (eds) Cell culture and somatic cell genetics of plants. Cell culture in phytochemistry, vol 4. Academic Press, San Diego, CA, pp 197–215

Yeoman MM, Yeoman CL (1996) Manipulating secondary metabolism in cultured plant cells. New Phytol 134:553–569

Yeoman MM, Miedzybrodzka MB, Lindsey K, Mclaughlan WR (1980) The synthetic potential of cultured plant cells. In: Sala F, Parisi B, Cella R, Ciferri O (eds) Plant cell cultures: results and perspectives. Elsevier/North Holland, Amsterdam, pp 327–343

Yu KW, Gao WY, Hahn EJ, Paek KY (2002) Jasmonic acid improves ginsenoside accumulation in adventitious root culture of Panax ginseng C.A. Meyer. Biochem Eng J 11:211–215

Yukimune Y, Tabata H, Higashi Y, Hare Y (1996) Methyl jasmonate-induced overproduction of paclitaxel and baccatin III in Taxus cell suspension cultures. Nature Biotechnol 14:1129–1132

Zárate R (1999) Tropane alkaloid production by Agrobacterium rhizogenes transformed hairy root cultures of Atropa baetica Willk. (Solanaceae). Plant Cell Rep 18:418–423

Zárate R, Yeoman MM (2001) Application of recombinant DNA technology to studies on plant secondary metabolism. In: Bender L, Kumar A (eds) From soil to cell—a broad approach to plant life. Giessen Electronic Library, GEB, pp 82–93. http://geb.uni-giessen.de/geb/ebooks_ebene2.php

Zárate R, el Jaber-Vazdekis N, Medina B, Ravelo AG (2006) Tailoring tropane alkaloid accumulation in transgenic hairy roots of Atropa baetica by over-expressing the gene encoding hyoscyamine 6β-hydroxylase. Biotechnol Lett 28:1271–1277

Zenk MH, Rueffer M, Kutchan TM, Galneder E (1988) Future trends of exploration for the biotechnological production of isoquinoline alkaloids. In: Yamada Y (ed) Applications of plant cell and tissue culture. Wiley, New York, pp 213–223

Zhang ZY, Zhong JJ (2004) Scale-up of centrifugal impeller bioreactor for hyperproduction of ginseng saponin and polysaccharide by high-density cultivation of Panax notoginseng cells. Biotechnol Prog 20:1076–1081

Zhang L, Ding R, Chai Y, Bonfill M, Moyano E, Oksman-Caldentey KM, Xu T, Pi Y, Wang Z, Zhang H, Kai G, Liao Z, Sun X, Tang K (2004) Engineering tropane biosynthetic pathway in *Hyoscyamus niger* hairy root cultures. Proc Natl Acad Sci U S A 101:6786–6791

Zhang C, Gong FC, Lambert GM, Galbraith DW (2005) Cell type-specific characterization of nuclear DNA contents within complex tissues and organs. Plant Methods 1:7

Zhang L, Yang B, Lu B et al (2007) Tropane alkaloids production in transgenic *Hyoscyamus niger* hairy root cultures over-expressing Putrescine N-methyltransferase is methyl jasmonate-dependent. Planta 225:887–896

Zhao Y, Sun W, Wang Y, Saxena PK, Liu CZ (2012) Improved mass multiplication of Rhodiola crenulata shoots using temporary immersion bioreactor with forced ventilation. Appl Biochem Biotechnol 166:1480–1490

Zhou M-L, Zhu X-M, Shao J-R, Tang Y-X, Wu Y-M (2011) Production and metabolic engineering of bioactive substances in plant hairy root culture. Appl Microbiol Biotechnol 90:1229–1239

Ziegler J, Facchini PJ (2008) Alkaloid biosynthesis: metabolism and trafficking. Annu Rev Plant Biol 59:735–769

Phytohormones and Growth Regulators 11

Phytohormones occupy a central position in the regulation of growth, and especially of differentiation of plants in general, as well as in cell and tissue culture systems. Although a wealth of literature exists on the reactions of cell and tissue cultures after a supply of one or the other growth regulator to the nutrient medium, our knowledge of an endogenous hormonal system of cultured cells is rather limited. In analogy to hormones in animals, phytohormones are defined as substances produced in some tissues at certain developmental stages of a plant and are then distributed by the vascular system, often exerting functions at remote tissues in very low concentrations. Actually, phytohormones are a kind of signal system to coordinate the growth and development of plants. In animal systems, hormone functions are generally rather specific and localized, whereas in plant systems, the phytohormones are rather unspecific, and often related to the physiological state, or the position of the target tissue or cells. Here, often a specific reaction can be induced by several of these substances, and a given phytohormone can induce several reactions. In Table 11.1, some examples are given for reactions of intact plants to the application of growth regulators. On the basis of such reactions, phytohormones are usually divided into five groups, i.e., auxins, gibberellins, and cytokinins, some gaseous compounds like ethylene, and a group associated predominantly with growth retardation and senescence, such as abscisic acid (ABA), and based on more recent data, Jasmonic acid and Brassinosteroids, and also salicylic acid. Both Jasmonic acid and salicylic acid are important in defense mechanisms in plants. Especially for auxins but also for cytokinins, many compounds of synthetic origin are available that are inductive to reactions characteristic of the group. A summary of recent developments in phytohormone research in general is given in a number of review articles in "Plant Biology 8, number 3, pp. 277–406" (2006).

The establishment of the phytohormone system involves quite different metabolic areas. Auxins are synthesized from tryptophan, i.e., an amino acid, gibberellins are diterpenes, cytokinins are products of nucleotide metabolism, and ethylene is synthesized from methionine, again an amino acid. The synthesis of abscisic acid originates from diterpene metabolism. Actually, phytohormones could be also

© Springer Nature Switzerland AG 2020

309

K.-H. Neumann et al., *Plant Cell and Tissue Culture – A Tool in Biotechnology*,

https://doi.org/10.1007/978-3-030-49098-0_11

Table 11.1 Some reactions of higher plants following the application of growth regulators

	Auxins+	Gibberellins	Cytokinins	Ethylene
Root formation	+	−	−	+
Breaking dormancy of storage organs	+	+	−	−
Growth of leaf area	−	−	−	+
Development of leaf buds	−	+	−	−
Internode length	−	+	+	−
Leaf senescence	+	−	−	+
Breaking dormancy of leaf buds	+	+	−	−
Shedding of flowers, fruit	+	−	−	+
Fruit growth	−	+	+	
Fruit ripening	−	−	−	+

+, positive reaction; −, negative reaction

classified as products of "secondary metabolism." However, secondary metabolism is characteristic of older cells with low or no division activities, but phytohormones are synthesized mostly in younger tissue with high cell division rates, i.e., meristems.

Phytohormones, and to some degree, also synthetic growth regulators occur in cells in (broadly speaking) three forms: first, as free molecules; secondly, as so-called conjugates, in which the compounds are bound to low molecular weight molecules like amino acids or carbohydrates; and thirdly, as molecules bound to high molecular structures (mainly peptides and proteins). Based on current ideas, the free molecules are the physiologically active compounds, whereas the conjugates with small molecules function as inactive transport forms. The molecules bound to high molecular structures could possibly function for immobile and inactive storage in cells, though no clear evidence for this is yet available. This classification is rather schematic, and deviations often occur, especially for cytokinins.

If a growth regulator molecule enters a cell, it either remains as free molecule and functions accordingly, or it will be conjugated with small molecules, or bound to bigger molecules and become inactivated. Eventually, the molecule can be broken down, or it passes unaltered through the cell. All these reactions are catalyzed by enzymes, and consequently the fate of the molecule will be determined by the molecular differentiation of a given cell.

The significance of growth regulators applied with the nutrient medium has been discussed earlier and will be dealt with later again. Here, reference will be made mainly to the endogenous hormonal system, for which again investigations using carrot cell cultures in the NL medium shall serve as example. Under standard conditions, root explants are cultured in continuous illumination (ca. 5000 lux). This means that light-sensitive IAA supplied to the nutrient medium will be photooxidized. Autoclaving of the nutrient medium, however, apparently has only small adverse effects on IAA (Table 11.2).

Twenty-four hours after application of ^{14}C-labeled IAA (side chain labeled), the isotope could be detected in several fractions of the tissue, as well as in the nutrient solution (Table 11.3). The study indicates a fast incorporation of ^{14}C of the acetate

Table 11.2 Influence of autoclaving, illumination, and transfer of explants on IAA concentration in the nutrient medium

	µg IAA/100 ml NL
Freshly prepared medium	200.0
Immediately after autoclaving	182.6
After 3 days in darkness	179.2
After 3 days of illumination, without explants	2.7
After 3 days of illumination, with carrot root explants	15.7

NL medium, see Table 3.3; Bender and Neumann (1978)

Table 11.3 Balance sheet of IAA in a system of cultured carrot root explants during the first 24 h of culture

Parameter	Value
Content in carrot root explants	0.1
Content in culture medium at t0	700.0
Uptake by explants	1.0 per hour
Synthesis by explants	2.5 per hour
Total breakdown	2.0 per hour
By photooxidation	1.4
By explants	0.6
IAA concentration in explants after 24 h in culture	1.4

µg/g fresh weight; NL medium, see Table 3.3, after Bender and Neumann (1978)

side chain of IAA into basic metabolism. In cultured cells, by far most of the ^{14}C can be found in the organic acid fraction. Evidently, a large proportion of IAA taken up is broken down by splitting off the acetate side chain. Some breakdown products can apparently be extruded to the nutrient solution. In any case, highest labeling was found in the fraction "acidic indoles," followed by the organic acid fraction.

In the nutrient medium after 12 days of culture, radioactive labeling was concentrated in two components that could not be detected in the cultures. It appears that these are breakdown products of IAA in the medium. In the tissue "acidic indole" fraction, ^{14}C in IAA was only second to an unknown component. Other compounds labeled with ^{14}C in the medium could not be detected in the cultures and therefore would be the products of destruction of IAA in the medium by photooxidation.

Immediately after initiation of culture, the explants produce ethylene to be excreted to the surrounding atmosphere. It would be some form of "wound ethylene" production, as observed on intact plants upon wounding. About 5 h after transfer of the explants to the culture vessel, a minimum can be observed (Fig. 11.1), followed by a steep increase up to the sixth day of culture. In the subsequent decrease, ethylene production on the 12th day is as low as that observed in the 20th hour. Ethylene production is determined by IAA, and also in cultured carrot tissue an application of IAA to the medium at that stage, and also later, promotes a strong increase in ethylene production (Fig. 11.1). Apparently, despite the low native

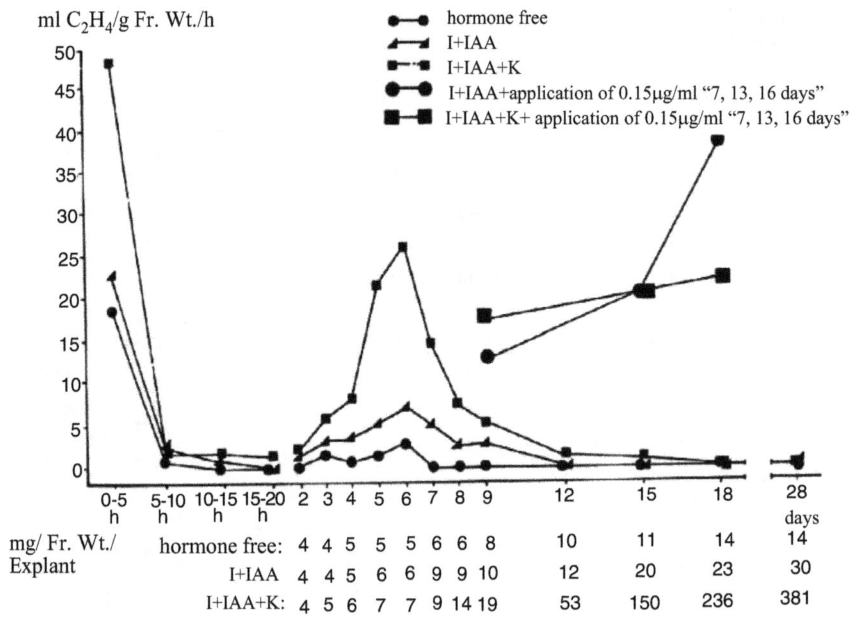

mg/ Fr. Wt./ Explant	0-5 h	5-10 h	10-15 h	15-20 h	2	3	4	5	12	15	18	28 days
hormone free:	4	4	5	5	5	6	6	8	10	11	14	14
I+IAA	4	4	5	6	6	9	9	10	12	20	23	30
I+IAA+K:	4	5	6	7	7	9	14	19	53	150	236	381

Fig. 11.1 Production of ethylene by cultured carrot tissue during a culture period of 28 days. *I* m-inositol, *K* kinetin

production level, the ability of the cultures to produce ethylene is maintained during the later periods of culture. At that stage under undisturbed conditions, the endogenous concentration of IAA is very low, which may be the cause of the low native production of ethylene.

It is interesting to see a correlation between the level of ethylene production on the sixth day and the growth regulator supplement to the nutrient medium. The highest production was observed for the treatment kinetin/IAA/inositol. At that stage, the growth of all treatments is essentially the same, but during a 4-week culture cycle, this treatment by far exceeds the other two in terms of fresh and dry weight, as well as cell number. This maximum of ethylene production occurs at the end of the lag phase of cell division activity. Still, growth activity during the following log phase seems predetermined already at that stage. Here, also IAA in the tissue is at its maximum, which could be related to the concurrent high ethylene production of the cultures. Apparently, ethylene production is simply an expression of a determination induced by the exogenous growth regulators (see also data on mineral nutrient uptake in Sect. 8.2.2).

In cultured carrot explants, considerable concentrations of 2iPA can be observed (Fig. 11.2). In many plant species, this is considered as a precursor of zeatin. In the carrot system, the 2iPA concentration increases continuously until the 12th day of culture, corresponding to the initiation of the log phase of growth, followed by a decrease. In the treatment without kinetin, the concentration of 2iP is very low throughout the culture cycle. Detailed investigations on the metabolism of

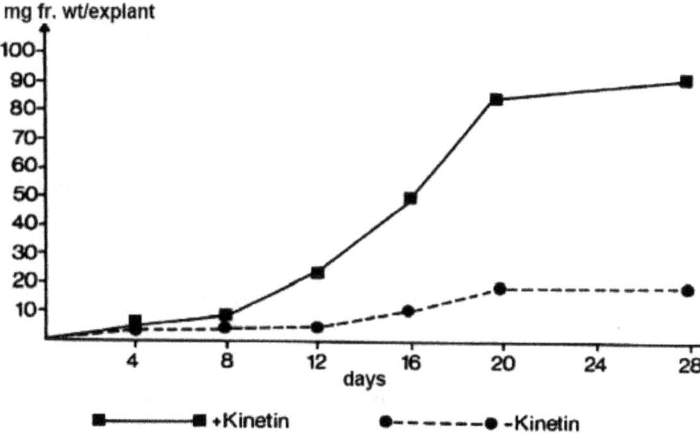

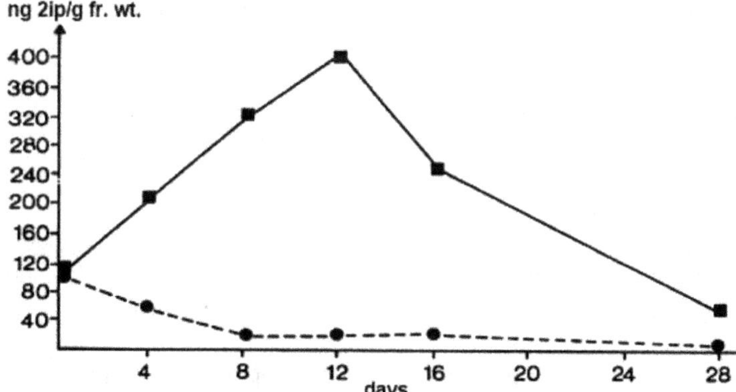

Fig. 11.2 Fresh weight and concentration of isopentenyladenine (IPA) of cultured explants of the secondary phloem of the carrot root as influenced by kinetin during 4 weeks of culture (NL, see Table 3.3; Stiebeling and Neumann 1987)

cytokinins in the carrot system are not available, but existing data again indicate a correlation between cytokinin concentration and the growth performance of the cultures.

Also in habituated cell cultures of tobacco, zeatin riboside (up to 230 pM/g f. wt.) and isopentenyl adenosine, but neither free Zeatin nor iPA were detected. Following an application of auxin, the concentration of the cytokinins is reduced, indicating a role as regulators of the endogenous hormone system for these growth regulators. A supplement of kinetin, however, increases the concentration of endogenous cytokinins, just as in the carrot system.

All this evidence clearly indicates that cultured cells are able to produce auxins, cytokinins, and ethylene—as discussed before, they are able to establish a characteristic endogenous hormonal system. The question remains, however, as to whether

Table 11.4 Influence of GA3 (50 ppm) and chlorcholine chloride (CCC) on the fresh weight of cultured carrot root explants

Nutrient medium	−GA3	+GA3
NL	15	19
NL+I+IAA+K	162	173
NL+I+IAA+K+50 ppm CCC	77	79
NL+I+IAA+K+500 ppm CCC	51	63

mg per explant, 21 days of culture, NL medium, see Table 3.3
I m-inositol, *K* kinetin

this system is an expression of the growth performance of the cultures, or a system to govern growth of those "meristems." For further information on this, experiments using inhibitors of the hormones or mutants with disturbed hormonal pathways of synthesis or function should be performed. Such methods were employed to determine the metabolism of cytokinins of *Arabidopsis* after determination of the nucleotide sequence of DNA. Here, key enzymes are isopentenyltransferases that catalyze the first step to transfer the isopentenyl group from dimethylallyl phosphate to ATP, ADP, or AMP and isopentenyl-ATP, isopentenyl-ADP, and isopentenyl-AMP as precursors of active cytokinins (Kakimoto 2001; Takei et al. 2001). In *Arabidopsis*, seven such transferases are expressed in root and shoot tissue. Cytokinin breakdown into adenine and an aldehyde is catalyzed by cytokinin oxydases and dehydrogenases (Houba-Herin et al. 1999; Morris et al. 1999).

Only scarce information is available on gibberellins in cultured cells. An application of GA3 to carrot cultures was without effect. An application of CCC (chlorcholine chloride), an inhibitor of gibberellin synthesis, reduced growth of carrot cultures considerably. This, however, could not be reversed by an application of GA3 (Table 11.4). This indicates that the missing reaction to a GA3 application is not due to a high native synthesis. Although an application of GA3 to intact carrot plants induces strong changes in growth and development, possibly cultured carrot cells may need one of the many other gibberellins. Another explanation of the reaction to a CCC application could be influences of this compound on parts of metabolism other than gibberellin synthesis. A growth promotion of gibberellins, however, was observed by its application to tobacco shoot explants.

Changes in the hormonal system of cultured petiole explants during somatic embryogenesis have been described in detail in Sect. 7.3. Here, only a short statement on abscisic acid will be given. Clear differences in the level of concentrations of IAA and in cytokinin activity were observed in embryogenic and non-embryogenic cultures, as were variations at various stages of embryo development (Table 11.5). The concentration of ABA in embryogenic strains is considerably elevated, compared to that of non-embryogenic strains (Rajasekaran et al. 1987), as has been described also for carrot cultures. The application of an inhibitor of ABA synthesis (fluoridon) reduces ABA concentration, and a reduction of embryogenic capacity was also observed.

Only few results are available on the uptake of growth regulators by cultured cells. For cultured tobacco cells, the uptake rate of NAA and 6-BA was proportional to the concentration in the nutrient medium. At identical concentrations, the uptake

Table 11.5 IAA concentration and cytokinin activity in cultured petiole explants of two carrot varieties at various stages of somatic embryogenesis

Variety	Lobbericher, non-embryogenic		Rotin, embryogenic	
	ng IAA/g f. wt.	Cytokinin act. (ng/g f. wt.)	ng IAA/g f. wt.	Cytokinin act. (ng/g f. wt.)
Stage 3, meristematic to embryogenic	146	25.6	524	53.6
Stage 4, four-cell stage	155	0	76	48.0
Stage 5, four-cell to torpedo stage	105	18.0	145	28.4

Analytic data were obtained on the same days of culture (for description of embryonic stages, see Sect. 7.3; Li and Neumann 1985)

of NAA was 10 times that of 6-BA. As described above for IAA, these two synthetic growth regulators are metabolized shortly after uptake, and only about 10% remains in the free, i.e., biologically active, state (Croes and Barendse 1986). Both regulators are then inactivated by the formation of conjugates of NAA with aspartate and BA as the 7-glucoside. Also unstable indole lactate forms are conjugated with aspartate. It has to be kept in mind that conjugates serve also as transport compounds for growth regulators, possibly from the peripheral cells to the more central core of the explants. Earlier reports indicate a conjugation of IAA with aspartate, from which it can later be released. A conjugation of BA from the medium as 9-riboside and in the 3- and 9-glucoside was reported for *Gerbera* shoot explants. The 9-riboside seems to be the precursor of the 3-riboside (Blakesley et al. 1986).

Finally, some remarks on hormone autotrophic cultures will be made. Early indications of the existence of such cultures are reports by Gautheret in the 1940s. More recently, the research group of Meins (Hasen et al. 1985, 1987; Ghosh and Pal 2013, 2014) has refocused on this problem. It was long considered that this autotrophy was due to a higher auxin synthesis of these cultures. Today, the evidence points more to a reduced auxin inactivation, e.g., by peroxidases, as cause of hormone autotrophy. As an example, a sugar beet cell culture strain after 2.4D application exhibits the same level of endogenous free IAA as that of the auxin autotrophic strain cultured without 2.4D. The activity of peroxidases involved in auxin breakdown is reduced in auxin autotrophic (habituated) cultures (Table 11.6). Similar results have been reported for an auxin autotrophic crown gall system of tobacco. Peroxidase activity was particularly high in the cell wall fraction.

Some preliminary results indicate that, like in intact plants, a diurnal rhythm in the concentration of free hormones exists in cultured cells (Stiebeling and Neumann 1987; Nessiem, unpublished data of our laboratory; see above). As an example, the concentration of cytokinins in the carrot system reaches a maximum at about 6 PM (unpublished results of our institute).

Kohli et al. (2013) suggested highly coordinated, dynamic nature of growth requirements plants perceive and react to various environmental signals in an interactive manner among various phytohormones. This leads to integrating the diverse input signals and readjusting growth as well as acquiring stress tolerance.

Table 11.6 Comparative investigations on IAA concentration, peroxidase concentration, and the concentration of "auxin protectors" in normal and habituated callus cultures of sugar beet (Coumans-Gilles et al. 1982)

	Normal cultures	Habituated cultures
IAA (ng/g fresh weight)	1393.00	1241.00
Peroxidase (µg/mg protein)		
Soluble fraction	1.90	0.73
Membrane fraction	7.27	3.45
Cell wall fraction		
Ionic	62.24	5.10
Covalent	1.50	1.46
"Auxin protectors" (percentage inhibition of IAA-oxidase per 12 min)	91.90	27.00

Besides, there are several examples of similar developmental changes occurring in response to distinct abiotic stress signals, which can be explained by the cross talk in phytohormone signaling. Kohli et al. (2013) highlighted the major phytohormone cross talks with a focus on the response of plants to abiotic stresses. The recent findings have made it increasingly apparent that such cross talk will also explain the extreme pleiotropic responses elicited by various phytohormones. Indeed, it would not be presumptuous to expect that in the coming years, this paradigm will take a central role in explaining developmental regulation.

Recent studies of the expression of genes controlling hormone synthesis showed expression of genes responsible for cytokinin (Cheng et al. 2013) and auxin (Bai et al. 2013) synthesis in cells cultivated in vitro, while auxin-induced organogenesis was accompanied by endogenous cytokinin production (Pernisova et al. 2009). Auxin synthesis, in turn, was decreased by a low level of cytokinins in vivo (Jones et al. 2010). It is obvious from these observations that the concentration of one hormone added to the nutrient medium may influence morphogenesis indirectly by inducing changes in the content of other hormones inside the cultivated cells. Thus, it is impossible to unravel the mechanisms of hormonal control of plant morphogenesis without identifying the location of the phytohormones and the qualitative assessment of their relative amounts, which is difficult to predict but necessary to study. Auxin plays a key role in differentiation of conductive tissues in vivo (Scarpella and Helariutta 2010; Ohashi-Ito et al. 2013) and has been shown to induce formation of xylem elements from isolated mesophyll cells (Fukuda 1997).

Seldimirova et al. (2016) studied the localization of endogenous zeatin and indole-3-acetic acid during simultaneous bud and root formation in calluses derived from immature embryos of wheat (*Triticum aestivum* L.). Strong immune staining for both hormones was detected in proliferating callus tissue, in developing meristematic centers and meristematic zones, and in the sites of shoot and root apex initiation. During further development, shoot apices with leaf primordia were heavily immune stained for zeatin, while immune staining for indole-3-acetic acid was more intense at the sites of incipient primordia and leaf primordial. However, immunostaining for both hormones reached a maximum in the root apex and gradually declined with increasing distance from the apex. This data suggests considerable similarity between patterns of hormone distribution in organs in vitro and in vivo. Thus, callus culture is a convenient and useful model for the study of fundamental biological questions such as how hormones regulate development.

Ermoshin et al. (2016) revealed indirect participation of genes of the HMG subfamily in growth and development of tobacco plants. Introduction of extra copy of the gene studied, as well as its downregulation with antisense RNAs, led to changes in the structure of phototrophic tissues of leaves and mesophyll density. This effect depends on the location of a leaf on the stem that is apparently connected with the gradient of phytohormone level along the shoot. The differences observed may also be due to either changes in the level of sterols in cellular membranes of the transgenic plants or in the level of Brassinosteroids in organs and tissues.

The role of acid invertase has also been suggested in morphogenesis (Maiti et al. 2011). Sengupta et al. (2011) reported the role of phenolic signals in shoot morphogenesis.

Seldimirova et al. (2016) studied the localization of endogenous zeatin and indole-3-acetic acid during simultaneous bud and root formation in calluses derived from immature embryos of wheat (*Triticum aestivum* L.). Strong immunostaining for both hormones was detected in proliferating callus tissue, in developing meristematic centers and meristematic zones (whose cells were shown to be involved in organ formation), and in the sites of shoot and root apex initiation. During further development, shoot apexes with leaf primordia were heavily immunostained for zeatin, while immunostaining for indole-3-acetic acid was more intense at the sites of leaf primordia initiation and incipient primordia themselves.

In the developing roots, immunostaining for both hormones reached a maximum in the root apex and gradually declined with increasing distance from the apex. Cells of developing procambial strands were also strongly stained for both zeatin and indole-3-acetic acid. These data suggest considerable similarity between patterns of hormone distribution in organs in vitro and in vivo. Thus, callus culture is a convenient and useful model for the study of fundamental biological questions such as how hormones regulate development.

Recent studies of the expression of genes controlling hormone synthesis showed expression of genes responsible for cytokinin (Cheng et al. 2013) and auxin (Bai et al. 2013) synthesis in cells cultivated in vitro, while auxin-induced organogenesis was accompanied by endogenous cytokinin production (Pernisova et al. 2009). Auxin synthesis, in turn, was decreased by a low level of cytokinins in vivo (Jones et al. 2010).

It is obvious from these observations that the concentration of one hormone added to the nutrient medium may influence morphogenesis indirectly by inducing changes in the content of other hormones inside the cultivated cells. Thus, it is impossible to unravel the mechanisms of hormonal control of plant morphogenesis without identifying the location of the phytohormones and the qualitative assessment of their relative amounts, which is difficult to predict but necessary to study. Auxin plays a key role in differentiation of conductive tissues in vivo (Scarpella and

Helariutta 2010; Ohashi-Ito et al. 2013) and has been shown to induce formation of xylem elements from isolated mesophyll cells (Fukuda 1997).

References

Bai B, Su YH, Yuan J, Zhang XS (2013) Induction of somatic embryos in Arabidopsis requires local YUCCA expression mediated by the down-regulation of ethylene biosynthesis. Mol Plant 6:1247–1260

Bender L, Neumann K-H (1978) Investigations on Indole-3-acetic acid metabolism of carrot tissue cultures. Z Pflanzenphysiol 88:209–217

Blakesley D, Lanten JR, Morgan R (1986) The metabolism of 6-benzylaminopurine and Zeatin in Gerbera shoots. In: Somers DA, Gengenbach DG, Biesboer DD, Hackett WP, Green CE (eds) Abstr. 6th Int. congress plant tissue and cell culture. University of Minnesota, Minneapolis, MN, p 375

Cheng ZJ, Wang L, Sun W, Zhang Y, Zhou C, Su YH, Li W, Sun T, Zhao XY, Li XG, Cheng Y, Zhao Y, Xie Q, Zhang XS (2013) Pattern of auxin and cytokinin responses for shoot meristem induction results from the regulation of cytokinin biosynthesis by auxin response factor 3. Plant Physiol 161:240–251

Coumans-Gilles MF, Kefers C, Coumans M, Gaspar T (1982) Auxin content and metabolism in auxin repairing and -non- requiring cultures. In: Fujiwara A (ed) Plant tissue culture. Japanese Association for Plant Tissue Culture, Tokyo, pp 197–198

Croes AF, Barendse GWM (1986) Uptake and metabolism of hormones as related to in vitro flower bud development in tobacco. In: Somers DA, Gengenbach DG, Biesboer DD, Hackett WP, Green CE (eds) Abstr. 6th Int. congress plant tissue and cell culture. University of Minnesota, Minneapolis, MN, p 299

Ermoshin AA, Kiseleva IS, Bortsova SA, Sanaeva YV, Alekseeva VV (2016) Morphological features of the transgenic tobacco plant shoot expressing the 3-hydroxy-3-methylglutagyl-CoA reductase (HMG1) gene in the direct and reverse orientations towards the promoter. Russ J Dev Biol 47(4):216–222

Fukuda H (1997) Tracheary element differentiation. Plant Cell 9:1147–1156

Ghosh S, Pal A (2013) Proteomic analysis of cotyledonary explants during shoot organogenesis in *Vigna radiata*. Plant Cell Tissue Organ Cult 115:55–68. https://doi.org/10.1007/s11240-013-0340-0

Ghosh S, Pal A (2014) Abscisic acid, one of the key determinants of *in vitro* shoot differentiation from cotyledons of *Vigna radiata*. Am J Plant Sci 5:704–713

Hasen CE, Meins F, Milani A (1985) Clonal and physiological variation in cytokinin content of tobacco cell lines differing in cytokinin requirements and capacity for neoplastic growth. Differentiation 29:1–6

Hasen CE, Meins F, Aebi R (1987) Hormonal regulation of zeatin ribosid accumulation by cultured tobacco cells. Planta 172:520–525

Houba-Herin N, Perthe C, d'Alayer J, Laloue M (1999) Cytokinin oxidase from Zea mays: purification C DNA cloning and expression in moss protoplast. Plant J 17:615–626

Jones B, Gunnerås SA, Petersson SV, Tarkowski P, Graham N, May S, Dolezal K, Sandberg G, Ljung K (2010) Cytokinin regulation of auxin synthesis in Arabidopsis involves a homeostatic feedback loop regulated via auxin and cytokinin signal transduction. Plant Cell 22:2956–2969

Kakimoto T (2001) Identification of plant cytokinin biosynthetic enzymes as dimethylallyldiphosphate: ATP/ADP isopentenyl transferases. Plant Cell Physiol 42:677–685

Kohli A, Sreenivasulu N, Lakshmanan P, Kumar PP (2013) The phytohormone crosstalk paradigm takes center stage in understanding how plants respond to abiotic stresses. Plant Cell Rep 32:945–957. https://doi.org/10.1007/s00299-013-1461-y

Li T, Neumann K-H (1985) Embryogenesis and endogenous hormone content of cell cultures of some carrot varieties (Daucus carota L.). Ber Dtsch Bot Ges 98:227–235

Maiti S, Chakraborty D, Kundu S, Paul S, Sengupta S, Pal A (2011) Developmentally regulated temporal expression and differential acid invertase activity in differentiating cotyledonary explants of mungbean [*Vigna radiata* (L.) Wilczek]. Plant Cell Tissue Organ Cult 107:417–425

Morris RO, Bilyeu KD, Laskey LG, Cheikh NN (1999) Isolation of a gene encoding a glycosylated cytokinin oxidase from maize. Biochem Biophys Res Commun 255:328–333

Ohashi-Ito K, Oguchi M, Kojima M, Sakakibara H, Fukuda H (2013) Auxin-associated initiation of vascular cell differentiation by LONESOME HIGHWAY. Development 140:765–769

Pernisova M, Klima P, Horak J, Válková M, Malbeck J, Souček P, Reichman P, Hoyerová K, Dubová J, Friml J, Zažímalová E, Hejátko J (2009) Cytokininsmodulate auxin-induced organogenesis in plants via regulation of the auxin efflux. Proc Natl Acad Sci U S A 106:3609–3614

Rajasekaran K, Heim MB, Davis GC, Carnes MG, Vasil IK (1987) Endogenous plant growth regulators in leaves and tissue cultures of Napier grass (Pennisetum purpureum Schum.). J Plant Physiol 130:13–25

Scarpella E, Helariutta Y (2010) Vascular pattern formation in plants. Curr Top Dev Biol 91:221–265

Seldimirova OA, Kudoyarova GR, Kruglova NN, Zaytsev DY, Veselov SY (2016) Changes in distribution of zeatin and indole-3-acetic acid in cells during callus induction and organogenesis in vitro in immature embryo culture of wheat. In Vitro Cell Dev Biol Plant 52:251–264

Sengupta S, Das S, Ghosh S, Pal A (2011) Phenolic signals and its perception leads to *in vitro* shoot induction in *Vigna radiata* cotyledonary explants. Int J Plant Dev Biol 5(1):37–41

Stiebeling B, Neumann K-H (1987) Identification and concentration of cytokinins in carrots (Daucus carota L.) as influenced by development and a circardian rhythem. J Plant Physiol 127:111–121

Takei K, Sakakiwara H, Sugiyama T (2001) Identification of genes encoding adenylate isopentenyl transferases, a cytokinin biosynthesis enzyme in Arabidopsis thaliana. J Biol Chem 276:26405–26410

Cell Division, Cell Growth, and Cell Differentiation

12

The basis of growth and development of all biological systems is cellular growth, cell division, and cell differentiation. Using cell and tissue culture opens up possibilities to investigate influences of nutrients and growth regulators on each of these phenomena, without confounding effects of remote tissue, as would be the case when using intact plants. Such studies should help to understand influences of these factors on cell life in more details. On the other hand, the use of cell cultures for specific aims, like the cloning of plants, production of secondary metabolites, gene technology, or plant breeding, requires a thorough understanding of these phenomena to optimize such procedures. Still, the interpretation of results of many such experiments remains mainly empirical, because of the lack of detailed knowledge of these systems. Attempts to integrate the results of histological, biochemical, and cytological assessments for interpretation of a given system are still quite rare.

In most systems, cellular growth seems to be rather independent of cell division activity. Nevertheless, under identical conditions in many cultural systems, there seems to be a close correlation between cell division activity and cell differentiation. Here, also a time factor has to be considered. Cellular growth shall be defined as dry weight increment per cell for a given time unit. At a high cell division rate, expressed as newly formed cells per time unit (or as cells produced per number of cells actively growing per time interval), the average dry weight per cell is usually reduced (Fig. 3.6, Sect. 3.7). This contradicts ideas expressed in botany text books not so long ago—for example, that cell division occurs as soon as a cell reaches a certain size. Apparently, cell division and cellular growth are to some extent regulated independently. Of course, intensive cell division requires a high production rate of cytoplasm and its ingredients, and for realization of one or the other line of cellular differentiation, optimal cellular growth is a prerequisite anyway.

As already pointed out for products of secondary metabolism, substances of commercial interest usually accumulate at higher concentration after the transition of the system from highly active cell division, a log phase, to a stationary phase. If secondary metabolism, and its characteristic enzyme complement, is regarded as a result of cell differentiation, to be produced and/or activated during a prolonged

© Springer Nature Switzerland AG 2020

K.-H. Neumann et al., *Plant Cell and Tissue Culture – A Tool in Biotechnology*,

https://doi.org/10.1007/978-3-030-49098-0_12

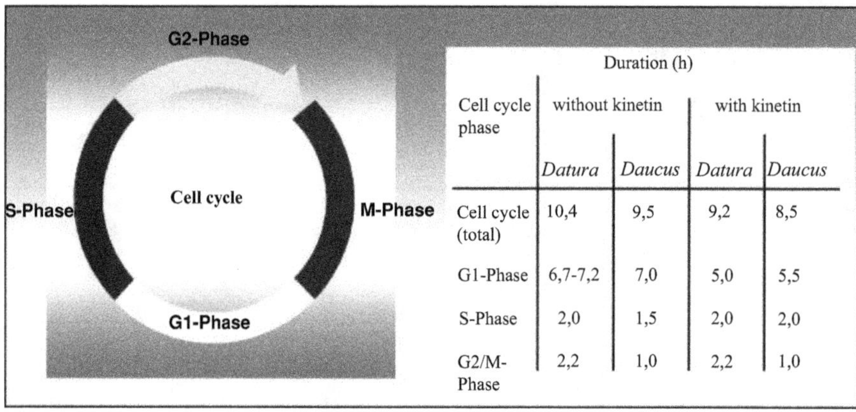

Cell cycle phase	Duration (h)			
	without kinetin		with kinetin	
	Datura	*Daucus*	*Datura*	*Daucus*
Cell cycle (total)	10,4	9,5	9,2	8,5
G1-Phase	6,7-7,2	7,0	5,0	5,5
S-Phase	2,0	1,5	2,0	2,0
G2/M-Phase	2,2	1,0	2,2	1,0

Fig. 12.1 Duration of the cell cycle and of its phases of meristematic cells of *Datura innoxia* and *Daucus carota* as influenced by kinetin

G1-phase of the cell cycle (see below), then again a prolongation of the cell cycle at reduced cell division activity is an example of opposite influences on the status of the culture of cell division and cell differentiation, here at a molecular level. Other examples already discussed include the induction of rhizogenesis only after transition of carrot cultures from a log phase of cell division, to a stationary phase with a longer G1-phase, or the induction of somatic embryogenesis in petiole explants after a transfer from a 2.4D-containing medium, which promotes high cell division activity, to an auxin-free medium resulting in a slowing down of cell division activity.

On the basis of characteristic biochemical processes, and the DNA content of the nucleus, the period between two mitoses, the interphase, can be roughly divided into three subphases (see Fig. 12.1). Usually, mitosis (M-phase) is followed by the G1-phase, appearing as a quiescent phase under the light microscope until, in preparation for the next M-phase, a doubling of DNA can be observed in the nucleus, usually lasting for a few hours. This is defined as the S-phase (phase of DNA synthesis). This terminates with twice as much DNA in the nucleus as in the zygote, i.e., 4 C-values can then be found for DNA in the nucleus. Until the initiation of mitosis, again a quiescent period can be observed under the light microscope. These two "quiescent" periods between processes observed in the nucleus, or gaps, are the G1-phase between mitosis and the initiation of the S-phase and the G2-phase between the end of the S-phase and the initiation of the M-phase. As will be shown later, the passage of a cell population through the cell cycle can be synchronized. Cell growth occurs mainly in the G1-phase—it is the actual working phase of a cell, with RNA and protein synthesis, and other activities. Some RNA and protein synthesis can also be observed during the G2-phase, which, however, mainly seems to serve to prepare for the next M-phase. Associated with cellular growth is molecular differentiation of the cells, as a basis for cytological differentiation. Again, the sequence of all steps is defined as the cell cycle. Variations in cell cycle length for

Table 12.1 Average duration (h) of the cell cycle of vegetative and flowering shoot tip cells

	Vegetative	Flowering
Triticum	41	22
Secale	50	31
Datura	36	26
Sinapis	157	25
Silene	20	10
Lupinus	48	34
Ranunculus	56	47
Epilobium	45	45

Generally, the cell cycle of flowering shoot tips is shorter than that of vegetative shoot tips. *Epilobium* is an exception (Lyndon 1990)

various species, and for two differentiation statuses of tissue (vegetative vs. flowering), can be seen in Table 12.1.

The duration of the M-phase, S-phase, and G2-phase seems to be characteristic for a given species. The duration of the G1-phase, however, seems to be rather specific for the differentiation of various tissues of a plant.

In meristematic tissue in general, about 30% of dividing cells can be found concurrently in roughly the same phase of the cell cycle. To characterize cytological or biochemical processes at each phase in more detail, synchronization has to be enhanced, for which several methods are available. Such methods use controlled changes of light and darkness, temperature shock, or an exact pulsation of cytokinins. A broadly applicable method uses FDU (fluorodeoxyuridine) as inhibitor of the synthesis of thymidine; as monophosphate, this is an essential nucleotide of DNA. This method was preferred in our own investigations. In the absence of thymidine, i.e., its phosphonucleotide, DNA synthesis is blocked, and the cells are not able to pass from the G1 to the S-phase to synthesize DNA. Consequently, cells will "pile up" in the G1-phase. A supplement of exogenous thymidine releases this blockage; the cells will simultaneously enter the S-phase, and this cell population will pass through later stages of the cell cycle synchronized. By this method, a synchronization of the cell cycle of a given cell population reaching 80 or 90% can be achieved and also maintained for at least one round of cell division. The stage of the cell cycle reached can be determined cytophotometrically by measurements of the DNA content of the nucleus on representative samples (see Chap. 6). Cells in G2-phase contain doubled amounts of DNA (4C). Other systems of synchronization employ the DNA polymerase inhibitor aphidicolin, which blocks cell cycle progression in the early S-phase, or propyzamide, which blocks the cell cycle at early mitosis (Nagata et al. 1992), or others. A review of various methods is given by Planchais et al. (2000).

The exposure time to FDU depends on the total length of the cell cycle. Its duration can be easily determined by double labeling using both [3]H- and ^{14}C-labeled thymidine. As an example, experiments to synchronize *Datura* cultures will be described (Blaschke et al. 1978). In this case, unsynchronized *Datura* cultures were labeled by supplying ^{14}C thymidine for 2 h, marking S-phase cells. This was followed by co-culture of the cell suspension with ^{3}H-labeled thymidine at 2-h intervals. After this, representative samples of cells were obtained at 30-min intervals, and historadioautograms were produced. The occurrence of double-labeled nuclei could be easily traced using an ordinary light microscope. This indicates a completion of the cell cycle from one S-phase to the next. As soon as double-labeled nuclei occur, the total length of the cell cycle can be calculated (Fig. 12.1). Even in cell cultures growing with a high rate of cell division, 5–10% of cells will leave the cell cycle at each turn and pass into a G0-phase.

Different rates of cell propagation described as being related to differentiation can be due either to variations in the length of the cell cycle or to differences in the number of cells of a given cell population actively engaged in cell division. A kinetin supplement results in a strong promotion of cell proliferation. In the *Datura* system, e.g., due to a kinetin supplement, cell cycle duration is reduced by 1–2 h, compared to the control without kinetin and growing under otherwise identical culture conditions. This shortening of the cell cycle evolves at the expense of the length of the G1-phase (Fig. 12.1). Furthermore, the percentage of cells leaving the active cell cycle to enter a G0-phase in the cytokinin-free control amounts to about 30%, compared to about 10% after kinetin supplementation. Consequently, the number of cells entering a G0-phase in the slowly growing control without kinetin should be increased, and here the number of cells actively engaged in cell division be reduced. Similar results were obtained for cultured carrot cells (Fig. 12.1). It is difficult to decide which of these two processes—changes in the length of the cell cycle, notably the G1-phase, or in the percentage of cells engaged in the cell cycle—is related to the formation of adventitious roots in the kinetin-free treatment, as described elsewhere (Chap. 3). The same question is posed to understand negative influences of a high proliferation rate of cell cultures on the concentration of compounds of secondary metabolism.

To understand the regulation of development in cell cultures in more detail, the relation of some aspects of the cell cycle to differentiation shall be considered. For a start, some data on rhizogenesis in cultured explants of the secondary phloem of the carrot taproot will be discussed as basis. As already shown earlier, a kinetin supplement to a nutrient medium containing IAA greatly increases growth of cultured tissue, mainly by cell division (Linser and Neumann 1968), and concurrently suppresses regeneration, e.g., adventitious root formation. Whereas those explants growing in a hormone-free nutrient medium show, besides growth mainly by cell expansion, also a few cell divisions (data not shown), those supplemented with IAA and m-inositol without kinetin show an approximate tripling of cell number per explant during a 3-week culture period, indicating a clear, though rather small log phase in cell number (Chap. 3, Fig. 3.6).

The samples supplemented additionally with kinetin are characterized by an extensive log phase of high cell division activity and a decrease in average cell size. Most remarkably, only the treatment containing IAA and m-inositol in the medium shows a differentiation of cells leading to the formation of adventitious roots from 14 days of culture onward. Both other treatments remain as morphologically undifferentiated cell material. Still, after about 4–5 weeks of culture in a

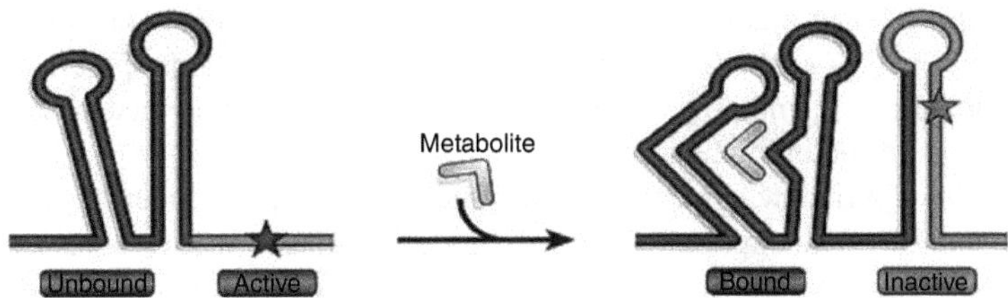

Fig. 12.2 Schematic view of a riboswitch

medium originally supplemented also with kinetin during a stationary phase of cell division, some roots appeared in some experiments. A similar course of cell number per explant can be observed if kinetin and IAA are replaced by the rather stable auxin 2.4D as the only growth substance, which strongly promotes growth by cell division at 0.2 ppm. Here also the formation of adventitious roots is prevented. Evidently, some general correlation between the possibility of root formation, i.e., differentiation, and cell division activity seems to exist.

As mentioned before, the number of cells produced per cultured explant during a given unit of time is due either to the number of cells engaged in active cell division or to the length of time between successive divisions, i.e., the duration of the cell cycle, or both. This could be demonstrated for a kinetin supplement resulting in a ca. 60–90 min reduction in the duration of the cell cycle for *Datura*, as well as for carrot cell cultures. This was due mainly to a reduction of the G1-phase of the cell cycle (see Fig. 12.1; Blaschke et al. 1978; Froese and Neumann 1997). Again, to understand cell differentiation, time has to be considered.

Although much information was lacking at the time, the following rough hypothesis was formulated many years ago to explain the correlation between cell division activity and differentiation (Neumann 1968; see also Neumann 1995, 2006): after completion of the M-phase, during the G1-phase genetic information is utilized in a hierarchical sequence until this is interrupted at the onset of the S-phase, and it depends on the length of the G1-phase which part of the genetic information potential can be activated during the cell cycle. Within these confines, other mechanisms like substrate activation of enzymes are in operation. Of significance could be the rather recent concept of riboswitches (Sudarsan et al. 2003), originally developed for the regulation of activity of mRNA of microbes (Fig. 12.2). Meanwhile, evidence is available for its occurrence also in higher plants like *Arabidopsis* and *Oriza*. Apparently, binding of small molecules to mRNA brings about allosteric changes in base-pairing near gene control elements, like ribosome binding sides, which could modulate pathways of RNA processing or expression also in eukaryotes. By and large, these noncoding regulatory mRNA domains directly precede the protein-encoding sequence. It is thought that a metabolite-sensing domain recognizes the substrate and induces a structural rearrangement of the mRNA either to initiate or to mask the gene expression signal and protein synthesis. The effectors of riboswitches can be metabolites like amino acids, vitamins, nucleotides, key coenzymes (Winkler 2005; Thore et al. 2006), or (as speculation) possibly also growth regulators like IAA or kinetin. The hierarchical activation of genetic information as predicted above possibly provides only the potential that is realized in protein synthesis by such riboswitch mechanisms, according to the general status of the individual cell.

The RNA of a riboswitch contains two functional domains: a metabolite-sensing domain and a gene-expression signal. These domains adopt interdependent conformations in response to the presence or absence of a particular metabolite. In the example illustrated in Fig. 12.2, the gene-expression signal is required for the initiation of protein synthesis. When the metabolite is absent, the metabolite-sensing domain adopts a conformation that reveals the gene-expression signal and enables protein synthesis to occur (indicated by the left star). When the metabolite binds, the ensuing structural reorganization leads to the sequestration of the gene-expression signal, shutting off protein production (indicated by the right star; Reichow and Varani 2006).

In our system, apparently, the information related to differentiation is localized later in this sequential activation of genetic information during the G1-phase. The initiation of the S-phase following a kinetin supplement results in the termination of the G1-phase, and consequently the potential to differentiation is here determined by kinetin. In plus kinetin treatments, it seems that the length of the G1-phase is simply not sufficient to activate the information required to bring about root or embryo development (see below). These ideas are illustrated by the following scheme further down.

Reski (2006) describes some aspects of the hormonal system during the cell cycle and of the two hormone groups particularly involved. The auxin levels remain more or less constant throughout, whereas strong fluctuations occur in cytokinin concentrations. Concentration maxima of cytokinins were observed immediately before the transition from one phase to the next and minima close to zero in between the phases (Hartig and Beck 2006). The kinetin in the medium in our system apparently simulates such a cytokinin maximum, to prematurely initiate the transition from the G1- to the S-phase. By doing so, the duration of the G1-phase is reduced. This seems to be of no obvious influence for the duration and transition for the other phases of the cell cycle (see Fig. 9.2). In addition to auxins and cytokinins, a third factor in the regulation of the cell cycle seems to be sucrose, which will be dealt with in another context.

Kinetin, cell cycle phases, and differentiation ("time hypothesis")
(*********=genetic information in G1, §=transition from one phase to the next)
§--------M--------§*-*-*-*-*G$_1$*-*-*-*-*§--------S-------§--------G$_2$-------§--------M----

Without kinetin: cell cluster, PEMs, rhizogenesis, somatic embryogenesis
§--------M--------§*-*-*G$_1$*-*-§--------S-------§--------G$_2$-------§--------M----
With kinetin: cell cluster, PEMs

The results reported below for cell suspension cultures related to somatic embryogenesis confirm this long-standing time hypothesis to some extent. Beside de novo synthesis of proteins, also protein breakdown occurs, and both would form the basis of a specific differentiation status of a given cell at a given time. Possibly also here a hierarchical regulation exists, but the precise mechanism is not yet known. Protein breakdown is related to the activity of the ubiquitin/26S proteasome pathway, in which proteins to be broken down are first phosphorylated by CDKs (cycline-dependent kinases) and then covalently attached to ubiquitins (consisting of about 70 amino acids). These complexes are recognized and eventually broken down by the proteasome, leaving the ubiquitins intact ready for reuse. Both processes, protein synthesis and protein degradation, could provide for a sequential changing of the composition of the protein moiety of the cell during its passage through G1.

As discussed above for callus cultures, a higher cell division activity in embryogenic carrot cell suspensions, resulting from a kinetin supplement to the nutrient solution, is associated with a reduction of the duration of the cell cycle by about 1 h (cf. Table 9.2; Froese and Neumann 1997), essentially due to a reduction of the duration of the G1-phase. Rhizogenesis and the development of embryos into plantlets were observed in the carrot system in the kinetin-free medium supplemented with IAA and m-inositol, but in the medium supplemented with kinetin (0.1 mg/l) in addition to IAA and m-inositol, no roots were produced, and embryo development did not proceed beyond the stage of PEMs (pre-embryogenic masses). It was assumed that also here a correlation between the duration of the G1-phase and the regeneration capacity exists. The determination of protein synthesis patterns during the prolonged G1-phase by labeling of proteins using synchronized cultures (FDU/thymidine system, Fig. 12.3) fed with ^{14}C leucine indicated the synthesis of 132 additional proteins in the minus kinetin cultures, not synthesized in the plus kinetin treatment presumably due to the immediate transition from the G1- to the S-phase following thymidine application (Table 12.2).

In both treatments in Table 12.2 in the labeling period 120 min, some proteins are missing that were synthesized in the 60-min period. Those missing proteins have apparently been broken down and are not resynthesized. To understand these data in terms of the working hypothesis formulated above, three mechanisms have to be considered, i.e., a mechanism to inactivate translational activity, a mechanism of protein breakdown, and a mechanism to regulate the performance of these in a hierarchical sequence. Whereas many details have recently became known for the former two mechanisms, this is only partly the case for the latter. Here, epigenetic control mechanisms should come into play (DNA methylation, DNA amplification, histone acetylation, and others).

It is assumed that coding mRNAs for these missing proteins are no longer active in the later labeling period, and the genes for these proteins are "closed" to producing new RNA molecules. The iRNA systems certainly play a role in this respect, an aspect that will not be discussed here in detail (for a review, see Xie et al. 2004). This, however, is in agreement with the predicted hierarchically organized sequential activity of genes to initiate the synthesis of stage-specific proteins. To date, two types of small interfering RNA molecules have been identified—micro-RNA (miRNA) and short interfering RNA (siRNA). Both function sequence specifically to suppress or inactivate posttranscriptional RNAs, and both regulate sequence specifically the nucleolytic activity of an RNA-induced silencing complex. Although the activity of siRNAs is related mostly to highly repeated sequences, and retroelements or

Experimental outline

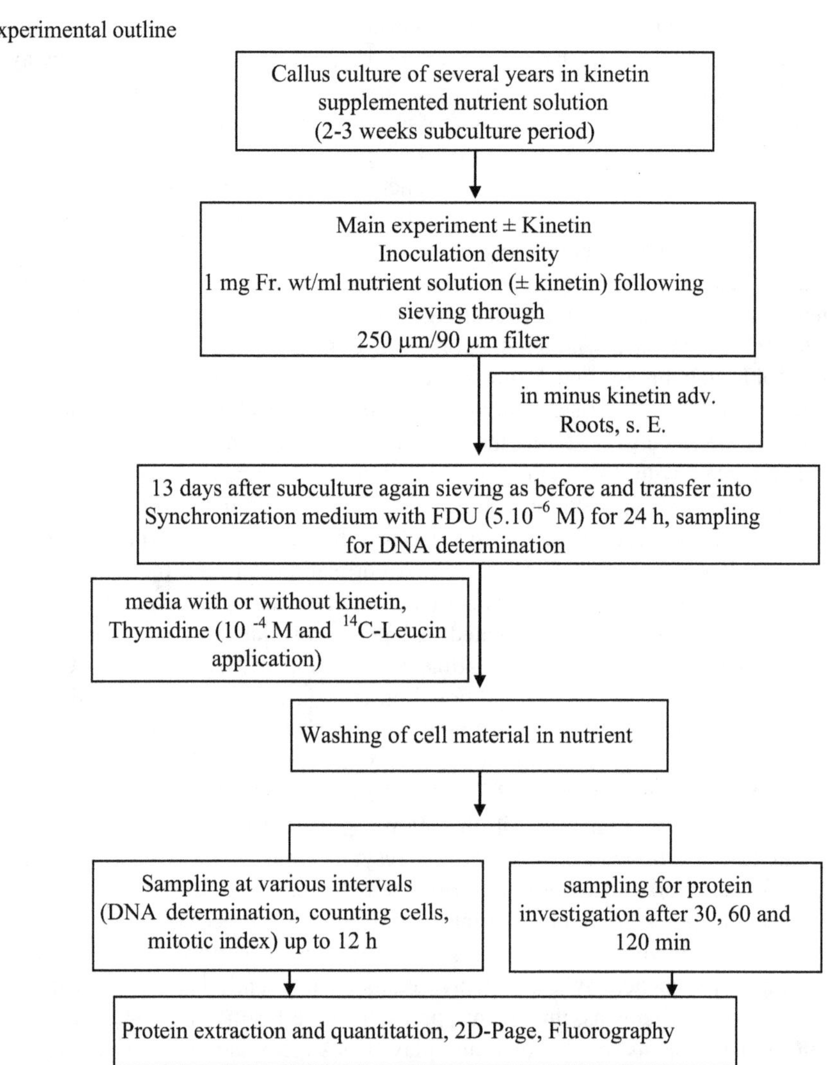

Fig. 12.3 Protocol for cell cycle synchronization and labeling with ^{14}C-leucine of an embryogenic *Daucus* cell suspension

transposons, reports on a few expressed genes are available (Hamilton et al. 2002; Mette et al. 2002). Another function seems to defend plants against virus infections.

The cell cycle is a highly conserved process apparently common to all eukaryotes (for a review, see Dewitte and Murray 2003). In the regulation of the hierarchical variations in the proteome during the G1-phase for protein breakdown, the cycline system certainly seems to play a central role, consistent with recent, more detailed findings. These will not be dealt with in depth here, and in addition to the remarks made above, the following should suffice as summary.

Table 12.2 Summary of ^{14}C-labeled protein spots (^{14}C from ^{14}C-labeled leucine) in soluble protein fractions at various stages of the cell cycle of synchronized carrot suspension cultures

		(−/+/id.)		(−/+/id.)	
Kinetin-free treatment					
Duration of labeling	30 min		60 min		120 min
Phase of cell cycle	G1		G1		G1
No. of labeled protein spots	14	(3/117/11)	128	(54/75/74)	149
+0.1 ppm Kinetin					
Phase of cell cycle	(G1) S		S		G2
No. of labeled protein spots	0	(0/43/0)	43	(18/26/25)	51

The values between the sampling times indicate either that a given labeled protein spot was missing at the subsequent sampling (−), or that it was an additional one (+), or that it occurred already at the preceding sampling (id.; Fröse 1993; Froese and Neumann 1997)

Summary of elements related to the Zeithypothese (time hypothesis), as described in the literature:

- Primary homeotic genes produce mRNA to synthesize homeotic peptides of the homeobox (ca. 70 amino acids). It is unclear what initiates the activity of homeotic genes.
- Homeotic peptides bind to DNAs of a family of secondary homeotic genes, presumably to the proper region, or enhancer elements of the target genes. These seem to serve as rather unspecific activators of the genes.
- To induce the synthesis of mRNA of these homeotic genes, transcription factors are required, such as zinc fingers or leucine zipper proteins. Where is the origin of these transcription factors, has to be asked? The proteins synthesized usually undergo some processing. Here, certainly the Golgi apparatus plays a role.
- Later, these proteins have to be broken down again to reach a protein complement specific for the next stage of the G1-phase. The labeling of proteins to be removed is achieved by the CDK, and actual breakdown follows by the proteasome complex.

A review of genes related to the cell cycle and development characterized to date is given by Arias et al. (2006).

- A prolongation of the cell cycle duration by G1 alone is not sufficient to promote differentiation. Similarly to an omission of kinetin, and as an example, in the case of iron deficiency, the number of cells produced by the cultured explants per unit time is reduced. However, average cell growth is also retarded, and no root formation could be observed (Neumann 1972). Beside the length of the G1-phase, optimal conditions for cell growth seem to be as important for differentiation, here restricted by iron deficiency.

Within this context, the fate of plastids in cycling cells should be addressed. Based on many investigations that show uncoupling of cell and plastid division, both are commonly considered to be independent processes. In a recent paper, however, a downregulation of members of the pre-replication complex has been demonstrated, and both nuclear DNA replication and plastid propagation seem to be influenced simultaneously (Raynaud et al. 2005).

As mentioned before, no differentiation was observed also in the hormone-free treatment. Here, only about two rounds of cell division occur in some areas of the explant, and then growth takes place only as cell expansion. Cell division activity is also strongly reduced in the treatment supplemented with IAA and m-inositol, compared to the plus kinetin treatment. In this case, however, a short log phase of cell division activity with a reduced duration of the cell cycle was observed in some areas of the cultures. During this log phase, apparently some dedifferentiation takes place in the dividing cells, later to become rhizogenic. This dedifferentiation may result in some reprogramming of the genetic machinery—and, in fact, of the cells in toto. In cells, a continuous protein turnover takes place (see above). If the hypothesis of the relation between the duration of the cell cycle, and the realization of genetic information has some meaning, then some proteins with coding positions later in the G1-phase, but present at the beginning of this short log phase with high cell division activity, could be removed from the cells by breakdown; due to the reduction of the G1-phase, these would not be replaced by newly synthesized molecules. To which extent RNAi could come into play has been discussed above. If the protein moiety of a cell represents its molecular differentiation status, then this should be responsible for changes in differentiation at the molecular level of cultured cells engaged in active cell division.

Reprogramming of the genetic machinery could be due to a change in DNA methylation (Loschiavo et al. 1989). Also in our system, an increase in methylation can be observed during the log phase of cell division, and only in the stationary phase of cell division was an amplification of some DNA sequences also found, usually associated with de novo differentiation. Here, DNA methylation is reduced again. As was shown later in other studies, this increase in methylation during the log phase of cell division is not directly related to kinetin application but rather due to the reduction in cell cycle duration (Dührssen and Neumann 1980; De Klerk et al. 1997; Arnholdt-Schmitt et al. 1998). Variation in the amplification of DNA sequences also requires mechanisms of DNA breakdown. Years ago, the occurrence of such "metabolic" DNA was reported in particular for differentiating cells, employing the pulse/chase technique (Schäfer et al. 1978).

A reduction in the duration of the cell cycle at the expense of the length of the G1-phase has also been reported for transgenic tobacco plants (Cockcroft et al. 2000). The transformation consisted in the integration of the cycline CycD2 originating from *Arabidopsis thaliana*. Compared to the wild-type control, the rate of cell division was enhanced due to a reduction of the G1-phase, and a faster cycling of the cells was inferred. The overall growth rate was increased at normal size of the meristems. The size of the cells was not affected and comparable to that of the untransformed control plants. The initiation of leaves was accelerated, as well as the development at all stages; flowering was 9–14 days earlier. As explanation, Cockcroft et al. (2000) proposed a general acceleration of metabolic processes due to genetic transformation during G1, though no biochemical data were made available to substantiate this conclusion. A preliminary and not really satisfying interpretation of the discrepancy in cell cycle regulation between the carrot system and the tobacco system could be the use of cell cultures for the former and of whole plants for the latter. At present, it is possibly more useful to ask questions than to forcibly try to give answers.

In summary, during recent years, molecular-based investigations on the regulation of the cell cycle of eukaryotes have strongly increased. A generally accepted hypothesis is not available yet. Again as discussed at length above, the basic mechanisms of the cell cycle are apparently highly conserved in all eukaryotes, and important regulators are heterodimeric serine–threonine protein kinases consisting of a catalytic CDK subunit and a cycline as activator. Apparently, phosphorylation and dephosphorylations of proteins and consequently changes in binding ability and catalytic properties are important processes in this regulation. These processes are governed by protein kinases consisting of a cycline and P34 as the catalytic unit. These cyclines occur mainly at the transition between various phases of the cell cycle, and by binding of P34 and the cycline (e30–50 kDa), proteolytic activity is initiated. After the transition, the cyclines are broken down, and P34 is no longer able to act as catalyzer.

In plants, eight different CDKs and 30 cycline genes are currently known, and these act at different stages of the cell cycle. The retinoplastoma protein, which represses the transition from the G1-phase to the S-phase, is phosphorylated and inactivated by a CDK/cycline complex, and DNA replication is initiated. Breakdown of proteins requires phosphorylation for these to be recognized as targets of ubiquitin-mediated proteolysis. Proteasomes are responsible for the degradation of the protein molecules, and these are ATP-dependent; for recognition, polyubiquitination of the protein substrate is required. A summary is given by Yanagawa and Kimura (2005). All this describes cell cycle regulation by removal of proteins, associated with changes in the composition of the proteome. Nothing is said on the activation of the synthesis of proteins de novo and its regulation. Only both processes—protein synthesis and protein breakdown—define the progression of the cell cycle and its function in cell division and cell differentiation.

The question arises to which extent the induction of programs of differentiation depends on a prior dedifferentiation of the original tissue brought about by cell division. Also here papers with controversial results can be found in the literature. Culture systems have been described in which direct initiation of a new path of differentiation without prior cell division is possible, as well as those for which a strict requirement for this seems to exist. Examples of the former situation are the differentiation of tracheids from parenchyma cells or direct somatic embryogenesis of subepidermal cells of cultured carrot petiole explants and for the latter rhizogenesis of carrot root callus cultures.

Xylogenesis, i.e., the differentiation of tracheid-like cells from parenchyma cells, is extensively characterized for mesophyll cells and protoplasts of *Zinnia elegans* (Kohlenbach et al. 1982). Here, cell culture systems consist of two cell types, one of which can be directly transformed into tracheid elements, whereas the other requires cell division before it can do so. In other culture systems, like *Helianthus tuberosus*, *Helianthus annuus*, and *Raphanus sativus*, cell division is always essential before tracheid differentiation. Using explants of immature tubers of *H. tuberosus*, trachea differentiation is still possible after X-ray irradiation, which inhibits cell division, whereas this is impossible for explants from mature tubers after irradiation. Possibly, the age of cells determines whether a dedifferentiation brought about by cell division is a prerequisite. Younger cells, before reaching a final stage of differentiation, are able to initiate tracheid differentiation directly. Differences in cellular age may possibly also be an explanation for the different reactions of *Zinnia* cells described above. Indeed, *Zinnia* explants were obtained from immature leaves, and here cells of different ages should occur, as in any other young leaf.

In many culture systems, xylogenesis requires, besides an exogenous auxin, also a cytokinin supplement. Moreover, ethylene, i.e., native ethylene, should play an important role. Thus, gazing of cultures with ethylene, or a supplement of ethrel that serves as a synthetic precursor of ethylene, increases the number of tracheid elements differentiated. This can be observed also after application of ACC (L-aminocyclopropane-1-carbonic acid), which is a natural precursor of ethylene, and xylogenesis can be inhibited with some ethylene inhibitors like aminoethoxyvinylglycine, Ag^+, Co^{++}, and Na-benzoate. Furthermore, a promotion of xylogenesis can be observed after the application of methionine, from which ethylene synthesis originates naturally (Roberts et al. 1982).

Particularly the significance of growth regulators for the differentiation of tracheids indicates a more direct influence of this group of compounds on development, in addition to their indirect function via influences related to cell division intensity and interphase duration. A maximum of ethylene production can be observed immediately before tracheid formation, and in *Phaseolus*, *Acer*, and *Populus* cultures, this stage coincides with the induction of phenylalanine lyase activity involved in xylane formation.

The investigations on tracheid differentiation discussed above indicate that some predetermination or competence of cells seems to exist that enables these to "transdifferentiate" (Umdifferenzierung) without prior cell division. Such a suppressed competence to various lines of differentiation, though camouflaged, could also play a role in the production of adventitious roots, shoots, or embryos. In the carrot petiole system, primary root formation is confined to the area between the conductive vessels and the glandular canals, shoot differentiation to the highly vacuolated parenchyma cells, and somatic embryogenesis to a subepidermal cell layer. These differentiation events occur in this sequence. It remains an open question whether the one or other of these events is a prerequisite for the remainder to proceed. As a first indication of cell differentiation, histological investigations reveal an increase in cytoplasm in the original vacuolated cells. After starting the culture, some time elapses before one or the other line of differentiation is realized, during which preparation proceeds for one or the other line of differentiation at the biochemical and cytological level. This state shall be defined as the induction phase.

During the induction phase, the area destined to follow one or the other line of differentiation (as described above) contains originally vacuolated cells now densely packed with cytoplasm, indistinguishable under a light microscope. Still, it can be assumed that the composition of the protein moiety will be different. If cultures are in a nutrient medium containing 2.4D as an auxin, then the development of somatic embryos will be blocked at the stage of pre-embryonic masses (PEMs), associated with a specific composition of proteins and other cell components that will be different from that of the original cells. Only a transfer into an auxin-free medium enables the development into somatic embryos to be completed, with a different composition of cells involved. Apparently, at the end of the induction phase, a control point exists that is regulated by auxins. Embryo development proceeds only after this control point is circumvented in an auxin-free medium or a medium with low auxin concentration.

The development of embryos from induced cells can be designated as the realization phase. The processes to differentiate root or shoots cannot be as clearly characterized. Still, all observations available to date indicate the existence of control points for these processes, too. Transdifferentiation can be defined in terms of three conditions: the competence of cells to pursue a specific line of differentiation (rhizogenesis, caulogenesis, somatic embryogenesis), its induction, and its realization (see above, e.g., Christianson 1987). A more detailed description of these hypothetical control points cannot be adequately provided here.

Suffice it to add that the concept of competence described above is difficult to explain on the basis of today's knowledge. If this concept is generalized beyond cultured petioles, then it can be hypothesized that most living cells of a tissue or an organ are concurrently preprogrammed to follow several lines of differentiation (subepidermal leaf cells/somatic embryos; parenchymatic cells near vascular bundles/root cells). It then depends on their position within the confines of the intact plant or on other environmental factors which of these possibilities will be realized. In cells of a different tissue, other lines of differentiation could be possible. As already shown by Skoog and Miller (1957) in the 1950s, here the ratio of growth regulators should be one factor involved. For further understanding, more knowledge on the genome organization of higher plants is required, in which also apomixes and similar phenomena should be considered.

The differentiation processes described above for cultured petiole explants were restricted to a specific tissue (direct somatic embryogenesis). As described before, induction of somatic embryogenesis, or of rhizogenesis can, however, occur also in other tissues. In this case, usually, a prior phase of rapid cell division is required, i.e., the formation of a callus (indirect somatic embryogenesis; De Klerk et al. 1997). It is during this phase of high division activity that the basis to produce the competence to this new line of differentiation is established. Again also here, first an increase in cell division activity results in a reduction in the length of the G1-phase. If a hierarchical activation of genetic information is assumed to exist (as described above), then the synthesis of proteins positioned later in G1, i.e., in the state now blocked due to the shortening of the G1-phase, and corresponding to the initiation of the S-phase, will be prevented. In living cells, as discussed before a continuous protein breakdown operates, some proteins to be synthesized later in the cell cycle will not be replaced by new molecules if those originally present have been removed by breakdown.

Consequently, the daughter cells will have a different protein complement than the mother cell, and if the protein complement is the molecular basis of cell differentiation, then daughter cell differentiation will no longer be identical with that of the mother cell. A transdifferentiation ("reprogramming") occurs for several rounds of cell division, at the end of which could arise the competence to rhizogenesis, or somatic embryogenesis.

All these ideas are quite speculative, and experimental evidence is still lacking for some. Such working hypotheses, however, are necessary to invoke new experimental approaches leading to advanced insight.

References

Arias RS, Filichkin SA, Strauss SH (2006) Divide and conquer: development and cell cycle genes in plant transformation. Trends Biotechnol 24:267–273

Arnholdt-Schmitt B, Schäfer C, Neumann KH (1998) Untersuchungen zum Einfluß physiologisch bedingter Genomvariationen auf RAPD-Analysen bei *Daucus carota*. Vortr Pflanzenzüchtung 43:107–113

Blaschke JR, Forche E, Neumann KH (1978) Investigations on the cell cycle of haploid and diploid tissue cultures of *Datura innoxia* Mill. and its synchronisation. Planta 144:7–12

Christianson ML (1987) Causal events in morphogenesis. In: Green CE, Somers DA, Hackett WP, Biesboer DD (eds) Plant tissue and cell culture. A.R. Liss, New York, pp 45–55

Cockcroft CE, Bart GW, den Boer JM, Healy S, Murray JAH (2000) Cyclin D control of growth rates in plants. Nature 405:575–579

De Klerk GJ, Arnholdt-Schmitt B, Lieberei R, Neumann KH (1997) Regeneration of roots, shoots and embryos: physiological, biochemical and molecular aspects. Biol Plant 39:53–66

Dewitte W, Murray JAH (2003) The plant cell cycle. Annu Rev Plant Biol 54:235–264

Dührssen E, Neumann KH (1980) Characterisation of satellite-DNA of *Daucus carota* L. Zeitschr Pflanzenphysiol 100:447–454

Froese C, Neumann KH (1997) The influence of kinetin on the protein synthesis pattern at some stages of the cell suspension cultures of *Daucus carota* L. Angew Bot 71:111–115

Fröse C (1993) Der Einfluß von Kinetin auf Wachstum, Zelldifferenzierung und Proteinsynthsemuster bei zellzyklussynchronisierten Suspensionskulturen der Karotte (*Daucus carota* L.). Dissertation, Justus Liebig Universität, Giessen

Hamilton A, Voinnet O, Chappell L, Baulcombe D (2002) Two classes of short interfering RNA in RNA silencing. EMBO J 21:4671–4679

Hartig K, Beck E (2006) Crosstalk between auxin, cytokinins, and sugars in the plant cell cycle. Plant Biol 8:1–8

Kohlenbach HW, Körber M, Lang H, Li L, Schöpke C (1982) Differentiation in suspension cultures of isolated protoplasts. In: Fujiwara A (ed) Plant tissue culture 1982. Japanese Association for Plant Tissue Culture, Tokyo, pp 95–96

Linser H, Neumann KH (1968) Untersuchungen über Beziehungen zwischen Zellteilung und Morphogenese bei Gewebekulturen von *Daucus carota* L. I. Rhizogenese und Ausbildung ganzer Pflanzen. Physiol Plant 21:487–499

Loschiavo F, Pitto L, Guiliano G, Torti G, Nuti-Ronchi V, Marazziti D, Vergara R, Orselli S, Terzi M (1989) DNA methylation of embryogenic carrot cell cultures and its variations as caused by mutation, differentiation, hormones and hypomethylating drugs. Theor Appl Genet 77:325–331

Lyndon RF (1990) Plant development: the cellular basis. Unwin Hyman, London

Mette MF, van der Winden J, Matzke M, Matzke AJ (2002) Short RNAs can identify can identify new candidate transposable element families in *Arabidopsis*. Plant Physiol 130:6–9

Nagata T, Nemoto Y, Hasezawa S (1992) Tobacco BY-2 cell line as the HeLa cell in cell biology of higher plants. Int Rev Cytol 132:1–30

Neumann K-H (1968) Der Beziehungen zwischen hormonalgesteuerter Zellteilungsgeschwindigkeit und Differenzierung, ein Beitrag zur Physiologie der Pflanzlichen Eetragsbildung. Habilitation, Justus Liebig Universitaet, Giessen

Neumann KH (1972) Untersuchungen über den Einfluß des Kinetins und des Eisens auf den Nukleinsäure- und Proteinstoffwechsel von Karottengewebekulturen. Zeitschr Pflanzenernäh Bodenkunde 131:211–220

Neumann KH (1995) Pflanzliche Zell- und Gewebekulturen. Eugen Ulmer, Stuttgart

Neumann K-H (2006) Some studies on somatic embryogenesis: a tool in plant biotechnology. In: Kumar A, Roy S, Sopory S (eds) Plant biotechnology and its applications in tissue culture. I.K. International, New Delhi, pp 1–14

Planchais S, Glab N, Inze D, Bergounioux C (2000) Chemical inhibitors: a tool for plant cell cycle studies. FEBS Lett 476:78–83

Raynaud C, Perennes C, Reuzeau C, Catrice O, Brown S, Bergounioux C (2005) Cell and plastid division are coordinated through the prereplication factor AtCDT1. Proc Natl Acad Sci USA 102:8216–8221

Reichow S, Varani G (2006) Structural biology—RNA switches function. Nature 441:1054–1055

Reski R (2006) Small molecules on the move: homeostasis, crosstalk, and molecular action of phytohormones. Plant Biol 8:277–280

Roberts LW, Baba S, Shrshraishi T, Miller AR (1982) Progress in cytodifferentiation under in vitro conditions. In: Fujiwara A (ed) Plant tissue culture 1982. Japanese Association for Plant Tissue Culture, Tokyo, pp 87–90

Schäfer A, Blaschke JR, Neumann KH (1978) On DNA metabolism of carrot tissue cultures. Planta 139:97–101

Skoog F, Miller CO (1957) Chemical regulation of growth and organ formation in plant tissues cultured in vitro. Symp Soc Exp Biol 11:118–140

Sudarsan N, Barrick JE, Breaker RR (2003) Metabolite-binding RNA domains are present in the genes of eukaryotes. RNA 9:644–647

Thore S, Leinbundgut M, Ban N (2006) Structure of the eukaryotic thiamine pyrophosphate riboswitch with its regulatory ligand. Science 312:1208–1211

Winkler WC (2005) Riboswitches and the role of noncoding RNAs in bacterial metabolic control. Curr Opin Chem Biol 9:594–602

Xie X, Johanson LK, Gustafson AM, Kasschau KD, Lellis AD, Zilberman D, Jacobsen SE, Carrington JC (2004) Genetic and functional diversification of small RNA pathways in plants. PLoS Biol 2:E104

Yanagawa Y, Kimura S (2005) Cell cycle regulation through ubiquitin/proteasome-mediated proteolysis in plants. JARQ 39:1–4

13.1 Somaclonal Variations

According to Leva et al. (2012), the term somaclone was coined to refer to plants derived from any form of cell culture, and the term "somaclonal variation" was coined to refer to the genetic variation among such plants. Phenotypic and genotypic variations are sometime shown in in vitro regenerated plants derived from organ cultures, calli, protoplasts, and somatic embryos (Orbovic et al. 2008; Machczyńska et al. 2015). However, somaclonal variation provides a valuable source of genetic variation for the improvement of crops through the selection of novel variants, which may show resistance to disease, improved quality, or higher yield (Unai et al. 2004). Classical plant breeders also generate genetic diversity within a species by exploiting a process called somaclonal variation, which occurs in plants produced from tissue culture, particularly plants derived from callus. Induced polyploidy and the addition or removal of chromosomes using a technique called chromosome engineering may also be used.

Somaclonal variation will be compounded with epigenetic effects. This makes the use for crop improvement more difficult, particularly in asexually propagated plants, since crossing and reselection would be required to sort out desired mutations from other alterations. In vitro selection is handicapped by atypical gene expression. The use of haploids derived from anther culture has found its best application in the "doubled haploids technique" which leads faster to homozygosity for more effective selection. Leva et al. (2012) and Sridevi and Giridhar (2014) obtained somaclonal variants in somaclonal variants of robusta coffee with reduced levels of cafestol and kahweol.

13.1.1 Ploidy Stability

Long-term cultures commonly show cells with ploidy levels higher than haploidy or diploidy besides aneuploidy. This ploidy instability observed in originally haploid or

© Springer Nature Switzerland AG 2020
K.-H. Neumann et al., *Plant Cell and Tissue Culture – A Tool in Biotechnology*,
https://doi.org/10.1007/978-3-030-49098-0_13

diploid cultures is actually a general characteristic of cell cultures. Cytogenetic stability, however, is a prerequisite for genetic manipulations, as well as for the use of such cultures for breeding purposes. Still, genetic instability also sometimes offers a chance to isolate genotypes with properties important for practical applications. However, the problem of genetic stability of the offspring arises remains to be addressed.

The epigenetic variations play an important role. Following Chaleff, these are defined as reactions of cultures maintained after removal of the stimulus that caused these reactions. In contrast to this, normal physiological reactions cease operation after removal of the stimulus.

The classical method to determine the ploidy level is to count chromosomes at the metaphase of mitosis. Cells in active divisions are mostly ploidy stable, whereas a broad scattering of DNA levels can be observed by cytophotometry in older cell cultures with a low division rate. In this case, reliable chromosome counts would reflect a higher genetic stability and homogeneity than really exist. This can be seen in Fig. 13.1 for a *Datura* cell suspension at the stationary growth phase, in which the amount of DNA was determined by means of cytophotometric measurements (see Chap. 6). Such inhomogeneities occur particularly in cultures of haploid origin. Cytophotometric methods, however, are usually not sensitive enough to detect aneuploidies, or the loss of sections of chromosomes. For this, chromosome analysis is still recommended.

As mentioned before for studies assessing cell-specific DNA contents, the so-called C-value is often used. 1C represents the DNA content of the haploid genome of the species in the G-phase of the cell cycle and 2C that of the G2-phase. The 1C-value can be easily obtained by cytophotometric determination of DNA content in the tetrads for standardization of the method. In carrots, for example, measurable (albeit small) differences in DNA content can be observed between varieties, and therefore it is recommended to perform such measurements for each variety separately.

In the G1-phase of the cell cycle, the DNA content of diploid cells is defined as 2C and in the G2-phase cells as 4C. In the S-phase, intermediate values are found. Values higher than those defined for the diploid genome can be also observed in cells of intact plants raised from seeds. Moreover, and despite some data scattering particularly for stem cells, comparing the profiles of DNA content of cells of intact plants, and of cell culture systems clearly demonstrates more uniformity in DNA content for the former.

In barley plants raised from microspores, a distinct inhomogeneity can be observed within a plant, with different ploidy levels for the leaves (Fig. 13.2). Under the conditions employed, direct somatic embryogenesis could not be induced. The plants were obtained from a callus produced by anther culture via separate shoot and root differentiation. The root tip of these plants consists of cell lines of different ploidy levels (Neumann 1995). This is not the case for plants raised by the bulbosum method. The same scattering of ploidy levels can be detected in the callus from which the plant was obtained. Apparently, the root primordia developed from a group of cells within the callus that contained cells of different C-values. In the

Fig. 13.1 DNA content in nuclei of cells of haploid and diploid origin in cell suspension cultures: at t0, and after the treatment described in Fig. 13.3, at t28 (Kibler and Neumann 1980) X axes: relative units, Y axes: percent of nuclei with a given DNA content

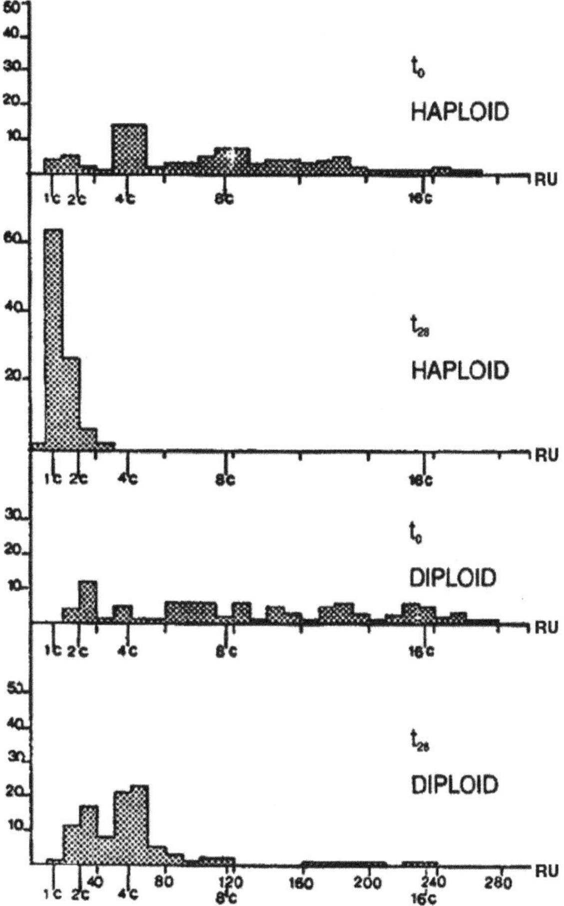

shoots of these plants, variations in C-values were recorded in the leaves, and infertile flowers were observed. If the initiation of shoot primordia was analogous to that of the root primordia considered above, then these leaf and flower trends would be explained.

These results have important implications. As we know today, plants derived from somatic embryogenesis develop from a single cell (see Sect. 7.3). Therefore, in experiments on the genome, or for use in breeding programs and the like, these plants would be by far preferable to those obtained from callus cultures with separate differentiation of roots and shoots. For some plant species, even nowadays, the term recalcitrant is used with respect to the induction of somatic embryogenesis. For these species, the following results could be of help in producing ploidy homogenous material (Figs. 13.3 and 13.4).

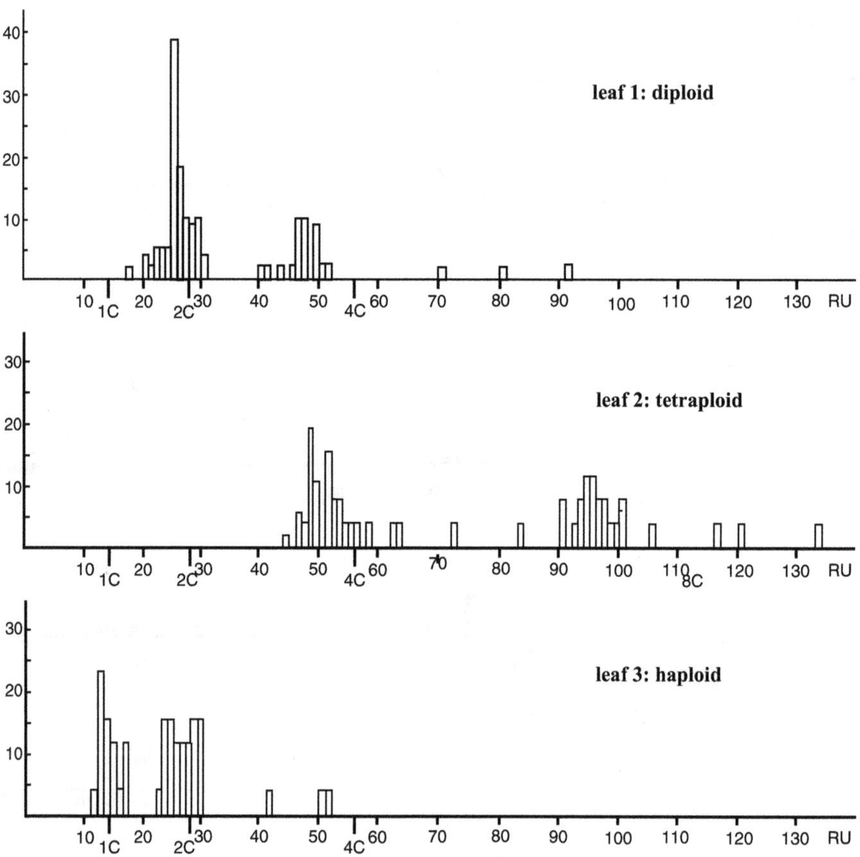

Fig. 13.2 C-value distribution in various leaves of a barley plant raised from the callus of anther cultures (Forche et al. 1979) *X* axes: relative units of DNA content in nuclei, *Y* axes: percent of nuclei with a given DNA content

The protocol is based on two assumptions: first, cells with the lowest ploidy level have the shortest cell cycle duration, and second, in cell suspensions, highest cell division activity takes place in small cell aggregates, comparable to the meristematic nests described for callus cultures. High cell division activity is supported by a kinetin supplement, and these cell aggregates are isolated by means of some sieving technique. The end result is ploidy stable *Datura* suspensions maintained for 3–4 weeks (Kibler and Neumann 1980; Neumann 1995). The application of this protocol to barley suspensions of haploid origin with a broad scattering of C-values in several successive subcultures resulted in ploidy homogenous cell material (Neumann 1995).

Procedure

1. ca. 5g of plated soft callus material (after 2 weeks subculture) are transferred to 250ml liquid medium (MS + Kinetin) in star flasks or Erlenmeyer flasks

2. Development of a dense cell suspension

3. after 6-7 days sieving of the suspension (250μm). The filtrate is used for further cultivation and the residue is discarded. Some times a second sieving (90μm) is required. Here the material on the filter is used for further processing.

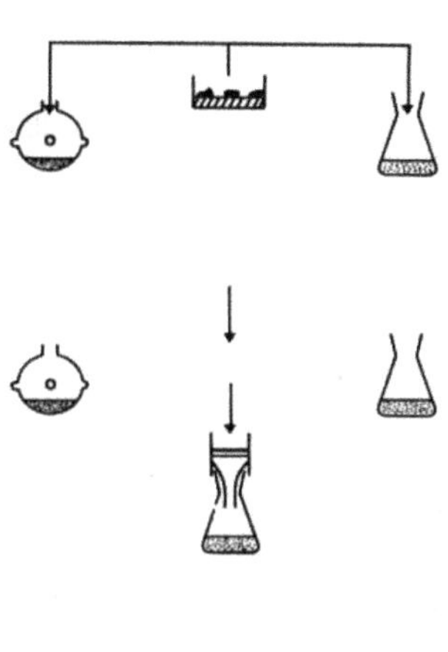

4. Sedimentation of cell material of the 250μm filtrates by centrifugation at 100 g

5. Mixture of the pellets to obtain a concentration of all suspension processed

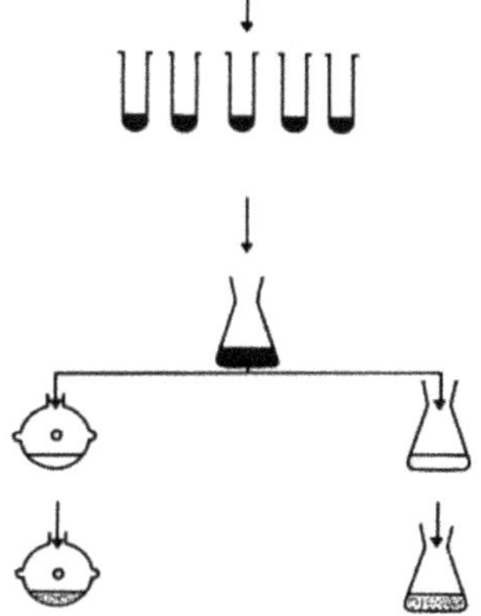

6. Transfer to fresh medium (MS + 2.4-D + kinetin) at a ratio of 1:5 cell density: $1\text{-}4\text{x}10^4$/ml

Main culture

7. 21-28 days, 22 °C, continuous illumination at ca. 400 Lux (Erlenmeyer flask or fermenter)

Fig. 13.3 Procedure to obtain ploidy homogenous cell cultures (Kibler and Neumann 1980)

13.1.2 Some More Somaclonal Variations

Following Larkin and Scowcroft (1981), scattering of ploidy levels (as described above), and some other epigenetic variations of cultured cells, can be categorized as somaclonal variations. These include mutations, chromosomal rearrangements, changes in chromosome structure, gene amplification, gene methylations, activation of transposons, exchange of sections of chromosomes, and others.

For a number of plant species (rice, wheat, maize, lettuce, tobacco, tomato, and rapeseed, lemon), so-called point mutations have been detected in plants raised from cell cultures. An early review on this topic was published by Scowcroft et al. already in 1987 (see also Orbovic et al. 2008; Leva et al. 2012). Plants with such mutations were obtained from the same callus as others are free of these, and it was speculated that these mutations were produced during cell culture.

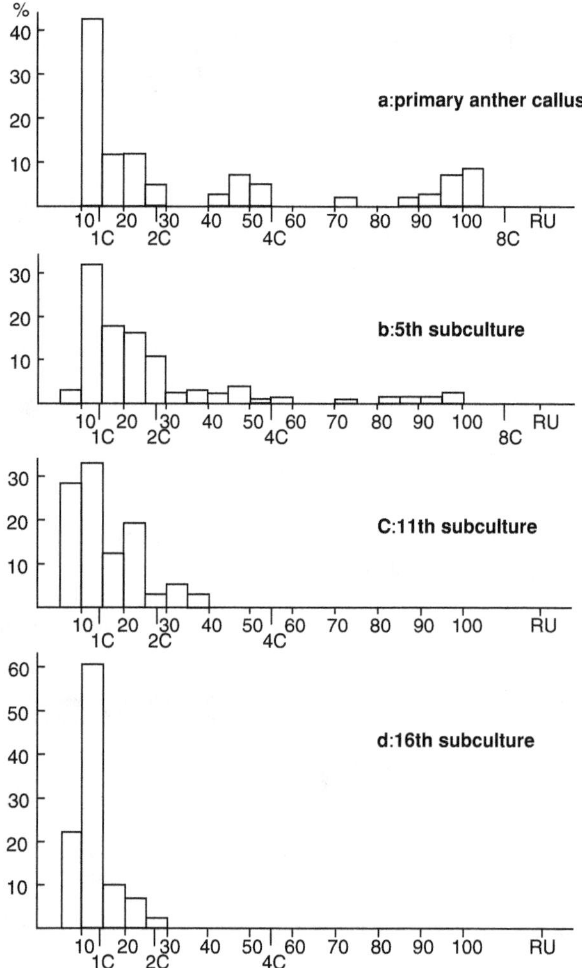

Fig. 13.4 Distribution of C-values in cell cultures of *Hordeum vulgare* during some subcultures with the application of the procedure described in Fig. 13.3. Until the 5th subculture, a transfer was performed at 4–6 week intervals and later every 2 weeks. *X* axes: DNA content in relative units, *Y* axes: percentage of nuclei with a given DNA content (Kibler and Neumann 1980)

This, however, cannot be regarded as a conclusive proof. However, Orbovic et al. (2008) analyzed genetic variability in various tissue culture-derived lemon plant populations using RAPD and flow cytometry. Usually, an original explant consists of 10,000–15,000 cells and more, and one cannot exclude the accidental existence of cells already mutated in the mother plant, which are later propagated in culture. Here, protoplast cultures and somatic embryogenesis should be a more suitable system.

In higher plants, the fraction of repeated DNA sequences amounts to about 40–60% of the total genome, and some sequences can occur in up to million copies. The fraction repetitive sequences with a moderate number of copies include genes for rRNA. In tissue culture-derived individual plants of *Triticale*, for a ratio of four fragments of an rRNA sequence to each other obtained by application of the restriction enzyme Tag 1, quantitative variations were detected by Brown and Lörz (1986). These changes were stable through meiosis. For interpretation of such results, it has to be considered that changes in the number of copies of genes can be also induced by phytohormones or herbicides (Widholm 1987). It remains unclear to which extent such changes are heritable. Notably, an amplification of some DNA stretches induced by GA3 applications to carrot plants, which resulted in a reduction of the diameter of the taproot, was not inherited.

Larkin et al. (1985) already described tissue culture-derived plants of 14 species in which changes in chromosome structure were observed. These included deletions, exchange of chromosome sections, isochromosome formations, inversions, DNA amplifications, and others. Furthermore, such changes were observed in 17 of 551 tissue culture-derived hexaploid wheat plants. In 14 plants, aneuploidy of a chromosome was detected, and four plants were euploid. These variations were interpreted as being due to the formation of isochromosomes or translocations within the genome.

The question has to be raised which are the causes of somaclonal variations and what makes plants derived from tissue cultures particularly prone to such changes.

The simple conventional understanding of genetic regulation of cellular life can be summarized as DNA is transcribed into RNA, which acts as template to synthesize proteins, these being responsible for essentially all processes occurring within the cell, or its reactions to the environment. As DNA nucleotide sequences become increasingly known, however, unexplainable inconsistencies with this central dogma appear. This raises questions like why apparently genetically identical twins are not really identical in some ways. Epigenetic factors seem to be responsible for these anomalies, and these may be associated with the so-called junk DNA (see above). It has to be kept in mind that only a few percent of our DNA codes for proteins through mRNA. For a long time, this junk DNA was considered as a byproduct of millions of years of evolution. This could still be true to some extent, but here at least part of the information for epigenetic factors could be localized. Somaclonal variations, often observed in cultured cell material, or embryos derived thereof, can be defined as epigenetic factors, and this may be a valuable system for studying the broad spectrum of epigenetics as such, and its regulation of cells.

During recent years, ever more evidence points to an epigenetic system of regulation of growth and development that seems to control gene activity. This includes DNA methylation, DNA amplification, histone acetylation, and others without changes in the nucleotide sequence. Changes in this system are also heritable. It remains to be seen whether these are part of a regulatory system above that of the classical DNA/RNA/protein system or rather largely independent factors related to the classical schema.

As mentioned elsewhere, during callus formation in vitro two of these epigenetic factors, i.e., DNA methylation and DNA amplification, were characterized in carrot cultures. These proved to be transient. During the logarithmic growth phase, DNA amplification was decreased, and DNA methylation was promoted. This was an indication of a rearrangement of epigenetic factors (or systems) at transfer from the original, rather quiescent root cells, to proliferating callus cells. At the stationary phase often associated with rhizogenesis, methylation was reduced, and the formation of amplified DNA sequences was increased again. Here, a presumably qualitatively different, new epigenetic system would have been established. Changes in methylation can also be induced by growth regulators, notably auxins (Loschiavo et al. 1989; Arnholdt-Schmitt 1993). Repeated elements are known to be preferably methylated (Arnholdt-Schmitt et al. 1995).

During such rearrangements, errors may occur in the reorganization of the epigenetic system, expressed as somaclonal variations. Such somaclonal variations are frequent in transgenic material. Transgenic plants are usually derived from tissue culture systems, and since such epigenetic changes can be heritable, the genome of transgenic plants is often rather unstable. This could contribute to answering the second part of the question posed above.

As answers to the first part of this question, DNA amplifications, transposons, and somatic reorganizations of the genome can be considered. The latter seems to involve mainly changes in polygenic traits. Furthermore, somatic crossing over, exchange of material of sister chromatides, variations in the methylation pattern of DNA, activation or inactivation of genes due to mutations of DNA sequences originally not coding, though associated with coding DNA stretches, and other aspects can be discussed. To date, however, there is insufficient experimental evidence published in the literature pointing to any one mechanism as the cause of somaclonal variations. The situation is not much different in responding to the second part of the question posed above. As early as the 1970s, D'Amato discussed the high plasticity of the genome of higher plants in tissue culture. Based on the availability of sufficient data already at that time, several mechanisms seemed to be possible, related mainly to the original material from which the explants were obtained. This may contain cells with DNA replications without concurrent division of the nucleus or eventually the cell. If such cells occur in an explant at the initiation of cell division, then when these cells divide, polyploidy will result. Such DNA extrareplications have been recorded in more than 80% of angiosperms screened. As already described, these can be more or less tissue specific and seem to be related to differentiation. In meristematic tissue, they are almost absent. In mature tissue like the *Phaseolus* suspensor, an amplification of DNA without concurrent division of the chromatids results in giant chromosomes (Nagl 1970). Consequently, ploidy homogeneity of cell cultures would depend at least partly on the developmental status of the tissue used to obtain explants for culture. The use of explants from meristematic areas should yield culture systems showing higher ploidy homogeneity.

Important are also reactions of the nucleus during callus inductions.

Here, endoreduplications are often observed and also unequal nucleus fragmentation due to multipolar spindle formation. As a result, cells with more than one nucleus are produced (D'Amato et al. 1980). Later, cell divisions will be associated with the loss of one or the other chromosome, eventually leading to aneuploidy. As already mentioned elsewhere, phytohormones seem to play a central role in these processes, particularly the auxin/cytokinin ratio.

In general, the variability of the genome increases with the age of cultures. This can be observed mainly for the occurrence of endoreplications without concurrent division of the nucleus, as well as of the cell. The extent of this influence depends on the hormonal supplement to the nutrient medium but also on nitrogen form and concentration, the osmotic pressure of the medium, and other factors. Thus, which cell type will find optimum conditions under which it can dominate will depend on the composition of the nutrient medium. An example is the influence of kinetin to enrich haploid cells in a suspension originally showing a broad scattering of C-values (see above).

A similar significance can be assigned to changes of the nutrient medium to induce differentiation. Thus, by changing the culture conditions, it is possible to selectively promote growth of certain genomes. By implication, careful genetic characterization, particularly of plant material used for plant breeding, or gene technology is recommended here.

As discussed earlier, somaclonal variation can cause problems for cell culture systems preserving germplasm for long duration through subculturing. These problems can be circumvented by the use of various methods of cryopreservation (Sect. 3.6).

As mentioned before, somaclonal variation can be exploited to select germ lines with properties beneficial for one or the other application. Here again the problem of stability of such traits arises. Although some changes in the genome were shown to be heritable, these changes were conferred mostly to areas of low significance for the growth and development of intact plants derived from cell suspensions. Similar to mutations induced by chemical treatments, or by X-rays, somaclonal variations are rather accidental. To improve the yield of crop plants, or the resistance to environmental stresses and attacks of pests, a controlled change of the genome is required. Success depends on understanding the physiological or biochemical basis of the process in the plant to be altered and its dependence on genetic factors. As will be described later, here gene technology comes into play.

An important aspect in tracking genomic changes induced either naturally as somaclonal variations or by artificial means such as X-rays is the method of selection. Generally, a positive method is applied through which plants with altered genomes are not or at least less affected in an environment hostile to the unchanged wild type (Figs. 13.5 and 13.6). An example is a stepwise increase of a toxin. Callus cultures, cell suspensions, or meristem cultures are suitable for an induction of mutations. In callus cultures, as well as in meristem cultures, the cells are not all exposed equally neither to the mutagen, nor to the same selection pressure. Consequently, a variable number of cells escape either the one or the other. Plants obtained from such cultures through regeneration of shoots and roots may be chimeras, as described for ploidy chimeras above. An alternative could be the use of protoplasts, which can be plated and selected after initiation of culture to be used to produce plants following somatic embryogenesis. Plating entails the transfer of protoplasts onto agar not yet fully hardened, and through further solidification of the agar, their position is subsequently fixed. The position of individual protoplasts on the plate can be marked, and their further development can be easily observed. Cultures of a given size (100s of cells) can be exposed to the selection pressure. The problems of selections will be addressed again after the discussion of gene technology. A

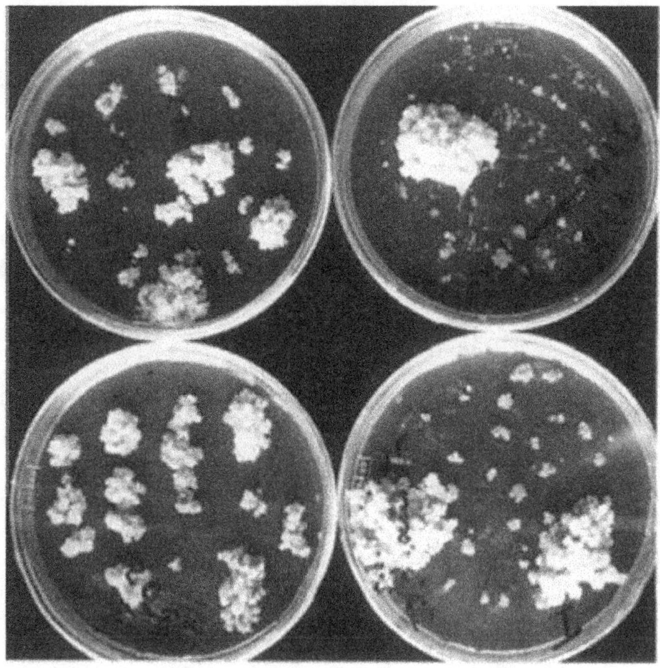

Fig. 13.5 Selective growth of barley cultures on an agar medium supplemented with a toxin (photograph by E. Forche)

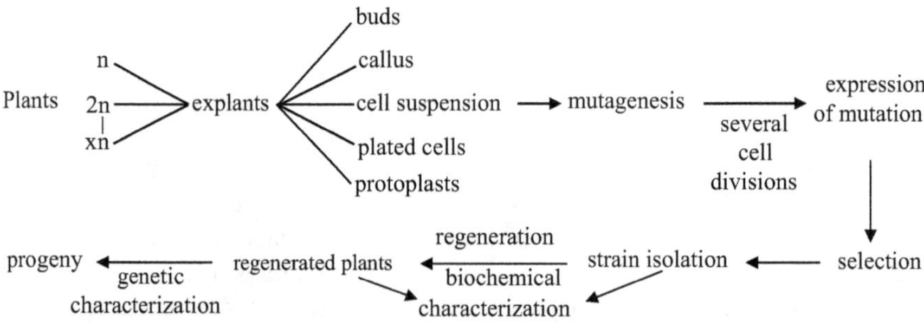

Fig. 13.6 Examples of a program to select for mutations

summary of work done on mutations before the advent of gene technology is given by Jacobs et al. (1987).

The discovery of many biochemical pathways in microbes was based on the use of deficiency mutants. Meanwhile, mutants of a few master plant systems, like *Arabidopsis* or rice, are employed for studying the metabolism of higher plants (metabolomics). Details on this are found in the sections dealing with metabolism and gene technology in this book.

13.2   DNA Fingerprinting and Characterization of Germplasm

A number of studies on DNA fingerprinting have been conducted in several laboratories in India, including those at the National Research Centre for DNA Fingerprinting (NRCDNAF) located at National Bureau of Plant Genetic Resources (NBPGR), New Delhi, and the Centre for DNA Fingerprinting and Diagnostics (CDFD), Hyderabad. In particular, NRCDNAF, which is the main center for DNA fingerprinting of plant genetic resources (PGRs) of India, has the following mandate:

- To identify suitable molecular markers for DNA profiling of some economically important plants
- Standardization of statistical methods for application of molecular marker techniques in variety identification and DUS (distinctness, uniformity, and stability) testing for patents and plant breeder's rights (PBRs)
- DNA fingerprinting of released and notified varieties, parental lines of hybrids and elite strains/genetic stocks of rice, wheat, maize, brassica, pulses (including chickpea), oilseeds, fibre crops, millets, etc. Among these PGRs, studies on DNA fingerprinting of the germplasm of banana, plantain, sesame, mungbean, soybean, chickpea, maize, rice, wheat, barley, mango, chili, tea, etc. have already been undertaken.

13.3   Estimations of Genetic Diversity

Several studies on molecular marker-based genetic diversity have also been conducted at NRCDNAF (New Delhi), Agharkar Research Institute (ARI, Pune) and National Chemical Laboratory (NCL, Pune), University of Delhi South Campus (UDSC, New Delhi), and The Energy Research Institute (TERI, New Delhi). These studies included several crops (e.g., tetraploid and hexaploid wheats, rice, brassica, millets) and a few tree species (e.g., neem, poplar, etc.)

The molecular marker systems, that were used, included RAPD, SSR, AFLP, SAMPL, etc. Das et al. (2007) studied phylogenetic relationships among the bamboo species by morphological characters and polymorphism analyses. However, there is a need to use maximum information collected through a variety of approaches, and more than one molecular marker systems should be used for getting reliable and useful estimates of genetic diversity.

13.4   Evaluation of Genetic Fidelity During Long-Term Conservation

Identity preservation of genetically modified crops has been reviewed by Doshi and Eudes (2008). Twenty-six countries (21 developing and 5 industrialized countries) planted 191.7 million hectares of biotech crops, which added 1.9 million hectares to the record of plantings in 2017. The continuous adoption of biotech crops by farmers worldwide indicates that biotech crops continue to help meet global challenges of hunger, malnutrition, and climate change. The production of genetically modified crops will increase from 112 million tons in 2015 to 130 million tons in the year 2021 (https://www.marketwatch.com/press-release/genetically-modified-food-gmo-market-size-is-expected-to-exhibit-32-billion-usd-by-2026-2019-08-08). With introduction of GM crops, there is a spectrum of alternative procurement strategy. Identity preservation is the process of maintaining the identity of different crops. Doshi and Eudes (2008) described in detail various components of IP system including testing of GM crops. Plant-made pharmaceuticals (PMPs) provide alternative method of providing antibiotics and other therapeutic proteins at a lower cost than current methods that use cultured mammalian cells (Thomas et al. 2002, see also Kumar and Sopory 2008). Faisal et al. (2012) tested genetic fidelity of *Rauvolfia serpentina*. Clones raised from synthetic seeds following 4 weeks of storage at 4 °C were assessed by using random amplified polymorphic DNA (RAPD) and inter-simple sequence repeat (ISSR) markers. All the RAPD and ISSR profiles from generated plantlets were monomorphic and comparable to the mother plant, which confirms the genetic stability among the clones. Fatima et al. (2013) generated RAPD and ISSR profiles from regenerated plantlets of *Withania somnifera* L. which were found to be monomorphic which thus confirms the genetic stability among the clones (see also Kumar and Shekhawat 2009 and Kumar and Sopory 2010).

13.5　Molecular Marker-Assisted Selection (MAS)

As reported by Francia et al. (2005), molecular marker-assisted selection (MAS) is an approach that has been developed to avoid the problems connected with conventional plant breeding changing the selection criteria from selection of phenotypes toward selection of genes, either directly or indirectly. According to Francia et al. (2005), several factors regulate MAS breeding:

1. Availability of cost-efficient and high-throughput genotyping methods.
2. Exploiting information derived from comparative genetic maps and genomic regions and conferring positive traits across syntenic species might be directly applied across species for the improvement of the trait.
3. Improving polygenic traits in a quick time frame and in a cost-effective manner, recent advances in MAS strategies.
4. Increasing amount of information on the topic which is available on public domain. Recent large-scale sequencing projects have produced a large amount of single-pass sequences of cDNAs from different plant species. The number of ESTs deposited in gene bank for wheat, maize, barley, and soybean has mounted to 562,000, 416,000, 380,000, and 342,000 sequences, respectively (http://www.ncbi.nlm.nih.gov/dbEST/). Because SSRs and SNP-based markers can be obtained quite easily from ESTs (Rafalski 2002), the development of molecular markers has recently shifted from anonymous DNA fragments to genes. Transcription-based genetic maps have thus been obtained from different crop

species including wheat (Gao et al. 2004), barley (Graner et al. 1999), and rice (Wu et al. 2002).

5. The availability of comprehensive cDNA and oligoncleotide arrays is now providing an option for the development of functional genomics-based strategies for the investigation of quantitatively inherited traits, using different strategies. Molecular markers are detectable in all stages of plant growth clearly and are not environmentally regulated and remain unaffected by the conditions in which the plants are grown. With the availability of an array of molecular markers and genetic maps, MAS has become possible both for traits governed by major genes and for quantitative trait loci (QTLs). Usefulness of a given molecular marker is dependent from its capability in revealing polymorphisms in the nucleotide sequence allowing discrimination between different molecular marker alleles.

These polymorphisms are revealed by molecular techniques such as restriction fragment length polymorphisms (RFLP), amplified fragment length polymorphisms (AFLP), microsatellite or simple sequence length polymorphisms (SSR), random amplified polymorphic sequences (RAPD), cleavable amplified polymorphic sequences (CAPS), single-strand conformation polymorphisms (SSCP), single-nucleotide polymorphisms (SNPs), and others (Mohan et al. 1997; Rafalski 2002; Kumar et al. 2014).

Genetic mapping of major genes and quantitative trait loci (QTLs) for many important agricultural traits is increasing the integration of biotechnology with the conventional breeding process. Traits related to disease resistance to pathogens and to the quality of some crop products are offering some important examples of a possible routinary application of MAS. For more complex traits, like yield and abiotic stress tolerance, a number of constraints have determined severe limitations on an efficient utilization of MAS in plant breeding, even if there are a few successful applications in improving quantitative traits. Recent advances in genotyping technologies together with comparative and functional genomicapproaches are providing useful tools for the selection of genotypes with superior agronomical performancies.

Thus, because different approaches are improving the strategies on which MAS rely on, an increased complementarity between molecular technologies and conventional breeding is expected in the near future for a more efficient improvement of the crop plants.

The first is functional association strategy: cDNA array can reveal that gene expression within a given tissue varies between genotypes differing for a given trait. Genetic mapping of identified candidate genes can then reveal congruency between the map position of the candidate gene and the presence of a QTL. The second strategy allows the identification of QTLs by a methodology called as eXtreme array mapping (XAM). This method can estimate the differences in allele frequency between pools of lines, selected for extreme phenotypes, by hybridization of total genomic DNA to a GeneChips (Wolyn et al. 2004).

Thus, because different approaches are improving the strategies on which MAS rely on, an increased complementarity between molecular technologies and conventional breeding is expected in the near future for a more efficient improvement of the crop plants.

13.6   Modes of Plant Gene Modification

There are a number of methods for modifying of plant genome listed as follows:

1. Classical breeding (wild crossing) Mutation classical chemical mutagens like EMS (*ethylmethane*-physical mutagens like x- or gamma rays *Sulphonate*) or NMH (*N*-nitroso-*N*-methylurea)
2. Somaclonal variation (in vitro manipulation) Chromosomal rearrangements: gene amplification and gene methylation (see Sect. 13.1.2)
3. Protoplast fusion (see Chap. 5)
4. Gene Transfer

13.6.1 Classical Breeding (Wild Crossing)

Plant breeding can be accomplished through many different techniques ranging from simply selecting plants with desirable characteristics for propagation to more complex molecular techniques. Classical plant breeding uses deliberate interbreeding (crossing) with themselves of closely or distantly related individuals to produce new crop varieties or lines with desirable properties (Jimenez-Lopez and Hernandez 2012). Plant breeding in certain situations may lead the domestication of wild plants. Almost all the domesticated plants used today for food and agriculture were domesticated in the centers of origin. Examples are the landraces of rice, *Oryza sativa* subspecies *indica*, which was developed in South Asia, and *Oryza sativa* subspecies *japonica*, which was developed in China.

Plants are crossbred to introduce traits/genes from one variety or line into a new genetic background. For example, a mildew-resistant pea may be crossed with a high-yielding but susceptible pea, the goal of the cross being to introduce mildew resistance without losing the high-yield characteristics. Classical breeding relies largely on homologous recombination between chromosomes to generate genetic diversity. The classical plant breeder may also make use of a number of in vitro techniques such as protoplast fusion, embryo rescue, or mutagenesis to generate diversity and produce hybrid plants that would not exist in nature.

Traits that breeders have tried to incorporate into crop plants in the last 100 years include: (1) increased quality and yield of the crop, (2) increased tolerance of environmental pressures (salinity, extreme temperature, drought), (3) resistance to viruses, fungi, and bacteria, and (4) increased tolerance to insect pests. Genetic modification of plants that can produce pharmaceuticals (and industrial chemicals), sometimes called pharmacrops, is a rather radical new area of plant breeding.

13.6.2 Mutagenesis

Mutagenesis deals with developing new strains wherein plants are subjected to radiation treatments or doused in toxic chemicals that randomly scramble genes to produce new traits. This finding was one of the most important breakthroughs in the history of genetics. The pioneering experiments by Muller (1927, 1928) have shown that it is possible to knock down the stability of genes.

In vitro mutagenesis is a process developed for the management of floral chimeric sectors and reported in chrysanthemum (Misra et al. 2003; Datta et al. 2005). In vitro mutagenesis through direct shoot regeneration from flower petals developed solid mutants without diplontic selection that too in a relatively short period of time.

Isolation of chimera and establishment of solid mutant are a lengthy process as it involves two-steps—first, in vivo mutagen treatment and second, in vitro regeneration of viable plants from mutated sectors. Chemical and physical mutagenesis has been used to increase genetic variability in crop plants. Chemical mutagens like EMS and DMS, radiation, and transposons are used to generate mutants with desirable traits to be bred with other cultivars. According to Osakabe and Osakabe (2015) gene function, genetic analysis has traditionally studied the effects of natural and artificial mutations induced by physical and chemical mutagens such as X-rays and ethyl methanesulfonate (EMS) (Sikora et al. 2011). These methods require screening to identify important traits as a consequence of mutations that are introduced randomly into the genome. Determining the effects of sequence-specific modifications on the host organism helps to identify gene function. However, performing site-specific mutagenesis in eukaryotes remains challenging. Other methods, such as posttranscriptional silencing of genes of interest using short interfering RNA (siRNA), have been utilized effectively, but gene knockdown by siRNA can be variable and incomplete.

More than 430 new varieties have been derived as mutants of rice (*Oryza sativa* L.) via the application of different mutagenic agents. Chemical mutagens such as ethyl methane sulphonate (EMS), diepoxybutane-derived (DEB), sodium azide, and irradiation (Gamma rays, X-rays, and fast neutrons) have been widely used to induce a large number of functional variations in rice and other crops (Benjavad Talebi et al. 2012). This new possibility was looked upon with great optimism. EMS produces primarily C/G to T/A transitions (Ashburner 1990). Today, mutation breeding is not anymore based only upon classical physical mutagens like x- or gamma rays or classical chemical mutagens like EMS or NMH but also upon variation that occurs during in vitro culture and has been termed "somaclonal variation"; see Sect. 13.1.2—(Skirvin 1978; Oono et al. 1984; Novak et al. 1988; Novak 1991; Sukekiyo and Kimura 1991).

The ray florets of chrysanthemums were treated with γ-radiation just after inoculation/culture them on regeneration medium, and the induced mutated cell was isolated through direct regeneration without any normal chimeric competition.

To understand gene function, genetic analysis has traditionally studied the effects of natural and artificial mutations induced by physical and chemical mutagens such as X-rays and ethyl methanesulfonate (EMS) (Sikora et al. 2011). Determining the effects of sequence-specific modifications on the host organism helps to identify gene function. However, performing site-specific mutagenesis in eukaryotes remains challenging. Other methods, such as posttranscriptional silencing of genes of interest using short interfering RNA (siRNA), have been utilized effectively, but gene knock down by siRNA can be variable and incomplete.

Recently, a new method has been developed that uses the type II prokaryote-specific adaptive immune system nuclease (CRISPR/Cas9). The CRISPR/Cas9 system is based on RNA-guided engineered nucleases (RGNs) that use complementary base pairing to recognize DNA sequences at target sites (Cong et al. 2013; Mali et al. 2013). For sequence-specific silencing, CRISPR RNA (crRNA) and trans-activating crRNA (tracrRNA) participate in target sequence recognition.

13.6.3 Gene Technology

The increased use of gene technology in biotechnology for numerous categories of common products and other important aspects (pharmaceuticals, food quality, agricultural chemicals, disease resistance, functional food, plant nutrition, etc.), as well as for basic investigations to understand gene regulation has an ever-increasing impact on our society. Presently plant breeding uses techniques of molecular biology to select, or in the case of genetic modification, to insert, desirable traits into plants.

Gene technology is a term that encompasses a wide range of techniques for genetic analyses based on the direct manipulation of DNA and the transfer of genes between different species. In the medical/pharmaceutical field, biotechnology signifies a dramatic change in the approach to drug discovery, research and development, diagnosis, and disease management. Examples of biopharmaceuticals, i.e., enzymes, or regulators of enzyme activity, hormones, or hormone-like growth factors, cytokines, vaccines, monoclonal antibodies, and gene transfer in humans can be discussed (Ritschel and Forusz 1994).

A transgenic organism carries, in all its cells, the foreign gene that was inserted by laboratory techniques. Each transgenic organism is produced by introducing cloned genes, composed of deoxyribonucleic acid (DNA) from microbes, animals, or plants, into plant and animal cells. Transgenic technology consists of methods that also enable the transfer of genes between different species. Transferring a gene from one source (plants, bacteria, fungi, insects) to another, especially across kingdoms (e.g., from an animal to a plant), requires a laboratory step that is often called gene splicing, or genetic modification (GM).

There are a number of methods for genetically modifying (or "transforming") plants. The most common three methods are outlined below. More details about these and other techniques can be found in a number of recent reviews (e.g., Barampuram and Zhang 2011; Hooykaas 2010; Rivera et al. 2012).

13.6.3.1 Genetic Modification

Genetic modification of plants is achieved by adding a specific gene or genes to a plant or by knocking out a gene with RNAi, to produce a desirable phenotype. Such plants resulting from adding a gene are often referred to as transgenic plants. If for genetic modification genes of the species or of a crossable plant are used under control of their native promoter, then they are called Cisgenic plants.

In order to genetically modify a plant, a genetic construct must be designed so that the gene to be added or removed will be expressed by the plant. To achieve this, a promoter to drive transcription and a termination sequence to stop transcription of the new gene and the gene or genes of interest must be introduced to the plant. A marker for the selection of transformed plants is also included. In the laboratory, antibiotic resistance is a commonly used marker: plants that have been successfully transformed will grow on media containing antibiotics; plants that have not been transformed will die. In some instances, markers for selection are removed by backcrossing with the parent plant prior to commercial release.

The majority of commercially released transgenic plants are currently limited to plants that have introduced resistance to insect pests and herbicides. Insect resistance is achieved through incorporation of a gene from *Bacillus thuringiensis* (Bt) that encodes a protein that is toxic to some insects. For example, the cotton bollworm, a common cotton pest, feeds on Bt cotton; it will ingest the toxin and die. Herbicides usually work by binding to certain plant enzymes and inhibiting their action. The enzymes that the herbicide inhibits are known as the herbicides target site. Herbicide resistance can be engineered into crops by expressing a version of target site protein that is not inhibited by the herbicide. This is the method used to produce glyphosate-resistant crop plants (Pollegioni et al. 2011).

13.6.3.2 Transformation Techniques

Methods for introducing new DNA into plant cells can be divided into two major categories: indirect and direct DNA delivery. In the former approach, genes of interest are introduced into the target cell via bacteria, usually *Agrobacterium tumefaciens* or (less commonly) *Agrobacterium rhizogenes* (Tzfira and Citovsky 2006). Using plant viruses to insert genetic constructs into plants is also a possibility, but the technique is limited by the host range of the virus. For example, cauliflower mosaic virus (CaMV) only infects cauliflower and related species. Another limitation of viral vectors is that the virus is not usually passed on the progeny, so every plant has to be inoculated.

Direct transformation introduces genetic material without an intermediate host. The most commonly used direct transformation methods are biolistic transformation and protoplast transformation (reviewed by Rivera et al. 2012).

Many different genes can influence a desirable trait in plant breeding. The use of tools such as molecular markers or DNA fingerprinting can map thousands of genes. This allows plant breeders to screen large populations of plants for those that possess the trait of interest. The screening is based on the presence or absence of a certain gene as determined by laboratory procedures, rather than on the visual identification of the expressed trait in the plant.

Indirect Gene Transfer

Virus-Based Techniques

Delivery of genome engineering reagents can be done by plant virus-based vectors. Both DNA viruses (bean yellow dwarf virus, wheat dwarf virus, and cabbage leaf curl virus) and RNA virus (tobacco rattle virus) have demonstrated efficient gene targeting frequencies in model plants (*Nicotiana benthamiana*) and crops (potato, tomato, rice, and wheat). This system can be useful especially in economically important crops that are difficult to transform by means of *Agrobacterium*-mediated transformation or particle bombardment. Nowadays, important viral both DNA and RNA virus vectors are available for plant genome engineering (Zaidi and Mansoor 2017).

Agrobacterium-Mediated Gene Transformation

Agrobacterium tumefaciens is a common soil bacterium that naturally causes gall formation on a wide range of plant species, including most dicotyledonous and some monocotyledonous species (Van Larebeke et al. 1974). The gall is induced by transfer of hormone-producing genes from the bacterial cell into the plant genome. The genes are carried on a circular DNA molecule found within the bacterial cell called a tumor-inducing (Ti) plasmid. During the infection process, only a section of the Ti plasmid known as the transfer DNA (T-DNA) is transferred to the plant. The infection and T-DNA transfer process of *A. tumefaciens* has been extensively studied. This natural process has been used to facilitate genetic modification of plants. *Agrobacterium*-mediated transformation is well established for dicotyledonous plants (flowering plants whose seeds produce two seed leaves when germinating; most herbaceous plants, trees, and bushes are dicotyledonous). However, most monocotyledonous plants (flowering plants that produce one seed leaf, e.g., grasses) are not natural hosts of *A. tumefaciens* (De Cleene and De Ley 1976), and until recently, transformation of monocots using *Agrobacterium* was difficult and unreliable (Sood et al. 2011). *A. tumefaciens* Ti plasmids have been produced that lack the genes responsible for gall formation (disarmed plasmids). Genes to be inserted into the plant are put into the T-DNA section of these disarmed plasmids. *A. tumefaciens* cells carrying such plasmids cannot produce a gall in an infected plant but will transfer the T-DNA sequence carrying the genes of interest into the plant cell where they stably integrate into the plant genome (Bevan 1984; Klee and Rogers 1989).

Nowadays, an artificial transfer DNA (T-DNA) binary system s in which T-DNA and vir genes are located on separate replicon (T-DNA) binary is developed. This system is consisting of a binary plasmid and a helper plasmid (disarmed plasmids) that together produce genetically modified plants. The binary vector contains two origin(s) of replication that leads to replication in both *E. coli* and in *Agrobacterium tumefaciens* host. Additionally, an antibiotic resistance is located on binary vector to select for the presence of the binary vector in bacteria.

Transformation with *Agrobacterium* can be achieved using a variety of plant tissues including protoplasts (isolated cells with their cell walls removed) and leaf discs. The plant tissue is incubated with the bacterium for a variable period to allow infection to occur and then transferred to a synthetic medium containing nutrients and hormones that promote the growth of plants from single transformed cells, as well as a selective agent such as an antibiotic to eliminate untransformed cells (Barampuram and Zhang (2011) Recent Advances in Plant Transformation. In: Birchler J. (eds) Plant Chromosome Engineering. Methods in Molecular Biology (Methods and Protocols), vol 701. Humana Press, Totowa, NJ). During this regeneration process, transformed cells usually first divide to form a mass of tissue called callus (callus normally forms at a wound or graft site on a plant) before formation of plantlets. Independent of the process of *Agrobacterium*-mediated transformation, the regeneration of callus can induce mutations or other changes in the plant cell genome, known as somaclonal variation. Alternatively, methods have been developed for *Agrobacterium*-mediated transformation using whole plants, such as vacuum infiltration and floral dip. These methods minimize somaclonal variation and result in more genetically uniform progeny. For the floral dip method (Clough and Bent 1998), flowers are dipped into an *Agrobacterium* culture, and the bacterium transforms the germline cells that make the female gametes. The seeds from these plants are grown and screened for the genetic modification, e.g., using an antibiotic resistance or herbicide tolerance marker.

For plants transformed via *Agrobacterium*, residual bacteria have been shown in some cases to survive through the transformation process and remain within regenerated GM plants (Barrett et al. 1997; Yang et al. 2006). *Agrobacteria* are usually eliminated when GM plants are grown from seed but may remain in plants that are propagated vegetatively, potentially allowing residual GM *Agrobacterium* to transfer genes to other bacteria. It has also been suggested that if *Agrobacteria* and fungi were both present at plant wound sites, natural gene transfer may occur from bacterium to fungus (Knight et al. (2010). Nguyen et al. (2018), evolutionary novelty in gravity sensing through horizontal gene transfer and high-order protein assembly. PLoS Biology. https://doi.org/10.1371/journal.pbio.2004920). For these reasons, *Agrobacterium*-mediated transformation techniques usually include stringent methods to prevent *Agrobacteria* surviving, as well as steps to screen for and eliminate any GM plants with remaining *Agrobacteria*, to minimize potential for gene transfer occurring in the environment.

In addition to transfer of the T-DNA sequence, small segments of flanking Ti plasmid sequence and *A. tumefaciens* chromosomal sequence may be transferred into the plant genome at a low frequency (Smith 1998; Ulker et al. 2008).

Since the application of *Agrobacterium*-mediated transformation to monocotyledonous species such as rice, maize, barley, sugarcane, and wheat, and also to animal cells, as has been recently reported also here the use of *Agrobacterium* is still in central focus. In summary, the main characteristics of the *Agrobacterium* system in these, as well as dicotyledonous species are:

- A rather high frequency of transformation
- Proper integration of the foreign gene into the host genome and low copy number of the gene inserted

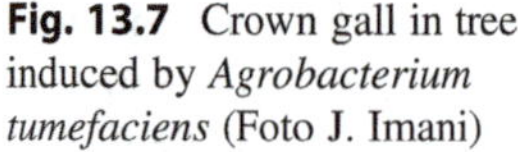

Fig. 13.7 Crown gall in tree induced by *Agrobacterium tumefaciens* (Foto J. Imani)

This results in most cases in a correct expression of the transgene itself. Because of its significance, it is necessary to give some details of this transformation process.

The first reliable method for plant transformation was based on a pathogen that attacks plants and causes crown gall disease—formation of galls at the—crown of a plant. The organism that causes this disease is *Agrobacterium tumefaciens*, a soil-borne plant pathogenic bacterium. The galls are produced at the site of infection and consist of a mass of undifferentiated cells, also known as tumors (Fig. 13.7).

Agrobacterium produces these tumors by transferring a piece of its DNA (T-DNA, transferred DNA) into the plant. This is a natural transfer of DNA from a prokaryote into an eukaryote. Plant transformation mediated by *A. tumefaciens* has become the most commonly used method for the introduction of foreign genes into plant cells and the subsequent regeneration of transgenic plants. The first evidence indicating this bacterium as the causative agent of crown gall goes back 100 years (Smith and Townsend 1907). Since that time, a large number of researchers have focused on the study of this neoplastic disease and its causative pathogen, and this for various reasons. During the first and most extensive period, scientific effort was devoted to disclosing the mechanisms of crown gall tumor induction. Then, in the 1970s, pathogenicity transferred between bacteria via conjugation, and evidence of

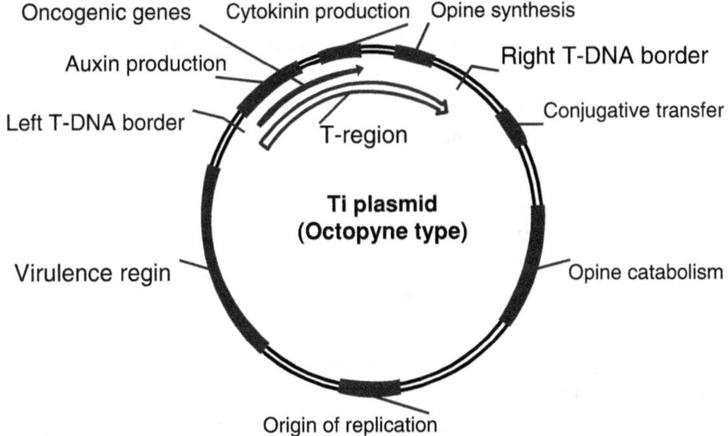

Fig. 13.8 Ti-plasmid of *Agrobacterium tumefaciens*

plasmid involvement was investigated, and finally in the 1970s–1980s, the Ti (tumor inducing) plasmid was characterized. It was apparent to researchers working with *A. tumefaciens* that this gram-negative soil bacterium co-opted normal plant cell metabolism by leaving a small portion (the transfer DNA or T-DNA) of its Ti-plasmid in the genome of an infected plant cell. Several research groups realized that if the genes normally present within the bacterium T-DNA were to be replaced with other genes, then one could obtain expression of the new genes in plant cells. However, the bacterium infected only one, or a few cells, and so not all cells of a plant harbored the new genes flanked by the T-DNA. Consequently, it was necessary to include a selectable marker in the T-DNA, so that only those cells that had taken up the engineered T-DNA, including the selectable marker, could be identified and allowed to grow.

When interacting with susceptible dicotyledonous plant cells, virulent strains of *A. tumefaciens* and *Agrobacterium rhizogenes*, another pathogen, have the exceptional ability to transfer a particular part (T-DNA) of their large plasmid, the Ti-plasmid (Fig. 13.8, >250 kb), into the nucleus of the infected cell, where it is then stably integrated into the host genome, and transcribed. This induces the diseases known as crown gall (*A. tumefaciens*) and hairy roots (*A. rhizogenes*). Two types of genes are responsible for tumor formation: the oncogenic genes, encoding for enzymes involved in the synthesis of auxins and cytokinins, and additionally T-DNA contains genes encoding for the synthesis of opines. These compounds, produced by the condensation of amino acids and sugars, are synthesized and excreted by the crown gall cells and consumed by *A. tumefaciens* as carbon and nitrogen sources. The synthesis of the opine form is dependent on the bacterial strain, e.g., for plasmids of C58 chromosomal background; it is the nopaline synthase that is responsible for nopaline formation from arginine, and it is agropine for strain EHA 105 with TiAch5 chromosomal background (Roger et al. 2000).

Mechanisms of gene transfer mediated by *A. tumefaciens*. The process of gene transfer from *Agrobacterium tumefaciens* into plant cells can be considered in terms of several steps: (1) infestation of bacteria, (2) induction of the bacterial virulence genes, (3) generation of the T-DNA transfer complex, and (4) integration of the T-DNA complex into the plant genome. An essential, and also the earliest step in tumor induction, is the infestation of plant cells by *A. tumefaciens* after its attachment to the plant cell surface (Matthysse 1986). Non-attaching mutants show a loss of the tumor-inducing capacity (Cangelosi et al. 1987; Thomashow et al. 1987; Bradley et al. 1997).

The polysaccharides of the *A. tumefaciens* cell surface are proposed to play an important role in the colonization of and also during the interaction with the host plant. The gene responsible for successful bacterium attachment to the plant cell is located at the chromosomal 20 kb locus.

Lipopolysaccharides (LPSs) are an integral part of the outer membrane and include the lipid membrane anchor and the antigen polysaccharide. *A. tumefaciens*, like other plant-associated *Rhizobiaceae* bacteria, produces also capsular polysaccharides (antigens) lacking lipid anchor, of strong anionic nature and tightly associated with the cell. There is some evidence indicating that capsular polysaccharides may play a specific role during the interaction with the host plant. In the particular case of *A. tumefaciens*, a direct attachment of wild-type bacteria to plant cells was observed.

This attachment region is composed of at least six essential operons (virA, virB, virC, virD, virE, virG) and two nonessential ones (virF, virH). The number of genes per operon differs.

VirA is a transmembrane dimeric sensor protein that detects signal molecules, mainly small phenolic compounds, released from wounded plants (Pan et al. 2003). The signals for VirA activation include acidic pH, phenolic compounds, such as acetosyringone (Winans 1992), and certain classes of monosaccharides that act synergistically with phenolic compounds (Ankenbauer et al. 1990; Cangelosi et al. 1990; Shimoda et al. 1990; Doty et al. 1996). VirA protein serves as periplasmic or input domain (important for monosaccharide detection), and two transmembrane domains act as a transmitter (signaling) and receiver (sensor; Chang and Winans 1992; Parkinson 1993). One of these transmembrane domains corresponds to the kinase domain and plays a crucial role in the activation of VirA, phosphorylating itself (Huang et al. 1990; Jin et al. 1990a, b) in response to signaling molecules from wounded plant sites.

VirG functions as a transcriptional factor regulating the expression of vir genes when it is phosphorylated by VirA (Jin et al. 1990a, b). The C-terminal region is responsible for the DNA binding activity, while the N-terminal is the phosphorylation domain and shows homology with the VirA receiver (sensor) domain.

The activation of vir systems also depends on external factors like temperature and pH. Virulence capacity is reduced by high temperature (exceeding 32 °C), because of a conformational change in the folding of VirA (Jin et al. 1993; Fullner et al. 1996). This indicates that the temperature for co-culture is crucial for genetic transformation. Actually, it should be between 21 and 28 °C.

The activation of vir genes results in the generation of single-stranded (ss) molecules representing the copy of the bottom T-DNA strand. Any DNA placed between T-DNA borders will be transferred to the plant cell as single-stranded DNA and integrated into the plant genome. The proteins VirD1 and VirD2 play a key role in this step, recognizing the T-DNA border sequences and nicking (cf. endonuclease activity) the bottom strand at each border. After endonucleotidic digesting, the rest of the VirD2 protein covalently attaches to the $5'$ end of the ssT strand. This association prevents exonucleolytic attack to the $5'$ end of the ssT strand (Dürrenberger et al. 1989) and distinguishes the $5'$ end as the leading end of the T-DNA transfer complex. T-DNA strand synthesis is initiated at the right border, and it proceeds in the $5'$–$3'$ direction until the termination process takes place. The left border may also act as a starting site for ssT strand synthesis, but the efficiency is much lower (Filichkin and Gelvin 1993). Inside the plant cell, the ssT-DNA complex is targeted to the plant nucleus after passing through the cell and nuclear membranes. For the export of VirE2 to the plant cell, VirE1 is essential. VirE2 contains two plant nuclear location signals (NLS) and VirD2 one (Bravo Angel et al. 1998). This indicates that both proteins probably play important roles once the complex is in the plant cell, mediating the complex uptake into the nucleus (Herrera-Estrella et al. 1990; Rossi et al. 1993; Tinland et al. 1995; Zupan et al. 1996). VirD2 and VirE2 also ensure that the DNA is efficiently transported into mammalian nuclei.

The final step of T-DNA transfer is its integration into the plant genome. The mechanism involved in the T-DNA integration has not yet been fully characterized. It has to be considered that integration occurs by some illegitimate recombination (Gheysen et al. 1989; Lehman et al. 1994; Puchta 1998). According to this model, pairing of a few bases, known as microhomologies, is required for a preannealing step between the T-DNA strand coupled with VirD2 and plant DNA. These homologies are very low and provide only a minimum specificity for the recombination process by positioning VirD2 for the ligation. The $3'$ end, or adjacent sequences of T-DNA find some low homologies with plant DNA, resulting in the first contact (synapses) between the T strand and plant DNA, and forming a gap in the $3'$–$5'$ strand of plant DNA. Displaced plant DNA is subsequently cut at the $3'$-end position of the gap by endonucleases and the first nucleotide of the $5'$ attaches to VirD2 pairs with a nucleotide in the top ($5'$–$3'$) plant DNA strand. The $3'$ overhanging part of T-DNA, together with the displaced plant DNA, are digested either by endonucleases, or by $3'$–$5'$ exonucleases. Then, the $5'$ attached to the VirD2 end and the other $3'$ end of the T strand (paired with plant DNA during the first step of integration process) join the nicks in the bottom plant DNA strand. Once the introduction of the T strand into the $3'$–$5'$ strand of the plant DNA is completed, a torsion followed by a nick into the opposite plant DNA strand is produced. This situation activates the repair mechanism of the plant cell, and the complementary strand is synthesized using the earlier inserted T-DNA strand as template (Tinland et al. 1995).

The first records on transgenic tobacco plant expressing foreign genes appeared at the beginning of the last decade, although many of the molecular characteristics of this process were unknown at that time (Herrera-Estrella 1983; Potrykus 1991).

Since that crucial turn in the development of plant science, a great progress in understanding of the *Agrobacterium*-mediated gene transfers to plant cells has been achieved. However, *A. tumefaciens* naturally infects only dicotyledonous plants, and many economically important plants are monocots, including cereals, and remained inaccessible for genetic manipulation for a long time.

Agrobacterium-mediated gene transfer into monocotyledonous plants was not possible until recently, when reproducible and efficient methodologies were established for rice (Cheng et al. 1998), banana, corn (Ishida et al. 1996), wheat (Cheng et al. 1997; Rasci-Gaunt et al. 2001; Wu and Lin 2003), sugarcane (Enríquez-Obregón et al. 1997, 1998; Arencibia et al. 1998), and barley (Tingay et al. 1997; Wang et al. 1998, 2001; Murray et al. 2001).

Many factors, including plant genotype, explant type, *Agrobacterium* strain, and binary vector, influence the *Agrobacterium*-mediated transformation of monocotyledonous plants. Inoculation and co-culture conditions are very important for the transformation of monocots. For example, antinecrotic treatments using antioxidants and bactericides, osmotic treatments, desiccation of explants before or after *Agrobacterium* infection, and inoculation and co-culture medium compositions had influences on the ability to recover transgenic monocots. The plant selectable markers used, and the promoters driving these marker genes, have also been recognized as important factors influencing stable transformation frequency. Extension of transformation protocols to elite genotypes and to more readily available explants in agronomically important crop species will be the challenge of the future. Further evaluation of genes stimulating plant cell division or T-DNA integration and of genes increasing the competency of plant cells to *Agrobacterium* may improve transformation efficiency in various systems (Cheng et al. 2004). In this book, we present updated information about the mechanisms of gene transfer mediated by *A. tumefaciens* and assessments for the application of this method to the transformation of monocotyledonous and dicotyledonous plants, as examples using barley and carrots, respectively, and as practiced in our own research program (see later). Transformation is currently used for the genetic manipulation of more than 120 species of at least 35 families, including the major economic crops, vegetables, fruit trees, as well as ornamental, medicinal, and pasture plants (Birch 1997), based on *Agrobacterium*-mediated, or direct transformation methods. The number of GM plant species increases continuously. The argument that some species cannot accept the integration of foreign DNA in their genome and lack the capacity to be transformed cannot be accepted, in view of the increasing number of species that have already been transformed; however, the establishment of an efficient tissue culture system still forms the basis for genetic manipulation.

As mentioned above, efficient methodologies for *Agrobacterium*-mediated gene transfer have been established mainly for dicotyledonous plants. To extend these to monocotyledonous plant species, it is important to account for critical aspects in the *Agrobacterium tumefaciens*–plant interaction and the cellular and tissue culture methodologies developed for these species. The suitable genetic material (bacterial strains, binary vectors, reporter, marker genes, and promoters) and the molecular biology techniques available in the laboratory are necessary considerations for selection of the DNA to be introduced. This DNA must be able to be expressed in the plants, enabling the identification of transformed plants in a selectable medium and using molecular biology techniques to test and characterize the transformation events (for a review, see Birch 1997). The optimization of *A. tumefaciens*–plant interaction is probably the most important aspect to be considered. It includes the integrity of the bacterial strain, its correct manipulation, and the study of reactions in wounded plant tissue, which may develop into a necrotic process in the wounded tissue, or affect the interaction and release of inducers or repressors of the *Agrobacterium* virulence system. The type of explants is also important and must be suitable for regeneration, enabling the recovery of whole transgenic plants. The successful establishment of a method for the efficient regeneration of one particular species is crucial for its transformation.

It is recommended to work firstly on the establishment of optimal conditions for gene transfer, through preliminary experiments on transient gene expression using reporter genes (Jefferson 1987), like the—green fluorescent protein (GFP; Tsien 1998; Chalfie and Kain 1998), DSRED Lux, and Bax. The proapoptotic protein Bax can serve as rapid gene answer by fast killing of plant cells (Eichmann et al. 2006; Technologie-Lizenz-Buero (TLB) der Baden-WuertingenschenHochschule GmbH, Germany). This is supported by the fact that *Agrobacterium*-mediated gene transfer is a complex process, and many aspects of the mechanisms involved still remain unknown. Transient expression experiments help to identify also the explants that may be used as targets for gene transfer, providing definitive evidence of successful transformation events, and correct expression of the transgene.

Preliminary studies also include the use of histologically defined tissues of different explants, and regeneration of whole plants, preferably by somatic embryogenesis (Sect. 7.3). Transient expression experiments may be directed to the regenerable tissue and cell. Optimization of transient activity is a futile exercise if experiments are conducted on non-regenerable tissues, or under conditions inhibiting regeneration, or altering the molecular integrity of the transformed cell. As mentioned before, *Agrobacterium*-mediated transfer introduces a smaller number of copies of foreign DNA per cell than is the case for particle bombardment or electroporation, but high efficiencies of stable transformation may be obtained even from cells without positive results in transient expression assays. These aspects are important in establishing an appropriate transformation procedure for any plant, particularly for those species categorized as recalcitrant. Cereals, legumes, and woody plants, which are difficult to transform or still remain untransformed today, can be included in this category. Many species originally considered for this category have nevertheless been transformed in recent years.

Direct Gene Transfer

Monocotyledons, as well as for dicotyledonous plants shown to be recalcitrant to *Agrobacterium*-based transformation, are transformed using direct method of genetic transformation (Shillito 1999; Sautter et al. 1991). The direct transformation methods include polyethylene glycol-mediated transfer (Ushiyama 1991), microinjection, protoplast and intact cell electroporation (Fromm et al. 1990; Lörz et al. 1985; Arencibia et al. 1995), and gene gun technology (Sanford et al. 1993).

Biolistic Approaches

Biolistic transformation (synonym: microprojectile bombardment, biolistics, particle bombardment, particle gun, gene gun, particle acceleration) is a powerful method and effective way for the creation of many transgenic organisms (plant, microbe, and mammalian cells) and most commonly used method to deliver DNA into plant tissues and callus. It is a rapid and very simple procedure, and it should facilitate the direct transformation of totipotent tissues such as pollen, embryos, meristems, and morphogenic cell cultures. This method is suitable for the transient gene expression in numerous tissues of different species and to be uniquely suitable for organelle transformation.

The development of the original particle gun started in 1984, and the concept was later patented by its inventors (Sanford et al. 1990). One of the early devices was based on acceleration of DNA-coated tungsten particles in a vacuum, to velocities that allowed penetration into biological tissues. After tungsten particles were shown to have phytotoxic effects (Russell et al. 1992), they were replaced with gold particles. The original gun design was followed by modified versions such as The He Biolistic Particle Delivery System licensed by DuPont (Hercules, CA, USA), the non-commercialized Accel Particle Gun (McCabe and Christou 1993), the Particle Inflow Gun (Finer et al. 1992), microtargeting devices designed for apical meristem transformation (Sautter et al. 1991), and the Helios gene gun (BioRad, Hercules, CA, USA) contrary to all other devices, the Helios gene gun does not require a vacuum chamber to hold the target tissue and can be used as a handheld device (Finer et al. 1999) to be effective regardless of species or tissue type. The damaging of DNA in cells, high copy number of intruded gene sequences, and the low transformation efficiency can be underlined as disadvantages of this method.

Current tools to assess the food safety of GE crops include extensive multisite and multiyear agronomic evaluations, compositional analyses, animal nutrition, and classical toxicology evaluations. In the 2000s, new methodologies were developed to allow, in theory, a holistic search for alterations in GE crops at different biological levels (transcripts, proteins, metabolites). These methodologies include cDNA microarrays, microRNA fingerprinting, proteome, metabolome, and toxicological profiling. The term—omics in relation to food and feed safety appeared for the first time in 2005 (Li et al. 2005). This review highlights the knowledge generated by recently published profiling studies regarding the effect of genetic modification itself, compared with environmental and intervariety variation, for major crops (44 studies) and for *Arabidopsis* (*Arabidopsis thaliana*) as a reference plant.

It should be mentioned that as far back as 1987, a report by the National Academy of Science (entitled Introduction of Recombinant DNA-Engineered Organisms into the Environment) had already stated that—there is no evidence that unique hazards exist in the use of recombinant DNA techniques or in the transfer of genes between unrelated organisms and—that the risk[s] are the same in kind as those associated with other genetic techniques.

Genetic Enrichment of Cereal Crops Via Alien Gene Transfer

Cereal crops were initially difficult to genetically engineer, mainly due to their recalcitrance to in vitro regeneration and their resistance to *Agrobacterium* infection. Herbicide and/or insect resistance are the traits that are often controlled by a single gene. However, in recent years, more complex traits, such as dough functionality in wheat and nutritional quality of rice, have been improved by the use of biotechnology. The current challenges for genetic engineering of plants will be to understand and control factors causing transgene silencing, instability, and rearrangement which are often seen in transgenic plants and are highly undesirable in lines to be used for crop development. Further improvement of current cereal cultivars is expected to benefit greatly from information emerging from the areas of genomics, proteomics, and bioinformatics.

However, it was soon realized that cereal transformation was problematic; in general, monocot cells and tissues were relatively recalcitrant to in vitro regeneration and did not respond to *Agrobacterium*-mediated transformation. Furthermore, some of the promoters that exhibited high activity in dicot transformation systems showed a low level of activity in monocot cells and tissues. At present, many of the problems initially encountered during the development of genetic transformation systems for cereals have been overcome, and transgenic rice, maize, wheat, and barley are now routinely produced in several laboratories.

The increased transformation frequencies for cereals have mainly been the result of systematic screenings of genotypes and explant tissues for suitability in transformation and regeneration systems,

- Improvement of direct DNA delivery systems, such as particle bombardment, and identification of useful scorable and selectable marker genes
- Adaptation of the *Agrobacterium*-mediated transformation system to cereals
- Reduced somaclonal variation by shortening the tissue culture period
- Optimization of codon usage, transcriptional, and translational signals to fit the monocot system

Wide Hybridization and Chromosome Manipulation

Wild relatives of cereal crops have genes for superior traits, some of which have been incorporated into these crops via interspecific and intergeneric hybridization (Jauhar and Chibbar 1999). Chromosome pairing between chromosomes of the alien donor and those of cereal crops is the key to such gene introgressions.

Cereal Transformation

Successful production of transgenic plants depends on the characteristics of the gene expression cassettes, the method chosen to deliver the DNA into recipient cells and the tissue culture, and selection techniques used to regenerate fertile plants from the transformed cells. In order to prove the transgenic nature of a plant, certain criteria must be fulfilled (Potrykus 1990). These qualifying factors require a detailed genetic phenotypic and molecular characterization of the primary transformant and its offspring. Stable transmission and expression of the transgenic phenotype in subsequent generations are not always achieved, but this represents an essential factor for development of commercial cultivars.

In Vitro Regeneration of Cereals

Protoplasts prepared from immature embryo-derived suspension cultures of pearl millet were among the first cereal cells shown to possess totipotency (Vasil and Vasil 1980).

Subsequently, regeneration protocols for immature embryos, young inflorescences, young leaves, anthers, microspores, and shoot apices were developed (Vasil 1987). Successful production of transgenic plants from callus derived from these tissues has been demonstrated for oat (Gless et al. 1998). Furthermore, a transformation system based on shoot meristem cultures, induced to produce axillary shoot meristems, has allowed recovery of transgenic plants of commercial cultivars of oat and barley (Zhang et al. 1999). The promising results from experiments using mature or germinating seeds as starting explant material indicate that these explant systems may soon replace those using immature tissue material.

Protoplasts prepared from immature embryo-derived suspension cultures of pearl millet were among the first cereal cells shown to possess totipotency (Vasil and Vasil 1980).

Subsequently, regeneration protocols for immature embryos, young inflorescences, young leaves, anthers, microspores, and shoot apices were developed (Vasil 1987). Successful production of transgenic plants from callus derived from these tissues has been demonstrated for oat (Gless et al. 1998). Furthermore, a transformation system based on shoot meristem cultures, induced to produce axillary shoot meristems, has allowed recovery of transgenic plants of commercial cultivars of oat and barley (Zhang et al. 1999). The promising results from experiments using mature or germinating seeds as starting explant material indicate that these explant systems may soon replace those using immature tissue material.

Growth of Donor Plants

- Wheat (*Triticum aestivum* L. var. bob white) plants are grown at 18 °C day and 14 °C night temperatures under a 16 h photo period (HQI lamps, 400W) in walk-in growth room.
- Collection and sterilization of wheat caryopses.
- Collect spikes from growth room plants. Embryos at the correct stage are usually found 12–14 days post anthesis.
- Remove the caryopses sterilize by rinsing in 70% ethanol for 5 min then soak for 15 min in (3% active chlorine) Na-hypochlorite for 20 min under gentle shaking
- Rinse the caryopses 3–4 times with sterile water.

Coating of gold particles with DNA (for six shots):

- Place 20 mg gold particles (0.6–1 µm) and 1ml 100% ethanol into a 1.5 ml Eppendorf. Sonicate for 2 min, pulse spin in a microfuge for 3 s, and remove the supernatant. Repeat this ethanol and wash twice more.

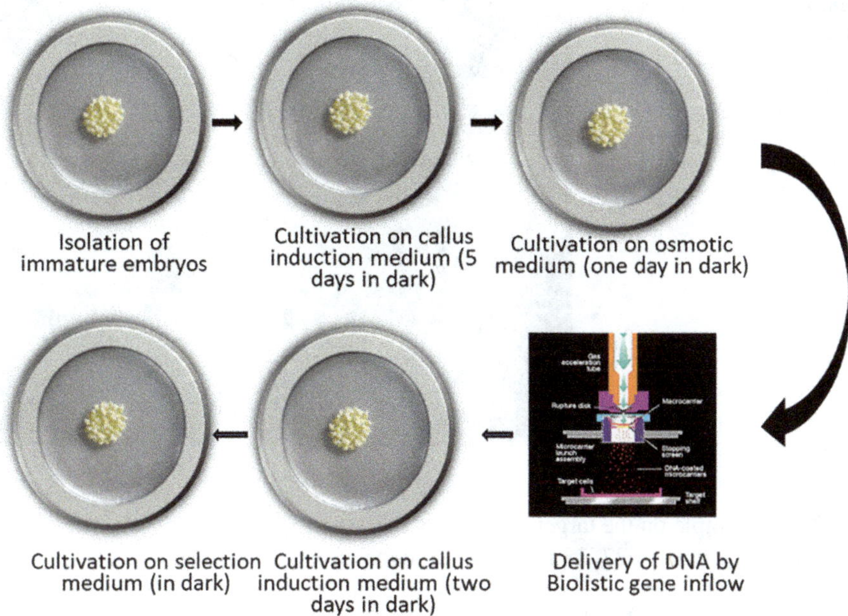

Fig. 13.9 Flow sheet indicating transformation by particle bombardment

- Add 1 ml sterile water, sonicate for 2 min, pulse spin in a microfuge for 3 s, and remove the supernatant. Repeat this step.
- Resuspend gold in 1 ml sterile water and divide into 50 µl aliquots.
- Add 5 µl DNA (1 µg/µl) vortex briefly.
- Place 50 µl 2.5 M $CaCl_2$ and 20 µl 0.1 M spermidine and vortex into the gold pellet the DNA-coated particles by centrifugation and discard the supernatant.
- Add 150 µl 100% ethanol to wash the particles and resuspend.
- Again, pellet the particles and discard the supernatant. Resuspend fully in 85 µl 100% ethanol and maintain on ice.

Delivery of DNA-Coated Gold Particles (Fig. 13.9)
The following settings are recommended for this procedure:

- Sterilize the gun's chamber and component parts with 70% (v/v) ethanol.
- Sterilize macro-carrier holders, macro carriers, stopping screens, and rupture discs by dipping in 100% ethanol and allow evaporating completely.
- Briefly vortex the coated gold particles, take 5 µl, place centrally onto the macro-carrier membrane, and allow drying.
- Place a stopping screen into the fixed nest. Invert the macro-carrier holder containing macro-carrier + gold particles/DNA and place over the stopping screen in the nest and maintain its position using the retaining ring. Mount the fixed nest assembly onto the second shelf from the top to give a gap of 2.5 cm.

Fig. 13.10 *GFP* expression in wheat embryos. Gold particles were coated with construct coating GFP gene sequences under control of 35 promoter. The picture was taken 48 h after particle bombardment

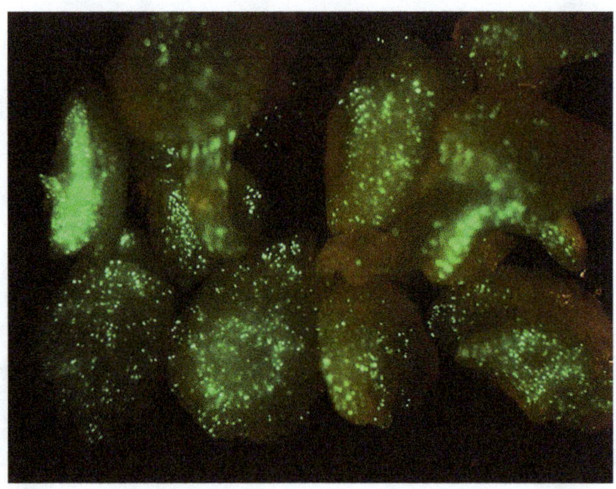

- Place a sample on the target stage (target distance 5.5 cm).
- Draw a vacuum of 27–28 in. Hg and fire the gun.

The efficient of DNA coating and firing the coated particle into wheat embryos is demonstrated in Fig. 13.10.

Comparison of Biolistics and *Agrobacterium*-Mediated Transformation Methods

Agrobacterium-based systems and biolistic methods share common features. The same explant tissues can be used as targets, the transformation frequencies are comparable (Stoger et al. 1998), and both techniques are genotype dependent (Lee et al. 1999).

However, delivering DNA using *Agrobacterium* as a vehicle has one major advantage over direct delivery techniques, as it allows insertion of a well-defined DNA fragment into the host genome. Some transgenic plants produced from bombarded explants have been reported to have very high transgene copy numbers and complex integration patterns (Register et al. 1994). However multiple copy inserts are undesirable in transgenic plants as these clustered inserts may increase the risk of gene silencing and DNA rearrangements (Kumpatla et al. 1997; Kumpatla and Hall 1998; Pawlowski and Somers 1998). In attempts to achieve simpler integration patterns, de Block et al. (1997) reported some success by expressing a nuclear-associated protein inhibitor. Another attempt to achieve simpler integration patterns was reported by Hansen and Chilton (1996), who developed a hybrid method between biolistics and *Agrobacterium*-mediated transformation, termed agrolistics. In their system, DNA to be delivered by particle bombardment is flanked by T-DNA border sequences and co-transferred with genes encodingVirD1 and VirD2, to simulate the *Agrobacterium*-mediated T-DNA integration into the genome.

Transgene rearrangements were initially reported to be more frequent in plants derived from biolistic methods than in plants derived from *Agrobacterium*-mediated transformations (Songstad et al. 1995). However, a recent study showed no difference in occurrence of transgene rearrangement or transgene integration pattern in plants produced by either method (Kohli et al. 1999). Although *Agrobacterium*-based DNA delivery techniques once appeared as the method of choice (Komari et al. 1998), their so-called superiority over direct delivery techniques remains to be demonstrated by studies on large populations of transgenic plants.

However, *Agrobacterium*-mediated transformation has remarkable advantages over these direct transformation methods. It reduces the copy number of the transgene, potentially leading to fewer problems with transgene co-suppression and instability (Koncz et al. 1994; Hansen et al. 1997). In addition, it is a single-cell transformation system and usually does not form chimeric plants, which are more frequent when direct transformation is used (Enríquez-Obregón et al. 1997, 1998). Two barley transformation systems *Agrobacterium* mediated and particle bombardment were compared in terms of transformation efficiency, transgene copy number, expression, inheritance, and physical structure of the transgenic loci, using fluorescence in situ hybridization (FISH). The efficiency of *Agrobacterium*-mediated transformation was double that obtained with particle bombardment. Whereas 100% of the *Agrobacterium*-derived lines integrated between one and three copies of the transgene, 60% of the transgenic lines derived by particle bombardment integrated more than eight copies of the transgene. In most of the *Agrobacterium*-derived lines, the integrated T-DNA was stable and inherited as a simple Mendelian trait. By contrast, transgene silencing was frequently observed in the T1 populations of the bombardment-derived lines (Travella et al. 2005). Because of genotype-dependent transformation by *A. tumefaciens*, however, the use of ballistic methods for recalcitrant genotypes is sometimes essential.

Still, as mentioned above, the advantages of *Agrobacterium*-mediated transformation of plant tissue are generally a low transgene copy number, minimal rearrangements, and higher transformation efficiency than for the direct DNA delivery techniques (Gelvin 1988; Pawlowski and Somers 1996).

Polyethylene Glycol (PEG)
The plant cell wall forms a major obstacle to transformation, so early direct transformation methods relied on protoplasts (plant cells with the cell wall removed typically by enzymatic digestion). Protoplast technology was initially restricted to certain dicotyledonous plants but subsequently became feasible for important cereal crops (Hooykaas 2010).

A number of methods have been developed for transformation of protoplasts (reviewed in Rivera et al. 2012), most of these being variations of methods used for animal cell transformation. For reasons of low cost and ease of use, the two most commonly used methods are chemical treatment with polyethylene glycol (in combination with calcium ions) and electroporation. Both of these methods change the permeability of the cell membrane, allowing entry of exogenous DNA. The use of protoplasts for transformation requires elaborate tissue culture steps to regenerate fertile adult plants and may introduce genetic instability and resulting somaclonal variation.

Other Transformation Methods

To date, particle bombardment and *Agrobacterium*-mediated DNA delivery remain the most commonly used techniques for transformation of cereals. However, transgenic cereal plants can also be obtained by electroporation of DNA into rice suspension-cultured microcolonies (Arencibia et al. 1998; Kim and Choi 1998), maize immature embryos or callus (D'Halluin et al. 1992), wheat scutella, and tritordeum inflorescence explants (He and Lazzeri 1998). Silicon carbide whisker-mediated techniques have also been successfully used to transform maize cell suspension cultures (Frame et al. 1994) and rice mature scutellar tissues (Matsushita et al. 1999). Besides this transfer of DNA into plant cells by imbibition, microinjection, macroinjection, electrophoresis, ultrasound treatment, laser treatment, or via pollen tube (reviews by Songstad et al. 1995; Barcelo and Lazzeri 1998) are other potential methods for cereal transformation. Some of these simple techniques require very little equipment and could have broad applications if developed.

The *Agrobacterium* virulence proteins VirD2 and VirE2 perform important functions for the transport into nuclei. The reconstituted complexes consisting of the bacterial VirD2, VirE2, and single-stranded DNA (ssDNA) in use in vitro for import into HeLa cell nuclei (Ziemienowicz et al. 1999) are inefficient. The import of ssDNA requires both VirD2 and VirE2 proteins, and a VirD2 mutant lacking its C-terminus nuclear localization signal is inefficient. The system described above was efficiently performed by plant cells. Here it was shown that VirE2 can be a protein without ssDNA, to be able to be imported into cell nuclei. The smaller ssDNA required only VirD, whereas the import of longer ssDNA additionally required VirE2. RecA, another ss DNA-binding protein, could substitute for VirE2 in the nuclear import of T-DNA, but not in earlier events of T-DNA transfer to plant cells (Ziemienowicz et al. 2001).

13.7 *Agrobacterium*-Mediated Transformation in Mono and Dicotyledonous Plants

The stable genetic transformation of major cereal crops, such as specific varieties of maize, rice, wheat, and barley, has changed from being solely laboratory vehicles for basic research to providing new varieties grown in large areas throughout the world (Casas et al. 1993; Dai et al. 2001; Hiei et al. 1997; Ritala et al. 1994; Somers et al. 1992; Tingay et al. 1997; Vasil et al. 1991; Zhao et al. 2000).

13.7.1 Generation of Transgenic Wheat Plants

Hayta et al. (2019) described an efficient and reproducible *Agrobacterium-mediated* transformation method for generation of hexaploid wheat (*Triticum aestivum* L.) plants using sculteum of wheat immature embryo cultivar. Wheat immature caryopes were collected from spikes at the early milk stage GS73 (Zadoks et al. 1974) and immature embryos 1–1.5 mm in diameter were separated under sterile condition and inoculated with *Agrobacterium tumeation* containing binary vector with GUS gene and control of rice ubiquitin 3 promoter (Fig. 13.11, see also Sect. 13.7.2). The transformation efficient was evaluated by means of GUS expression in transformed lines in Fig. 13.12.

13.7.2 Generation of Transgenic Barley Plants

Recent work on *Agrobacterium*-mediated genetic transformation of monocotyledonous plant species has focused on the use of the so-called super-binary vector systems, i.e., binary vectors carrying a DNA fragment from the *A. tumefaciens* virulence region (Komari et al. 1996; Torisky et al. 1997).

Using highly virulent strains of *Agrobacterium* and improved vectors, much progress in the transformation efficiency of cereals has been made (Tingay et al. 1997; Wang et al. 1998, 2001; Murray et al. 2001; Rasci-Gaunt et al. 2001; Wu and Lin 2003). Nowadays, quite routinely applied transformations of cereals are still rather time-consuming and cost-intensive. For transformation, the strongly constitutive ubiquitin and CaMV35S promoters are used. At present, a shortage exists of suitable regulatable promoters for the expression of defense-associated genes (Stuiver and Custers 2001).

Fig. 13.11 (a–c) Selection of wheat spikes and immature embryos at the correct stage, (d) isolated immature embryo, (e) immature embryos in Eppendorf tube containing 1 ml co cultivation media, (f) immature embryos on co-cultivation medium, (g) scutelum with the embryonic axis removed before transferring to resting medium, (h) callus induction on Selection 1 and Selection 2 media, (i) transformed callus starting to green and produce small shoots, (j) regenerated shoots with visibly strong roots, (k)putative transgenic wheat plant transferred to culture tube showing strong root system in hygromycin containing medium, (l) transgenic wheat plants before transferring to soil (modified after Hayta et al. 2019)

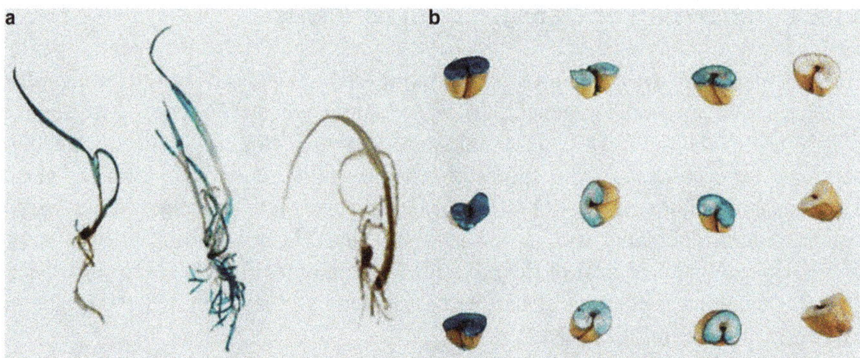

Fig. 13.12 (a) Transgenic plants showing GUS expression after being on rooting medium and Fielder (non-transformed) tissue culture control on the right. (b) Segregating T1 generation seeds showing GUS expression in three independent transgenic lines (Source: Hayta, S, Smedley, M.A, Demir, S.U. et al. An efficient and reproducible *Agrobacterium*-mediated transformation method for hexaploid wheat (*Triticum aestivum* L.) Plant Methods 15, 121 (2019) https://doi.org/10.1186/s13007-019-0503-z Open access This article is distributed under the terms of the Creative Commons Attribution 4.0 International License (http://creativecommons.org/licenses/by/4.0/), which permits unrestricted use, distribution, and reproduction in any medium, provided you give appropriate credit to the original author(s) and the source, provide a link to the Creative Commons license, and indicate if changes were made. The Creative Commons Public Domain Dedication waiver (http://creativecommons.org/publicdomain/zero/1.0/) applies to the data made available in this article, unless otherwise stated)

For constitutive overexpression and for tagging expression, we cloned a functional cDNA fusion of the green fluorescent protein (GFP) and HvBI-1 into the binary vector pLH6000 (http://www.dna-cloningservice.de/lh-vectors.htm; DNA Cloning Service, Hamburg, Germany), maintaining the original cauliflower mosaic virus 35S promoter (CaMV35S). For barley transformation, pLH6000 CaMV35S:: GFP, as well as pLH6000 CaMV35S:GFP-HvBI-I were then introduced into *A. tumefaciens* strain AGL1 (Lazo et al. 1991) by Gene Pulser Xcell™ Electroporation Systems (Bio Rad, Germany).

GFP-BI-1 fusion protein accumulates in nuclear membranes (Matthews et al. 2001; Deshmukh et al. 2006).

The barley cultivar Golden Promise was grown in a growth chamber or in the greenhouse at 22 °C, 60% relative humidity, and a photoperiod of 16 h (150 μmol/ s per m^2 photon flux density). Stable genetic transformation of barley was performed as described by Tingay et al. (1997; see flow sheet in Fig. 13.13, and Table 13.1). Twelve to 14 days after anthesis, immature kernels were surface sterilized for 3 min with 70% ethanol, and 20 min with a sodium hypochlorite solution containing 3% active chlorine, followed by rinsing 3 times with sterile distilled water. Excised immature embryos were then infected with *A. tumefaciens*. The callus induced grew on.

Murashige and Skoog medium containing 50 mg hygromycin B/liter (Roche, Germany). Established calli were then subcultured on regeneration medium

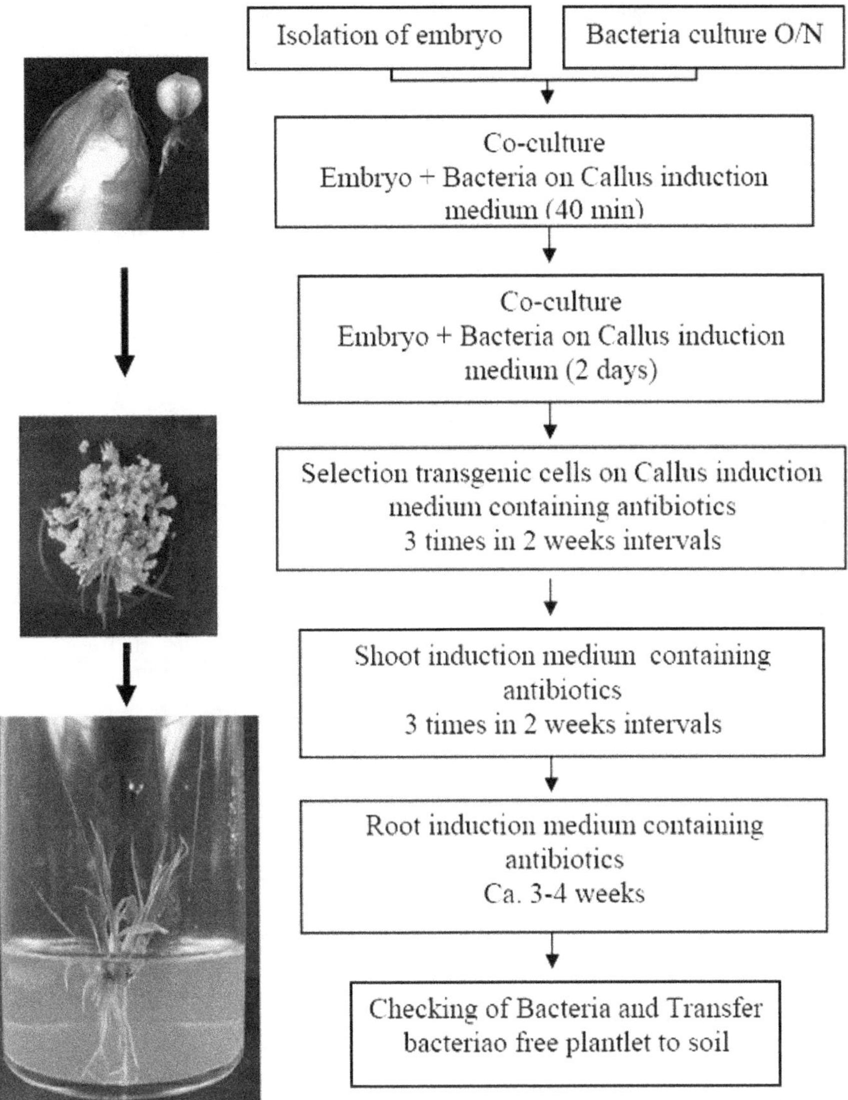

Fig. 13.13 Flow sheet indicating *Agrobacterium*-mediated genetic transformation of barley immature embryo and their regeneration into the whole plant

supplemented with 25 mg hygromycin B/l, until rooted plantlets could be transferred to soil. Timentin (150 mg/L) was applied until tests for the presence of bacteria proved negative. The GFP reporter was visualized with either a standard fluorescence microscope or the confocal laser scanning microscope TCS SP2 AOBS (excitation: laser line 488 nm, emission: 500–540 nm; Leica Microsystems, Bensheim, Germany; Schultheiss et al. 2005; Deshmukh et al. 2006; Fig. 13.14).

Table 13.1 Different culture media used for barley transformation

Component	Amount
Barley callus induction medium (1 l) MS stock (Duchefa M0221) 4.3 g	
$CuSO_4 \times 5H_2O$	1.2 mg (5 µM)
Maltose	30 g
Thiamine-HCl	1 mg
Myo-inositol	250 mg
Casein hydrolysate	1 g
L-Proline	690 mg
Dicamba	2.5 mg
pH: 5.9, filter sterilization	
Phytoagar	5 g
Barley shoot induction medium (1 L)	
MS stock (NH_4NO_3-free)	2.7 g
$CuSO_4 \times 5H_2O$	1.2 mg (5 µM)
NH_4NO_3	165 mg
Maltose	62 g
Thiamine-HCl	0.4 mg
Myo-inositol	100 mg
Glutamine	150 mg
BAP (benzylamine purine)	1 mg
pH: 5.6, filter sterilization	
Phytoagar for liquid medium 5 g Barley root induction medium (1 L)	
MS stock	2.15 g
$CuSO_4 \times 5H_2O$	0.6 mg (2.5 µM)
Maltose	15 g
Thiamine-HCl	0.5 mg
Myo-inositol	125 mg
Casein hydrolysate	0.5 g
L-Proline	345 mg
pH: 5.9, filter sterilization phytoagar for liquid medium 5 g	

13.7.3 Stable Root Transformation System for the Functional Study of Proteins in Barley

Introduction: A major drawback in the functional characterization of plant proteins supporting or restricting microbial root colonization of monocots is the lack of appropriate root transformation systems. In order to study the impact of barley proteins on the outcome of mutualistic and pathogenic root–microbe interactions, a reliable and fast stable root transformation system (STARTS) for barley roots was established (Imani et al. 2011). They presented a suitable method for stable root transformation of barley, which is as robust as conventional transformation methods but is highly time efficient in the generation of transformed roots (Fig. 13.15).

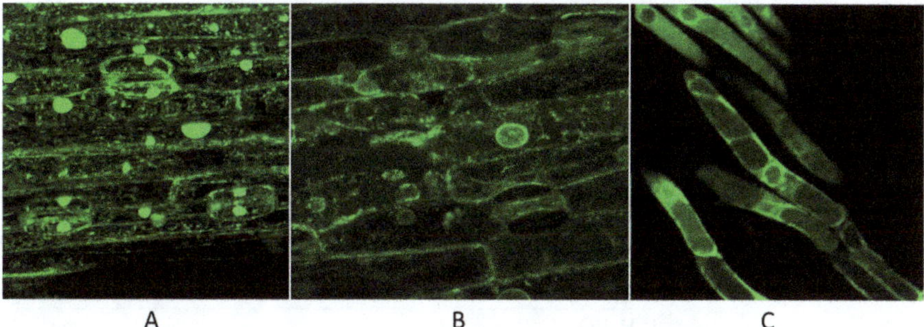

Fig. 13.14 (**a**) GFP protein accumulates mainly in the nucleus of epidermis leaf cells of barley. (**b**) GFP-BI-1 fusion protein accumulates in the nuclear membrane of epidermis leaf cells. (**c**) GFP-BI-1 fusion protein accumulates in the nuclear membrane of root cells of barley

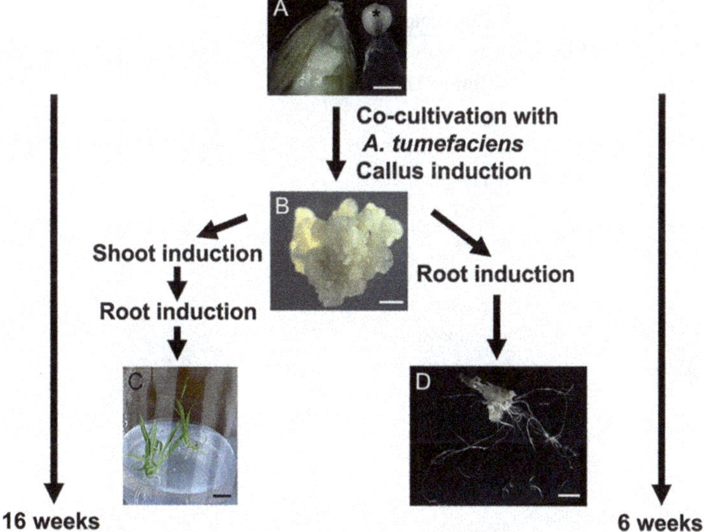

Fig. 13.15 Comparison of the conventional transformation method and the stable root transformation system (STARTS) scutella (*) are isolated from immature embryos without the embryo axis (**a**) (see Fig. S1 for details on the preparation of scutella) and co-cultured with *Agrobacterium tumefaciens* for transformation. Calli are regenerated from transformed scutella cells. (**b**) For conventional transformation, calli are subjected to shoot and thereafter to root induction medium to obtain T0 plants (**c**) For STARTS, calli (**b**) are subjected to root induction medium to regenerate roots from individual, transformed cells (**d**) Transformed roots are obtained within 6 weeks in comparison with the conventional transformation procedure by which heterozygous root material is obtained after 16 weeks. Notably, each transgenic root regenerated by STARTS represents an independent integration event. The multiplicity of transgenic events and, as a result, regenerated roots diminishes the occurrence of integration-associated phenotypes during functional studies. A higher amount of heterozygous roots regenerated from single shoots by conventional transformation approaches are required to obtain an equally confident result

Table 13.2 Composition of barley root induction medium (1 L)

$CaCl_2 \times 2H_2o$	295 mg
KH_2PO_4	170 mg
KNO_3	2200 mg
$MgSo4 \times 7H_2O$	310 mg
$NaH_2PO_4 \times H_2O$	75 mg
$(NH_4)_2SO_4$	67 mg
NH_4NO_3	600 mg
$CoCl_2 \times 6H_2O$	0.025 mg
$CuSO_4 \times 5H_2O$	0.025 mg
H_3BO_3	3 mg
$MnSO_4 \times H_2O$	5 mg
$Na_2MoO_4 \times 2H_2O$	0.25 mg
$ZnSO_4 \times 7H_2O$	5 mg
Fe-citrat $\times 5H_2O$	20 mg
FeEDTA	28 mg
Nicotinamide	1 mg
Pyridoxine HCL	1 mg
Thiamine HCL	10 mg
Arginine	25 mg
Asparagine	50 mg
Asparaginic acid	30 mg
Glutamine	120 mg
Proline	50 mg
Threonine	25 mg
Pepton from casein	125 mg
Myo inositol	100 mg
Liquid endosperm of coconut fruits (LE)	25 ml
Glucose	7 g
Sucrose	20 g
Charcoal[a]	1 g

[a]Added after adjustment of pH 5.3

Media was filter sterilized and 6 g L^{-1} autoclaved phyto agar (Duchefa, the Netherlands) was added

The barley scutellum tissue was separated from immature embryo of cv. Golden Promise. For transformation the scutellum was immediately co-cultivated with *Agrobacterium tumefaciens* containing desired genes. Two days after transformation, the scutellum was transferred into barley callus induction medium (Tingay et al. 1997) including the antibiotics. 2–3 weeks after co-cultivation calli were formed from the transformed cells; then they were transferred into the modified root induction medium (modified after Jensen 1983; Table 13.2) containing a charcoal concentration of 1 g/L, and roots were developed within 2 weeks.

This system can be used for overexpression of desired gene in barley roots. This is demonstrated for of the *green fluorescent protein (GFP)* using transformation vector pLH6000-Ubi::GFP (Fig. 13.16). The GFP overexpression was obtained in barley

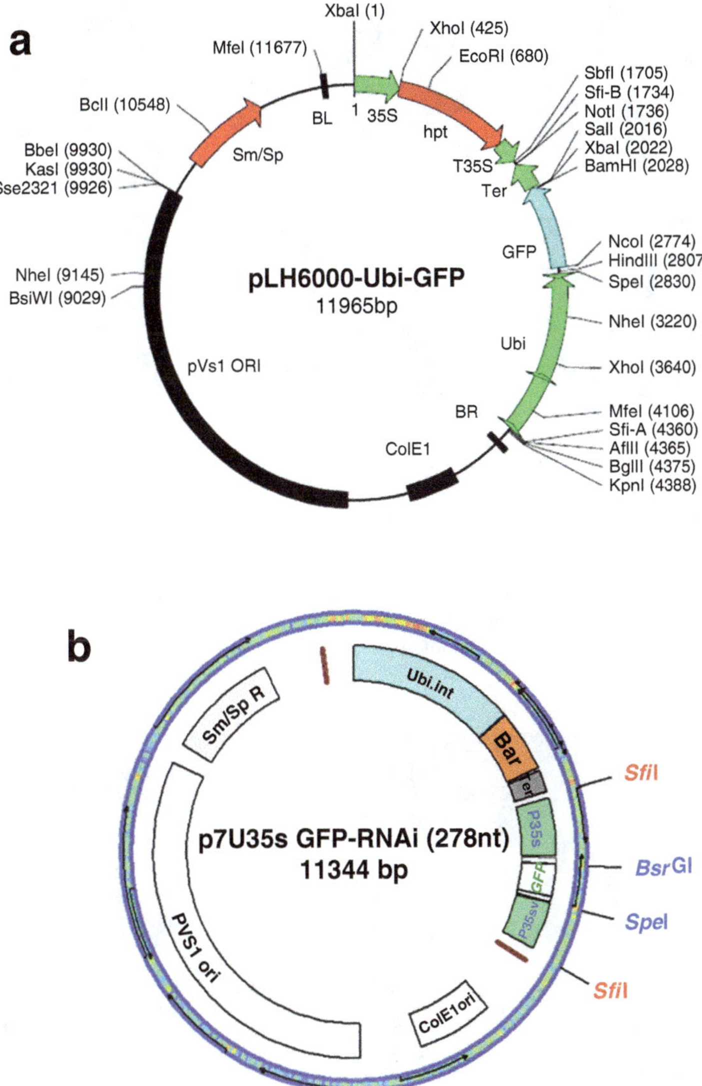

Fig 13.16 Maps of plant transformation vectors (DNA Cloning Service Hamburg, Germany). (**a**) vector for GFP overexpression in plant, (**b**) vector contains inverted repeat promoter for generation of dsGFP-RNAi in plant

roots (Fig. 13.17). (a) Scutellum was separated without the embryo axis. (b) 2 days after transformation, GFP expression is observed partially in the scutellum cells. 1–2 weeks after co-cultivation, expression of GFP was observed in the dividing scutellum cells. (c) 2–3 weeks later callus was formed from the transformed cells. (d) Root development on rooting medium. (e) Excitation with UV light (GFP2 filter) under the fluorescence microscope exhibited GFP uniform green fluorescence from all roots emerging from a single callus.

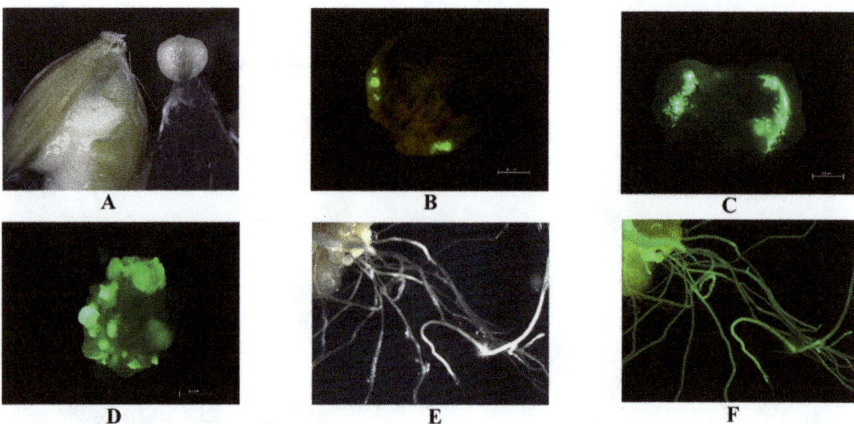

Fig. 13.17 Overexpression of synthetic green fluorescent protein (sGFP) in barley roots by STARTS (Fig. 13.15) (**a**) Scutellum separated without the embryo axis 2 days after transformation by co-cultivation with *Agrobacterium tumefaciens*. (**b**) Fluorescent image of (**a**) showing sGFP expression in cells of the scutellum. (**c**) Expression of sGFP is observed in dividing scutellum cells by 1–2 weeks after co-cultivation. (**d**) Callus is formed from sGFP-transformed scutellum cells by 2–3 weeks after co-cultivation. (**e**, **f**) sGFP expressing roots regenerated from single sGFP-transformed callus cells. sGFP was visualized by excitation at 450–490 nm, and emission was detected at 510–530 nm using a stereofluoresence microscope (Leica, Germany, http://www.leica-microsystems.com) Bars = 2 mm (**a–d**) and 2 cm (**e**, **f**)

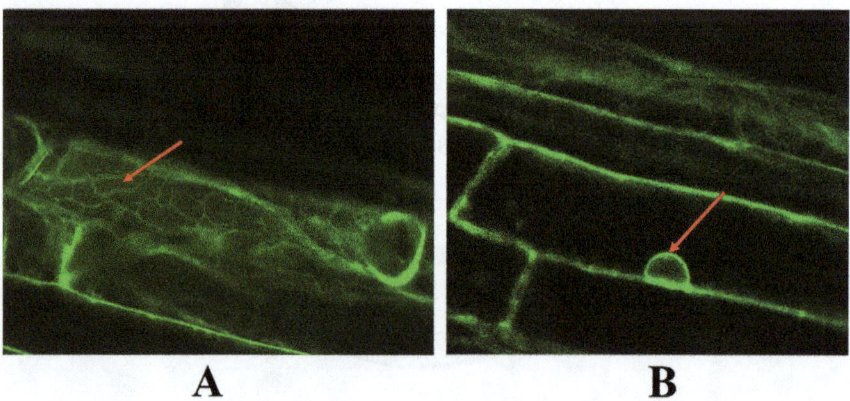

Fig. 13.18 mGFP-ER protein localization in the stably transformed barley roots. (**a**) mGFP-ER protein localization in the ER; (**b**) mGFP-ER protein localization in the nuclear envelope

Furthermore, the STARTS was proven successfully for protein localization studies by overexpressing *GFP* fused to an ER retention signal. The ER localization was subsequently determined by confocal laser scanning microscopy (Fig. 13.18). For more details, see Imani et al. (2011).

Because of shortening the time required for root transformation, the STARTS system allows us to screen a higher number of candidate genes before investing the time to grow whole plants for screening resistance. It is will also reveal more information about root defenses to soilborne fungal pathogens (Imani et al. 2011; Okubara 2016; Okubara and Imani unpublished).

13.7.4 Carrot Cell Culture and Transformation Procedures

The *Agrobacterium*-mediated transformation protocols differ from one plant species to the other, and within species, from one cultivar to the other. Therefore, the optimization of *Agrobacterium*-mediated transformation methodologies requires the consideration of several factors that can be determined in the successful transformation of one species. The optimization of *Agrobacterium*-plant interaction in competent cells from different tissues and the development of a suitable method for regeneration from transformed cells are needed. In this book, we present the carrot as model for genetic manipulation in a successful establishment of a routine protocol for *Agrobacterium*-mediated transformation of dicotyledons. Here, we have optimized the integration rate of a desired gene into the plant genome.

The basis for a successful production of transgenic carrot plants, as an example for dicots, has been formed by lengthy experience in the field of somatic embryogenesis and cell cycle synchronization, as well as extensive work on the differentiation and DNA organization of this species. One example will be the integration of a surface protein of hepatitis B (SHBs), another the integration of a second gene for phosphoenolpyruvate carboxylase (PEPCase; Sect. 9.2.2), *Hordeum vulgare* Bax inhibitor 1 (HvBI-1), and some attempts to target foreign genes (ROL genes; see Sect. 7.3) for integration into the genome and gene studies related to plant pathogen resistance.

The B5 medium (Gamborg et al. 1968; Sect. 3.4) for carrot cell suspensions originally derived from cultured petiole explants was supplemented with 2.26 μM 2.4D (dichlorophenoxy acetic acid) to promote cell division. For synchronization of the cell cycle, an FDU/thymidine system was employed, as described earlier (see Sect. 7.3 and Fig. 12.1; Blaschke 1977; Blaschke et al. 1978; Neumann 1995; Froese and Neumann 1997). By these means, 80–90% of the dividing cultured carrot cells can be synchronized. In short, to arrest cycling cells at the transition from Gl- to the S-phase, fluorodesoxyuridine (FDU, 10^{-6} M) was added to the nutrient solution for 24 h to inhibit the synthesis of thymidine and consequently of DNA replication. Thereafter, the cells were transferred to a fresh nutrient medium supplemented with 10^{-5} M thymidine to initiate the transition of the synchronized cultures from the late G1-phase into the S-phase for DNA replication.

In order to make use of an advantageous preferential integration of foreign DNA into replicating DNA, a suspension of *A. tumefaciens* containing the plasmid pGE2 with the GUS construct (Fig. 13.19) was simultaneously supplied to the synchronized carrot cell suspension, together with thymidine. After 48 h of co-culture, cefotaxime (500 μg/ml) was added to remove excess *Agrobacterium*

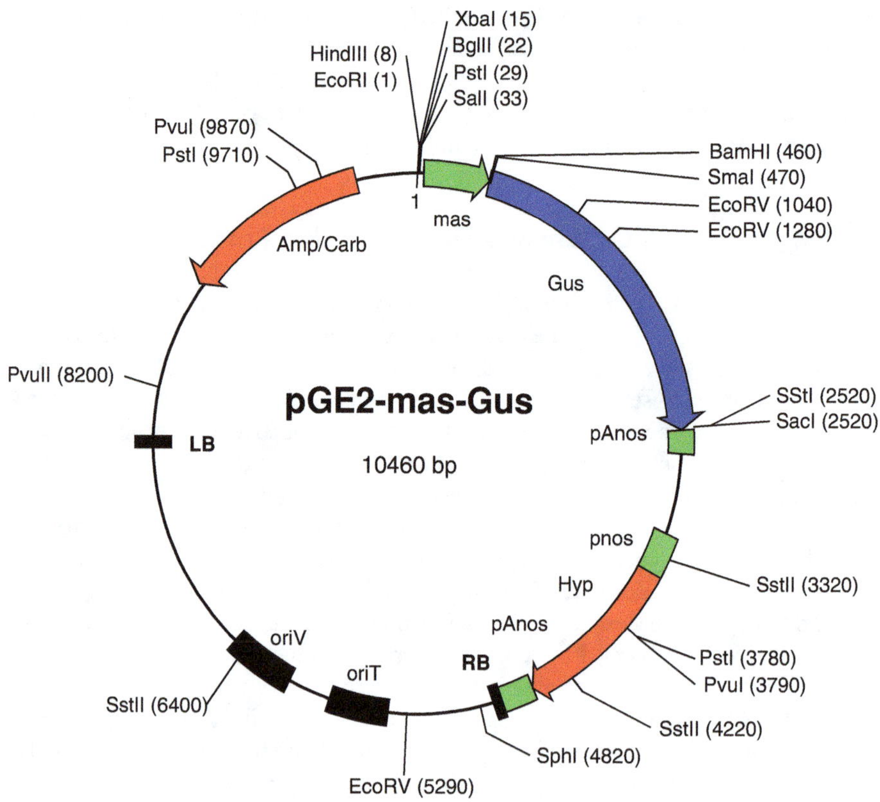

Mas =	Manopine synthase promote
pAnos =	Poly(A) tail (termiater) of nopaline systase gene
Pnos =	Nopaline synthase promoter
GUS =	Reporter gene (glucuronidase)
Hyp =	Hygromycin resistant gene
RB =	Right border
LB =	Left border

Fig. 13.19 Schematic representation of pGE2 MAS::GUS binary vector. Downstream of the auxin-inducible promoter MAS, the reporter gene GUS (glucuronidase) was inserted

(Fig. 13.20). After transformation, the cells were either subcultured in a B5 medium, again with 2.26 μM 2.4D for proliferation, or transferred into a hormone-free medium to enable the development of somatic embryos (Li and Neumann 1985). The germinated embryos were raised to intact plants growing in soil. Integration of foreign genes into the carrot genome was confirmed by genomic Southern blotting (Imani et al. 2002).

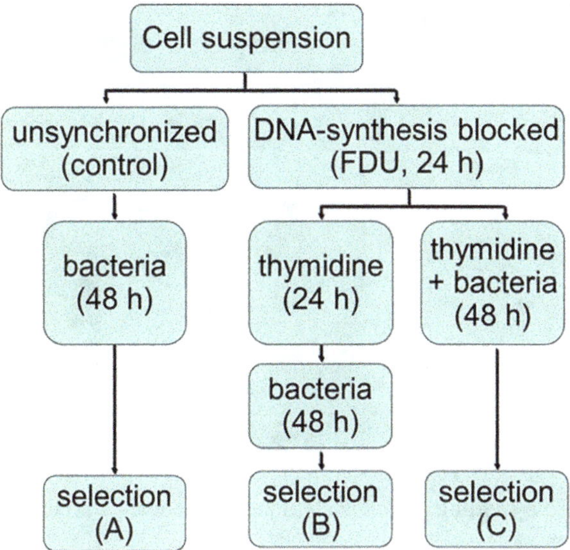

Fig. 13.20 Flow sheet indicating the integration of cell cycle synchronization in the experiments on genetic transformation

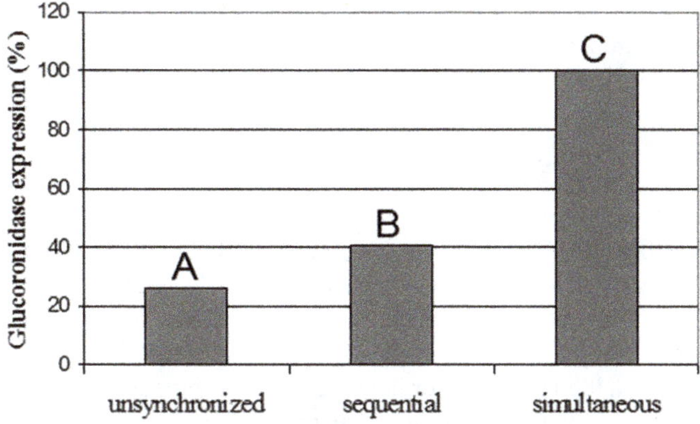

Fig. 13.21 Influence of cell cycle synchronization (FDU/thymidine system) on the expression of foreign gene (MAS::GUS). As indicator, X-Gluc was used as substrate. (**a**) unsynchronized control; (**b**) application of *A. tumefaciens* containing the construct 24 h after thymidine supplement; (**c**) simultaneous application of *A. tumefaciens* containing the genetic construct together with thymidine to initiate the S-phase (equals 100%). The expression of beta-glucuronidase was quantified by a densitometric procedure with the software program SIS

The integration efficiency for GUS was determined by the X-Gluc reaction, resulting in the production of a blue color proportional to the number and expression activity of transformed cells. Comparing the intensity of the color in the cultures shown in Fig. 13.21, it is evident that the integration of the MAS::GUS constructs is

Fig. 13.22 Somatic embryos derived from stable transformed carrot cells. The blue dye indicates the auxin distribution in embryos during their development

greatly enhanced by adding *A. tumefaciens* simultaneously with thymidine to initiate DNA replication.

Adding the bacteria 24 h after the application of thymidine results in a markedly reduced reaction, indicating a lower level of transformation, probably due to loss of cell cycle synchronization, and a reduced number of cells passing through the S-phase. Quantification of transgene expression with densitometric techniques shows a fourfold increase in gene expression of synchronized cell cultures, compared to unsynchronized cultures (Figs. 13.20 and 13.21).

In such transformed cell suspensions, the development of somatic embryos (Fig. 13.22) was initiated. The embryos were then raised to intact plantlets and transferred to soil in the greenhouse.

After approximately 3 months of culture in soil, in various parts of the plants, a blue color evidently indicated the long-lasting expression of β-glucuronidase. Optimization of transformation and protein extraction techniques in our laboratory, followed by the screening of several transgenic cell lines, revealed significant differences in expression levels of proteins of interest. This helped to increase the concentration of SHBs protein to 15 μg/g in carrots (Ph.D. Thesis of H. Lorenz 2006). Thus, transgenic carrots may have a high potential for the production of various oral vaccines.

Based on this method, *Cyclamen persicum* Mill. embryogenic cell suspension cultures were transformed. A high rate of transformation was obtained when *A. tumefaciens* was added concurrently with thymidine (Imani et al. 2007; Fig. 13.23). The regeneration of transformed cyclamen plants via somatic embryogenesis has been described by Winkelmann et al. (2000).

13.7.5 Uses of Transgenes to Increase Host Plant Resistance to Plant Pathogens

As is known, an infection by plant pathogens leads to plant cell death as self-protection (programmed cell death, PCD), which should inhibit the spreading of

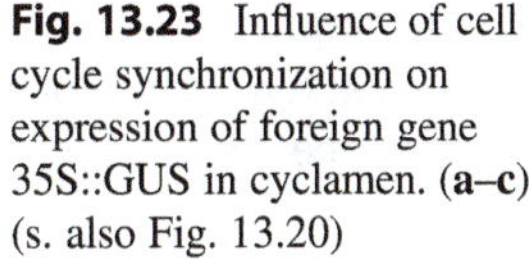

Fig. 13.23 Influence of cell cycle synchronization on expression of foreign gene 35S::GUS in cyclamen. (**a–c**) (s. also Fig. 13.20)

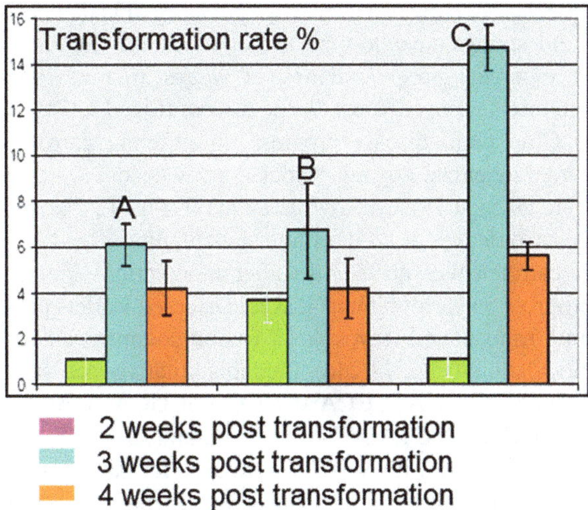

fungi within the plant. PCD is a mechanism to remove aged, unwanted, damaged, or infected cells from multicellular organisms. It is under genetic control and must be tightly regulated to avoid false ultimate decisions and diseases. In the interaction of plants with pathogenic microbes, PCD appears to play different roles depending on the lifestyle of the disease-inducing agent. Among cell death regulator proteins, only a few are structurally and functionally conserved across eukaryotic kingdoms. One of these PCD regulator proteins is the BAX inhibitor-1 (BI-1), an endoplasmic reticulum membrane protein that can suppress diverse kinds of PCD. BI-1 controls heterologous BAX-induced cell death, hypersensitive reaction, and abiotic stress-induced cell death in plants (reviewed by Hückelhoven 2004).

Gray mold caused by *Botrytis cinerea* is a severe disease for many dicotyledonous crop plants, and it also occurs as postharvest disease in carrots. Owing to its necrotrophic lifestyle, *B. cinerea* destroys plant tissue by secretion of hydrolytic enzymes, host nonspecific toxins, and reactive oxygen species (ROS; Von Tiedemann 1997; Gronover et al. 2001; Tenberge et al. 2002; Siewers et al. 2005). The species is difficult to control by chemical means, because pesticide resistance occurs rapidly in fungal populations (Staub 1991). *B. cinerea* might be a target for genetically engineered resistance.

To restrict pathogenesis of the necrotroph *Botrytis cinerea* (carrot root pathogen), we generated the carrot (*Daucus carota* ssp. *sativus*) cultivar Rotin expressing HvBI-1 protein under control of constitutive CaMV35S and inducible MAS promoter (Imani et al. 2006). From the corresponding lines, we regenerated plants by tissue culturing and somatic embryogenesis, under selective conditions according to Imani et al. (2002). For pathogen challenge, excised leaves from 7- to 8-week-old plants were inoculated with agar blocks containing *B. cinerea* (strain 7890).

Symptom development on excised leaves was examined in three independent experiments. Symptom development progressed slowly until 4 days after inoculation. However, runaway spreading of leaflet necrosis and overgrowth by gray mold started on wild-type leaves as of the 5th day after inoculation. Symptom development progressed until 4 weeks after inoculation, when the leaves were necrotic, and overgrown by the fungus (Fig. 13.17).

Gray mold disease progress on different carrot genotypes. (wild-type carrot leaves inoculated by agar blocks (arrowheads) at the central leaf axis. Photographs were taken at 28 days after inoculation (Fig. 13.24a) MAS::GUS-MAS:HvBI-1 line 2, carrot leaves at 28 days after inoculation (Fig. 13.24b) CAMV35S::HvBI-1 line 2, carrot leaves at 28 days after inoculation (Fig. 13.24c). Counting of disease progress on carrot pinnate leaves. Diseased leaflets per composite leaf were counted on 10 inoculated leaves at 14 (white columns) and 21 (gray columns) days after inoculation (Fig. 13.24d). Semiquantitative one-step reverse transcription PCR (primers, 5′-GGATCCCAACGCGAGCGCAGGACAAGC-3′, 5′-GTCGACGCGG TGACGGTATCTACATG-3′, 55 °C annealing temperature) to estimate expression levels of HvBI-1 in excised leaves used for pathogen challenge. RNA was extracted from excised leaves after 5 days on agar plates and DNAse digested. PCR products were separated in gels. Expected product size was 850 bp. rRNA bands are shown to demonstrate the amount of RNA used as template (Fig. 13.24e). Leaves from the same batch as in Fig. 13.24f were used for GUS-activity histochemistry.

Together, HvBI-1 can delay or prevent diseases caused by hemibiotrophic and necrotrophic fungal pathogens in carrots. This supports the assumption that factors promoting biotrophic fungal growth concurrently cause inhibition of fungi with a necrotrophic lifestyle phase. To address the problem that overexpression of PCD inhibitors in plants might confer undesired side effects on biotrophs, one strategy could be expression of BI-1 under control of a necrotrophy-specific and/or a tissue-specific promoter.

13.7.6 Transgenic Carrot: Potential Source of Edible Vaccines

Transgenic plants have a high scientific and economical potential for the production of foreign proteins of biomedical importance. Since the early 1990s, it has been shown that transgenic plants could express viral and bacterial antigens, with preservation of their immunogenic properties. Such plants could therefore serve as an inexpensive vaccine, because it would not be necessary to make a large capital investment into a production facility for a vaccine. These vaccines stimulate the immune response at the mucosal level and thus would be particularly effective against diseases. For example, enough antigens for one dose of hepatitis B vaccine could be produced in unprocessed plant material at a cost of US$ 0.005. This advantage of plant-derived vaccines is important, because it can lead to a much less complicated vaccine development decision-making process. This represents an alternative for the production of vaccines against infectious diseases of worldwide importance, e.g., hepatitis B. Moreover, transgenic plants expressing sufficient levels of antigens bear the potential of oral delivery of antigenic proteins as edible

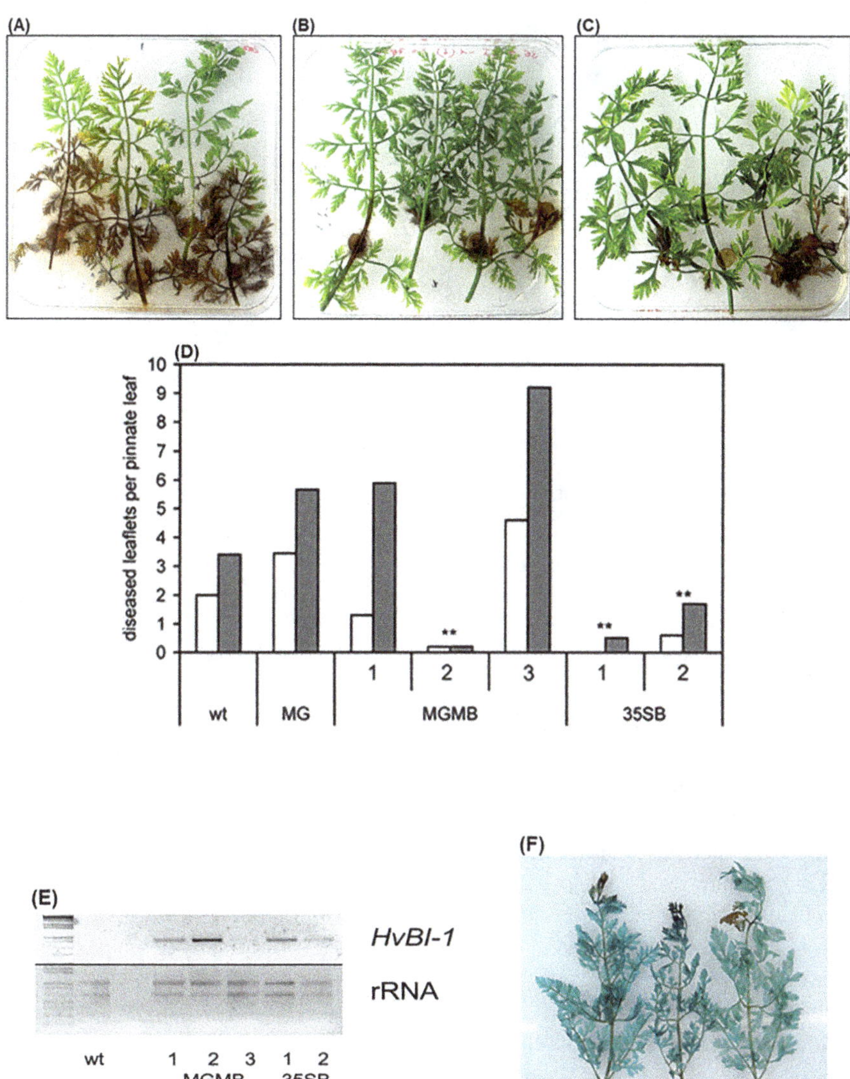

Fig. 13.24 Gray mold disease progress on different carrot genotypes. (**a**) Wild-type carrot leaves inoculated by agar blocks (arrowheads) at the central leaf axis. (**b**) MAS::GUS-MAS:HvBI-1 line 2, carrot leaves. (**c**) CAMV35S::HvBI-1 line 2. (**d**) Counting of disease progress on carrot pinnate leaves. Asterisk mark lines that were significantly different from wild type at both times of evaluation (Student's *t*-test, $P < 0.01$) (**e**). Semiquantitative one-step reverse transcription PCR, (**f**) GUS-activity in leaves of Wt = wild type; MG = MAS::GUS; MGMB = MAS::GUS-MAS: HvBI-1; 35SB = CAMV35S::HvBI-1; Numbers 1, 2, and 3 designate independent transgenic lines

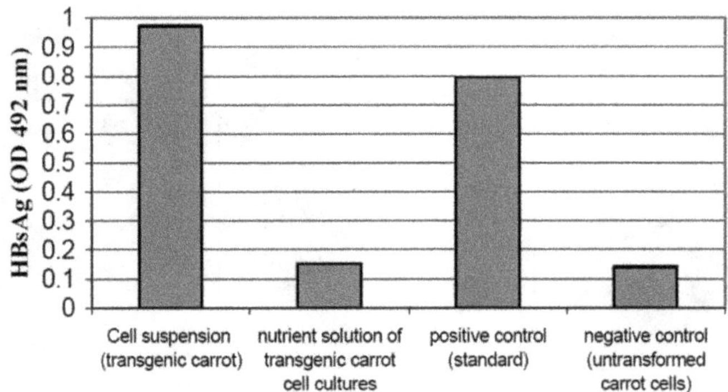

Fig. 13.25 HBV surface protein expression detected by HBsAg-specific ELISA. Proteins extracted from cells, as well as from the nutrient solution of transgenic carrot suspension cultures, were subjected to HBsAg ELISA. The sample extracted was 2 g fresh weight in 1000 µl buffer, of which 100 µl was used for the test. Again, 100 µl of nutrient medium was used

vaccines, if the antigen is resistant against the gastric passage. The hepatitis B virus (HBV) is distributed worldwide, with an estimated 350 million persistent carriers (Maynard et al. 1986). HBV infections are responsible for a high proportion of the world cases of cirrhosis, are the cause of up to 80% of all cases of hepatocellular carcinoma (HCC), and are directly responsible for more than one million deaths each year (report of WHO meeting, 1993). Major pathways for transmission of HBV include parental exposure to the blood or other infective body fluids, transmission from mother to infant, as well as sexual transmission. In order to minimize HBV infections, large-scale immunization is required (Chen et al. 1988, 1996). Current hepatitis B vaccines contain the major or small hepatitis B virus surface protein (SHBs) and are produced in transgenic yeast. Vaccines may also be produced in genetically engineered higher plants. There are several reports on transgenic plants producing biomedically relevant molecules, such as antibodies in tobacco plants (Mason et al. 1992, 1996; Tsuda et al. 1998), and potatoes (Ehsani et al. 1997). Furthermore, a plant-derived edible vaccine against HBV expressed in lettuce and lupine has been reported (Kapusta et al. 1999). More recently, successful clinical tests on humans, using raw potatoes after incorporation of SHBS, have been carried out (Thanawala et al. 1995). For vaccination purposes, however, HBV surface proteins should preferably be produced in an edible plant that is easy to store and to transport and that can be consumed without cooking. Thus, carrots (*Daucus carota* L, ssp. *sativus*) seem to be very suitable for the production of HBV vaccines.

For production of major hepatitis B virus surface protein, we linked the MAS promoter to SHBs. In carrot cell suspensions, as well as in roots of mature transgenic carrot plants, the hepatitis B virus surface protein was produced (Imani et al. 2002; Figs. 13.25 and 13.26). Mannopine synthase (mas) promoter of *Agrobacterium tumefaciens* is a strong inducible promoter, active in roots and leaves, as well as in callus of carrot (Imani et al. 2002) and tobacco, and induced by wounding, auxins, and cytokinins (Langridge et al. 1989). Because of its inducibility by exogenous

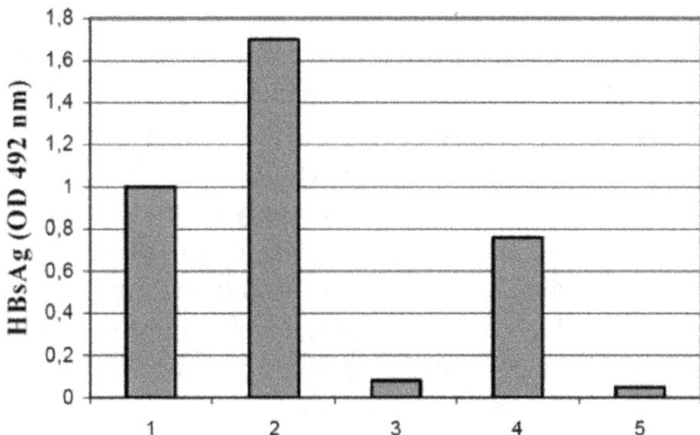

Fig. 13.26 HBsAg expression in callus cultures (liquid media) derived from the secondary phloem of taproots of transformed carrot, as influenced by the application of 11.4 µM IAA. In addition to the DNA for the virus protein, the transgenic plants contain the auxin-sensitive MAS promoter. NL3: liquid medium according to Neumann (1966, 1995) 1 NL3 for 3 weeks of culture in an IAA (11.4 µM) and kinetin (0.45 µM) liquid medium; 2 as 1, but after 2 weeks of culture, 11.4 µM IAA was additionally supplemented to the liquid medium for 1 week; 3 non-transformed carrot; 4 positive control (Fa. DADE Behring, Germany); 5 negative control (Fa. DADE Behring, Germany)

application of auxins, we tested its suitability for engineering HBsAg in carrot. The HBV concentration depends on the form of application of NAA to enhance the activity of the MAS promoter and the production of HBsAg. The following experiments using callus cultures are encouraging.

As control for the function of an additional application of an auxin in the production of the viral protein, callus cultures were initiated in explants of the secondary phloem of the taproots of transgenic and of untransformed carrot plants of the same developmental stage in a liquid medium (Steward et al. 1952; Neumann 1995). The cultured explants were bathed in the nutrient medium containing IAA; consequently, the effect of the auxin on the production of the viral protein is much higher than after an application to intact plants growing in soil (Fig. 13.13).

Production levels of SHBs antigen in carrot, tobacco, tomato, banana, and other plants were compared. Expression of SHBs was reported to be rather low, at 25 ng/g f. wt. in carrot (Kumar et al. 2007). In order to reach high transformation rates as a basis for strain selection, an effective method to transform carrot suspension cells is required. The percentage of cells successfully transformed by *Agrobacterium* is usually very low. In this report, we show that in carrot (*Daucus carota* L, ssp. *sativus*), cell suspension transformation efficiency was strongly improved by using cell cycle synchronized cells (Chap. 12). Fluorodeoxyuridine (FDU, Sigma, Germany) was added for 24 h to inhibit thymidine synthesis. This blocked the cell cycle at the transition from the Gl- to the S-phase. Then, the block was released by applying thymidine. A high rate of transformation was obtained when *A. tumefaciens* was added concurrently with thymidine. As examples of efficient and long-term foreign gene expression in transgenic cells, the reporter enzyme β-glucuronidase (GUS) was used as model. The GUS gene was linked to an inducible mannopine synthase promoter (MAS) from the Ti-plasmid of *A. tumefaciens*. In carrot cell suspensions containing the gus gene, the corresponding GUS protein was produced. In roots of mature transgenic carrot plants generated via somatic embryogenesis and raised in soil, as well as in callus cultures derived thereof, the GUS protein was also produced (Imani et al. 2002).

Here, a synchronization of the cell cycle strongly promoted transformation. From the cell suspension, we regenerated transgenic carrot plants following somatic embryogenesis. Carrot cell suspensions were first transformed with a MAS::GUS construct, which was used as control for the functioning of the system. Based on the experience gained in this model system, the SHBs gene was successfully introduced.

13.8   RNA Interference (RNAi) Technology

Gene silencing by means of RNA interference (RNAi) technology offers the possibility to inhibit expression of entire gene families. This aim can be achieved by selecting nucleotide regions of high similarity often found within coding regions. In contrast, selecting less conserved 50- or 30-untranslated regions allows inhibition of individual members of gene families. Although systemic silencing has been shown for non-plant target genes (i.e., viruses), silencing of endogenous genes seems to be cell autonomous (Ossowski et al. 2008).

RNA interference, first coined by Andrew Fire and Craig C. Mello in 1998, is the phenomenon of posttranscriptional gene silencing in eukaryotes. It was first discovered in nematode *Caenorhabditis elegans* that double-stranded RNA (dsRNA) could lead to sequence-specific gene silencing (Fire et al. 1998). It occurs by introduction of dsRNA (double-stranded RNA), siRNA (short interfering RNA), or miRNA (micro RNA) precursor molecules into cytoplasm of eukaryotic cells. A protein Dicer of Rnase III family, around 200 kDa, which contains ATPase, RNA helicase, and PAZ domain, two catalytic RNase III domains, and a C-terminal dsRNA-binding domain (dsRBD) cleaves dsRNA or Pre-miRNA into 22–25 bp nt duplex with a 2 nt 3′ overhang at each end (Bernstein et al. 2001; MacRae et al. 2006). Nucleotide duplexes formed as a result of cleavage by dicer are known as siRNA, while those from Pre-miRNA are called miRNA. One of the strands of the duplex, called guide strand, is incorporated into a RNA-induced silencing complex (RISC), a multiprotein complex. This guide strand is used by RISC as a template to recognize complementary mRNA, through Watson–Crick base pairing and degrades them using its catalytic component Argonaute protein.

13.8.1 MicroRNA (miRNA)

The first microRNA (miRNA) to be identified was *lin-4* from *Caenorhabditis elegans* (Chalfie et al. 1981; Ambros 1989) as a result of a genetic screening to find out defects in the temporal control of postembryonic development (Hannon and He 2004). MicroRNAs are a family of small noncoding RNAs, 21–25-nt long. They are derived from longer hair pin RNA precursors. Primary miRNA or Pri mRNA are transcribed from endogenous genome. Their origin may lie in non-coding genes or introns of protein coding genes. Pri-miRNA are processed by enzyme drosha into ~70-nt pre-miRNAs inside the nucleus. Pre-miRNA is then exported outside the nucleus into the cytoplasm by exportin 5, where they are processed into miRNA: miRNA duplexes by enzyme dicer (an RNAse III enzyme). And later one of the duplex strands called the guide strand is incorporated into RISC and takes part in posttranscriptional gene silencing (Fig. 13.25).

Double-stranded RNA (dsRNA) introduced in *Drosophila* in vitro system to trigger sequence specific gene silencing were found to be processed into 21–23 nt long RNA segments (Zamore et al. 2000). It was also shown that synthetically synthesized 21-nt RNA mediate RNA interference in Mammalian cells (Tuschl et al. 2001). They were called short interfering RNA or siRNA. Apart from targeted gene silencing, it was shown that siRNA can induce sequence-specific promoter methylation in plants *Arabidopsis* and tobacco, thereby acting as transacting silencing signals (Matzke et al. 2009). siRNA also play part in heterochromatin formation in yeast (Grewal et al. 2002; Grewal and Moazed 2003). Figure 13.27 describes a model for origin and posttranscriptional gene silencing by siRNA.

Silenced phenotypes can persist for generations in *Caenorhabditis elegans* (Plasterk et al. 2006), mice (Rassoulzadegan et al. 2006), *Moniliophthora perniciosa* (Cascardo et al. 2009) and in (Abdellatef et al. 2015). Silenced phenotypes can persist for generations in *Caenorhabditis elegans* (Plasterk et al. 2006), mice (Rassoulzadegan et al. 2006), *Moniliophthora perniciosa* (Cascardo et al. 2009), and in the aphid *Sitobion avenae* (Abdellatef et al. 2015).

We have demonstrated the use of hairpin RNAs targeted against GFP in GFP expressing transgenic barley (Schultheiss et al. 2005; Imani et al. 2011). A plasmid was constructed that contained a 280 nt inverted repeat of the coding sequence of GFP, cloned into the vector p7U-35S-RNAi (Fig. 13.16). For GFP silencing, subsequently this vector was transferred into GFP expressing barley. Different degrees of silencing were observed in the transformants (Fig. 13.28).

RNAi-Based Pest Control

Despite an incomplete understanding of RNAi mechanisms, RNAi-based crop protection strategies have been widely used in laboratory settings to control insects (Zotti et al. 2018) (Fig. 13.29).

Of the various approaches tested, both expression of dsRNA in transgenic crop plants and direct application of dsRNA as an insecticide appear promising for use in the field (Koch and Kogel 2014). However, many practical aspects of dsRNA application need further investigation. For example, the rules guiding dsRNA design to maximize its stability and efficacy are not well understood (Scott et al. 2013; Fellmann and Lowe 2014; Yu et al. 2016). Assessing the safety aspects of these pest management strategies, particularly the potential for side effects on non-target organisms, also is of critical importance.

The RNAi-mediated plant protection from insect was represented in a review by Liu et al. (2020) (Fig. 13.30).

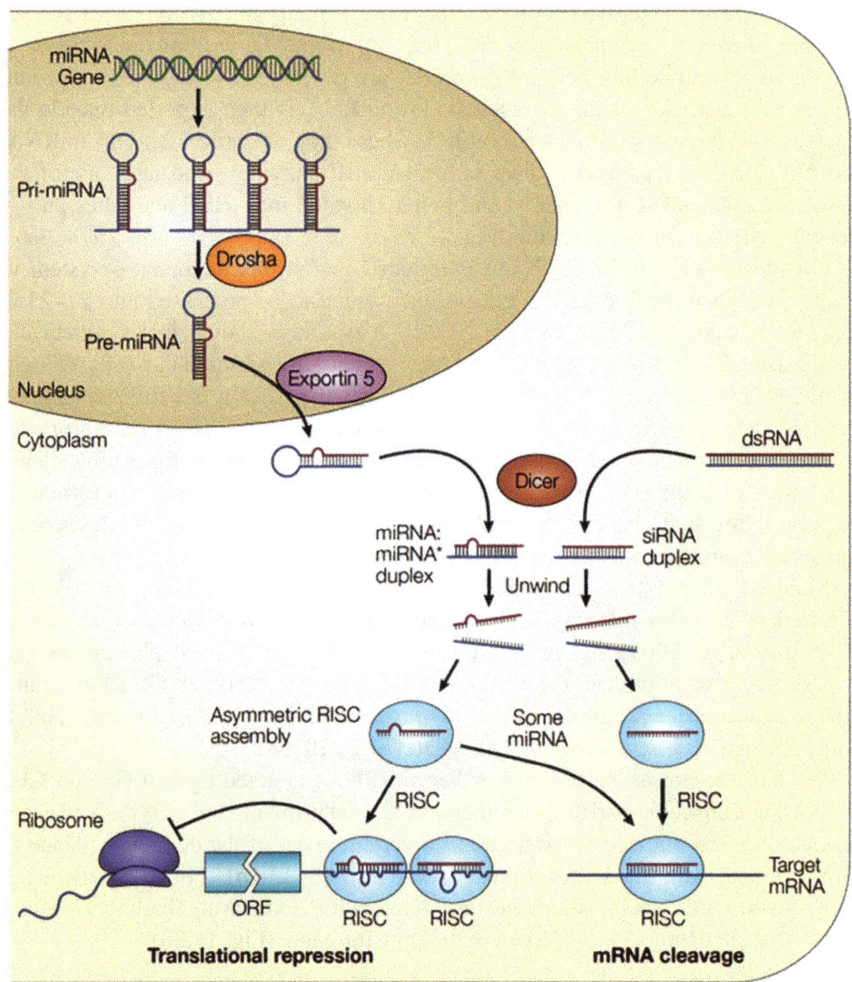

Fig. 13.27 Model for origin and posttranscriptional gene silencing of microRNAs and short interfering RNAs (Hannon and He 2004) (Source: Hannon, G., He, L. (2004). MicroRNAs: small RNAs with a big role in gene regulation. *Nat Rev Genet 5,* 522–531 https://doi.org/10.1038/nrg1379 license number 4802031047134 dated April 4, 2020)

13.8.2 Generation of Hypoallergenic Carrot by Means RNAi

Carrot (*Daucus carota*) allergy is also a common phenomenon in patients with birch pollen allergy. The carrot allergen protein Dau c 1 and denominated Dau c 1.02 was identified (Ballmer-Weber et al. 2005). This isoform showed only 50% amino acid sequence identity with the known isoform Dau c 1 (then denominated 1.01), and limited cross-reactivity was observed between both isoforms in some patients. Because limited cross-reactivity to Bet v 1 was also observed at the T-cell level (Bohle et al. 2006), it was concluded that Dau c 1.02 may represent a partially

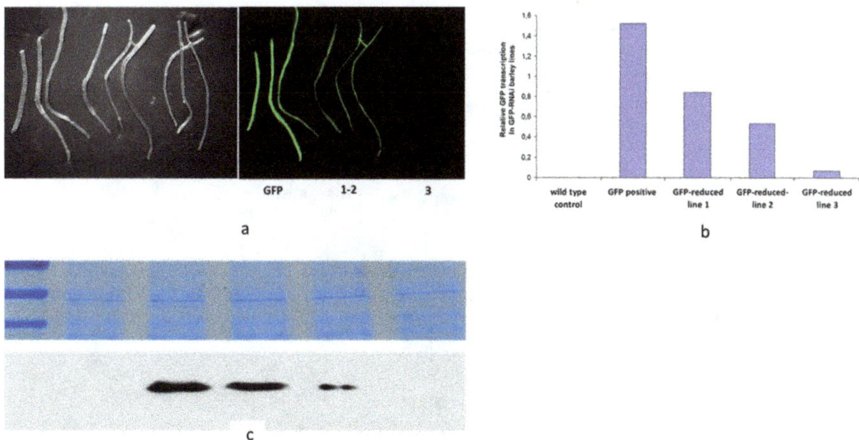

Fig. 13.28 GFP silencing effect confirmation. (**a**) The two types of the GFP silencing effects observed under the fluorescence microscope. (**b**) QPCR identification for the degrees of three types of silencing roots. (**c**) Western blot confirmation of GFP in silenced roots. Silencing efficiencies were ca. 40–100%, Lines 1–3. QPCR result showed the silencing effects were correlative with phenotype observed under fluorescence microscope. (**b**) In addition, the protein level of the independently transformed root were identified by Western blot using GFP monoclonal antibody (**c**)

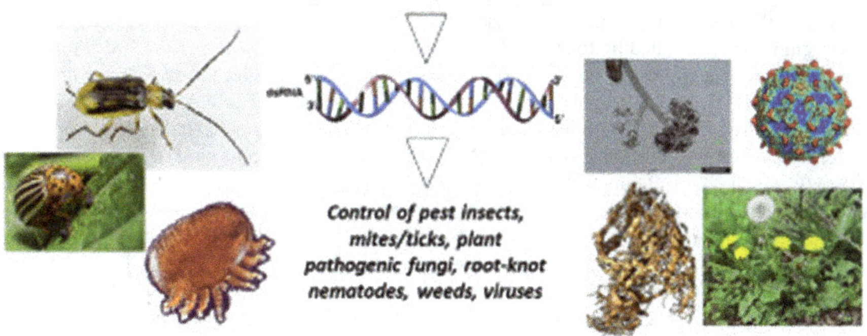

Fig. 13.29 RNA interference technology in crop protection against arthropod pests, pathogens and nematodes RNA interference technology in crop protection against arthropod pests, pathogens, and nematodes (Source: Moises Zotti Ericmar Avila dos Santos Deise Cagliari Olivier Christiaens Clauvis Nji Tizi Taning Guy Smagghe (2018) RNA interference technology in crop protection against arthropod pests, pathogens and nematodes. Pest Management Science 74: 1239–1250. Reproduced with license number 4802500366920 dated April 5)

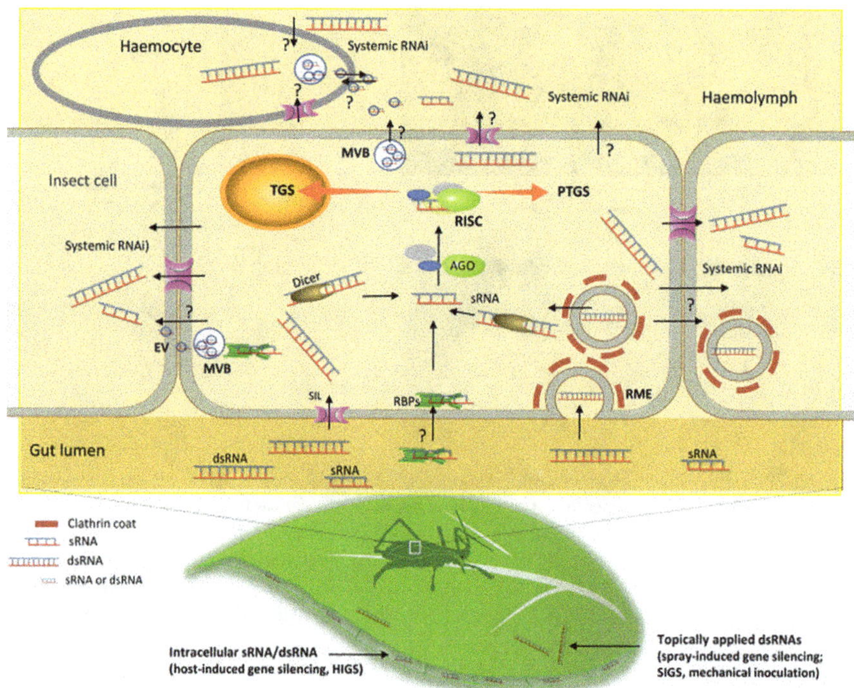

Fig. 13.30 Schematic representation of RNAi-mediated plant protection from insect predation. For host-induced gene silencing (HIGS), dsRNA is produced in planta from a transgene that either generates an RNA hairpin sequence or contains a set of inverted promoters (bottom left). Alternatively, chemically synthesized dsRNA can be delivered to the plant by direct application to the leaf (bottom right) or via several other strategies depicted in Fig. 13.2. Ingestion of dsRNA/sRNA present on the plant surface and/or inner tissues by an insect predator leads to their presence in the insect gut lumen (indicated by the white box on the insect and the expanded view above). Uptake from the insect gut appears to be accomplished primarily by clathrin-dependent receptor-mediated endocytosis (RME), which preferentially imports long dsRNA over sRNA (Saleh et al. 2006); additionally, SID1-like (SIL)-proteins may play a role. Studies of mammalian and plant–microbe systems raise the possibility that dsRNA/sRNA uptake also might occur via extracellular vesicles (EVs) (Cai et al. 2018; Baldrich et al. 2019). RNA-binding proteins (RBPs) may support dsRNA/sRNA transfer, although their role is not well understood. For many insects, uptake of long dsRNA triggers RNAi more efficiently than short sRNA. However, since HIGS is triggered by in planta expression of dsRNA, which is largely processed into sRNA by plant Dicer-like (DCL) proteins, the size requirement may be variable and/or an sRNA uptake mechanism(s) may exist. After entering the cells lining the insect's gut, dsRNA is processed by DICER, and the resultant sRNA, like any sRNA imported directly from the gut lumen, is loaded into specific members of the ARGONAUTE (AGO) protein family. Once an RNA-induced silencing complex (RISC) has been generated, the guide strand of the sRNA mediates binding of the RISC complex to complementary RNA targets. Target recognition subsequently leads to posttranscriptional gene silencing (PTGS) in the insect cell cytoplasm via degradation of the mRNA target or inhibition of its translation or to transcriptional gene silencing (TGS) in the nucleus (depicted as an orange sphere) via chromatin modifications. In some insects, RNAi can spread systemically via a process that is poorly understood. Analyses of cell-to-cell dsRNA movement in *C. elegans*, as well as in mammalian and plant–microbe systems, have suggested roles for SID-1, vesicle transport, and/or endocytosis. Thus, SIL proteins, RME, and EVs, which are secreted after multivesicular bodies (MVB), fuse with the plasma membrane, are depicted as possible mechanisms for moving dsRNA/sRNA between insect cells. In addition, a

pollen-independent stimulus to the immune system. In contrast to Bet v 1-related allergy to apple, food allergy caused by Dau c 1 tends to be more severe, with approximately 50% of patients experiencing systemic reactions (Ballmer-Weber et al. 2001). Dau c 1 represents the major allergen in carrot, and approximately 60% of the allergic patients are mono-sensitized to this protein.

An RNAi strategy was followed to generate carrots with reduced Dau c 1 expression. The two isoforms Dau c 1.01 and Dau c 1.02 share approximately 50% amino acid identity and 66% nucleotide identity (Ballmer-Weber et al. 2005). Empirically, the degree of similarity was not sufficient to allow silencing of both isoforms by one RNAi construct. Thus, a 253-bp cDNA fragment of Dau c 1.01 (position nucleotides [nt] 1–253) and a 267-bp cDNA fragment of Dau c 1.02 (position nt 56–323) were inserted separately in sense and antisense orientation into the plant transformation vector pK7GWIWG2 (II) (Fig. 13.31).

The RNAi constructs were transformed separately into carrot cells by *Agrobacterium tumefaciens*-mediated gene transfer, and transgenic plants were generated on kanamycin containing medium via somatic embryogenesis as described by Imani et al. (2002).

The results show successful respective inhibition of Dau c 1.01 and 1.02 expression in transgenic carrot plants by constitutive expression of Dau c 1.01- and Dau c 1.02-specific hairpin RNAs (hnRNA). On the basis of a quantitative PCR analysis (qPCR) and an immunoblot assay, accumulation of Dau c 1.01 and 1.02 was strongly reduced compared with untransformed carrots (Fig. 13.32).

When challenged with carrot extracts from Dau c 1.01- and 1.02-silenced plants, patients with carrot allergy had strongly reduced skin prick test (SPT) reactivity, indicating a reduced allergenic reactivity in vivo caused by suppression of Dau c 1 expression (for detailed information, see Peters et al. 2010; Wangorsch et al. 2011).

13.9 Precise Genome Editing Tools

Biological studies with RNAi transgenic plants are highly reliant on the stability of knockdown construct and constantly high silencing efficiency of RNAi lines in successive generations. Contrary to conventional mutagenesis techniques where a gene function is completely abolished and irreversible in mutants, RNAi techniques have highly variable efficacies depending on the gene silenced, regions within specific genes, and even within plants carrying identical constructs (Wang et al. 2005). Additionally, even after silencing using RNAi system, some amount of transcript is left over that might influence biological function or in some cases

Fig. 13.30 (continued) hypothetical role for hemocytes has been included, since these cells play a critical role in the systemic antiviral RNAi response in Drosophila (Source: Liu S Jaouannet M Dempsey D Imani J Coustau C et. al. (2020) RNA-based technologies for insect control in plant production. Biotechnology Advances. 39 (107463): 1-13. This is an open-access article distributed under the terms of the Creative Commons CC-BY license, which permits unrestricted use, distribution, and reproduction in any medium, provided the original work is properly cited)

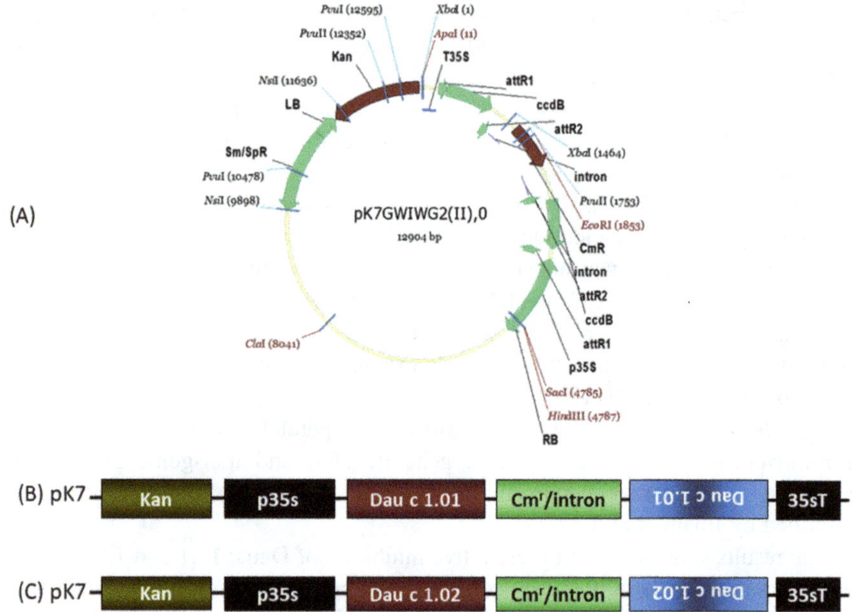

Fig. 13.31 (**a**) Schematic representation of the T-DNA region inserted into plant transformation RNAi vector pK7GWIWG2 (II). (**b**) 253 nt fragment of Dau c 1.01 and 267 nt fragment of Dau c 1.02 (**c**) were inserted separately in sense and antisense orientation into vector pK7GWIWG2 (II) by means of a lambda reconstruction (LR) recombination reaction (Invitrogen) resulting in plasmids pK7GW_Dau c 1.01_RNAi and pK7GW_Dau c 1.02_RNAi (Source: Karimi, Mansour et al. 2002. GATEWAY™ vectors for *Agrobacterium*-mediated plant transformation. Trends in Plant Science, Volume 7, Issue 5, 193–195 License number 4801510859121 April 3, 2020)

reduced transcript levels might not be enough to generate a phenotype. Besides, the knockdown efficiency with identical constructs may range from anything between 0 and 90% reduction in transcript levels that might make comparison of biological assays and data analysis difficult. The loss in silencing might be a concern if experiments are planned with this later generation of transgenic plants and might affect production, identification, and use of homozygous lines altogether. Additionally, the common methods for gene manipulation are random and lead some time to unwanted gene disruption, position effect, and multiple copies of inserts.

13.9.1 Genome Manipulation Using the DNA Repair System

DNA Damage, Repair Mechanisms, and Consequences
DNA repairing mechanisms exist in organisms. That is a process by which a cell identifies and corrects damage to the DNA molecules that encode its genome. DNA

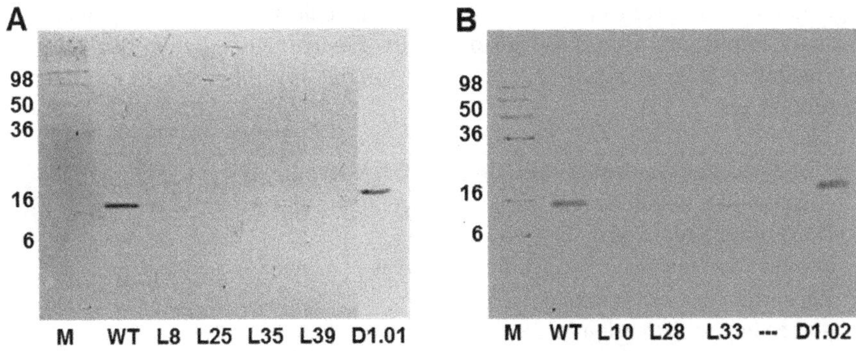

Fig. 13.32 Immunodetection of Dau c 1 in protein extracts from carrot root (cv Rodelika) of wild type (WT, mix of 8 mock samples) and Dau c 1.01_RNAi transgenic lines 8, 25, 35, and 39 (0.8 lg/ lane) by using monoclonal antibody M3 directed against Dau c 1.0104 (**a**) and Dau c 1.02_RNAi transgenic lines 10, 28, and 33 (0.1 lg/lane) by monoclonal antibody M9 directed against Dau c 1.0201 (**b**) Recombinant rDau c 1.0104 (D1.01) and rDau c 1.0201 (D1.02), 5 ng each, were used as positive controls. M SeeBlue_Plus2 marker (numbers on the left side represent molecular weights of marker proteins)

repairing mechanisms are based on base excision, nucleotide excision, and mismatch repair. If the DNA is not repaired, it can lead to an irreversible state of dormancy (senescence), cell suicide (persistent DNA damage: apoptosis or PCD), or unregulated cell division (tumor). There are two main types for sources of DNA damage: (1) endogenous damage caused by attack of reactive oxygen on spontaneous mutation, e.g., replication errors and process of oxidative deamination—(2) exogenous damage is caused by viruses; ultraviolet, radiation from the sun, X-rays, and gamma rays; hydrolysis or thermal disruption; certain plant toxins; and human-made mutagenic chemicals, especially aromatic compounds that act as DNA inserting agents: cancer chemotherapy and radiotherapy.

Genome Manipulation Using the DNA Repair System
Creation of a double-strand break at a specific location within a genome by using ZFN, TALEN, and CRISPR-CAS system.

- DNA will be repaired.
- Looking for mutants in which the repair was performed.
- Repairing performs incorrectly:
  - Leading to disruption of the targeted gene
  - Insertion and/or deletion (indel)

Genome Manipulation Using the DNA Repair System
DNA repair provides a forum for the comprehensive coverage of cellular responses to DNA damage in living cells to identify and correct damage to the DNA molecules that encode its genome. The sources of DNA damage are (1) endogenous damage,

e.g., attack by reactive oxygen (process of oxidative deamination) and spontaneous mutation (replication errors) and (2) exogenous damage caused by viruses; ultraviolet radiation from the sun, X-rays, and gamma rays; hydrolysis or thermal disruption, etc.

Precise Genome Manipulation Using the DNA Repair System
- Creation of a double-strand break at a specific location within a genome
- DNA will be repaired.
- Looking for mutants in which the repair was performed.
- Repairing performs incorrectly:
  – Leading to disruption of the targeted gene

Double-Strand Break Repair Pathways
Genome modification requires an engineered nuclease to create double-strand breaks (DSB) at pre-defined targets which then trigger cellular DNA repair mechanism. This can lead to genome modification depending on the DNA repair pathway and the presence of a repair template. There are two known DSB repair pathways, non-homologous end joining (NHEJ) and homologous recombination (HR) NHEJ in most instances leads to random insertions or deletions (indels) at the repair site (Fig. 13.29). The joining of DNA ends in NHEJ is led by DNA ligase IV. In the case DSB generates overhangs, NHEJ can also introduce gene insertions or precise gene modifications with a double-stranded DNA fragment with compatible overhangs (Fig. 13.29) (Cristea et al. 2013; Maresca et al. 2013; Bortesi and Fischer 2015). In cases where a DNA template is available with sequence homologous to the separated ends, the HR repairs the DNA damage, and this process can be used to introduce precise gene modifications (Fig. 13.33) or insertions (Bortesi and Fischer 2015).

De Souza (2011) and Osakabe and Osakabe (2015) elaborated on genome editing. There are three principal classes of engineered nucleases. Zinc finger nucleases (ZFNs) consist of a DNA-binding domain, derived from zinc finger proteins, linked to the nuclease domain of the restriction enzyme Fok-1. Like their parent nuclease, ZFNs must dimerize to cleave DNA. The DNA-binding domain, which can be designed (in principle) to target any genomic location of interest, is a tandem array of Cys2His2 zinc fingers, each of which recognizes 3 nt in the target DNA sequence. By linking together multiple fingers (the number varies: three to six fingers have been used per monomer in published studies), ZFN pairs can be designed to bind to genomic sequences 18–36 nt long. When two ZFN monomers bind, in inverse orientation, with an optimal spacing of 5–7 nt, the resulting dimeric nuclease cleaves the DNA between the binding sites (Fig. 13.34).

Presently Precision Tools for DNA Editing Include
- Zing finger nuclease (ZFN)
- Transcription activator-like effector (TALE)
- Clustered regularly interspaced short palindromic repeats (CRISPR) (Fig. 13.35)

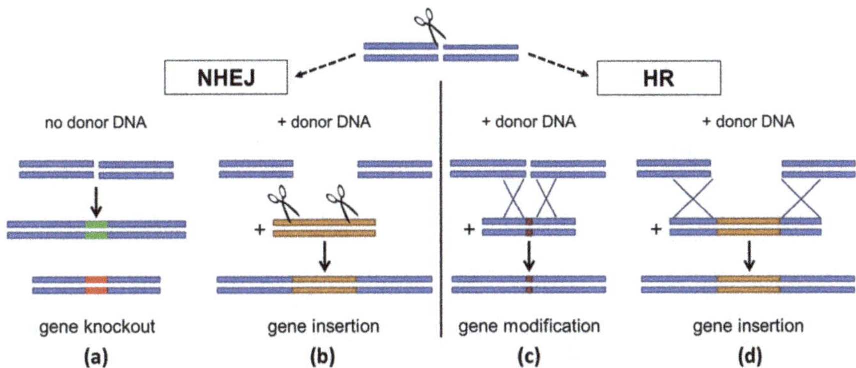

Fig. 13.33 Gene editing with site-specific nucleases. (a) DSB repair by NHEJ leads to random base–pair insertion (green) or deletion (red), resulting in gene knockout by disruption. (b) In case a donor DNA is present and also cut by the same nuclease to produce compatible overhangs, then NHEJ can lead to gene insertion. (c) HR using a donor DNA template can be used to achieve precise nucleotide substitutions or (d) gene insertions. (Adapted from Bortesi and Fischer 2015) (Source: Bortesi L, and Fischer R. (2015). The CRISPR/Cas9 system for plant genome editing and beyond. Biotechnology Advances 33: 41-52 License number 4801510103295 dated April 3)

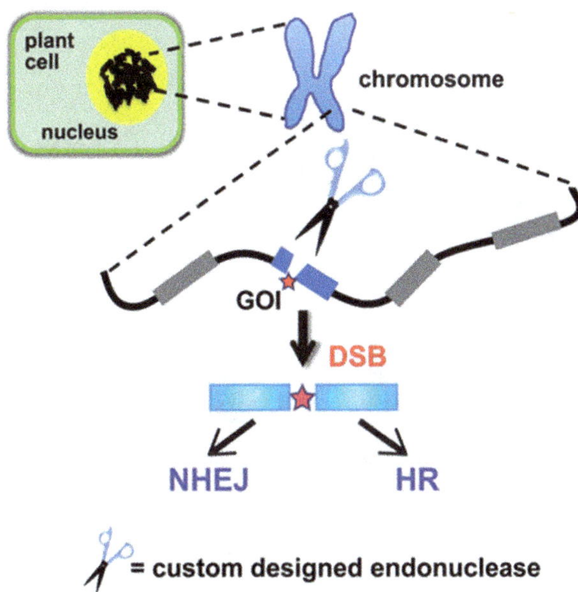

Fig. 13.34 Targeted genome editing in plants using engineered nucleases (GEEN) Engineered nucleases are used to induce a double-stranded DNA break (DSB) at a specified locus of the gene of interest (GOI) DSBs are repaired by either non-homologous end joining (NHEJ) or homologous recombination (HR) (Source: Yuriko Osakabe, Keishi Osakabe, Genome Editing with Engineered Nucleases in Plants, *Plant and Cell Physiology*, 56: 389–400, https://doi.org/10.1093/pcp/pcu170 printed with permission license number 4802011209977 granted April 4, 2020 attached)

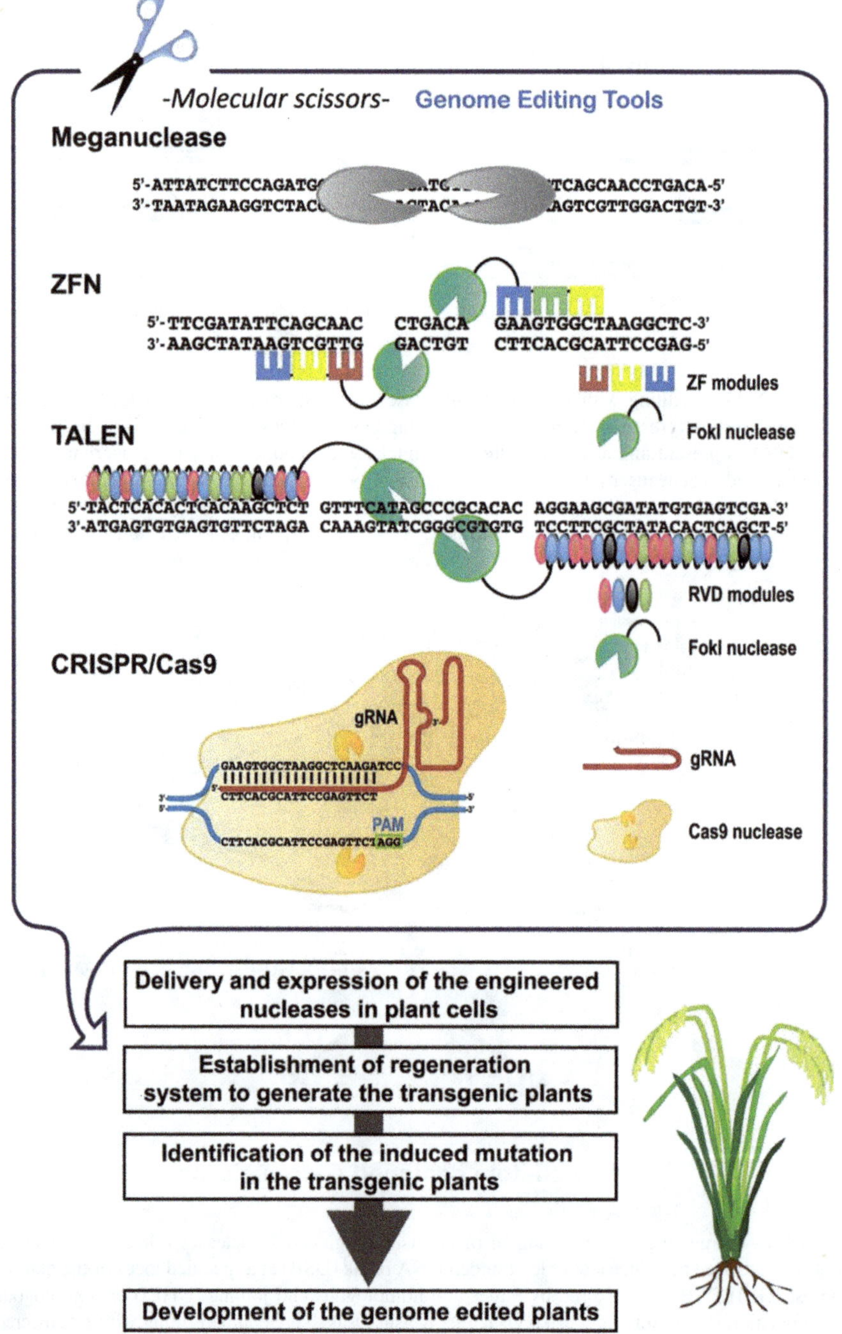

Fig. 13.35 Current methods in GEEN and an overview of the workflow of plant genome editing. Homing endonucleases/meganucleases (EMNs) recognize long (approx 20 bp) DNA sequences. FokI nuclease is used as the DNA cleavage domain in ZFN, which binds the target DNAs via engineered C2H2-zinc finger (ZF) domains and in TALEN that recognize the targets by the

Transcription activator-like effector nucleases (TALENs) have an overall architecture similar to that of ZFNs, with the main difference that the DNA-binding domain comes from TAL effector proteins, transcription factors from plant pathogenic bacteria. The DNA-binding domain of a TALEN is a tandem array of amino acid repeats, each about 34 residues long. The repeats are very similar to each other; typically, they differ only at two positions (amino acids 12 and 13, called the repeat variable di-residue, or RVD). Each RVD specifies preferential binding to one of the four possible nucleotides, meaning that each TALEN repeat binds to a single base pair. TAL effector DNA binding is mechanistically less well understood than that of zinc finger proteins, but their seemingly simpler code could prove very beneficial for engineered nuclease design. TALENs also cleave as dimers, have relatively long target sequences (the shortest reported so far binds 13 nt per monomer), and appear to have less stringent requirements than ZFNs for the length of the spacer between binding sites.

As reported by Zhang et al. (2013) the past few years, the use of sequence-specific nucleases for efficient targeted mutagenesis has provided plant biologists with a powerful new approach for understanding gene function and developing new traits. These nucleases create DNA double-strand breaks at chromosomal targeted sites that are primarily repaired by the non-homologous end joining (NHEJ) or homologous recombination (HR) pathways. NHEJ is often imprecise and can introduce mutations at target sites resulting in the loss of gene function. In contrast, HR uses a homologous DNA template for repair and can be employed to create site-specific sequence modifications or targeted insertions (Moynahan and Jasin 2010).

Zhang et al. (2013) further reported that in spite of being developed only 3 years ago, TALENs have become the reagent of choice for efficiently modifying eukaryotic genomes in a targeted fashion (Baker 2012; Liu et al. 2012; Tong et al. 2012).

Like ZFNs, TALENs are composed of an engineered array of DNA-binding domains fused with a nonspecific FokI nuclease domain (Christian et al. 2010). Two TALEN monomers bind target DNA sequences allowing the FokI domains to dimerize and cleave the sequence between the two recognition sites.

Each repeat in the DNA-binding domain of a TALEN recognizes 1 nt in the target in a largely context-independent fashion, making the engineering of nucleases with new DNA sequence specificities easier and more reliable than ZFNs and meganucleases (Reyon et al. 2012).

Fig. 13.35 (continued) engineered TALEs composed of RVD domains derived from the plant pathogen *Xanthomonas*. The CRISPR/Cas9 system utilizes RNA-guided engineered nucleases (RGNs) that use a short guide RNA (gRNA) to recognize DNA sequences at the target sites (Source: Yuriko Osakabe, Keishi Osakabe, Genome Editing with Engineered Nucleases in Plants, Plant and Cell Physiology. 56: 389–400, https://doi.org/10.1093/pcp/pcu170 printed with permission license number 4802011209977 granted April 4, 2020 attached)

13.9.2 Zinc Finger Nuclease

Zinc finger nucleases (ZFNs) are chimeric proteins composed of a highly specific integral zinc finger DNA-binding domain (ZF-DBD) zinc ion forms the protein for proper 3D confirmation. By taking advantage of endogenous DNA repair machinery, zinc finger nuclease can be used to precisely alter the genomes of higher organisms. Zinc finger nucleases (ZFNs) enable scientists to gene knockout (KO), gene knock in, and point mutations in the genome of cells. Zinc finger domains can be engineered to target desired DNA sequences that enable zinc finger nucleases to target unique sequences within complex genomes. An artificial restriction enzyme can be fused with a zinc finger DNA-binding domain to a DNA-cleavage domain. An individual zinc finger binds specifically to a three-base pair DNA sequence. For recognition of longer sequences (8 or 9 zinc fingers) stringing several different fingers together to make a protein. For DNA break zinc finger DNA-binding domain (ZF-DBD) linked to a nonspecific *Fok*I-nuclease cleavage domain that is only active as a dimer. Fok-1-nonspecific DNA nuclease found in *Flavobacterium okeanokoites*, Fok1 has DNA-binding domain which recruits Fok1 to cleavage site. The sequence recognition sites for specific zinc finger motifs and that Fok1 cleaves nonspecific DNA, activation is mediated my DNA binding and dimerization. Scientists design zinc finger to recognize specific sequence in gene fused with Fok1 gene sequences; Fok1 require 6nt of space between recognition sequence for best cleavage function. DNA repair pathway leads to mutation in non-homologous end joining (NHEJ) and homologous recombination (HR) NHEJ in most instances leads to random insertions or deletions (indels) at the repair site.

13.9.3 Meganucleases

As reported by Zhang et al. (2013) in the past few years, the use of sequence-specific nucleases for efficient targeted mutagenesis has provided plant biologists with a powerful new approach for understanding gene function and developing new traits. These nucleases create DNA double-strand breaks at chromosomal targeted sites that are primarily repaired by the non-homologous end joining (NHEJ) or homologous recombination (HR) pathways. NHEJ is often imprecise and can introduce mutations at target sites resulting in the loss of gene function. In contrast, HR uses a homologous DNA template for repair and can be employed to create site-specific sequence modifications or targeted insertions (Moynahan and Jasin 2010).

Zhang et al. (2013) further reported that in spite of being developed only 3 years ago, TALENs have become the reagent of choice for efficiently modifying eukaryotic genomes in a targeted fashion (Baker 2012; Liu et al. 2012; Tong et al. 2012). Like ZFNs, TALENs are composed of an engineered array of DNA-binding domains fused with a nonspecific FokI nuclease domain (Christian et al. 2010). Two TALEN monomers bind target DNA sequences allowing the FokI domains to dimerize and cleave the sequence between the two recognition sites (Supplemental Fig. 1A). Each repeat in the DNA-binding domain of a TALEN recognizes 1 nt in the target in a largely context-independent fashion, making the engineering of nucleases with new DNA sequence specificities easier and more reliable than ZFNs and meganucleases (Reyon et al. 2012). Zhang et al. (2013) reported the first study demonstrating highly efficient targeted knockouts in multiple genes in the model monocot species Brachypodium.

13.9.4 Transcription Activator-Like Effector (TALE): A Precision Tool for DNA Editing

Transcription activator-like effectors (TALEs) are secreted by plant–pathogenic *Xanthomonas* bacteria into plant cells where they act as transcriptional activators and, hence, are major drivers in reprogramming the plant for the benefit of the pathogen. Their DNA-binding domain forms a right-handed supercoil structure wrapping around the DNA and mediating sequence-specific binding to the promoter of plant genes. Different *Xanthomonas* strains possess different repertoires of TALEs (Nivina et al. 2016; Erkes et al. 2017) (Fig. 13.36).

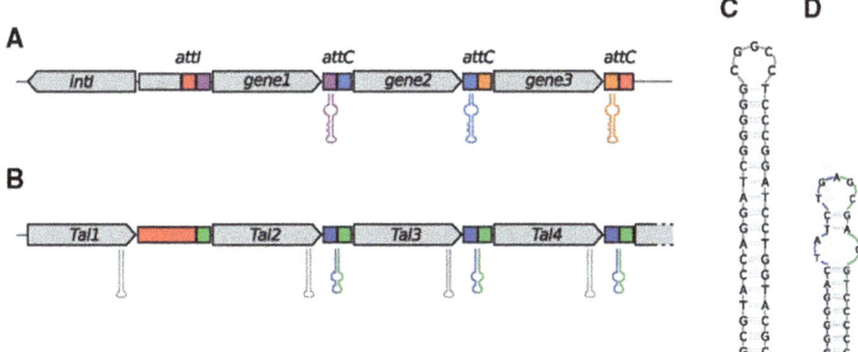

Fig. 13.36 An integron-like mechanism may be responsible for the reassortment of TALE genes. (a) Organization of gene clusters formed by integrons, including the hairpin structures at attC sites, adapted from Nivina et al. (2016). (b) Schematic arrangement of TALE genes in clusters with short spacer regions in Xoc. Putative hairpin structures may be found at the C-terminal region of TALE genes (gray) as well as in the spacer regions (green/blue). (c) Secondary structure of the C-terminal hairpin as predicted by Mfold. (d) Secondary structure of the hairpin in the spacer region (Source: Erkes A, Reschke M, Boch J, Grau J. Evolution of Transcription Activator-Like Effectors in Xanthomonas oryzae. Genome Biol Evol. 2017;9(6):1599–1615. doi:10.1093/gbe/evx108 This is an Open Access article distributed under the terms of the Creative Commons Attribution Non-Commercial License (http://creativecommons.org/licenses/by-nc/4.0/), which permits non-commercial re-use, distribution, and reproduction in any medium, provided the original work is properly cited)

Relationship Between a Host and a Pathogen

Dynamic: each modifies the activities and functions of the other. The outcome depends on virulence of the pathogen and relative degree of resistance or susceptibility of the host. There are two broad qualities of pathogenic bacteria underlie the means by which they cause disease: (1) Invasiveness, which means colonization (adherence and initial multiplication), production of extracellular substances which facilitate invasion (invasins), and ability to bypass or overcome host defense mechanisms. (2) Toxigenesis, which means production of toxins such as exotoxins (released from bacterial cells and may act at tissue sites removed from the site of bacterial growth) and endotoxins released from growing bacterial cells and cells that are lysed as a result of effective host defense (e.g., lysozyme).

Transcription activator-like (Tal) effectors are found in *Xanthomonas* and *Ralstonia*, bacterial strains that cause major crop diseases.

Up to 26 TAL members/strain:

- Proteins are injected into plants via a type III secretion system.
- Induce expression of plant genes.
- Bacterial pathogen proteins used to rewire transcription of host plants upon infection.

A Model for Sequence Specificity (Fig. 13.37)

Resistance (R) proteins, a class of plant immune receptors that ediate recognition of pathogen-derived avirulence (Avr) proteins, are a well-studied facet of the plant defense system.

Xanthomonas campestris pv. *vesicatoria* (Xcv):

- Type III secretion (T3S) system to inject an arsenal of about 20 effector proteins into the host cytoplasm (promotion of virulence R protein-mediated defense (HR resulting PCD).
- One Avr protein that R proteins recognize is AvrBs3, a member of a *Xanthomonas* family of highly conserved proteins.
  - The central region of AvrBs3 consists of 17.5 tandem near-perfect 34 –amino acid repeat units that determine avirulence specificity.
  - AvrBs3 contains nuclear localization signals (NLSs), acidic transcriptional activation domain (AD), and induces host gene transcription.

13.9.5 CRISPR-Cas9 System for Gene Knockout

Khatodia et al. (2016) reviewed the current knowledge about CRISPR/Cas9 genome editing tool. The targeted plant genome editing using sequence-specific nucleases has a great potential for crop improvement to meet the increasing global food

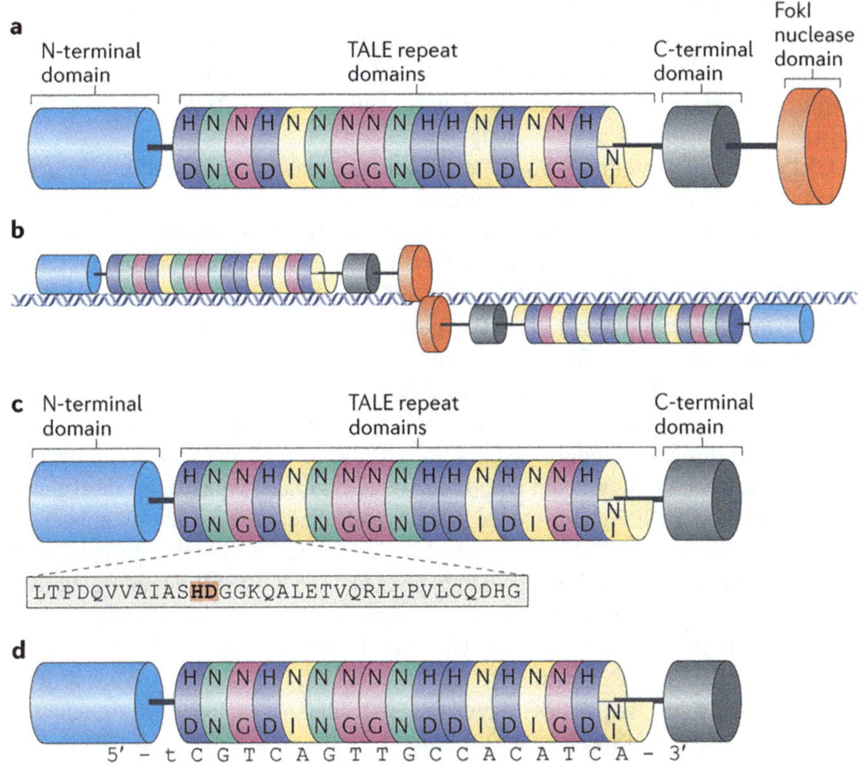

Nature Reviews | Molecular Cell Biology

Fig. 13.37 (**a**) Schematic diagram of a TALEN. TALE repeats are shown as colored cylinders with a final carboxy-terminal truncated "half" repeat. Letters inside each repeat represent the two hypervariable residues. TALE-derived amino- and carboxy-terminal domains required for DNA-binding activity are shown as longer blue and grey cylinders, respectively. The non-specific nuclease domain from the FokI endonuclease is shown as a larger orange cylinder. (**b**) TALENs bind and cleave as dimers on a target DNA site. Note that the TALE-derived amino- and carboxy-terminal domains flanking the repeats may make some contacts to the DNA. Cleavage by the FokI domains occurs in the "spacer" sequence that lies between the two regions of the DNA bound by the two TALEN monomers. (**c**) Schematic diagram of a TALE-derived DNA-binding domain. The amino acid sequence of a single TALE repeat is expanded below with the two hypervariable residues highlighted in orange and bold text. (**d**) TALE-derived DNA-binding domain aligned with its target DNA sequence. Note the matching of repeat domains to single bases in the target site according to the TALE code. Also note the presence of a 5′ thymine preceding the first base bound by a TALE repeat (Source: Joung JK, Sander JD. (2013) TALENs: a widely applicable technology for targeted genome editing. Nat Rev Mol Cell Biol.;14(1):49–55. doi:10.1038/nrm3486 Reproduced with leicence number 4802931229208 dated 6th April)

demands and to provide sustainable productive agriculture system (Liu et al. 2013). Traditionally, the crops were being improved by the conventional and mutation plant breeding techniques, which are now getting constrained by the declining of existing genetic variation of plants, hampering the production for future feeding (Chen and Gao 2014). There is an urgent need for efficient crop improvement strategies with novel genome editing techniques like CRISPR/Cas9 system, which can improve the existing important functions or make new valuable products (Zhang and Zhou 2014).

CRISPR-Cas9 system: A novel technique for plant genome editing targeted genome engineering is one of the alternatives to classical breeding and generation of transgenic plants. Even though mechanisms like RNAi-mediated gene silencing are used widely to study gene functions, they have limitations like variation in knock down levels and reduction in knock down efficiency in successive generations. In view of this, several alternatives have been developed to obtain complete gene silencing (knockout) zinc finger nucleases (ZFNs) and transcription activator-like effector nucleases (TALENs) can be used for targeted mutagenesis of genomes at specific loci. The major drawback of these systems is the laborious target site selection and design procedures leading to development of alternative approaches. One such new technology is the type II clustered regularly interspaced short palindromic repeats (CRISPR) interference system, part of adaptive immunity in bacteria and archaea (Sorek et al. 2013). It is a naturally occurring microbial nuclease system protecting the bacteria against invading phages. The CRISPR locus contains a combination of CRISPR-associated genes that encode a bacterial endonuclease Cas9 and two short noncoding RNA elements known as CRISPR RNA (crRNA) and trans- activating crRNA (tracrRNA). The noncoding pre-crRNA consist of an array of palindromic sequences (direct repeats) interspaced by short stretches of non-repetitive spacers (Sorek et al. 2013; Jinek et al. 2012; Cong et al. 2013). The Cas9 protein is a large monomeric DNA nuclease containing RuvC and HNH homologous nuclease domains. These two domains cleave the noncomplementary and complementary strands, respectively, to generate a blunt cut in the target DNA (Jinek et al. 2012). The double-stranded breaks (DSBs) disrupt gene function by forming premature stop codons or through mutations inserted by non-homologous end joining (NHEJ) pathway or homology-directed repair (Cong et al. 2013). The cleavage of target DNA takes place in four sequential steps. In the first step, transcription of the two noncoding RNAs, the pre-crRNAarray and tracrRNA take place. This is followed by hybridization of tracrRNA to the direct repeat or palindromic region of the pre-crRNA followed by processing of pre-crRNA into mature crRNAs containing individual spacer sequences. In the next step, Cas9 endonuclease is directed to the specific target sequence by the mature crRNA: tracrRNA complex via Watson–Crick base pairing between the spacer on the crRNA and the protospacer on the target DNA next to the protospacer adjacent motif (PAM), an additional requirement for target recognition. Finally, Cas9 endonuclease recognizes and creates a double-stranded break within the protospacer region of the target DNA molecule (e.g, in a bacteriophage 24 genome, Cong et al. 2013). Functional portions of crRNA and tracrRNA can be combined to give rise to a chimeric single guide RNA or sgRNA which along with Cas9 forms a targeted RNA-guided endonuclease (Mussolino and Cathomen 2013; Jinek et al. 2012). The Cas9 endonuclease could be easily redirected to different target sites by modifying the sequence of a chimeric single guide RNA (sgRNA) complexed with the enzyme (Jinek et al. 2012). Multiplexing can be achieved by combining Cas9 expression with multiple guide RNAs targeting different loci in the target genome (Cong et al. 2013) thereby reducing the costs and speeding up generation of organisms with multiple, targeted mutations. Thus, RNA-guided endonuclease seems to combine the efficiency of ZFNs and TALENs with a much simpler design process, as target site selection is determined solely by base complementarity to the guide RNA, and the protein does not require reengineering for each new target site (Fig. 13.38).

For targeted mutagenesis in plants using CRISPR, plant codon-optimized version of Cas9 from the bacterium *Streptococcus pyogenes* was used (Shan et al. 2013; Miao et al. 2013). The second important component, a synthetic RNA chimera created by fusing crRNA with 25 tracrRNA known as single-guide RNA (sgRNA) is required to form a complex with Cas9 nuclease for target recognition. The guide sequence located at the 5′ end of sgRNA determines DNA target specificity. The guide sequence is usually about 20 bp long (Jinek et al. 2012). The corresponding DNA target is also 20 bp long followed by PAM sequence (NGG). Contrary to mammalian systems, plant-guide sequences are of lengths varying from (N)19–22 NGG as against the stringent (N)20 NGG existing in mammalian systems (Shan et al. 2013; Miao et al. 2013; Feng et al. 2014). The plant sgRNAs are driven by type III RNA polymerase promoters, such as wheat U6 and rice U3. They have stringent requirements for transcription start sites to be—G or—A, for U6 or U3 promoters, respectively. Therefore, the guide sequences follow the consensus G(N)19–22 NGG for the U6 promoter and A(N)19–22 NGG for the U3 promoter, where the first G or A may or may not pair up with the target DNA sequence (Shan et al. 2013; Miao et al. 2013; Feng et al. 2014). Transient assays help in rapid screening and optimization of a method. In plants, protoplast transformation and leaf tissue transformation using the agro infiltration method have been used to test targeted mutagenesis by CRISPR/Cas9 system. Protoplast assay is good for achieving gene coexpression from separate plasmids, even though the protoplast isolation may be time-consuming and prone to contamination. Mutations with efficiency of 15% were detected 18 h after protoplast cultivation. Target mutation efficiencies were estimated by band intensities (Li et al. 2013). The induced mutations may be detected by PCR-restriction enzyme digestion assay. Cas9 nuclease usually cuts the target DNA about 3 bp away from the PAM and can be used to identify mutation in the target region which has a restriction site adjacent to the PAM motif. Repair of a DSB in protospacer region by the error-prone NHEJ pathway results in mutations that disrupt the restriction site. These mutations are detected by PCR amplification of genomic DNA using primers specific for the target region and digesting resulting amplicons with the restriction enzyme (Li et al. 2013). Cloning and sequencing of these uncut bands revealed indels in the targeted gene. SgRNA with a length of 20 nt of sequence complementarity to the OsPDS had the highest frequency and mutation efficiency (Li et al. 2013). Rice phytoene desaturase gene (OsPDS) was knocked out using Cas9 plasmid and sgRNA expression plasmids bombarded into rice calli resulting in biallelic mutations and some homozygous mutations carrying the same 1 nt insertion. Albino and dwarf phenotype confirmed disruption of OsPDS (Li et al.

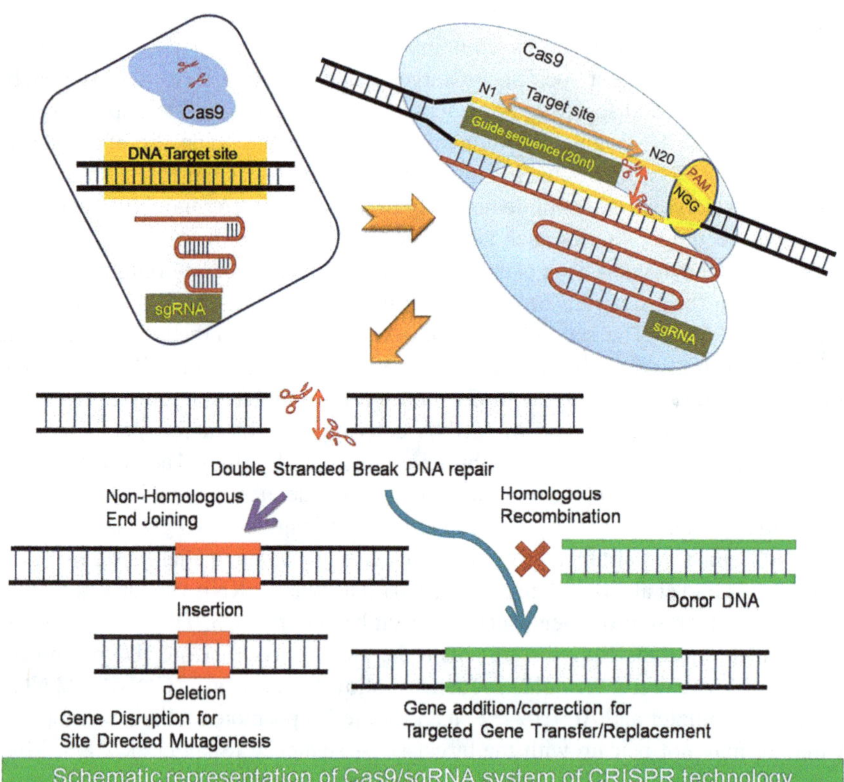

Fig. 13.38 The basic strategy of Cas9/sgRNA system. The Cas9 is a RNA-guided endonuclease consists of two nuclease domains, namely, HNH and RuvC. The target specificity of Cas9 depends upon the guide sequence (20 nt) short guide RNA (sgRNA). The target sites must lay immediately 5′ of a PAM (protospacer adjacent motif) sequence of the form N20-NGG (or N20-NAG). The Cas9 nuclease induces double-stranded breaks (DSB) at the target site which can be repaired either by non-homologous end-joining method or homologous recombination by cellular system which results in gene disruption by indels or gene addition/correction, respectively (Source: Khatodia S, Bhatotia K, Passricha N, Khurana SM, Tuteja N. (2016) The CRISPR/Cas Genome-Editing Tool: Application in Improvement of Crops. Front Plant Sci.;7:506. Published 2016 Apr 19. doi:10.3389/fpls.2016.00506. This is an open-access article distributed under the terms of the Creative Commons Attribution License (CC BY). The use, distribution, or reproduction in other forums are permitted, provided the original author(s) or licensor are credited and that the original publication in this journal is cited, in accordance with accepted academic practice)

2013). CRISPR/Cas9 system application in plant cells was demonstrated by DGU. US reporter assay, where DSB generated is repaired through single strand annealing, thus restoring the GUS activity that led to strong GUS staining spots in rice calli (Miao et al. 2013). Endogenous genes in rice chlorophyll a oxygenase 1 (CAO1) gene and LAZY1 gene were knockedout selectively using the CRISPR/Cas technology. Loss-of-function mutant cao1defective synthesis of chlorophyll b (Chl b) showed a pale green phenotype and loss-of-function mutant of LAZY1 gene, exhibited a tillerspreading phenotype which was observed after tillering stage. Sequencing analysis on these lines using gene-specific primers showed mutations in specific regions confirming the disruption of the respective genes (Miao et al. 2013).

13.9.6 Structure and Functional Mechanism of Cas9 Nuclease

Crystal structure of Cas9 nuclease provides insight into its conserved architecture that recognizes and cleaves DNA targets. Cas9 has a bilobed structure: a large recognition lobe (REC) and a small nuclease lobe (NUC) (Jinek et al. 2014). REC lobe consists of REC1, REC2, and a long α-helical arginine-rich domain, called Bridge Helix (BH) NUC has two nucleic acid-binding grooves, a wide major groove and a narrow minor groove, both are located within the REC and NUC lobes. NUC lobe has two nuclease domains, RuvC and HNH and a PAM-interacting domain (PI domain) (Hsu et al. 2014; Nishimasu et al. 2014). Cas9 protein resides in an auto-inhibitory mode, gets activated on loading of single guide RNA (sgRNA) and mediates target specific cleavage of DNA in a sequential manner (Fig. 13.39) (Belhaj et al. 2015).

The CRISPR (clustered regularly interspaced short palindromic repeats)-Cas9 (CRISPR-associated nuclease 9) system is a versatile tool for genome engineering that uses a guide RNA (gRNA) to target Cas9 to a specific sequence.

13.10   CRISPR/Cas9 for Plant Genome Editing

For achieving site-directed DNA breaks, zinc finger nucleases (ZFNs) and transcription activator-like effector nucleases (TALENs) have shown promising results. Both use a dimeric *Fok*1 nuclease for creating DNA breaks in genome (Smith et al. 2000; Christian et al. 2010). Recently experimentations have started with type II clustered regularly interspaced short palindromic repeats (CRISPR)/Cas9 (CRISPR-associated) system from *Streptococcus pyogenes* to induce precise mutations in plant genome (Jinek et al. 2012). Generating CRISPR constructs is easy and fast making it more attractive for genome editing (Bortesi and Fischer 2015). CRISPR/Cas9 system involves introduction of two components, the Cas9 protein and a guide RNA (gRNA), into the target cell (genome) to be mutated (Fig. 13.40). The gRNA defines the desired target and consists of a 20 bp nt sequence that has sequence similarity to the target gene or part of genome. The single gRNA (sgRNA) is driven by promoters from either *U3* or *U6* small nuclear RNA (snRNA) encoding genes. Preferred transcription initiation site for *U6* is 'G' and 'A' for *U3* promoters (Fig. 13.40). As a result, guide sequence follows the consensus $G(N)_{19-22}NGG$ for the *U6* promoter and $A(N)_{19-22}NGG$ for the *U3* promoter, where the first nucleotide G or A may or may not pair up with the target DNA sequence (Fig. 13.40) (Shan et al. 2013; Miao et al. 2013; Feng et al. 2014). The Cas9 protein induces the double-stranded DNA breaks (DSBs) by recruiting the gRNA. An important requirement for

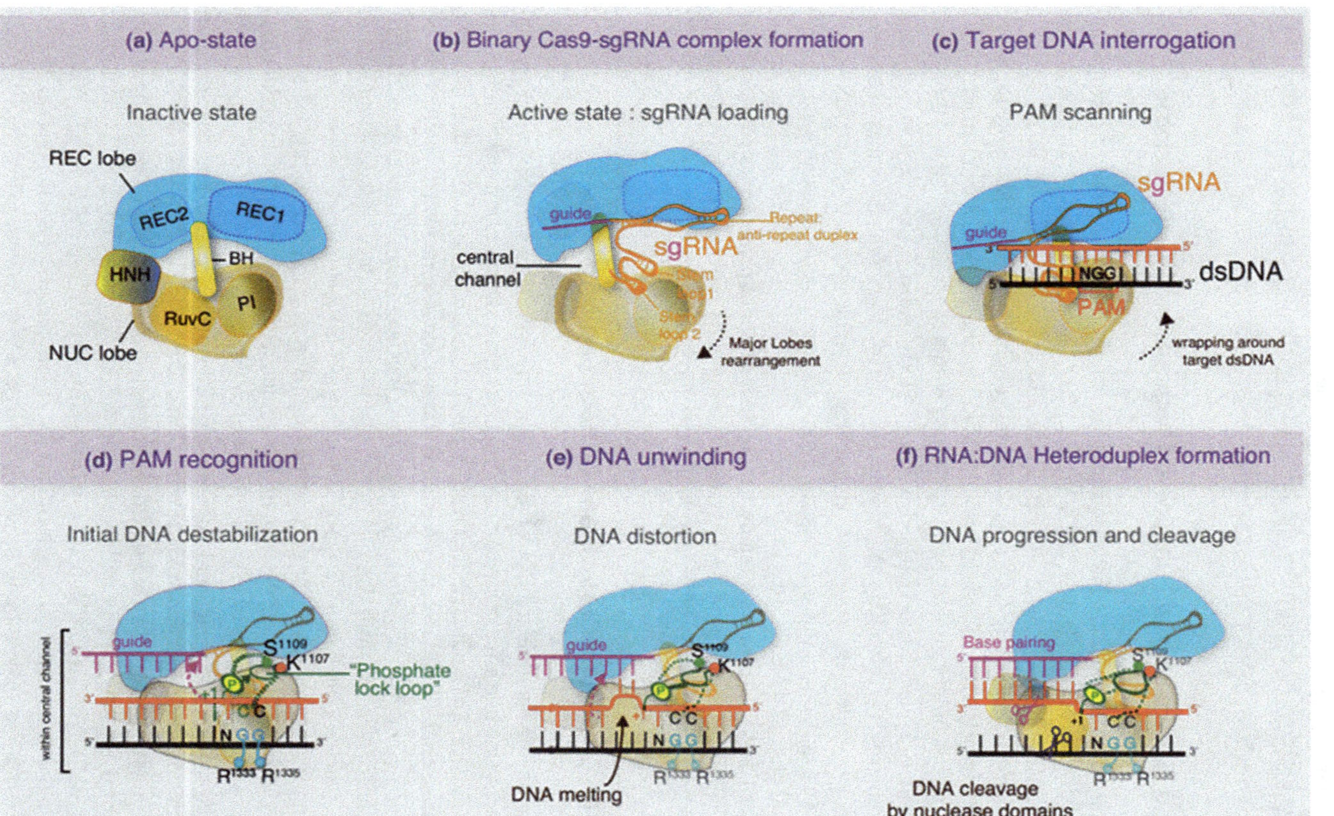

Fig. 13.39 The mechanism of DNA cleavage mediated by RNA-guided Cas9 nuclease. (**a**) Cas9 has a bilobed structure: a large recognition lobe (REC), and a small nuclease lobe (NUC) Cas9 protein remains in an auto-inhibitory mode. (**b**) Loading of 19–20 bp sgRNA activates Cas9 to form a Cas9-sgRNA complex

Fig. 13.39 (continued) with a modified central channel for placement of a RNA-DNA heteroduplex. (**c**) The NUC domain reorients leading to wrapping of Cas9 around DNA to look for a PAM motif. (**d**) Recognition of PAM by the PI domain (a part of NUC lobe) leads two arginine residues (R^{1333} and R^{1335}) of major grooves to read out the GG dinucleotide of PAM. The two lysine and serine residues (K^{1107} and S^{1109}) of the minor groove interact with the PAM sequence on the complimentary strand to create phosphate lock and destablize the DNA. (**e**) Target DNA further basepairs with the seed region (8–12 bp) of sgRNA, leading to melting of DNA upstream of (1–2 bp) of PAM. (**f**) The RNA-DNA heteroduplex forms, and the seed region further pairs up with the complementary DNA leading to further separation of DNA. Then both complementary and noncomplementary DNA strands are cleaved at +3 position, upstream of PAM by HNH and RuvC nuclease domains (part of NUC lobe) (adapted from Belhaj et al. 2015; Fig. 13.39) (Source: Belhaj K, Chaparro-Garcia A, Kamoun S, Patron NJ and Nekrasov V (2015) Editing plant genomes with CRISPR/Cas9. Curr. Opin. Biotechnol 32:76–84.printed with License number 4807521046305)

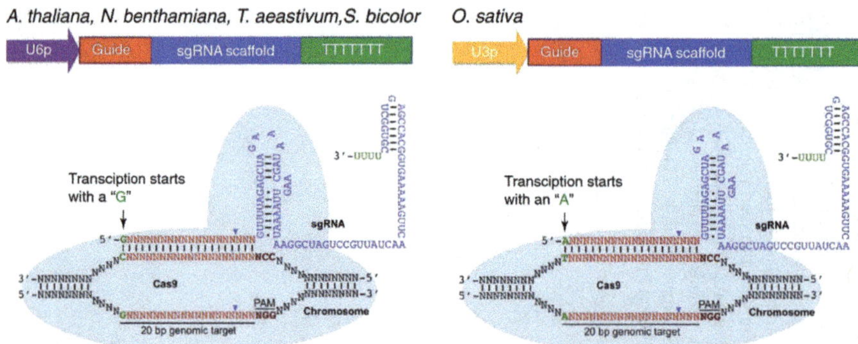

Fig. 13.40 SgRNA target recognition. The first nucleotide for transcription initiation is 'G' in case of *U6* promoters and 'A' for *U3* promoters (Source: Belhaj K, Chaparro-Garcia A, Kamoun S, Nekrasov V. (2013) Plant genome editing made easy: targeted mutagenesis in model and crop plants using the CRISPR/Cas system. Plant Methods.;9(1):39. Published 2013 Oct 11. doi:10.1186/1746-4811-9-39. This is an open-access article distributed under the terms of the Creative Commons Attribution License (http://creativecommons.org/licenses/by/2.0), which permits unrestricted use, distribution, and reproduction in any medium, provided the original work is properly cited)

DNA cleavage is the presence of a conserved protospacer adjacent motif (PAM), usually carrying the sequence 5′-NGG-3′ downstream of the target DNA (Gasiunas et al. 2012; Jinek et al. 2012), but sometimes can have NAG (Hsu et al. 2013). The DSBs can be repaired by the cellular NHEJ pathway resulting in insertions and/or deletions (indels), thus disrupting the target locus. When these mutations occur, or are induced in coding region of a gene, it leads to frameshift mutation and in most cases generate a gene knockout (Bortesi and Fischer 2015).

Since 2013, the CRISPR/Cas9 system has been successfully applied for gene editing in a wide range of plants such as *Arabidopsis*, *Nicotiana benthamiana*, rice wheat, and barley (Li et al. 2013; Nekrasov et al. 2013) and food crops rice, sorghum, and tomato (Jiang et al. 2013; Miao et al. 2013; Brooks et al. 2014; Zhang and Zhou 2014). Editing of *Mlo* alleles in wheat produced powdery mildew-resistant plants (Wang et al. 2014). Recently CRISPR/Cas9 system was used to create gene mutations in barley. Lawrenson et al. (2015) *exploited the wheat promoter of the TaU6* snRNA *gene for* Cas9-mediated DNA editing of *HvPM19*, which encodes an ABA-inducible plasma membrane protein. Mutation frequencies of 10–23% were observed in the T0 generation, and induced mutations were transmitted to T2 plants independently of the T-DNA construct. Kapusi et al. (2017) used the Cas9 system to disrupt a barley endo-*N*-acetyl-β-D-glucosaminidase (ENGase) gene by employing the *OsU6* promoter to drive the sgRNA, reaching a Cas9-induced mutation frequency of 78%. However, both CRISPR tool had their own drawbacks. While Lawrenson et al. (2015) enriched mutations using restriction enzymes to identify mutated plants, Kapusi et al. (2017) analyzed a large number of explants to identify Cas9-positive plants.

Genome editing uses engineered nucleases as powerful tools to target specific DNA sequences to edit genes precisely in the genomes of both model and crop plants, as well as a variety of other organisms.

According to Osakabe and Osakabe (2015), numerous examples of successful genome editing now exist. Genome editing uses engineered nucleases as powerful tools to target specific DNA sequences to edit genes precisely in the genomes of both model and crop plants, as well as a variety of other organisms. The DNA-binding domains of zinc finger (ZF) proteins were the first to be used as genome editing tools, in the form of designed ZF nucleases (ZFNs).

More recently, transcription activator-like effector nucleases (TALENs), as well as the clustered regularly inter-spaced short palindromic repeats/Cas9 (CRISPR/Cas9) system, which utilizes RNA–DNA interactions, have proved useful. A key step in genome editing is the generation of a double-stranded DNA break that is specific to the target gene. This is achieved by custom-designed endonucleases, which enable site-directed mutagenesis via a non-homologous end joining (NHEJ) repair pathway and/or gene targeting via homologous recombination (HR) to occur efficiently at specific sites in the genome. This review provides an overview of recent advances in genome editing technologies in plants and discusses how these can provide insights into current plant molecular biology research and molecular breeding technology.

Genome editing using engineered nucleases (GEEN) is an effective genetic engineering method that uses "molecular scissors," or artificially engineered nucleases, to target and digest DNA at specific locations in the genome. The engineered nucleases induce a double-stranded DNA break (DSB) at the target site that is then repaired by the natural processes of homologous recombination (HR) or non-homologous end joining (NHEJ) (Figs. 13.33 and 13.34). Numerous examples of successful genome editing have now been reported which provide insights into current plant molecular biology research and molecular breeding technology (Osakabe and Osakabe 2015). Currently, four types of engineered nucleases are used for genome editing: engineered homing endonucleases/meganucleases (EMNs), zinc finger nucleases (ZFNs), transcription activator-like effector nucleases (TALENs), and CRISPR (clustered regularly interspaced short palindromic repeats)/Cas9 (CRISPR-associated)9. In particular, TALEN and CRISPR/Cas9 are now used widely in various organisms.

13.11 Selectable Marker Genes

The production of transgenic plants usually requires the use of selection marker genes, which enables the selection of genetically modified cells, and their regeneration into whole plants. For this reason, genes coding for antibiotic resistance are frequently used. These genes have no intentional function in the genetically modified organism and no agronomic or other value in agriculture.

Current methods of generating transgenic plants employ a—selectable marker gene, which is transferred together with any other gene of interest usually on the

T-DNA between its borders. The presence of a suitable marker is necessary to facilitate the detection of genetically modified plant tissue during its development.

There are two types of marker genes:

1. Selectable markers to protect the organism against a selective agent, based on antibiotic resistance, herbicide tolerance, and metabolic marker genes.
2. Markers for screening. A marker for screening will make cells containing the gene "look" different.

The most frequently used marker genes in GM plants are the kanamycin resistance gene (neomycin phosphotransferase II, NPT II) from the bacterial transposable element Tn5 and the hygromycin resistance gene associated with hygromycin phosphotransferase (hpt).

Hygromycin B and kanamycin are inhibitors of RNA translation. Herbicide-tolerant marker genes are screenable selectable markers, which make the identification of transformed progeny efficient and fast. The Bar and pat genes have been widely used as selectable markers for plant transformation and are ideal for large- or small-scale screening to identify transformants that can easily be done in the greenhouse or in the field. Also herbicide screening has been used successfully to identify transformants. Transgenic cells and plants expressing the pat (phosphinothricin acetyltransferase isolated from soil bacteria) gene are able to degrade the herbicide agent phosphinothricin (glufosinate). This gene is resistant to the herbicides basta, bialaphos, and ignite.

Metabolic marker genes are:

2-DOG system
2-deoxyglucose is a plant growth inhibitor. Transgenes (2-deoxyglucose-6-phosphate-phosphatase) can grow on medium containing 2-Dog.—Palatinose system

Usually, plant cells are not able to assimilate palatinose. Therefore, if this is the carbohydrate source in the medium, no growth occurs. On the other hand, the transgenes containing palatinase can grow on medium containing palatinose, by converting it into fructose + glucose.

Mannose system
Mannose cannot be metabolized in plants. Transgenes with mannose6-phosphate-isomerase can grow on medium containing mannose.

13.12 Reporter Genes

In order to identify transformed cells or plants that have been growing on a selective medium, it is necessary to have an easily assayable reporter gene. The most useful reporter genes encode an enzyme activity not found in the organism being studied. A number of these genes are currently being used (cf. uidA, GFP, DsRED, Luc, LuxA/LuxB).

13.12.1 β-Glucuronidase (GUS)

The *E. coli* β-glucuronidase (GUS) is one of the most popular reporter genes currently in use. The protein has a molecular weight of 68,200 and appears to function as a tetramer. It is very stable and will tolerate many detergents, widely varying ionic conditions, and general stress. It is most active in the presence of thiol-reducing agents such as β-mercaptoethanol or DTT. It may be assayed at any physiological pH, with an optimum between 5.2 and 8.0. The GUS gene, like the GFP, can be used for overexpression (Fig. 13.41), and promoter activity assay (Fig. 13.42). It can be used also in gene fusion. This means that the GUS coding sequence is under the direction of the controlling sequence of another gene.

Agrobacterium containing some GUS plasmids show significant GUS activity. This seems to be due in part to read through transcription from the gene in which the GUS coding region might be located. *Agrobacterium* without these constructs shows little, if any detectable GUS activity. In order to solve this problem, one laboratory has constructed GUS genes carrying an intron, which has to be processed before expression takes place. This totally eliminates expression in any untransformed system (Vancanneyt et al. 1990).

13.12.1.1 Procedures for Assay of GUS Gene Expression Histochemical Assay

The best substrate currently available for the histochemical localization of β-glucuronidase activity in tissues and cells is 5-bromo-4-chloro-3-indolyl glucuronide (X-Gluc; see Table 13.3). The substrate works very well, giving a blue precipitate by incubation at 37 °C for 8–16 h at the site of enzyme activity. There

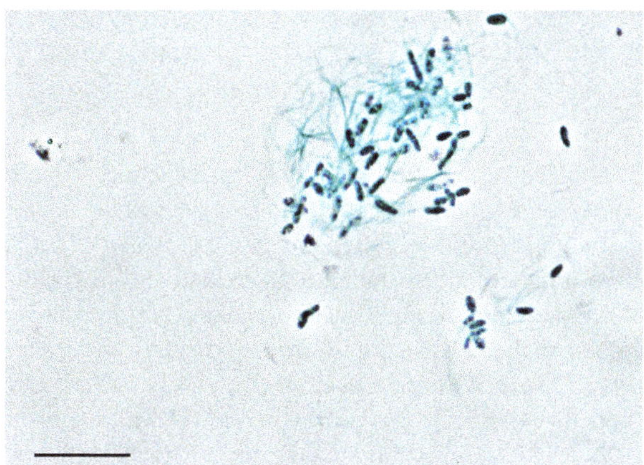

Fig. 13.41 Endomytotic bacteria *Rhizobium radiobacter* (RrF4, Sharma et al. 2008) expressing GUS protein (Bar scale 10 μm) (Source: Sharma M., Schmid M., Rothballer M., Hause G., Zuccaro A., Imani J., et al. (2008). Detection and identification of bacteria intimately associated with fungi of the order Sebacinales. *Cell Microbiol.* 10 2235–2246. 10.1111/j.1462-5822.2008.01202.x (Free access))

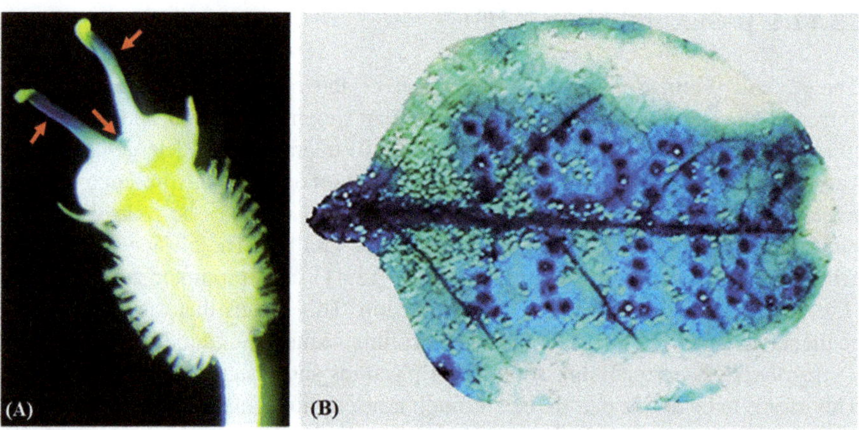

Fig. 13.42 Promoter study using GUS reporter gene. (**a**) The auxin activity (blue color, arrow) could be detected on the base of the anthers in the filament in transgenic carrot flower (Sect. 13.7.3, Fig. 13.19). (**b**) GUS activity in wound-induced tobacco leaf containing MAS::GUS construck. The letter *IPAZ* indication of Institute of Phytopathology and Applied Zoology were punched by a needle and 1 h later inoculated in substrate X-Gluc (13.12.1.1). The blue color is indication of the inducibility of MAS promoter by the wound

Table 13.3 Compositions of X-Gluc buffer for GUS assay (10 ml; store at −20 °C after filter (0.2 μ) sterilizing)

Component	Amount
N,N-Dimethylformamide	10–15 drops
X-Gluc	10 mg
0.1 M Phosphate buffer (KH$_2$PO$_4$/K$_2$HPO$_4$), pH 7.0	9.8 ml
5 mM potassium ferricyanide	100 μl
5 mM potassium ferrocyanide	100 μl
Triton X-100	10 μl

are numerous variables that affect the quality of histochemical localization, including all aspects of tissue preparation and fixation, as well as the reaction itself.

The product of glucuronidase action on X-Gluc is not colored. Indeed, the indoxyl derivative produced must undergo an oxidative dimerization to form the insoluble, highly colored indigo dye. This dimerization is stimulated by atmospheric oxygen and can be enhanced by using an oxidation catalyst such as a K$^+$ ferricyanide/ferrocyanide mixture. Without a catalyst, the results are often very good, but one has to consider the possibility that nearby peroxidases may enhance the apparent localization of glucuronidase. Fixation conditions will vary with the tissue and its permeability to the fixative. Glutaraldehyde can be used; it does not easily penetrate leaf cuticle but is immediately available to stem cross sections. Formaldehyde seems to be a more gentle fixative than glutaraldehyde and can be used for longer periods of time.

Whole tissues, callus, suspension culture cells, and protoplasts or whole plants or plant organs can be stained, but the survival of stained cells is not certain. After staining, clearing the tissue with 70% ethanol seems to improve the contrast in many cases.

13.12.1.2 Fluorometric Assay

Fluorometry is preferred over spectrophotometry because of its greatly increased sensitivity and wide dynamic range. Although spectrophotometric substrates for GUS are available, GUS activity in solution is usually measured with the fluorometric substrate 4-methylumbelliferyl-b-D-glucuronide (MUG). The assay is highly reliable and simple to use.

So far, the bacterial β-glucuronidase (GUS) gene has been the most common scorable marker used for plant transformation. However, GUS-stained tissues and cells cannot be used for plant regeneration, and therefore the GUS marker is not useful for selection purposes.

Production of green fluorescent protein (GFP) activity, anthocyanin pigmentation, or luciferase activity can be used as visual markers to identify transformed cells and tissues of plants. Detection of GFP activity or anthocyanin production is relatively simple, whereas assay of luciferase activity requires treatment of the tissues with a substrate and use of a rather expensive equipment to detect emitted photons. To increase the activity produced from GFP, the GFP gene has been engineered by gene shuffling (Stemmer 1994) to produce an enzyme that has a 42-fold higher activity than the wild-type enzyme (Crameri et al. 1996). Further modifications to the GFP gene have been made to adapt the codon bias to that of monocot plants (Pang et al. 1996). Recently, Vain et al. (1998) demonstrated the use of GFP for visual selection of transformed rice plants. So far, GFP has been the most promising scorable marker for identification of transformed cells and tissues in cereals.

13.12.2 Green Fluorescent Protein (GFP)

In addition to fluorescent stains, also fluorescent proteins are used. The most prominent of these is the green fluorescent protein (GFP; Chalfie et al. 1994; Tsien 1998; Chalfie and Kain 1998). Chalfie and Tsien received the 2008 Noble Prize in Chemistry for this discovery.

A general overview of this topic is given by Tsien (1998). This rather unusual molecule was detected in the 1960s in the luminescent jellyfish *Aequorea victoria*. The protein emits a green fluorescent in UV light. In modified forms, it has been used to make biosensors, and many animals have been created that express GFP as a proof of concept that a foreign gene can be expressed throughout a given organism. Reporter plants carrying GFP are used to analyze cytological events or to localize selected proteins. GFP is coupled to the gene to be analyzed. In most cases, the target protein is not disturbed, and GFP can be used as a chemical—flash light. The position of the protein to be investigated can be localized by the use of a fluorescent microscope. This provides the possibility not only to localize the protein but also to gain a fair estimation of its concentration. Additionally, it is possible to analyze new—switches for genes, i.e., promoters, and their activation in an organism. These very interesting properties have been used to analyze many physiological and biochemical problems. Meanwhile, besides GFP, some other fluorescent proteins, like DsRed of corals, have become known and are being used in molecular biology.

13.12.2.1 Variants of GFP

As mentioned above, the green fluorescent protein (GFP) has become an invaluable tool for pure and applied biological research. The inert nature of the protein, and the many potential uses of GFP, including the possibility to observe the behavior of proteins in living cells, were quickly recognized, and triggered the development of novel GFP variants with improved characteristics. An intensive search was initiated for other autofluorescent proteins, which fluoresce at different wavelengths than that of GFP. Mutagenesis of the wild-type gene yielded improved variants such as enhanced GFP (EGFP), as well as color variants such as the cyan (CFP) and yellow (YFP) fluorescent proteins. A recent paper has described a family of fluorescent proteins related to GFP. The most useful of these newly discovered proteins is DsRed, which is derived from the coral *Discosoma*. DsRed has an orange–red fluorescence with an emission maximum at 583 nm. It has a high quantum yield and is photostable. These characteristics make DsRed an ideal candidate for fluorescence imaging, particularly for multicolor experiments involving GFP and its variants. A codon-optimized version of DsRed is now available under the name DsRed1. The red fluorescent protein DrFP583 (DsRed) is the first true monomer, mRFP1, derived from the *Discosoma* sp. fluorescent protein (Geoffrey et al. 2000; Bevis and Glick 2002, Figs. 13.43 and 13.44).

Dendra

The first representative of a new class of photoactivatable fluorescent proteins that is capable of pronounced light-induced spectral changes is the monomeric variant Dendra from octocorals (*Dendronephthya* sp, red activatable), suitable for protein labeling. Dendra is capable of 1000- to 4500-fold photoconversion from the green to the red fluorescent states in response to either blue or UV light. It demonstrates high photostability of the activated state and can be photoactivated by a common, marginally phototoxic, 488-nm laser line. The suitability of Dendra for protein labeling and to quantitatively study the dynamics of fibrillarin and vimentin in mammalian cells is recommended (Gurskaya et al. 2002).

Proapoptotic Protein Bax

For promoter studies based on rapid gene answer by the fast killing of the plant cell, the proapoptotic protein Bax also seems suitable (Eichmann et al. 2006; Technologie-Lizenz-Buero der BadenWuertembergischen Hochschule GmbH, Germany). Figure 13.45 shows collapsed barley epidermal cells expressing Bax protein. Here, the leaves were transformed, via ballistic delivery of expression vectors, into single epidermal cells of barley leaf segments, according to a transient

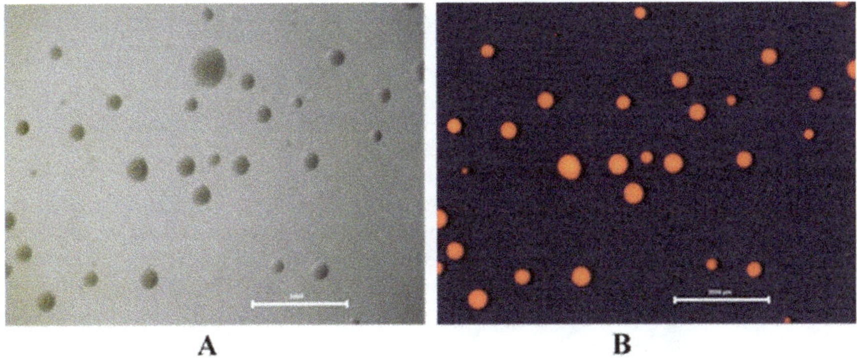

Fig. 13.43 Endomytotic bacteria Rhizobium radiobacter (RrF4, Sharma et al. 2008) expressing red fluorescent protein DsRed protein imaged with MZ16F, Germany. (**a**) Bright field. (**b**) Excitation with 488 nm and emission recorded at 620–670 nm (DsRed) (Sharma M., Schmid M., Rothballer M., Hause G., Zuccaro A., Imani J., et al. (2008). Detection and identification of bacteria intimately associated with fungi of the order Sebacinales. *Cell Microbiol.* 10 2235–2246. 10.1111/ j.1462-5822.2008.01202.x (Free access))

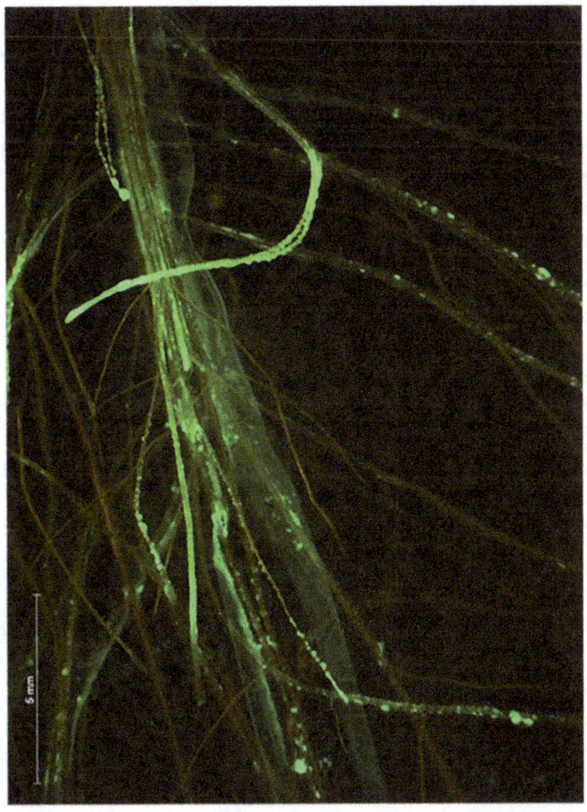

Fig. 13.44 Clonization of endomytotic bacteria Rhizobium radiobacter (RrF4, Sharma et al. 2008) expressing green fluorescent protein (GFP) imaged with Leica MZ16F, Germany (Kumar et al. unpublished)

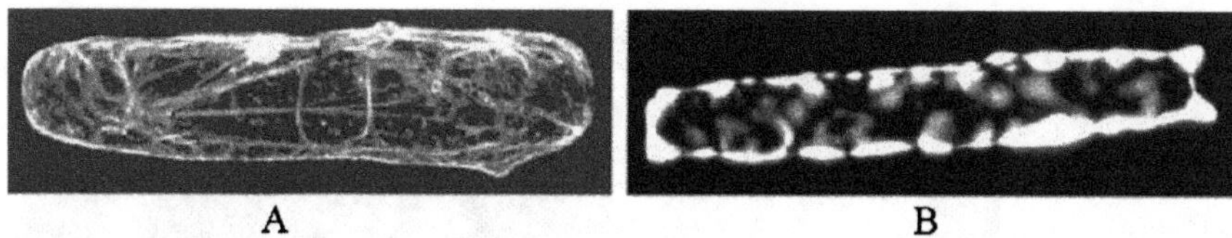

Fig. 13.45 Single-cell expression of a BAX gene from a mouse induces cell death in barley. (**a**) Confocal laser scanning whole cell projection of barley epidermal cells transiently expressing green fluorescent protein (GFP) as control. (**b**) The barley epidermal cell expressing BAX shows arrest of cytoplasmic streaming and induced fragmentation of the cytoplasm already 10 h after transformation (courtesy of Dr. Eichmann)

transformation protocol originally developed for wheat (Schweizer et al. 1999) Both GFP and the Bax gene were under control of the CaMV35S promoter (Fig. 13.45).

13.13 Antibiotics Resistance Genes

(ABR Genes, Restricted Use Recommended)
On April 2, 2004, a group of experts of the European Union (GMO panel of the European Nutritional Office EFSA) responsible for security problems in—green technology accepted a report on the use of antibiotics-resistant marker genes in gene technologically transformed plants. These experts declined a general prohibition and recommended a differential and cautious use. The use of some antibiotics-resistant markers would be prohibited, and the use of some others would be restricted. For the nptII marker gene employed in most transformed plants, no changes were recommended. Horizontal gene transfer from a transformed plant into microorganisms is very unlikely; still, the experts considered this a possibility. The frame of considerations was set by accepting the notion, as basis, that even in the case of such an unlikely horizontal transfer, human and environmental health would not be negatively influenced. The following criteria were applied:

- The medical significance of an antibiotic, its present use in human and veterinary medicine, and its efficiency to control some infectious diseases
- The natural distribution of antibiotics resistances in microorganisms of soil and water but also in the digestive systems of humans and other mammals

The experts recommended a release of transformed organisms into the environment only if they contain ABR marker genes occurring at large under natural conditions and inhibiting the action of only those antibiotics not in use in medicine.

As any other group of experts of EFSA, the GMO panel consists of independent and highly esteemed scientists. This group of scientists will usually be consulted in granting permission to grow transformed plants in the EU and present scientific recommendations. Decisions, however, are made by the political institutions—the

EU Commission, EU Counsel, and EU Parliament. Not all antibiotics resistance genes are similar. The GMO panel of EFSA categorizes antibiotics-resistant genes into three groups.

Group 1

These are ABR genes occurring at large in natural microbial communities. These antibiotics have either no or very limited significance for human or veterinary medicine (nptII gene for kanamycin resistance or hph gene for hygromycin resistance) example: nptII gene: this gene was abundantly used for many years to label transgenic plants. It was originally isolated from a transposon (jumping gene). It transmits resistance to several antibiotics, e.g., kanamycin, or neomycin. These are used only on patients not able to tolerate any other antibiotics. Kanamycin can have strong side effects. For marker genes of this group, it is assumed that their use in transgenic plants will have no influence on the group's already existing distribution in the environment. The EFSA experts recognized no arguments to restrict the use of these ABR genes. It was recommended to permit unlimited use of GM plants carrying these ABR marker genes for field experiments, as well as for commercial farming.

Group II

It is assumed that marker genes of this group used in transgenic plants would hardly have an influence on their present distribution. If an influence on the health of humans and animals exists, it would only be small. The EFSA experts recommended using these markers only in field experiments, but not in crops for the market. ABR genes are widely distributed in natural populations of microorganisms. Antibiotics are still prescribed to control specific diseases. To these belong the ampr gene (resistance to ampicillin), the aadA gene (resistance to streptomycin), and the Cmr gene (resistance to chloramphenicol).

Example: ampr gene: this gene transmits resistance to the antibiotic ampicillin. It originates from *E. coli* bacteria and is used in approved transgenic plants (Bt 176 corn). Generally, ampicillin is only rarely prescribed, to cure certain infectious diseases.

Group III

These are ABR genes transmitting resistance to antibiotics of great significance in controlling diseases in humans. Even if the effect of these antibiotics is not impaired by use in GM plants, this is not approved as a precaution, and the use of these genes should be avoided. The EFSA panel recommended that GM plants with such marker genes should be used neither for experimental purposes nor for commercial production.

Example: Npt gene: this gene transmits resistance to the antibiotic amikacin, an important storage antibiotic effective in controlling a number of infectious diseases.

In the near future, a prohibition to use antibiotics in transgenic plants can be expected.

One alternative could be the use of genes of D-amino acid-oxidase (dao1; Erikson et al. 2004) as selection markers:

I. D-Alanine and D-serine

$$\text{D-Alanine} + O_2 \xrightarrow[\text{Catalase}]{\text{Lactic Acid Dehydrogenase}} \text{Pyruvate} + NH_3$$

$$\text{Pyruvate} + \beta\text{-NADH} \xrightarrow{\text{Lactic Acid Dehydrogenase}} \text{Lactate} + \beta\text{-NAD}$$

where β-NADH is β-nicotinamide adenine dinucleotide, reduced form, and β-NAD is β-nicotinamide adenine dinucleotide, oxidized form.

II. D-valine and D-isoleucine do not affect plants but are metabolized by dao1 into keto acids, which inhibit plant growth.

13.14 Elimination of Marker Genes

Production of transgenic plants usually requires the use of selection marker genes, which enable the selection of genetically modified cells, and their regeneration into whole plants. For this purpose, as discussed above, genes coding for antibiotic resistance are frequently used. However, there is no agronomic or other value of the selection marker genes for the use of transgenic plants in agriculture. The presence of antibiotics resistance markers in transgenic plants intended for human or animal consumption may also be a cause of concern. Fears have been expressed that such genes may be transferred horizontally to microorganisms of the gut flora of man or animals, leading to the spread of antibiotics resistances in pathogenic microorganisms. Though extensive studies have failed to detect a measurable risk of this occurrence, many biotechnologists view the negative publicity related to the presence of unnecessary marker genes as sufficient reason to warrant their removal.

We present some common methods for the elimination of undesirable marker genes, which have already been successfully used in plant and animal cells.

13.14.1 Cre-lox Recombination-Based Systems

A group of site-specific recombinase enzymes that catalyze recombination at the specific target sequences have received considerable attention for the manipulation of heterologous genomes in vivo. Of these, the recombinase protein from bacteriophage P1 (Cre is a 38 kDa) mediates intramolecular (excisive or inversional) and intermolecular (integrative) site-specific recombination between loxP sites (locus of

X-ing over); it consists of two 13-bp inverted repeats separated by an 8-bp asymmetric spacer region (see review by Sauer 1993).

One molecule of Cre binds per inverted repeat or two Cre molecules line up at one loxP site. The recombination occurs in the asymmetric spacer region. Those eight bases are also responsible for the directionality of the site. Two loxP sequences, in opposite orientation to each other, invert the intervening piece of DNA, and two sites of same orientation dictate excision of the intervening DNA between the sites, leaving one loxP site behind. This precise removal of DNA can be used for the elimination of an endogenous gene or activation of a transgene (Fig. 13.46).

The advantage of this system is the automatic elimination during seed production—for example, when a seed-specific promoter is used. The following generation should therefore be marker gene-free.

Homologous recombination is efficient in chloroplasts (Corneille et al. 2001).

13.14.2 Ac/Ds System

The Nobel Prize winner Barbara McClintock in 1949 challenged the thesis that genes remain fixed at particular sites in a chromosome. She showed that genes can change their position and can move to other chromosomes. These—jumping genes or—transpones occur in many organisms. This phenomenon plays an important role in biology, because it contributes to the genetic variability of organisms. The transposons that have been most comprehensively characterized are those of the—Ac/Ds family. A transposon contains a gene for a particular enzyme

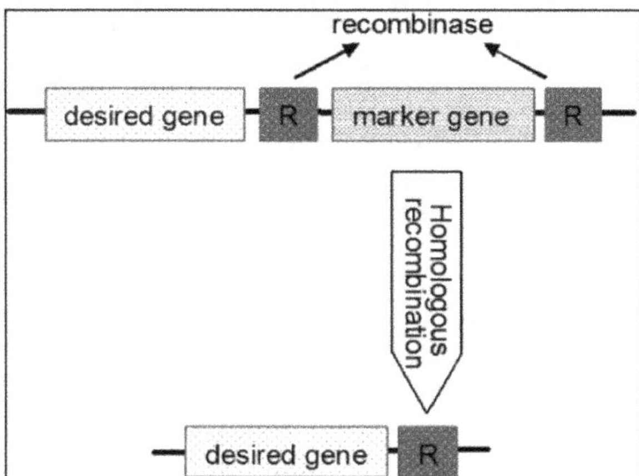

Fig. 13.46 General strategy for the excision of selectable marker genes. Between two identical sequence motives (R) that are recognized by a site-specific recombinase, the selectable marker gene is inserted into the transformation vector and used for the selection of transgenic plant cells. After expression of the corresponding recombinase, the marker gene is excised from the plant genome. Alternatively, recombination between the homologous overlaps could also result in marker gene elimination

(Ac transposase, activator element), which recognizes certain signals (Ds sequences, dissociation element) in the DNA, cuts pieces out of the DNA at these points, and reintegrates these into the genetic material at a different, unpredictable site. This ability can be exploited to remove undesirable marker genes from genetically modified plants (Fig. 13.47).

The marker-free GMP can be then selected by progeny segregation. The advantage of this system is not only to unlink the marker gene but also to create a series of plants with different transgene loci from one original transformant, which is particularly highly valued if recalcitrant plants have to be transformed. This repositioning enables expression of the transgene at different genomic positions and consequently at different levels. However, as segregation of the transgene and marker is required, and transposons tend to jump into linked positions, this approach is definitely more time-consuming than Cre-lox recombination-based systems.

13.14.3 Double Cassette System

The selectable marker in the transgenes also prevents retransformation with additional genes with the same selection procedure. Two cotransformation approaches are comparable: one *Agrobacterium* strain with one plasmid bearing two T-DNAs (double cassette) and two *Agrobacterium* strains each with one plasmid (one with the target gene and one with the selection gene) It is thus desirable to create marker-free transgenic plants. This was successfully achieved in tobacco, rice, and barley with binary—double cassette vectors, which carry two separate cassettes on the same plasmid, each bracketed by a left and a right T-DNA border sequence (Fig. 13.48; Komari et al. 1996; Slafer et al. 2002). One cassette contains the gene encoding βglucoronidase (uidA), and the other a hygromycin or kanamycin resistance gene.

Both T-DNA segments were co-transferred with a frequency of about 50% in the plants mentioned above. As the two cassettes were frequently inserted into different chromosomes, or chromosome arms, the resistance gene segregated independently from the marker gene in the T2 generation.

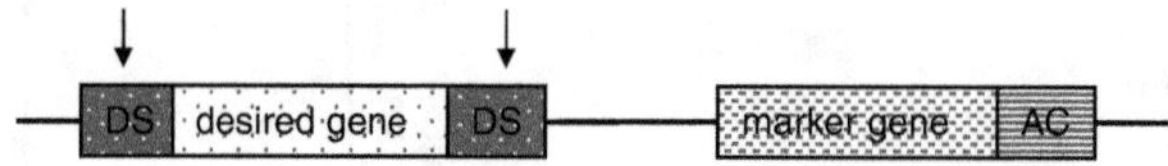

Fig. 13.47 Separation of two DNA sections after integration in the plant genome. The Ac transposase separates the gene construct at the sites annotated D (arrows). Then, the gene between the D sequences is moved to a different part of the genome and integrated there at random. As soon as the marker and target genes are located on different chromosomes, they are separated

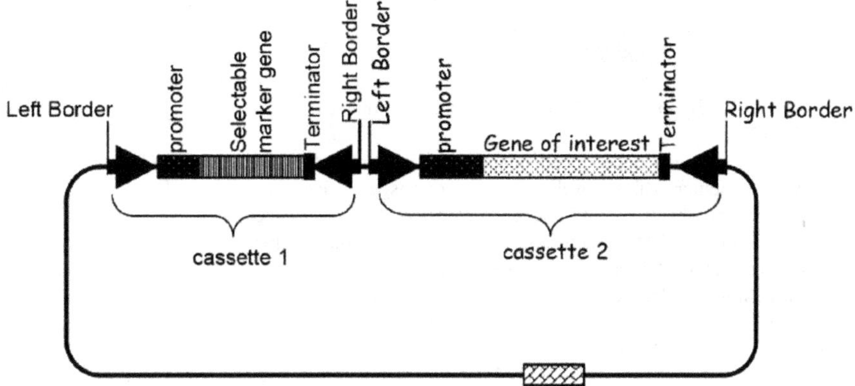

Fig. 13.48 Double cassette binary vector

13.15 Concluding Remarks

Agrobacterium tumefaciens is more than only the causative agent of crown gall disease affecting dicotyledonous plants. It now is also the natural system for the introduction of foreign genes into plants, enabling its genetic manipulation. Similarities have been found between T-DNA and conjugal transfer systems. They are evolutionarily related and apparently evolved from a common ancestor.

Although the gene transfer mechanisms remain largely unknown, great progress has been made in the practical implementation of transformation protocols for both dicotyledonous and monocotyledonous plants. Particularly important is the extension of this single-cell transformation methodology to monocotyledonous plants. This advance has biological and practical implications. Firstly, there are advantages of *A. tumefaciens*-mediated gene transfer over the direct transformation methods, which originally were the only available for genetic manipulation of economically important crops such as cereals and legumes. Secondly, it has been demonstrated that T-DNA is transferred to dicotyledonous and monocotyledonous plants by an identical molecular mechanism. This confirmation implies that potentially any plant species can be transformed by this method, if a suitable transformation protocol is established.

References

Abdellatef E, Will T, Koch A, Imani J, Vilcinskas A, Kogel KH (2015) Silencing the expression of the salivary sheath protein causes transgenerational feeding suppression in the aphid Sitobion avenae. Plant Biotechnol J. https://doi.org/10.1111/pbi.12322

Ambros V (1989) A hierarchy of regulatory genes controls a larva-to-adult developmental switch in C. elegans. Cell 57:49–57

Ankenbauer RG, Cangelosi GA, Nester EW (1990) Sugars induce the Agrobacterium virulence genes through a periplasmic binding protein and a transmembrane signal protein. Proc Natl Acad Sci USA 87:6708–6712

Arencibia A, Molina P, De la Riva G, Selman-Housein G (1995) Production of transgenic sugarcane (*Saccharum officinarum* L.) plants by intact cell electroporation. Plant Cell Rep 14:305–309

Arencibia A, Gentinetta E, Cuzzoni E, Castiglione S, Kohli A, Vain P, Leech M, Christou P, Sala F (1998) Molecular analysis of the genome of transgenic rice (*Oryza sativa* L.) plants produced via particle bombardment or intact cell electroporation. Mol Breed 4:99–109

Arnholdt-Schmitt B (1993) Rapid changes in amplification and methylation pattern of genomic DNA in cultured carrot root explants (*Daucus carota* L.). Theor Appl Genet 85:793–800

Arnholdt-Schmitt B, Heterich S, Neumann KH (1995) Physiological aspects of genome variability in tissue culture. 1. Growth phase—dependent differential DNA methylation of the carrot genome (*Daucus carota* L.) during primary culture. Theor Appl Genet 91:809–815

Ashburner M (1990) *Drosophila*: a laboratory handbook. Cold Spring Harbor Laboratory Press, Cold Spring Harbor, NY

Baker M (2012) Gene-editing nucleases. Nat Methods 9:23–26

Baldrich P, Rutter BD, Karimi HZ, Podicheti R, Meyers BC, Innes RW (2019) Plant extracellular vesicles contain diverse small RNA species and are enriched in 10- to 17-nucleotide "tiny" RNAs. Plant Cell 31:315

Ballmer-Weber BK, Wuethrich B, Wangorsch A, Fötisch K, Altmann F, Vieths S (2001) Carrot allergy: double-blinded, placebocontrolled food challenge and identification of allergens. J Allergy Clin Immunol 108:301–307

Ballmer-Weber BK, Wangorsch A, Bohlen B, Kaul S, Kündig T, Fötisch K, van Ree R, Vieth S (2005) Component-resolved in vitro diagnosis in carrot allergy: does the use of recombinant carrot allergens improve the reliability of the diagnostic procedure? Clin Exp Allergy 35:970–978

Barampuram S, Zhang ZJ (2011) Recent advances in plant transformation. Methods Mol Biol 701:1–35

Barcelo P, Lazzeri PA (1998) Direct gene transfer: chemical, electrical and physical methods. In: Lindsey K (ed) Transgenic plant research. Harwood Academic, Amsterdam, pp 1–34

Barrett C, Cobb E, McNicol R, Lyon G (1997) A risk assessment study of plant genetic transformation using *Agrobacterium* and implications for analysis of transgenic plants. Plant Cell Tissue Organ Cult 47:135–144

Belhaj K, Chaparro-Garcia A, Kamoun S, Nekrasov V (2013) Plant genome editing made easy: targeted mutagenesis in model and crop plants using the CRISPR/Cas system. Plant Methods 9 (1):39

Belhaj K, Chaparro-Garcia A, Kamoun S, Patron NJ, Nekrasov V (2015) Editing plant genomes with CRISPR/Cas9. Curr Opin Biotechnol 32:76–84

Benjavad Talebi A, Benjavad Talebi A, Shahrokhifar B (2012) Ethyl methane sulphonate (EMS) induced mutagenesis in Malaysian rice (cv. MR219) for lethal dose determination. Am J Plant Sci 3:1661–1665

Bernstein E, Caudy AA, Hammond SM, Hannon GJ (2001) Role for a bidentate ribonuclease in the initiation step of RNA interference. Nature 409:363–366

Bevan M (1984) Binary *Agrobacterium* vectors for plant transformation. Nucleic Acids Res 12:8711–8721

Bevis BJ, Glick BS (2002) Rapidly maturing variants of the *Discosoma* red fluorescent protein (DsRed). Nat Biotechnol 20:83–87

Birch RG (1997) Plant transformation: problems and strategies for practical application. Annu Rev Plant Physiol Plant Mol Biol 48:297–326

Blaschke JR (1977) Histologische, cytochcmische und biochemische Untersuchungen zur Charakterisierung des Kinetineinflusses auf die Entwicklung haploider und diploider Kalluskulturen von *Datura innoxia* Mill. Ph.D. thesis, Justus Liebig University, Giessen

Blaschke JR, Forche E, Neumann KH (1978) Investigations on the cell cycle of haploid and diploid tissue cultures of *Datura innoxia* Mill. and its synchronisation. Planta 144:7–12

Bohle B, Zwölfer B, Heratizadeh A, Jahn-Schmid B, Antonia YD, Alter M, Keller W, Zuidmeer L, van Ree R, Werfel T, Ebner C (2006) Cooking birch pollen-related food: divergent consequences for IgE- and T cell-mediated reactivity in vitro and in vivo. J Allergy Clin Immunol 118:242–249

Bortesi L, Fischer R (2015) The CRISPR/Cas9 system for plant genome editing and beyond. Biotechnol Adv 33(1):41–52

Bradley LR, Kim JS, Matthysse AG (1997) Attachment of *Agrobacterium tumefaciens* to carrot cells and Arabidopsis wound sites is correlated with the presence of a cell-associated, acidic polysaccharide. J Bacteriol 179:5372–5379

Bravo Angel AM, Hohn B, Tinland B (1998) The omega sequence of VirD2 is important but not essential for efficient transfer of the T-DNA by *Agrobacterium tumefaciens*. Mol Plant Microbe Interact 11:57–63

Brooks C, Nekrasov V, Lippman ZB, Eck JV (2014) Efficient gene editing in tomato in the first generation using the CRISPR/Cas9 system. Plant Physiol 166(3):1292–1297

Brown PTH, Lörz H (1986) Methylation changes in the progeny of tissue culture derived maize plants. In: Somers DA, Gengenbach BG, Biesboer DD, Hackett WP, Green CE (eds) Abstr VI Int. congress plant tissue and cell culture. University of Minnesota, Minneapolis, MN, p 261

Cai Q, Qiao L, Wang M, He B, Lin FM, Palmquist J, Huang SD, Jin H (2018) Plants send small RNAs in extracellular vesicles to fungal pathogen to silence virulence genes. Science 360:1126–1129

Cangelosi GA, Hung L, Puvanesarajah V, Stacey G, Ozga DA, Leigh JA, Nester EW (1987) Common loci for *Agrobacterium tumefaciens* and Rhizobium meliloti exopolysaccharide synthesis and their role in plant interaction. J Bacteriol 169:2086–2091

Cangelosi GA, Ankenbauer RG, Nester EW (1990) Sugars induce the Agrobacterium virulence genes through a periplasmic binding protein and a transmembrane signal protein. Proc Natl Acad Sci USA 87:6708–6712

Casas AM, Kononowicz AK, Zehr UB, Tomes DT, Axtell JD et al (1993) Transgenic sorghum plants via microprojectile bombardment. Proc Natl Acad Sci U S A 90:11212–11216

Cascardo JCM, Dias CV, Pungartnik C, Pirovani CP, Santos ACCD, Santos SC, Sena JAL, Valle RR, Vincentz M (2009) dsRNA-induced gene silencing in *Moniliophthora perniciosa*, the causal agent of witches' broom disease of cacao. Fungal Genet Biol 46:825–836

Chalfie M, Kain S (eds) (1998) GFP, green fluorescent protein: properties, applications, and protocols. Wiley-Liss, New York

Chalfie M, Horvitz HR, Sulston JE (1981) Mutations that lead to reiterations in the cell lineages of C. elegans. Cell 24:59–69

Chalfie M, Tu Y, Euskirchen G, Ward WW, Prasher DC (1994) Green fluorescent protein as a marker for gene expression. Science 263:802–805

Chang CH, Winans SC (1992) Functional roles assigned to the periplasmic, linker and receiver domains of the *Agrobacterium tumefaciens* VirA protein. J Bacteriol 174:7033–7039

Chen K, Gao C (2014) Targeted genome modification technologies and their applications in crop improvements. Plant Cell Rep 33:575–583. https://doi.org/10.1007/s00299-013-1539-6

Chen DS, Chen HMH, Sung JL, Hsu ST, Kuo YT, Lo KJ, Shih YT (1988) The Hepatitis Steering Committee, and the Hepatitis Control Committee. Control of hepatitis B virus infection in a hyperendemic area: a mass immunoprophylaxis programme in Taiwan. Viral Hepatitis and Liver Disease, A.R. Liss, New York, pp 971–976

Chen HL, Chang MH, Ni NY, Lee PI, Lee CY, Chen DS (1996) Seroepidemiology of hepatitis-B virus infection in children: ten years of mass vaccination in Taiwan. JAMA 276:906–908

Cheng M, Fry JE, Pang SZ, Zhou HP, Hironaka CM, Duncan DR, Conner W, Wan YC (1997) Genetic transformation of wheat mediated by *Agrobacterium tumefaciens*. Plant Physiol 115:971–980

Cheng X, Sardana R, Kaplan H, And Altosaar I (1998) *Agrobacterium* transformed rice plants expressing synthetic *cryIA(b)* and *cryI(c)* genes are highly toxic to striped stem borer and yellow stem borer. Proc Natl Acad Sci USA 95:2767–2772

Cheng M, Lowe BA, Spencer TM, Ye X, Armstrong CHL (2004) Factors influencing *Agrobacterium*-mediated transformation of monocotyledonous species. In Vitro Cell Dev Biol Plant 40:31–45

Christian M, Cermak T, Doyle EL, Schmidt C, Zhang F, Hummel A, Bogdanove AJ, Voytas DF (2010) Targeting DNA double-strand breaks with TAL effector nucleases. Genetics 186:757–761

Clough SJ, Bent AF (1998) Floral dip: a simplified method for Agrobacterium-mediated transformation of Arabidopsis thaliana. Plant J 16:735–743

Cong L, Ran FA, Cox D, Lin S, Barretto R, Habib N, Hsu PD, Wu X, Jiang W, Marraffini LA, Zhang F (2013) Multiplex genome engineering using CRISPR/Cas systems. Science 339 (6121):819–823

Corneille S, Lutz K, Svab Z, Maliga P (2001) Efficient elimination of selectable marker genes from the plastid genome by the CRE-lox site-specific recombination system. Plant J 27:171–178

Crameri A, Whitehorn EA, Tate E, Stemmer WPC (1996) Improved green fluorescent protein by molecular evolution using DNA shuffling. Nat Biotechnol 14:315–319

Cristea S, Freyvert Y, Santiago Y, Holmes MC, Urnov FD, Gregory PD et al (2013) In vivo cleavage of transgene donors promotes nuclease-mediated targeted integration. Biotechnol Bioeng 110(3):871–880

D'Amato F, Bennici A, Cionini PG, Baroncelli S, Lupi MC (1980) Nuclear fragmentation followed by mitosis as mechanism for wide chromosome number variation in tissue cultures, its implications for plant regeneration. In: Sala F, Parisi B, Cella R, Ciferri O (eds) Plant cell cultures: results and perspectives. Elsevier/North Holland, Amsterdam, pp 67–72

D'Halluin K, Bonne E, Bossut M, De Beuckeleer M, Leemans J (1992) Transgenic maize plants by tissue electroporation. Plant Cell 4:1495–1505

Dai S, Zheng P, Marmey P, Zhang S, Tian W, Chen S, Beachy RN, Fauquet C (2001) Comparative analysis of transgenic rice plants obtained by agrobacterium-mediated transformation and particle bombardment. Mol Breed 7:25–33

Das M, Bhattacharya S, Basak J, Pal A (2007) Phylogeneticrelationships among the bamboo species as revealed by morphologicalcharacters and polymorphism analyses. Biol Plant 51 (4):667–672

Datta SK, Misra P, Mandal AKA (2005) Mutagenesis—a quickmethod for establishment of solid mutant in chrysanthemum. Curr Sci 88:155–158

De Block M, Debrouwer D, Moens T (1997) The development of a nuclear male sterility system in wheat. Expression of the *barnase* gene under the control of tapetum specific promoters. Theor Appl Genet 95:125–131

De Cleene M, De Ley J (1976) The host range of crown gall. Bot Rev 42:389–466

De Souza N (2011) Primer: genome editing with engineered nucleases. Nat Methods 9(1):27–27. https://doi.org/10.1038/nmeth.1848

Deshmukh S, Hückelhoven R, Schäfer P, Imani J, Sharma M, Weiss M, Waller F, Kogel K-H (2006) The root endophytic fungus *Piriformospora indica* requires host cell death for proliferation during mutualistic symbiosis with barley. Proc Natl Acad Sci USA 103:18450–18457

Doshi MK, Eudes F (2008) Identity preservation in genetically modified crops. In: Kumar A, Sopory S (eds) Recent advances in plant biotechnology and its applications. I.K. International, New Delhi, pp 303–329

Doty SL, Yu NC, Lundin JI, Heath JD, Nester EW (1996) Mutational analysis of the input domain of the VirA protein of *Agrobacterium tumefaciens*. J Bacteriol 178:961–970

Dürrenberger F, Crameri A, Hohn B, Koukolikova-Nicola Z (1989) Covalently bound VirD2 protein of *Agrobacterium tumefaciens* protects the T-DNA from exonucleolytic degradation. Proc Natl Acad Sci USA 86:9154–9158

Ehsani P, Khabiri A, Domansky NN (1997) Polypeptides of hepatitis B surface antigen produced in transgenic potato. Gene 190:107–111

Eichmann R, Dechert C, Kogel K-H, Hückelhoven R (2006) Transient over-expression of barley BAX Inhibitor-1 weakens oxidative defense and MLA12-mediated resistance to *Blumeria graminis* f.sp. *hordei*. Mol Plant Pathol 7(6):543–552

Enríquez-Obregón GA, Vázquez-Padrón RI, Prieto-Samsónov DL, Pérez M, Selman-Housein G (1997) Genetic transformation of sugarcane by *Agrobacterium tumefaciens* using antioxidants compounds. Biotecnol Apl 14:169–174

Enríquez-Obregón GA, Vázquez-Padrón RI, Prieto-Sansonov DL, De la Riva GA, Selman-Housein G (1998) Herbicide resistant sugarcane (*Saccharum officinarum* L.) plants by *Agrobacterium*-mediated transformation. Planta 206:20–27

Erikson O, Herzberg M, Näsholm T (2004) A conditional marker gene allowing both positive and negative selection in plant. Nat Biotechnol 22(4):455–458

Erkes A, Reschke M, Boch J, Grau J (2017) Evolution of transcription activator-like effectors in Xanthomonas oryzae. Genome Biol Evol 9(6):1599–1615. https://doi.org/10.1093/gbe/evx108

Faisal M, Alatar AA, Ahmad N, Anis M, Hegazy AK (2012) Assessment of genetic fidelity in Rauvolfia serpentina plantletsgrown from synthetic (encapsulated) seeds following in vitrostorage at 4°C. Molecules 17:5050–5061

Fatima N, Ahmad N, Anis M, Ahmad I (2013) An improved in vitro encapsulation protocol, biochemical analysis and genetic integrity using DNA based molecular markers in regenerated plants of Withania somnifera L. Ind Crops Prod 50:468–477

Feng Z, Mao Y, Xu N, Zhang B, Wei P, Yang DL, Wang Z, Zhang Z, Zheng R, Yang L, Zeng L, Liu X, Zhu JK (2014) Multigeneration analysis reveals the inheritance, specificity, and patterns of CRISPR/Cas-induced gene modifications in *Arabidopsis*. Proc Natl Acad Sci USA 111:4632–4637

Fellmann C, Lowe SW (2014) Stable RNA interference rules for silencing. Nat Cell Biol 16:10–18

Filichkin SA, Gelvin SB (1993) Formation of a putative relaxation intermediate during T-DNA processing directed by *Agrobacterium tumefaciens* VirD1/D2 endonuclease. Mol Microbiol 8:915–926

Finer JJ, Vain P, Jones MW, McMullen MD (1992) Development of the particle inflow gun for DNA delivery to plant cells. Plant Cell Rep 11:323–328

Finer JJ, Finer KR, Ponappa T (1999) Particle bombardment mediated transformation. In: Hammond J, McGarvey P, Yusibov V (eds) Plant biotechnology. Springer, Berlin, pp 59–80

Fire A, Xu S, Montgomery MK, Kostas SA, Driver SE, Mello CC (1998) Potent and specific genetic interference by double-stranded RNA in *Caenorhabditis elegans*. Nature 391:806–811

Forche E, Foroughi B, Mix A, Neumann KH (1979) Untersuchungen zur Ploidieverteilung in Gerstenpflanzen (*Hordeum vulgare* L.) aus Antherenkulturen. Zeitschr Pflanzenzücht 83:222–235

Frame BR, Drayton PR, Bagnall SV, Lewnau CJ, Bullock WP, Wilson HM, Dunwell JM, Thompson JA, Wang K (1994) Production of fertile transgenic maize plants by silicon carbide whisker-mediated transformation. Plant J 6:941–948

Francia E, Tacconi G, Crosatti C, Barabaschi D, Bulgarelli D, Dall'Aglio E, Valè G (2005) Marker assisted selection in crop plants. Plant Cell Tissue Organ Cult 82(3):317–342

Froese C, Neumann KH (1997) The influence of Kinetin on the protein synthesis pattern at some stages of the cell suspension cultures of *Daucus carota* L. Angew Bot 71:11–115

Fromm ME, Morrish F, Armstrong C, Williams R, Thomas J, Klein TM (1990) Inheritance and expression of chimeric genes in the progeny of maize plants. Bio/technology 8:833–839

Fullner KJ, Lara JC, Nester EW (1996) Pilus assembly by Agrobacterium T-DNA transfer genes. Science 273:1107–1109

Gamborg OL, Miller RA, Ojima K (1968) Nutrient requirements of suspension cultures of soybean root cultures. Exp Cell Res 50:151–158

Gao LF, Jing RL, Huo NX, Li Y, Li XP, Zhou RH, Chang XP, Tang JF, Ma ZY, Jia JZ (2004) One hundred and one new microsatellite loci derived from ESTs (EST-SSRs) in bred wheat. Theor Appl Genet 108:1392–1400

Gasiunas G, Barrangou R, Horvath P, Siksnys V (2012) Cas9-crRNA ribonucleoprotein complex mediates specific DNA cleavage for adaptive immunity in bacteria. Proc Natl Acad Sci USA 109(39):E2579–E2586

Gelvin SB (1988) The introduction and expression of transgenes in plants. Curr Opin Biotechnol 9:227–232

Geoffrey SB, Zacharias DA, Tsien RY (2000) Biochemistry, mutagenesis, and oligomerization of DsRed, a red fluorescent protein from coral. Proc Natl Acad Sci USA 97:11984–11989

Gheysen G, Villarroel R, Van Montagu M (1989) Illegitimate recombination in plants: a model for T-DNA integration. Genes Dev 5:287–297

Gless C, Lörz H, Jähne-Gärtner A (1998) Transgenic oat plants obtained at high efficiency by microprojectile bombardment of leaf base segments. J Plant Physiol 152:151–157

Graner A, Streng S, Kellermann A, Schiemann A, Baner E, Waugh R, Pellio B, Ordon F (1999) Molecular mapping of the rym5 locus encoding resistance to different strains of the barley yellow mosaic virus complex. Theor Appl Genet 98:285–290

Grewal IS, Moazed D (2003) Heterochromatin and epigenetic control of gene expression. *Science* 301:798–802

Grewal SIS, Hal IM, Kidner C, Martienssen RA, Teng G, Volpe TA (2002) Regulation of heterochromatic silencing and histone H3 lysine-9 methylation by RNAi. Science 297:1833–1837

Gronover CS, Kasulke D, Tudzynski P, Tudzynski B (2001) The role of G protein alpha subunits in the infection process of the gray mold fungus *Botrytis cinerea*. Mol Plant Microbe Interact 14 (11):1293–1302

Gurskaya NG, Verkusha VV, Shcheglov AS, Staroverov DB, Hamilton RJ, Voinnet O, Chappell L, Baulcombe D (2002) Two classes of short interfering RNA in RNA silencing. EMBO J 21:4671–4679

Hannon GJ, He L (2004) MicroRNAs: small RNAs with a big role in gene regulation. Nat Rev Genet 5:522–531

Hansen G, Chilton MD (1996) Agrolistic transformation of plant cells: integration of T-strands generated *in planta*. Proc Natl Acad Sci USA 93:14978–14983

Hansen G, Shillito RD, Chilton MD (1997) T-strand integration in maize protoplasts after codelivery of a T-DNA substrate and virulence genes. Proc Natl Acad Sci USA 94:11726–11730

Hayta S, Smedley MA, Demir SU, Blundell R, Hinchliffe A, Atkinson N, Harwood WA (2019) An efficient and reproducible agrobacterium-mediated transformation method for hexaploid wheat (Triticum aestivum L.). Plant Methods 15:121. https://doi.org/10.1186/s13007-019-0503-z

He GY, Lazzeri PA (1998) Analysis and optimization of DNA delivery into wheat scutellum and tritordeum inflorescence explants by tissue electroporation. Plant Cell Rep 18:64–70

Herrera-Estrella L (1983) Transfer and expression of foreign genes in plants. PhD thesis, Laboratory of Genetics, Gent University, Belgium

Herrera-Estrella A, Van Montagu M, Wang K (1990) A bacterial peptide acting as a plant nuclear targeting signal: the amino-terminal portion of Agrobacterium VirD2 protein directs the b-galactosidase fusion protein into tobacco nuclei. Proc Natl Acad Sci USA 87:9534–9537

Hiei Y, Komari T, Kubo T (1997) Transformation of rice mediated by Agrobacterium tumefaciens. Plant Mol Biol 35:205–218

Hooykaas P (2010) Plant transformation. In: *Encyclopedia of life sciences (ELS)*. Wiley, New York

Hsu PD, Scott DA, Weinstein JA, Ran FA, Konermann S, Agarwala V, Li Y, Fine EJ, Wu X, Shalem O et al (2013) DNA targeting specificity of RNA-guided Cas9 nucleases. Nat Biotechnol 31:827–832

Hsu PD, Lander ES, Zhang F (2014) Development and applications of CRISPR-Cas9 for genome engineering. Cell 157(6):1262–1278. https://doi.org/10.1016/j.cell.2014.05.010

Huang Y, Morel P, Powell B, Kado CI (1990) VirA, a coregulator of Ti-specified virulence genes, is phosphorylated in vitro. J Bacteriol 172:1142–1144

Hückelhoven R (2004) BAX inhibitor-1, an ancient cell death suppressor in animals and plants with prokaryotic relatives. Apoptosis 9:299–307

Imani J, Berting A, Nitsche S, Schaefer S, Gerlich WH, Neumann KH (2002) The integration of a major hepatitis B virus gene into cell-cycle synchronized carrot cell suspension cultures and its expression in regenerated carrot plants. Plant Cell Tissue Organ Cult 71:157–164

Imani J, Baltruschat H, Stein E, Jia J, Vogelsberg J, Kogel K-H, Hueckelhoven R (2006) Expression of barley BAX inhibitor-1 in carrots confers resistance to Botrytis cinerea. Mol Plant Pathol 7 (4):279–284

Imani J, Doil A, Winkelmann T, Kogel K-H (2007) Enhancement of genetic transformation rates in plants using cell cycle synchronized suspension cultures. In: Abstr Int Conf Plant Transformation Technologies (PTT), 4–7 February 2007, Wien, p 62

Imani J, Li L, Schäfer P, Kogel KH (2011) STARTS—a stable root transformation system for rapid functional analyses of proteins of the monocot model plant barley. Plant J. https://doi.org/10.1111/j.1365-313X.2011.04620.x

Ishida Y, Saito H, Ohta S, Hiei Y, Komari T, Kumashiro T (1996) High efficiency transformation of Maize (Zea mays L.) mediated by Agrobacterium tumefaciens. Nat Biotechnol 14:745–750

Jacobs M, Nigruti I, Dirks R, Cammaerts R (1987) Selection programs for isolation and analysis of mutants in plant cell cultures. In: Green CE, Somers DA, Hackett WP, Biesboer DD (eds) Plant tissue and cell culture. A.R. Liss, New York, pp 243–264

Jauhar PP, Chibbar RN (1999) Chromosome-mediated and direct gene transfers in wheat. Genome 42:570–583

Jefferson RA (1987) Assaying chimeric genes in plants: the GUS gene fusion system. Plant Molec Biol Rep 5:387–405

Jensen CJ (1983) Producing haploid plants by chromosome elimination. In: Cell and tissue culture techniques for cereal crop improvement. Science Press, Beijing, pp 55–79

Jiang W, Bikard D, Cox D, Zhang F, Marraffini LA (2013) RNA-guided editing of bacterial genomes using CRISPR-Cas systems. Nat Biotechnol 31(3):233–239. https://doi.org/10.1038/nbt.2508

Jimenez-Lopez JC, Hernandez MC (2012) Biochemical testing methodes for agricultueran and food safety. In: Jimenez-Lopez JC (ed) Biochemical testing. IntechOpen. https://doi.org/10.5772/38873

Jin S, Prusti RK, Roitsch T, Ankenbauer RG, Nester EW (1990a) The VirG protein of Agrobacterium tumefaciens is phosphorylated by the autophosphorylated VirA protein and this is essential for its biological activity. J Bacteriol 172:4945–4950

Jin S, Roitisch T, Christie PJ, Nester EW (1990b) The regulatory VirG protein specifically binds to a cis acting regulatory sequence involved in transcriptional activation of Agrobacterium tumefaciens virulence genes. J Bacteriol 172:531–562

Jin S, Song Y, Pan S, Nester EW (1993) Characterization of a virG mutation that confers constitutive virulence gene expression in Agrobacterium tumefaciens. Mol Microbiol 7:555–562

Jinek M, Chylinski K, Fonfara I, Hauer M, Doudna JA, Charpentier E (2012) A programmable dual-RNA-guided DNA endonuclease in adaptive bacterial immunity. Science 337:816–821

Jinek M, Jiang F, Taylor DW, Sternberg SH, Kaya E, Ma E, Anders C, Hauer M, Zhou K, Lin S, Kaplan M, Iavarone AT, Charpentier E, Nogales E, Doudna JA (2014) Structures of Cas9 endonucleases reveal RNA-mediated conformational activation. Science 343:1247997

Joung JK, Sander JD (2013) TALENs: a widely applicable technology for targeted genome editing. Nat Rev Mol Cell Biol 14(1):49–55. https://doi.org/10.1038/nrm3486

Kapusi E, Corcuera-Gómez M, Melnik S, Stoger E (2017) Heritable genomic fragment deletions and small indels in the putative ENGase gene hvinduced by CRISPR/Cas9 in barley. Front Plant Sci 8:540. https://doi.org/10.3389/fpls.2017.00540

Kapusta J, Modelska A, Figlerowicz M, Pniewski T, Letellier M, Lisowa O, Yusibov V, Koprowski H, Plucienniczak A, Legocki AB (1999) A plant derived edible vaccine against hepatitis B virus. FASEB J 13:1796–1799

Karimi M, Inzé D, Depicker A (2002) GATEWAY vectors for Agrobacterium-mediated plant transformation. Trends Plant Sci 7:193–195

Khatodia S, Bhatotia K, Passricha N, Khurana SMP, Tuteja N (2016) The CRISPR/Cas genome-editing tool: application inimprovement of crops. Front Plant Sci 7:506. https://doi.org/10.3389/fpls.2016.00506

Kibler R, Neumann KH (1980) On cytogenetic stability of cultured tissues and cell suspensions of haploid and diploid origin. In: Sala F, Parisi B, Cella R, Ciferri O (eds) Plant cell cultures: results and perspectives. Elsevier/North Holland, Amsterdam, pp 59–65

Kim JC, Choi SJ (1998) Transformation system of rice suspension-cultured microcolonies by electroporation. J Plant Biol 41:193–200

Klee HJ, Rogers SG (1989) Plant gene vectors and genetic transformation: plant transformation systems based on the use of Agrobacterium tumefaciens. In: Schell J, Vasil IK (eds) Cell culture and somatic cell genetics of plants, vol 6. Academic Press, San Diego, pp 1–23

Knight CJ, Bailey AM, Foster GD (2010) Investigating Agrobacterium-mediated transformation of Verticillium albo-atrum on plant surfaces. PLoS ONE 5:1–5

Koch A, Kogel K-H (2014) New wind in the sails: improving the agronomic value of crop plants through RNAi-mediated gene silencing. Plant Biotechnol J 12:821–831

Kohli A, Griffiths S, Palacios N, Twyman RM, Vain P, Laurie DA, Christou P (1999) Molecular characterization of transforming plasmid rearrangements in transgenic rice reveals a recombination hotspot in the CaMV 35S promoter and confirms the predominance of microhomology mediated recombination. Plant J 17:591–601

Komari T, Hiei Y, Saito Y, Murai N, Kumashiro T (1996) Vectors carrying two T-DNA for cotransformation of higher plants mediated by Agrobacterium tumefaciens and segregation of transformants free from selection markers. Plant J 10(1):165–174

Komari T, Hiei Y, Ishida Y, Kumashiro T, Kubo T (1998) Advances in cereal gene transfer. Curr Opin Plant Biol 1:161–165

Koncz C, Németh K, Redei GP, Scell J (1994) Homology recognition during T-DNA integration into the plant genome. In: Paszkowski J (ed) Homologous recombination and gene silencing in plants. Kluwer, Dordrecht, pp 167–189

Kumar A, Sopory S (eds) (2008) Recent advances in plant biotechnology and its applications. I.K. International, New Delhi

Kumar A, Shekhawat NS (2009) Plant tissue culture and molecular markers: their role in improving crop productivity. New Delhi, I.K. International, 688 pp

Kumar A, Sopory S (eds) (2010) Applications of plant biotechnology: in vitro propagation, plant transformation and secondary metabolite production. New Delhi, I.K. International, 606 pp

Kumar S, Ganapathi TR, Bapat VA (2007) Production of hepatitis B surface antigen in recombinant plant systems: an update. Biotechnol Prog 23:532–539

Kumar A, Sharma M, Basu SK, Asif M, Li X, Chen X (2014) Plant molecular breeding: perspectives from plant biotechnology and marker assisted selection. Am J Soc Hum 4:177–189

Kumpatla SP, Hall TC (1998) Recurrent onset of epigenetic silencing in rice harboring multi-copy transgene. Plant J 14:129–135

Kumpatla SP, Teng W, Buchholz WG, Hall TC (1997) Epigenetic transcriptional silencing and 5-azacytidine-mediated reactivation of a complex transgene in rice. Plant Physiol 115:361–373

Langridge WHR, Fitzgerald KJ, Koncz C, Scheu J, Szalay A (1989) Dual promoter of Agrobacterium tumefaciens mannopine gyotbase genes is regulated by plant growth hormones. Proc Natl Acad Sci USA 80:3214–3223

Larkin PJ, Scowcroft WR (1981) Somaclonal variation—a novel source of variability from cell cultures for plant improvement. Theor Appl Genet 60:197–214

Larkin PJ, Brettell RIS, Ryan SA, Davies PA, Palotta WA, Scowcroft WR (1985) Somaclonal variation: impact on plant biology and breeding strategies. In: Zaitlin M, Day P, Hollaender A, Wilson CM (eds) Biotechnology in plant science: relevance to agriculture in the eighties. Academic Press, Orlando, FL, pp 83–100

Lawrenson T, Shorinola O, Stacey N, Li C, Østergaard L, Patron N, Uauy C, Harwood W (2015) Induction of targeted, heritable mutations in barley and Brassica oleracea using RNA-guided Cas9 nuclease. Genome Biol 16:258. https://doi.org/10.1186/s13059-015-0826-7

Lazo GR, Stein PA, Ludwig RA (1991) A DNA transformation-competent Arabidopsis genomic library in *Agrobacterium*. Biotechnology 9:963–967

Lee SH, Shon YG, Lee SI, Kim CY, Koo JC, Lim CO, Choi YJ, Han CD, Chung CH, Choe ZR, Cho MJ (1999) Cultivar variability in the Agrobacterium-rice cell interaction and plant regeneration. Physiol Plant 107:338–345

Lehman CW, Trautman JK, Carroll D (1994) Illegitimate recombination in Xenopus: characterization of end-joined junctions. Nucleic Acid Res 22:434–442

Leva AR, Petruccelli R, Rinaldi LMR (2012) Somaclonal variation in tissue culture: a case study with olive. In: Leva A, Rinaldi LMR (eds) Recent advances in plant in vitro culture. IntechOpen, pp 123–150. https://doi.org/10.5772/50367

Li T, Neumann KH (1985) Embryogenesis and endogenous hormone content of cell cultures of some carrot varieties (Daucus carota L.). Ber Deut Bot Ges 98:227–235

Li X, Huang KL, Zhu BZ, Tang MZ, Luo YB (2005) Potentiality of omics techniques for the detection of unintended effects in genetically modified crops. J Agric Biotechnol 13:1082–1088

Li W, Yuan JS, Stewart CN (2013) Advanced genetic tools for plant biotechnology. Nat Rev Genet 14:781–793. https://doi.org/10.1038/nrg3583

Liu J, Li C, Yu Z, Huang P, Wu H, Wei C, Zhu N, Shen Y, Chen Y, Zhang B, Deng WM, Jiao R (2012) Efficient and specific modifications of the drosophila genome by means of an easy TALEN strategy. J Genet Genomics 39:209–215

Liu W, Yuan JS, Stewart CN (2013) Advanced genetic tools for plant biotechnology. Nat Rev Genet 14:781–793. https://doi.org/10.1038/nrg3583

Liu S, Jaouannet M, Dempsey DA, Imani J, Coustau C, Kogel KH (2020) RNA-based technologies for insect control in plant production. Biotechnol Adv 39:107463

Lorenz H (2006) Replikation von drei Säuger-Hepadnaviren im amerikanischen Waldmurmeltier (Marmota monax) und Expression der viralen Oberflächenproteine in transgenen Pflanzen. GEB, Giessener Elektronische Bibliothek, Dissertation, Fachbereichs Veterinärmedizin, Justus Liebig Universität, Giessen

Lörz H, Baker B, Schell J (1985) Gene transfer to cereal cells mediated by protoplast transformation. Mol Gen Genet 199:473–497

Loschiavo F, Pitto L, Guiliano G, Torti G, Nuti-Ronchi V, Marazziti D, Vergara R, Orselli S, Terzi M (1989) DNA methylation of embryogenic carrot cell cultures and its variations as caused by mutation, differentiation, hormones and hypomethylating drugs. Theor Appl Genet 77:325–331

Machczyńska J, Zimny J, Bednarek PT (2015) Tissue culture-induced genetic and epigenetic variation in triticale (× Triticosecale spp. Wittmack ex A. Camus 1927) regenerants. Plant Mol Biol 89:279. https://doi.org/10.1007/s11103-015-0368-0

MacRae IJ, Zhou K, Li F, Repic A, Brooks AN, Cande WZ, Adams PD, Doudna JA (2006) Structural basis for double-stranded RNA processing by Dicer. Science 311:195–198

Mali P, Aach J, Stranges PB, Esvelt KM, Moosburner M, Kosuri S, Yang L, Church GM (2013) CAS9 transcriptional activators for target specificity screening and paired nickases for cooperative genome engineering. Nat Biotechnol 31:833–838

Maresca M, Lin VG, Guo N, Yang Y (2013) Obligate ligation-gated recombination (ObLiGaRe): custom-designed nuclease-mediated targeted integration through nonhomologous end joining. Genome Res 23(3):539–546

Mason HS, Lam DMK, Amtzen CJ (1992) Expression of hepatitis B surface antigen in transgenic plants. Proc Natl Acad Sci USA 89:11745–11749

Mason HS, Ball JM, Shi JJ, Jiang X, Estes MK, Amtzen CJ (1996) Expression of Norwalk virus capsid protein in transgenic tobacco and potato and its oral immunogenicity in rice. Proc Natl Acad Sci USA 93:5335–5340

Matsushita J, Otani M, Wakita Y, Tanaka O, Shimada T (1999) Transgenic plant regeneration through silicon carbide wiskermediated transformation of rice (Oryza sativa L.). Breed Sci 49:21–26

Matthews PR, Wang MB, Waterhouse PM, Thornton S, Fieg SJ, Gubler F, Jacobsen JV (2001) Marker gene elimination from transgenic barley, using co-transformation with adjacent 'twin T-DNAs' on a standard Agrobacterium transformation vector. Mol Breed 7:195–202

Matthysse AG (1986) Initial interactions of Agrobacterium tumefaciens with plant host cells. Crit Rev Microbiol 13:281–307

Matzke M, Kanno T, Daxinger L, Huettel B, Matzke AJ (2009) RNA-mediated chromatin-based silencing in plants. Curr Opin Cell Biol 21(3):367–376

Maynard JE, Kaue MA, Aller JM, Halder CH (1986) Control of hepatitis B by ünmunization: global perspectives. Viral hepatitis and Ihrer disease. A.R. Liss, New York, pp 967–969

McCabe D, Christou P (1993) Direct DNA transfer using electrical discharge particle acceleration (Accell technology). Plant Cell Tissue Organ Cult 33:227–236

Miao J, Guo D, Zhang J, Huang Q, Qin G, Zhang X, Wan J, Gu H, Qu L-J (2013) Targeted mutagenesis in rice using CRISPR-Cas system. Cell Res. https://doi.org/10.1038/cr.2013

Misra P, Datta SK, Chakrabarty D (2003) Mutation in flowercolour and shape of chrysanthemum by using gamma-radiation. Biol Plant 47:153–156

Mohan M, Nair S, Bhagwat A, Krishna TG, Yano M, Bhatia CR, Sasaki T (1997) Genome mapping, molecular markers and marker-assisted selection in crop plants. Mol Breed 3:87–103

Moynahan ME, Jasin M (2010) Mitotic homologous recombination maintains genomic stability and suppresses tumorigenesis. Nat Rev Mol Cell Biol 11:196–207

Muller HJ (1927) Artificial transmutation of the gene. Science 66:84–87

Muller HJ (1928) The problem of genic modification. In: Proc. 5th Int. Congress of Genetics, Berlin, pp 234–260

Murray F, Bishop D, Mathews P, Jacobsen J (2001) Improving barley transformation efficiency. In: Proceedings of the 10th Australian Barley Technical Symposium

Mussolino C, Cathomen T (2013) RNA guides genome engineering. Nat Biotechnol 31:208–209. https://doi.org/10.1038/nbt.2527

Nagl W (1970) Karyologische Anatomie des Endosperms von Phaseolus coccineus. Plant Syst Evol 128:566–557

Nekrasov V, Staskawicz WD, Jones J, Kamoun S (2013) Targeted mutagenesis in th model plant Nicotiana benthamiana using Cas9 RNA-guided endonuclease. Nat Biotechnol 31:691–693. https://doi.org/10.1038/nbt.2655

Neumann KH (1966) Wurzelbildung und Nukleinsäuregehalt bei Phloem-Gewebekulturen der Karottenwurzel auf synthetischem Nährmedium. Congr Coll Univ Liege 38:96–102

Neumann KH (1995) Pflanzliche Zell-und Gewebekulturen. Verlag Eugen Ulmer, Stuttgart, p 304

Nguyen TA, Greig J, Khan A (2018) Evolutionary novelty in gravity sensing through horizontal gene transfer and high-order protein assembly. PLoS Biol. https://doi.org/10.1371/journal.pbio.2004920

Nishimasu H, Ran FA, Hsu PD et al (2014) Crystal structure of Cas9 in complex with guide RNA and target DNA. Cell 156(5):935–949. https://doi.org/10.1016/j.cell.2014.02.001

Nivina A, Escudero JA, Vit C, Mazel D, Loot C (2016) Efficiency of integron cassette insertion in correct orientation is ensured by the interplay of the three unpaired features of attC recombination sites. Nucleic Acids Res 44(16):7792–7803

Novak FJ (1991) Mutation breeding by using tissue culture techniques. In: Gamma Field Symposia No.30. Inst. of Radiation Breeding, NIAR, MAFF, Japan, pp 23–32

Novak FJ, Daskalov S, Brunner H, Nesticky M, Afza R, Dolezelova M, Lucretti S, Herichova A, Hermelin T (1988) Somatic embryogenesis in maize and comparison of genetic variability induced by gamma radiation and tissue culture techniques. Plant Breed 101:66–79

Okubara P (2016) You ain't seen nothing yet. Can new biotech methods tackle Rhizoctonia? Wheat life. Volume 59 Number 09. www.wheatlife.org

Oono K, Okuno K, Kawai T (1984) High frequency of somaclonal mutations in callus culture of rice. In: Gamma Field Symposia No.23. Inst. of Radiation Breeding, NIAR, MAFF, Japan, pp 71–94

Orbovic V, Calovic M, Viloria Z, Nielsen B, Gmitter F, Castle W, Grosser J (2008) Analysis of genetic variability in various tissue culturederived lemon plant populations using RAPD and flow cytometry. Euphytica 161:329–335

Osakabe Y, Osakabe K (2015) Genome editing with engineered nucleases in plants. Plant Cell Physiol 56(3):389–400

Ossowski S, Schwab R, Weigel D (2008) Gene silencing in plants using artificial microRNAs and other small RNAs. Plant J 53:674–690. https://doi.org/10.1111/j.1365-313X.2007.03328.x

Pan SQ, Charles T, Jin S, Wu ZL, Nester EW (2003) Preformed dimeric state of the sensor protein VirA is involved in plant-Agrobacterium signal transduction. Proc Natl Acad Sci USA 90:9939–9943

Pang SZ, DeBoer DL, Wan Y, Ye G, Layton JG, Neher MK, Armstrong CL, Fry JE, Hinchee MAW, Fromm ME (1996) An improved green fluorescent green protein gene as a vital marker in plants. Plant Physiol 112:893–900

Parkinson JS (1993) Signal transduction schemes of bacteria. Cell 73:857–871

Pawlowski WP, Somers DA (1996) Transgene inheritance in plants genetically engineered by microprojectile bombardment. Mol Biotechnol 6:17–30

Pawlowski WP, Somers DA (1998) Transgenic DNA integrated into the oat genome is frequently interspersed by host DNA. Proc Natl Acad Sci USA 95:12106–12110

Peters S, Imani J, Mahler V, Foetisch K, Kaul S, Paulus KE, Scheurer S, Vieths S, Kogel KH (2010) Reduced allergenicity of carrot roots harvested from Dau c 1.01 and Dau c 1.02-silenced transgenic carrot plants. Transgenic Res 20:547–556. https://doi.org/10.1007/s11248-010-9435-0

Plasterk RHA, Brunschwig K, Marcel T, Müller F, Okihara KL, Vastenhouw NL (2006) Long-term gene silencing by RNAi. Nature 442:882

Pollegioni L, Schonbrunn L, Siehl D (2011) Molecular basis of glyphosate resistance: different approaches through protein engineering. FEBS J 278(16):2753–2766. https://doi.org/10.1111/j.1742-4658.2011.08214.x

Potrykus I (1990) Gene transfer to cereals: an assessment. Bio/technology 8:535–542

Potrykus I (1991) Gene transfer to plants: assessment of published approaches and results. Annu Rev Plant Physiol Plant Mol Biol 42:205–225

Puchta H (1998) Repair of genomic double-strand breaks in somatic cells by one-side invasion of homologous sequences. Plant J 13:331–339

Rafalski JA (2002) Applications of single nucleotide polymorphisms in crop genetics. Curr Opin Plant Biol 5:94–100

Rasci-Gaunt S, Riley A, Cannell M, Barcelo P, Lazzeri PA (2001) Procedures allowing the transformation of a range of european elite wheat (Triticum aestivum L.) varieties via particle bombardment. J Exp Bot 52:865–878

Rassoulzadegan MR, Cuzin F, Gillot I, Gounon P, Grandjean V, Vincent S (2006) RNA-mediated non-mendelian inheritance of an epigenetic change in the mouse. Nature 441:469–474

Register JC, Peterson DJ, Bell PJ, Bullock WP, Evans EJ, Frame B, Greenland AJ, Higgs NS, Jepson I, Jiao S, Lewnau CJ, Sillik JM, Wilson HM (1994) Structure and function of selectable and non-selectable transgenes in maize after introduction by particle bombardment. Plant Mol Biol 25:951–961

Reyon D, Tsai SQ, Khayter C, Foden JA, Sander JD, Joung JK (2012) FLASH assembly of TALENs for high-throughput genome editing. Nat Biotechnol 30:460–465

Ritala A, Aspegren K, Kurtén U, Salmenkallio-Marttila M, Mannonen L, Hannus R, Kauppinen V, Teeri T, Enari T-M (1994) Fertile transgenic barley by particle bombardment of immature embryos. Plant Mol Biol 24:317–325

Ritschel WA, Forusz H (1994) Chronopharmacology. A review of drugs studied. Meth Find Exp Clin Pharmacol 16:57–75

Rivera AL, Gomez-Lim M, Fernandez F, Loske AM (2012) Physical methods for genetic plant transformation. Phys Life Rev 9:308–345

Roger P, Hellens E, Edwards A, Leyland NR, Bean S, Mullineaux M (2000) pGreen: a versatile and flexible binary Ti vector for Agrobacterium-mediated plant transformation. Plant Mol Biol 42:819–832

Rossi L, Hohn B, Tinland B (1993) Vir D2 protein carries nuclear localization signals important to transfer of T-DNA to plants. Mol Gen Genet 239:345–353

Russell JA, Roy MK, Sanford JC (1992) Physical trauma and tungsten toxicity reduce the efficiency of biolistic transformation. Plant Physiol 98:1050–1056

Saleh M-C, van Rij RP, Hekele A, Gillis A, Foley E, O'Farrell PH, Andino R (2006) The endocytic pathway mediates cell entry of dsRNA to induce RNAi silencing. Nat Cell Biol 8:793

Sanford JC, Wolf ED, Allen NK (1990) Method for transporting substances into living cells and tissues and apparatus therefor. US Patent #4945050

Sanford JC, Smith FD, Russell JA (1993) Optimizing the biolistic process for different biological applications. Methods Enzymol 217:483–509

Sauer B (1993) Manipulation of transgenes by site-specific recombination: use of cre recombinase. Methods Enzymol 225:890–900

Sautter C, Waldner H, Neuhaus-Url G, Galli A, Niehaus G, Potrykus I (1991) Micro targeting: high efficiency gene transfer using a novel approach for the acceleration of microparticles. Bio/technology 9:1080–1085

Schultheiss H, Hensel G, Imani J, Sonnewald U, Kogel K-H, Kumlehn J, Hückelhoven R (2005) Ectopic expression of constitutively activated RACB in barley enhances susceptibility to powdery mildew and abiotic stress. Plant Physiol 139(1):353–362

Schweizer P, Pokorny J, Abderhalden O, Dudler R (1999) A transient assay system for the functional assessment of defense-related genes in wheat. Mole Plant Microbe Interact 12:647–654

Scott JG, Michel K, Bartholomay LC, Siegfried BD, Hunter WB, Smagghe G, Zhu KY, Douglas AE (2013) Towards the elements of successful insect RNAi. J Insect Physiol 59(12):1212–1221

Shan Q, Wang Y, Li J, Zhang Y, Chen K, Liang Z, Zhang K, Liu J, Xi JJ, Qiu JL et al (2013) Targeted genome modification of crop plants using a CRISPR-Cas system. Nat Biotechnol 31:686–688. https://doi.org/10.1038/nbt.2650

Sharma M, Schmid M, Rothballer M, Hause G, Zuccaro A, Imani I, Kämpfer P, Schäfer P, Hartmann A, Kogel K-H (2008) Detection and identification of bacteria intimately associated with fungi of the order Sebacinales. Cell Microbiol 10(11):2235–2224

Shillito R (1999) Methods of genetic transformation: electroporation and polyethylene glycol treatment. In: Vasil IK (ed) Molecular improvement of cereal crop. Kluwer Academic, London, pp 9–20

Shimoda N, Toyoda-Yamamoto A, Nagamine J, Usami S, Katayama M, Sakagami Y, Machida Y (1990) Control of expression of Agrobacterium tumefaciens genes by synergistic actions of phenolic signal molecules and monosaccharides. Proc Natl Acad Sci USA 87:6684–6688

Siewers V, Viaud M, Jimenez-Teja D, Collado IG, Gronover CS, Pradier JM, Tudzynski B, Tudzynski P (2005) Functional analysis of the cytochrome P450 monooxygenase gene bcbot1 of Botrytis cinerea indicates that botrydial is a strain-specific virulence factor. Mol Plant Microbe Interact 18(6):602–612

Sikora P, Chawade A, Larsson M, Olsson J, Olsson O (2011) Mutagenesis as a tool in plant genetics, functional genomics, and breeding. Int J Plant Genomics 314829:1–13. https://doi.org/10.1155/2011/314829

Skirvin RM (1978) Natural and induced variation from tissue culture. Euphytica 27:241–266

Slafer GA, Molina-Cano JL, Savin R, Araus JL, Romagosa I (2002) Barley science: recent advances from molecular biology to agronomy of yield and quality. Food Product Press, New York

Smith N (1998) More T-DNA than meets the eye. Trends Plant Sci 3:85. https://doi.org/10.1016/S1360-1385(98)01206-0

Smith EF, Townsend CO (1907) A plant tumor of bacterial origin. Science 25:671–673

Smith N, Singh SP, Wang MB, Stoutjesdijk PA, Green AG, Waterhouse PM (2000) Total silencing by intron-spliced hairpin RNAs. Nature 407:319–320

Somers DA, Rines HW, Gu W, Kaeppler HF, Bushnell WR (1992) Fertile, transgenic oat plants. Nat Biotechnol 10:1589–1594

Songstad DD, Somers DA, Griesbach RJ (1995) Advances in alternative DNA delivery techniques. Plant Cell Tissue Organ Cult 40:1–15

Sood P, Bhattacharya A, Sood A (2011) Problems and possibilities of monocot transformation. Biol Plant 55:1–15

Sorek R, Martin Lawrence C, Wiedenheftnheft B (2013) CRISPR-mediated adaptive immune systems in bacteria and archaea. Annu Rev Biochem 82:237–266. https://www.annualreviews.org/doi/abs/10.1146/annurev-biochem-072911-172315

Sridevi V, Giridhar P (2014) Establishment of somaclonal variants of robusta coffee with reduced levels of cafestol and kahweol. In Vitro Cell Dev Biol Plant 50(5):618–626

Staub T (1991) Fungicide resistance: practical experience with antiresistance strategies and the role of integrated use. Annu Rev Phytopathol 29:421–442

Stemmer WPC (1994) Rapid evolution of a protein in vitro by DNA shuffling. Nature 370:389–391

Steward PC, Mapes MO, Mears K (1952) Investigation on growth and metabolism of plant cells. I. New techniques for the investigation of metabolism, nutrition and growth in undifferentiated cells. Am J Bot 16:57–77

Stoger E, Williams S, Keen D, Christou P (1998) Molecular characteristics of transgenic wheat and the effect on transgene expression. Transgenic Res 7:463–471

Stuiver MH, Custers JH (2001) Engineering disease resistance in plants. Nature 865:968

Sukekiyo Y, Kimura Y (1991) Somaclonal variation in protoplast-derived rice plants. In: Gamma Field Symposia No.30. Inst. of Radiation Breeding, NIAR, MAFF, Japan, pp 43–58

Tenberge KB, Beckedorf M, Hoppe B, Schouten A, Solf M, von den Driesch M (2002) In situ localization of AOS in host–pathogen interactions. Microsc Microanal 8(Suppl 2):250–251

Thanawala Y, Yang YF, Lyons P, Mason HS, Arntzen C (1995) Immunogenicity of transgenic plant-derived hepatitis B surface antigen. Proc Natl Acad Sci USA 92:3358–3361

Thomas BR, Van Deynze A, Bradford KJ (2002) Production of therapeutic proteins in plants. Agricultural Biotechnology in California Series, Publication 8078

Thomashow MF, Karlinsey JE, Marks JR, Hurlbert RE (1987) Identification of a new virulence locus in Agrobacterium tumefaciens that affects polysaccharide composition and plant attachment. J Bacteriol 169:3209–3216

Tingay S, McElroy D, Kalla R, Fieg S, Wang M, Thornton S, Brettell R (1997) Agrobacterium tumefaciens-mediated barley transformation. Plant J 11:1369–1376

Tinland B, Schoumacher F, Gloeckler V, Bravo AM, Angel M, Hohn B (1995) The Agrobacterium tumefaciens virulence D2 protein is responsible for precise integration of T-DNA into the plant genome. EMBO J 14:3585–3595

Tong C, Huang G, Ashton C, Wu H, Yan H, Ying Q-L (2012) Rapid and cost-effective gene targeting in rat embryonic stem cells by TALENs. J Genet Genomics 39:275–280

Torisky RS, Kovacs L, Avdiushko S, Newman JD, Hunt AG, Collins GB (1997) Development of a binary vector system for plant transformation based on supervirulent *Agrobacterium tumefaciens* strain Chry5. Plant Cell Rep 17:102–108

Travella S, Ross SM, Harden J, Everett C, Snape JW, Harwood WA (2005) A comparison of transgenic barley lines produced by particle bombardment and Agrobacterium-mediated techniques. Plant Cell Rep 23:780–789

Tsien RY (1998) The green fluorescent protein. Annu Rev Biochem 67:509–544

Tsuda S, Yoshioka K, Tanaka T, Iwata A, Yoshikawa A, Watanabe Y, Okada Y (1998) Application of the human hepatitis B virus core antigen from transgenic tobacco plants for serological diagnosis. Vox Sang 74:148–155

Tuschl T, Elbashir SM, Harborth J, Lendeckel W, Yalcin A, Weber K (2001) Duplexes of 21-nucleotide RNAs mediate RNA interference in cultured mammalian cells. Nature 411:494–498

Tzfira T, Citovsky V (2006) Agrobacterium-mediated genetic transformation of plants: biology and biotechnology. Curr Opin Biotechnol 17:147–141

Ulker B, Li Y, Rosso MG, Logemann E, Somssich IE, Weisshaar B (2008) T-DNA-mediated transfer of Agrobacterium tumefaciens chromosomal DNA into plants. Nat Biotechnol 26:1015–1017

Unai E, Iselen T, de Garcia E (2004) Comparison of characteristics of bananas (Musa sp.) from the somaclone CIEN BTA-03 and its parental clone Williams. Fruit 59:257–263

Ushiyama K (1991) Large scale culture of ginseng. In: Komamine A, Misawa M, DiCosmo FE (eds) Plant cell culture in Japan. CMC, Tokyo, p 97

Vain P, Worland B, Kohli A, Snape JW, Christou P (1998) The green fluorescent protein (GFP) as a vital screenable marker in rice transformation. Theor Appl Genet 96:164–169

Vancanneyt G, Schmidt R, O'Connor-Sanchez A, Willmitzer L, Rocha-Rosa M (1990) Construction of an intron-containing marker gene: splicing of the intron in transgenic plants and its use in monitoring early events in Agrobacterium mediated plant transformation. Mol Gen Genet 220:245–250

Van Larebeke N, Engler G, Holsters M, Van den Elsacker S, Schilperoort RA, Schell J (1974) Large plasmid in Agrobacterium tumefaciens essential for crown gall-inducing ability. Nature 252 (5479):169–170. https://doi.org/10.1038/252169a0

Vasil IK (1987) Developing cell and tissue culture systems for the improvement of cereal and grass crops. J Plant Physiol 128:193–197

Vasil V, Vasil IK (1980) Isolation and culture of cereal protoplasts. II. Embryogenesis and plantlet formation from protoplasts of Pennisetum americanum. Theor Appl Genet 56:97–99

Vasil IK (1991) Plant tissue culture and molecular biology as tools for understanding plant development and plant improvement. Curr Opin Biotechnol 2:158–163

Von Tiedemann A (1997) Evidence for a primary role of active oxygen species in induction of host cell death during infection of bean leaves with Botrytis cinerea. Physiol Mol Plant Pathol 50:151–166

Wang L, Haeusler RA, Good PD, Thompson M, Nagar S, Engelke DR (2005) Silencing near tRNA genes requires nucleolar localization. J Biol Chem 280(10):8637–8639. https://doi.org/10.1074/jbc.C500017200

Wang MB, Upadhyaya NM, Li Z, Waterhouse PM (1998) Improved vectors for Agrobacterium tumefaciens-mediated transformation of monocot plant. Acta Hortic 461:401–407

Wang MB, Abbott DC, Upadhyaya NM, Jacobsen JV, Waterhouse PM (2001) Agrobacterium tumefaciens-mediated transformation of an elite Australian barley cultivar with virus resistance and reporter genes. Aust J Plant Physiol 28:149–156

Wang Y, Cheng X, Shan Q, Zhang Y, Liu J, Gao C, Qiu JL (2014) Simultaneous editing of three homoeoalleles in hexaploid bread wheat confers heritable resistance to powdery mildew. Nat Biotechnol 32:947–951

Wangorsch A, Weigand D, Peters S, Mahler V, Fötisch K, Reuter A, Imani J, DeWitt AM, Kogel KH, Lidholm J, Vieths S, Scheurer S (2011) Identification of a Dau c PRPlike protein (Dau c 1.03) as a new allergenic isoform in carrots (cultivar Rodelika). Clin Exp Allergy 42:156–166. https://doi.org/10.1111/j.1365-2222.2011.03900.x

Widholm JM (1987) Selected gene amplification, a review. In: Proc Int Botanical Congr, Berlin, pp 3-09-7

Winans SC (1992) Two way chemical signalling in Agrobacterium-plant interactions. Microbiol Mol Biol Rev 56:12–31

Winkelmann T, Hohe A, Pueschel AK, Schwenkel HG (2000) Somatic embryogenesis in Cyclamen persicum Mill. Curr Topics Plant Biol 2:51–62

Wolyn DJ, Borevitz JO, Loudet O, Scwartz C, Maloof J, Ecker JR, Berry CC, Chory J (2004) Light-response quantitative trait loci identified by composite interval mapping and eXtreme array mapping in Arabidopsis thaliana. Genetics 167:907–917

Wu J, Lin L (2003) Enhancement of taxol production and release in Taxus chinensis cell cultures by ultrasound, methyl jasmonate and in situ solvent extraction. Appl Microbiol Biotechnol 62:151–155

Wu J, Maehara T, Shimokawa T, Yamamoto S, Harada C, Takazaki Y, Ono N, Mukai Y, Koike K, Yazaki J, Fujii F, Shomura A, Ando T, Kono I, Waki K, Yamamoto K, Yano M, Matsumoto T, Sasaki T (2002) A comprehensive rice transcript map containing 6591 expressed sequence tag sites. Plant Cell 14:525–535

Yang M, Ewald D, Wang Y, Liang H, Zhen Z (2006) Survival and escape of Agrobacterium tumefaciens in triploid hybrid lines of Chinese white poplar transformed with two insect-resistant genes. Acta Ecol Sin 26:3555–3356

Yu X-D, Liu Z-C, Huang S-L, Chen Z-Q, Sun Y-W, Duan P-F, Ma Y-Z, Xia L-Q (2016) RNAi-mediated plant protection against aphids. Pest Manag Sci 72:1090–1098

Zadoks JC, Chang TT, Konzak CF (1974) A decimal code for the growth stages of cereals. Weed Res 14:415–421

Zaidi SS, Mansoor S (2017) Viral vectors for plant genome engineering. Front Plant Sci 8:539. https://doi.org/10.3389/fpls.2017.00539

Zamore PD, Bartel DP, Tuschl T, Sharp PA (2000) RNAi: double-stranded RNA directs the ATP-depenent cleavage of mRNA at 21 to 23 nucleotide intervals. Cell 101:25–33

Zhang L, Zhou Q (2014) CRISPR/Cas technology: a revolutionary approach for genome engineering. Sci China Life Sci 57:639–640. https://doi.org/10.1007/s11427-014-4670-x

Zhang S, Cho MJ, Koprek T, Yun R, Bregitzer P, Lemaux PG (1999) Genetic transformation of commercial cultivars of oat (Avena sativa L.) and barley (Hordeum vulgare L.) using in vitro shoot meristematic cultures derived from germinated seedlings. Plant Cell Rep 18:959–966

Zhang Y, Shan Q, Wang Y, Chen K, Liang Z, Li J, Gao C (2013) Rapid and efficient gene modification in rice and brachypodium using TALENs. Mol Plant 6:1365–1368

Zhao ZY, Cai T, Tagliani L, Miller M, Wang N, Pang H, Rudert M, Schroeder S, Hondred D, Seltzer J, Pierce D (2000) Agrobacterium-mediated sorghum transformation. Plant Mol Biol 44:789–798

Ziemienowicz A, Gorlich D, Lanka E, Hohn B, Rossi L (1999) Import of DNA into mammalian nuclei by proteins originating from a plant pathogenic bacterium. Proc Natl Acad Sci USA 96 (7):3729–3733

Ziemienowicz A, Merkle T, Schoumacher F, Hohn B, Rossi L (2001) Import of Agrobacterium T-DNA into plant nuclei: two distinct functions of VirD2 and VirE2 proteins. Plant Cell 13:369–383

Zotti M, Dos Santos EA, Cagliari D, Christiaens O, Taning CNT, Smagghe G (2018) RNA interference technology in crop protection against arthropod pests, pathogens and nematodes. Pest Manag Sci 74(6):1239–1250. https://doi.org/10.1002/ps.4813

Zupan JR, Citovsky V, Zambryski PC (1996) Agrobacterium VirE2 protein mediates nuclear uptake of single-stranded DNA in plant cells. Proc Natl Acad Sci USA 93:2392–2397

Summary of Some Physiological Aspects in the Development of Plant Cell and Tissue Culture

14

This summary will be divided into two parts, one dealing with physiological problems of cultured cells in vitro, the other with research results obtained using cell cultures as model systems to investigate physiological problems of plants in general. For both, some examples including also some of our own research activity will be presented and discussed.

In vitro cultured cells are isolated from the physiological situation and the control of the original plant from which they were isolated for culture, and at isolation from the donor plant, a wound is set by mechanical or enzymatic means. In particular the former drew the attention of Haberlandt, White, and other pioneers in the field. For the latter, some results from our own research program will be given first for primary explants. Using transgenic carrot plants containing the auxin-sensitive MAS promoter linked to the GUS gene as reporter to obtain primary explants, it could be shown that, within a few hours after setting, the wound adjacent cells show an increase in IAA concentration (see Chap. 11). IAA promotes the synthesis of ethylene, and consequently, ethylene is produced by the freshly isolated explants. The level of ethylene production soon decreases to a minimum, followed by an increase to a maximum value again after 5–6 days of cultivation. Whereas no clear influences of growth substances like IAA or kinetin in the medium on ethylene could be observed immediately after wounding—presumably, ethylene production was a mere wound reaction—distinct responses to these substances could be observed at the later maximum. Still, the responses to IAA (2 ppm) were rather low, compared to those following an additional supplement of kinetin (0.1 ppm). The number of cells produced by this treatment during a subsequent culture period greatly exceeded that of the IAA treatment, or the hormone-free control. Apparently, quite early in vitro, the cells of the cultured explants are stimulated by the kinetin supplement, to physiological responses responsible for the later growth performance, whatever its biochemical basis may be. At this period of the cultural cycle, also a strong increase in the uptake of potassium and phosphorus can be observed. At least for cultured carrot petiole explants at this period of the cultural cycle, strong increases in the concentrations of endogenous IAA and endogenous cytokinin 2-iP were recorded,

© Springer Nature Switzerland AG 2020

K.-H. Neumann et al., *Plant Cell and Tissue Culture – A Tool in Biotechnology*,
https://doi.org/10.1007/978-3-030-49098-0_14

reaching maximum values after about 10 days of culture. Apparently, soon after wounding during isolation in culture, an endogenous hormonal system is induced that, however, is not sufficient to promote considerable growth. This endogenous system would be supplemented by growth regulators from the nutrient medium, until a sufficient number of meristematic cells are produced to provide an endogenous hormonal system to sustain growth for a few weeks, according to the conditions in the medium at the beginning of the culture.

Somatic embryogenesis should serve as an example on possible consequences resulting from the isolation of cells from the framework of control of the intact plant, and the transfer into a suitable environment. Often, too much significance is attributed to such an isolation. If intact carrot seedlings are cultured partly submerged in an auxin-supplemented nutrient solution, somatic embryogenesis is induced in a great number of cells in the petiole, leaf lamina, or hypocotyl. This is, in particular, the case at elevated salt concentrations. Another example is the induction of this process in callus cells developed on germinating seeds in an auxin-containing medium. For these systems, evidently, neither wounding nor the isolation of explants from the intact plant is required to induce somatic embryos. Not all cells can be induced to this process, and as histological studies have indicated, e.g., in cultured petioles, a layer of small subepidermal cells are the origin of the embryos, and even here not every cell is responsive. In these protocols, only a suitable environment provided by an auxin-supplemented medium is sufficient to induce this developmental program. The auxin somehow disturbs the original hormonal system of the cells, and a new pattern develops. This was shown for cultured petiole explants in which the originally high ABA concentration decreased during culture of about 2 weeks, IAA reached a maximum after 7 days, followed by a steep decline, and cytokinins, originally quite low at t0, showed a small maximum 10–12 days after initiation of culture. The peak in the concentration of IAA coincided with the formation of adventitious roots, and that of cytokinins with the formation of cytoplasm-rich subepidermal cells as the origin of somatic embryos. Since the production of somatic embryos could also be observed on the petioles of intact plants of some carrot strains produced by the fusion of protoplasts of domestic and wild carrots (both transgenic) growing in an inorganic agar medium, even an exogenous auxin supply seems not to be an indispensable requirement for the induction of this adventitious developmental program. Here, apparently due to a specific genetic setup, a physiological situation is created that promotes the development of somatic embryos in competent cells without exogenous interference.

In embryogenic cell suspensions, the supplement of 0.1 ppm kinetin prevents the formation of somatic embryos. Concurrently, the duration of the G1-phase of the cell cycle of basically embryogenic cells is reduced by about 90 min. During this period, the synthesis of about 120 proteins is initiated in a hierarchical sequence in the kinetin-free medium, but which are not synthesized in the kinetin treatment. The synthesis of some of these proteins is terminated shortly after initiation. To date, no attempt has been made to identify these proteins, and it remains to be seen which significance can be attributed to one or the other of these protein moieties for the development of the embryos.

Due to space limitation, only a few examples can be given for the second key physiological aspect of the development of plant cell and tissue culture, i.e., to serve as model systems in plant sciences. A more detailed discussion of this topic was published previously. Here, first the work by the Skoog group at the University of Wisconsin needs to be mentioned, who introduced the concept of cytokinins, and the significance of the ratios of phytohormones for the differentiation and development of plants, based on their landmark investigations employing callus cultures of tobacco. The Steward group at Cornell University in Ithaca originally aimed at investigating the accumulation of minerals using fast- and slow-growing cells. For this, they used carrot root explants growing fast on a coconut milk supplement, and comparatively slowly without this supplement. The outcome was the unexpected observation of somatic embryogenesis. The results dealing with salt accumulation have recently been summarized by F.C. Steward himself, who stressed the significance of the intensity of the growing process to understand salt accumulation. Two broad physiobiochemical areas are of uppermost importance for the process of photosynthesis, i.e., the movements of the stomata, and the assimilation of CO_2 by the cells of the leaves. Experimentally, it is difficult to separate these two using intact leaves. Here, photoautotrophic cell cultures lacking stomata could be useful to study influences on the metabolic machinery of assimilating cells, without the confounding effect of stomata. By using such photoautotrophic cultures it was shown, e.g., that an excess of sucrose—as may occur due to pathological disturbances of the conducting systems of intact plants—strongly increases the fixation of CO_2 by a partial C4 fixation pathway, due to an enhancement of PEPCase activity. At the same time, RuBisCO activity and concentration, and consequently CO_2 fixation via the Calvin cycle are reduced. Such investigations could be helpful to influence the photosynthetic system by gene technological means.

When Haberlandt embarked on the use of plant cell cultures about 100 years ago, he was interested in knowing whether plant cells are able to grow in isolation from the intact plant. At that time, the cell theory of Schleiden and Schwann was only a few decades old, and his experimental question was therefore within the basic framework of general botany to understand the physiology of cells.

Going through textbooks on botany or plant physiology, like "Strasburger's Lehrbuch der Botanik" and others, the general impression cannot be avoided that the contribution of cell culture systems to understand general phenomena in these fields is rather meager, although thousands of scientists have been working with cell and tissue culture systems. One reason for this may be the early intensive search for practical applications with commercial implications starting in the 1960s, particularly for components of secondary metabolism—with, however, limited success, considering the input. Sensu Haberlandt, cells in a suspension are "autonomous organisms," with all genetic information of the species or strain. Which of these are realized depends on the environment at large. It is not enough to know and manipulate the enzymatic reaction chain to synthesize the desired product of secondary metabolism. At least as important are properties of membranes for its transport within the cell, and its accumulation and storage. The pH of various compartments of cells has to be considered, as well as other cytological aspects.

The study of basic problems relating cultured cells to fundamental aspects of plant cells, in general, has been a stepchild of research, certainly in terms of funding, but also otherwise. Evidently, we still do not know enough about the physiology of intact plants, and of cultured cells and tissue to close this gap, and meaningfully integrate insights gained in the latter field in attempting to solve problems of general plant physiology. This may soon change. Indeed, the currently emerging application of gene technology to agriculture requires cell and tissue culture systems as essential tool—today, progress is not feasible without.

Index

© Springer Nature Switzerland AG 2020
K.-H. Neumann et al., *Plant Cell and Tissue Culture – A Tool in Biotechnology*,
https://doi.org/10.1007/978-3-030-49098-0